SPACE TELESCOPE SCIENCE INSTITUTE

SYMPOSIUM SERIES: 9

Series Editor S. Michael Fall, Space Telescope Science Institute

THE COLLISION OF COMET SHOEMAKER-LEVY 9 AND JUPITER

T0186161

SPACE
TELESCOPE
SCIENCE
INSTITUTE

THE COLLISION OF COMET SHOEMAKER-LEVY 9 AND JUPITER

IAU Colloquium 156

Proceedings of the Space Telescope Science Institute Workshop, held in Baltimore, Maryland
May 9–12, 1995

Edited by

KEITH S. NOLL
Space Telescope Science Institute, Baltimore, MD 21218, USA

HAROLD A. WEAVER
Applied Research Corporation, Landover, MD 20785, USA

PAUL D. FELDMAN
John Hopkins University, Baltimore, MD 21218, USA

Published for the
Space Telescope Science Institute

CAMBRIDGE
UNIVERSITY PRESS

CAMBRIDGE UNIVERSITY PRESS
Cambridge, New York, Melbourne, Madrid, Cape Town, Singapore, São Paulo

Cambridge University Press
The Edinburgh Building, Cambridge CB2 2RU, UK

Published in the United States of America by Cambridge University Press, New York

www.cambridge.org
Information on this title: www.cambridge.org/9780521561921

First published 1996
This digitally printed first paperback version 2006

A catalogue record for this publication is available from the British Library

ISBN-13 978-0-521-56192-1 hardback
ISBN-10 0-521-56192-2 hardback

ISBN-13 978-0-521-03162-2 paperback
ISBN-10 0-521-03162-1 paperback

Contents

Participants

Mike A'Hearn	University of Maryland
Stephanie Abrams	Golden Gaters Productions
John Africano	Rockwell Power Systems
Claude Arpigny	Université de Liege
Tadashi Asada	Kyushu International University
Sushil Atreya	University of Michigan
Johannes Babion	Universitäts-Sternwarte
G. Ballester	University of Michigan
Don Banfield	Cornell University
Ed Barker	NASA Headquarters
Harry Bates	Towson State University
Kelly Beatty	Sky & Telescope
Reta Beebe	New Mexico State University
Bruno Bézard	Observatoire de Paris
Gordon Bjoraker	NASA/Goddard Space Flight Center
Carlo Blanco	Universita di Catania
William Blass	University of Tennessee
Jirí Borovicka	Astronomical Institute, Czechoslovakia
Tim Brook	Jet Propulsion Laboratory
Michael Brown	University of Arizona
Scott Budzien	Naval Research Laboratory
David Buhl	NASA/Goddard Space Flight Center
Marc Buie	Lowell Observatory
Robert Carlson	Jet Propulsion Laboratory
Kenneth Carpenter	NASA/Goddard Space Flight Center
Zdenek Ceplecha	Astronomical Observatory
Nancy Chanover	New Mexico State University
Clark Chapman	Planetary Science Institute
Paul Chodas	Jet Propulsion Laboratory
Klim Churyumov	Kiev University
John Clarke	University of Michigan
Anita Cochran	University of Texas at Austin
William Cochran	University of Texas at Austin
Regina Cody	NASA/Goddard Space Flight Center
Michael Collins	Naval Research Laboratory
J. E. P. Connerney	NASA/Goddard Space Flight Center
Barney Conrath	NASA/Goddard Space Flight Center
Ron Cowen	Science News
Thomas Cravens	University of Kansas
David Crawford	Sandia National Laboratories
David Crisp	Jet Propulsion Laboratory
Jacques Crovisier	Observatoire de Paris—Meudon
Catherine de Bergh	Observatoire de Paris
Imke de Pater	University of California, Berkeley
Drake Deming	NASA/Goddard Space Flight Center
Michael DiSanti	NASA/Goddard Space Flight Center
George Dulk	DESPA, Observatoire de Meudon
Steve Edberg	Jet Propulsion Laboratory

Participants

Scott Edgington	University of Michigan
Thérése Encrenaz	DESPA, Observatoire de Meudon
Irene Engle	United States Naval Academy
Sergio Fajardo-Acosta	Penn State Erie
Kelly Fast	NASA/Goddard Space Flight Center
Paul Feldman	Johns Hopkins University
Daniel Fischer	Skyweek
Brendan Fisher	University of Rochester
Alan Fitzsimmons	Queen's University
Michael Flasar	NASA/Goddard Space Flight Center
Michael Foulkes	The British Astronomical Association
James Friedson	Jet Propulsion Laboratory
Oliver Funke	Universitat Bonn
Patrick Galopeau	Observatoire de Meudon
Don Gavel	University of California, LLNL
Diane Gilmore	Space Telescope Science Institute
Galen Gisler	Los Alamos National Laboratory
Jay Goguen	Jet Propulsion Laboratory
Ed Grayzea	University of Maryland
Thomas Greathouse	Northern Arizona University
Caitlin Griffith	Northern Arizona University
Arie Grossman	University of Maryland
Joseph Hahn	University of Notre Dame
Michael Haken	University of Maryland
Doyle Hall	Johns Hopkins University
Doug Hamilton	Max Planck Institut für Kernphysik
Heidi Hammel	Massachusetts Institute of Technology
Joseph Harrington	NASA/Goddard Space Flight Center
Hitoshi Hasegawa	ASTEC, Inc.
John Hillman	NASA/Goddard Space Flight Center
Thomas Hockey	University of Northern Iowa
Richard Hook	ST-ECF/ESO
Walter Huebner	Southwest Research Institute
Donald Hunten	University of Arizona
Andrew Ingersoll	California Institute of Technology
W.-H. Ip	Max Planck Institut für Aeronomie
Kandis Lea Jessup	University of Michigan
Torrence Johnson	Jet Propulsion Laboratory
Robert Joseph	University of Hawaii
Dave Kary	University of California, Santa Barbara
Janos Kelemen	Konkoly Observatory, Hungary
Dick Kerr	Science Magazine
Kate King	CNN
Kenneth Kissell	University of Maryland
Valerij Kleshchonok	Astronomical Observatory, Ukraine
Roger Knacke	Penn State Erie
Theodor Kostiuk	NASA/Goddard Space Flight Center
Nikolay Kotsarenko	Kiev University
Arunav Kundu	University of Maryland
Pierre-Olivier Lagage	Service d'Astrophysique, France

Rob Landis	Space Telescope Science Institute
Bryan Laubscher	Amparo Corporation
Miska Le Louarn	NASA/Goddard Space Flight Center
Yolande Leblanc	DESPA, Observatoire de Meudon
David Leckrone	NASA/Goddard Space Flight Center
Emmanuel Lellouch	Observatoire de Paris—Meudon
Peter Leonard	University of Maryland
David Levy	University of Arizona
Sanjay Limaye	University of Wisconsin, Madison
Carey Lisse	NASA/Goddard Space Flight Center
Timothy Livengood	NASA/Goddard Space Flight Center
Mario Livio	Space Telescope Science Institute
Dmitrij Lupishko	Kharkov State University
James Lyons	California Institute of Technology
Mordecai-Mark Mac Low	University of Chicago
Karen Magee-Sauer	Rowan College of New Jersey
William Maguire	NASA/Goddard Space Flight Center
Stephen Mahan	University of Tennessee
Mark Marley	New Mexico State University
Terry Martin	Jet Propulsion Laboratory
Clifford Matthews	University of Illinois at Chicago
Lucy McFadden	University of Maryland
Melissa McGrath	Space Telescope Science Institute
Peter McGregor	Mt. Stromlo & Siding Spring Observatories
William McKinnon	Washington University
Vikki Meadows	Jet Propulsion Laboratory
Jennifer Mills	Cambridge, MA
Alexander Morozhenko	National Academy of Sciences
Julianne Moses	Lunar & Planetary Institute
Michael Mumma	NASA/Goddard Space Flight Center
Chan Na	University of Colorado
Ivan Nemtchinov	Institute for Dynamics of Geosphers RAS
Philip Nicholson	Cornell University
Malcolm Niedner	NASA/Goddard Space Flight Center
Keith Noll	Space Telescope Science Institute
Miles O'Brien	CNN
Carlos Olano	Instituto Argentino de Radioastronomia
Hari Om Vats	Physical Research Laboratory, India
Jose Luis Ortiz	Jet Propulsion Laboratory
Glenn Orton	Jet Propulsion Laboratory
Renée Prangé	Université Paris XI
Wayne Pryor	University of Colorado
David Rabinowitz	Carnegie Institution of Washington
Terrence Rettig	University of Notre Dame
Jessica Reynolds	University of Denver
Andrea Richichi	Osservatorio Astrofisico di Arcetri
John Rogers	University of Cambridge
Francoise Roques	DESPA, Observatoire de Meudon
Frank Roylance	Baltimore Sun
Tamara Ruzmaikina	University of Arizona

Leslie Sage	Nature Magazine
Agustin Sanchez-Lavega	Universidad Del Pais Vasco
Sho Sasaki	University of Tokyo
Takehiko Satoh	NASA/Goddard Space Flight Center
Paul Schenk	Lunar & Planetary Institute
David Schneider	Scientific American
James Scotti	University of Arizona
Zdenek Sekanina	Jet Propulsion Laboratory
Kazuhiro Sekiguchi	National Astronomical Observatory of Japan
Scott Severson	University of Chicago
Surja Sharma	University of Maryland
Carolyn Shoemaker	Northern Arizona University
Eugene Shoemaker	Lowell Observatory
Leonid Shulman	Main Astronomical Observatory, Ukraine
Amy Simon	New Mexico State University
Ed Smith	Space Telescope Science Institute
Derck Smits	HARTRAO
John Spencer	Lowell Observatory
Alex Storrs	Space Telescope Science Institute
Darrell Strobel	Johns Hopkins University
Richard Strom	Netherlands Foundation for Research in Astronomy
Michael Summers	Naval Research Laboratory
Bunji Suzuki	Misato Technological High School
Krishna Swamy	Tata Institute of Fundamental Research
Isshi Tabe	Kanagawa, Japan
Satoru Takeuchi	National Astronomical Observatory of Japan
Gonzalo Tancredi	Depto. Astronomía, Fac. Ciencias, Uruguay
Kate Tobin	CNN
Laurence Trafton	University of Texas at Austin
Pedro Valdés Sada	NASA/Goddard Space Flight Center
Hugh Van Horn	National Science Foundation
John Wang	University of Maryland
Qi Wang	Purple Mountain Observatory
Sichao Wang	Purple Mountain Observatory
Jun-ichi Watanabe	National Astronomical Observatory of Japan
Hal Weaver	Space Telescope Science Institute
Paul Weissman	Jet Propulsion Laboratory
Dennis Wellnitz	University of Maryland
Richard West	European Southern Observatory
Robert West	Jet Propulsion Laboratory
Peter Wilson	Cornell University
Maria Womack	Penn State Behrend
Laura Woodney	University of Maryland
Xingfa Xie	NASA/Goddard Space Flight Center
Padma Yanamandra-Fisher	Jet Propulsion Laboratory
Roger Yelle	NASA/Ames Research Center
Kevin Zahnle	NASA/Ames Research Center
Heqi Zhang	Purple Mountain Observatory
Jiaxiang Zhang	Purple Mountain Observatory
Xizhen Zhang	DRAO, Canada

Hongnan Zhou	Nanjing University
Xun Zhu	Johns Hopkins University
Krzysztof Ziolkowski	Space Research Center, Poland

Preface

In a cosmic sense, the collision of the ninth periodic comet discovered by the team of Carolyn and Gene Shoemaker and David Levy with the planet Jupiter was unremarkable. The history of the solar system, indeed its very genesis, has been marked by countless such events. The cratered surfaces of planetary bodies are a testament to this ubiquitous phenomenon; even the Earth's ephemeral surface records the continued action of this elemental process in impact craters and in the fossil record.

In human terms, on the other hand, the impact of Comet Shoemaker-Levy 9's 20-odd fragments into Jupiter was an unprecedented event of global significance. After a year of planning and preparation, the largest astronomical armada in history focussed on the planet Jupiter in July 1994. News of each successively more astonishing image or spectrum was broadcast with almost instantaneous speed over the world's increasingly sophisticated computer communications network. Astronomers were, for a time, to be found on daily newscasts and the front pages of newspapers. For a week in July, the world looked up from its normal preoccupations long enough to notice, and to ponder, the awesome beauty of the natural world and the surprising unpredictability of the universe.

Still one more perspective on this event remains. What has science gained from the terabytes of images, lightcurves and spectra obtained over the entire range of the electromagnetic spectrum? Do we now have a better understanding of comets, of the atmosphere of Jupiter, of the effect of high-velocity impacts on planetary atmospheres? What has been deduced about SL9's size, composition, or structure? Do we fully understand the sequence of events recorded in our data?

The Hubble Space Telescope played a central role in the initial characterization of SL9 and in recording unique images and spectra of the impacts and impact sites during the week of July 14–21, 1994. It is fitting that a meeting conceived as a first attempt to synthesize answers to the above, and many more, questions was held as part of the STScI May symposium series (and under the auspices of the International Astronomical Union as Colloquium 156). The meeting held in Baltimore in May 1995 at the Space Telescope Science Institute and the Johns Hopkins University attracted more than 200 scientists who heard 16 invited reviews, participated in 9 workshops, and discussed over 140 contributed posters. Some questions were answered, and new ones raised.

This volume is composed of sixteen chapters from the invited speakers on the topics reviewed at the meeting. Most chapters have been updated to include the latest results presented at the 1995 Division of Planetary Sciences Meeting. Two of the chapters have authors who did not give the review talk at the meeting and one, the first chapter, is written by the chairs of one of the workshops. The result is the first comprehensive synthesis of the SL9 event covering everything from its prehistory to the dissipation of the debris clouds. Though the first, this will probably not be the last volume to review this significant event. We hope that this volume will provide the starting point for future reviewers who will report on the deepening of our understanding of this example of one of the universe's fundamental processes.

Special thanks are due to a large number of people who helped support the symposium. Foremost, we thank the scientists at the Space Telescope Science Institute who were willing to commit the resources of our large annual meeting to this topic well *before* July 1994 when the dramatic results of the impacts were not imagined. The Johns Hopkins University gracefully provided its facilities for this large meeting and also added financial support. Dr. Michael A'Hearn led the science organizing committee and was instrumental in garnering the support of the IAU. Dr. Alex Storrs performed excellent service as head of the local organizing committee, along with many others. Cheryl Schmidt made the organization of this largest May symposium yet look easy with her usual skill. Finally, Sharon Toolan cheerfully and skillfully turned a collection of manuscripts into a beautiful and coherent volume.

Keith S. Noll, Harold A. Weaver, Paul D. Feldman
Space Telescope Science Institute
Baltimore, Maryland
May, 1995

The orbital motion and impact circumstances of Comet Shoemaker-Levy 9

By PAUL W. CHODAS and DONALD K. YEOMANS

Jet Propulsion Laboratory, California Institute of Technology, 4800 Oak Grove Drive,
Pasadena, CA 91109, USA

Two months after the discovery of comet Shoemaker-Levy 9 came the astonishing announcement that the comet would impact Jupiter in July 1994. Computing the orbital motion of
this remarkable comet presented several unusual challenges. We review the pre-impact orbit
computations and impact predictions for SL9, from the preliminary orbit solutions shortly after
discovery to the final set of predictions before the impacts. The final set of predicted impact
times were systematically early by an average of 7 minutes, probably due to systematic errors
in the reference star catalogs used in the reduction of the fragments' astrometric positions. The
actual impact times were inferred from the times of observed phenomena for 16 of the impacts.
Orbit solutions for the fragments were refined by using the actual impact times as additional
data, and by estimating and removing measurement biases from the astrometric observations.
The final orbit solutions for 21 fragments are tabulated, along with final estimates of the impact
times and locations. The pre-breakup orbital history of the comet was investigated statistically,
via a Monte Carlo analysis. The progenitor nucleus of SL9 was most likely captured by Jupiter
around 1929 ± 9 years. Prior to capture, the comet was in a low-eccentricity, low-inclination
heliocentric orbit entirely inside Jupiter's orbit, or, less likely, entirely outside. The ensemble
of possible pre-capture orbits is consistent with a group of Jupiter family comets known as the
quasi-Hildas.

1. Introduction

The late-March 1993 discovery of multiple comet Shoemaker-Levy 9 by Carolyn and
Gene Shoemaker and David Levy set in motion an extraordinary international effort to
study the evolution of a remarkable cometary phenomenon and to witness its ultimate
collision with Jupiter (Shoemaker *et al.* 1993). From the beginning, it was clear that
the orbital dynamics of this comet were unique. It had spectacularly split into ~ 20
fragments, most likely because of tidal disruption during a recent very close approach to
Jupiter. Preliminary orbit computations soon confirmed the close approach, and revealed
the surprising fact that the comet was actually in orbit about the planet (Marsden 1993b).
Even more extraordinary news came several weeks later, when further orbit computations
suggested that the comet would likely collide with Jupiter in late July 1994 (Nakano 1993,
Yeomans and Chodas 1993a). Early calculations indicated that the collision would take
place on the far side of the planet as viewed from the Earth, but the precise location
was very uncertain. After the comet emerged from solar conjunction in December 1993,
important new astrometric measurements were added to the data set, and the predicted
impact locations moved much closer to the limb of Jupiter, although they were still on
the far side (Yeomans and Chodas 1993d). During the months leading up to the impacts,
increasingly more accurate predictions of the impact times and locations were computed
and distributed electronically to the astronomical community. These predictions made
it possible for the extraordinary impact events to be well recorded by an unprecedented
array of ground-based and space-based instruments.

Orbital computations for comet Shoemaker-Levy 9 (referred to as SL9 hereafter) presented several challenges beyond what is normally the case for comets and asteroids.

Because the comet was in orbit about Jupiter and was heading for an impact, new parameters such as jovicentric positions and velocities in various reference frames, jovicentric orbital elements, impact times, and impact locations had to be computed. Since the comet had fragmented into a string of nuclei with no obvious bright central condensation to use as a reference point, astrometric measurements and orbit computations were referenced to the mid-point of the string, which was rather ill-defined. Eventually, the mid-point was abandoned in favor of tracking the approximately 20 individual fragments, requiring that orbit computations and impact predictions be repeated for each nucleus. Determining the orbits for some of the fainter fragments was difficult, since very little astrometric data were available for these poorly observed objects. Some of the fragments disappeared completely as the comet evolved, while others split. Proper identification of fragments was a problem, as they were sometimes mislabeled in the astrometric data. Detective work was required to sort out their true identities. Even Mother Nature conspired to add confusion, when a telescope observing SL9 from Kitt Peak in Arizona was unknowingly shaken during the January 17, 1994 earthquake in southern California. The effect of the earthquake on these astrometric observations was detected only through the resulting large orbit residuals.

Accurately determining the motion of the fragments close to the July 1994 impacts offered additional computational challenges. The need for accurate impact predictions required the modeling of the perturbative effects of the Galilean satellites and Jupiter's oblateness. Also, as the fragments approached the planet, their motion became very non-linear. The fact that our software used a variable integration step size, and that the partial derivatives required in the orbital differential correction process were numerically integrated along with the comet's motion, rather than being approximated using finite differences, allowed us to refine the orbit solutions right up to the times of impact.

In the next section, we review the pre-impact orbit computations and impact predictions for SL9, from the preliminary orbit solutions shortly after discovery to the final set of predictions before the impacts. We then discuss post-impact analyses, indicating how the observed impact phenomena were interpreted, and how the actual impact times were inferred. We give a compilation of the times of key events in the observed light curves. Following this, we describe how the orbit solutions were improved after the impacts, by using the actual impact times as additional data, and by removing measurement biases from the astrometric observations. We tabulate our final orbit solutions for 21 fragments, in both heliocentric and jovicentric forms. Next, we present our final estimates of the impact times, locations, and geometries, as derived from the final orbit solutions. Finally, we discuss the pre-breakup orbital history of the comet, which we have investigated statistically using a Monte Carlo analysis. We give our estimate of when the comet was likely captured by Jupiter, and characterize SL9's possible pre-capture heliocentric orbits.

2. Pre-impact orbital analyses and impact predictions

The early orbital analyses of SL9 were based on the supposition that the comet had broken up during a recent close approach to Jupiter. The circumstantial evidence was strong: SL9 had split into a large number of fragments in a well organized geometry, and it was currently situated only 4 degrees from the largest planet in the Solar System. Tidal disruption during an approach to within the Roche limit of a large perturbing body is a common mechanism for cometary splitting. Several comets have been known to split after close approaches to the Sun, and one, P/Brooks 2, is known to have split after approaching Jupiter to ~ 2 Jupiter radii (R_J) in July, 1886 (Sekanina and Yeomans

1985). Thus, the supposed breakup scenario would not be unprecedented. It was far from a certainty, however, as comets have been seen to split spontaneously, when nowhere near a large body. The day after the announcement of the comet's discovery, B. G. Marsden published a very preliminary orbit solution in which he used the assumption of a close passage by Jupiter (Marsden 1993a). His computations suggested that the comet's close approach to the planet had been at a distance of 0.04 AU in late July 1992, surprisingly accurate considering how little data were used in the solution. It would be many weeks before enough astrometric data became available to confirm that the comet had indeed made an extremely close approach to Jupiter on July 7, 1992, at a distance of only 1.3 R_J from the center of the planet.

Computing the orbit of SL9 in the first month or so after its discovery was very difficult. Few astrometric measurements were available, and the presence of nearby Jupiter introduced a large nonlinearity into the orbit computations. Furthermore, SL9 had no single central condensation to serve as a reference point for astrometric measurements. Since the individual nuclei were not easily resolvable by many observers, the convention was adopted to measure only the center of the train of nuclei, the mid-point of the bar (Marsden 1993b). This simplifying assumption greatly facilitated astrometry for many observers, especially amateurs, who provided a large fraction of the early measurements. We certainly would not have learned as much as we did about the orbit of SL9 as quickly as we did without this convention. However, the center of the train was rather ill-defined, and different observers placed it at different points in the train, according to the extent of the train each could see. Moreover, as the length of the train grew, errors in locating its center also grew.

A week after his first orbit solution, Marsden (1993b) obtained an improved solution which indicated a surprising new result: SL9 appeared to be in orbit about Jupiter. This was also not unprecedented. Carusi *et al.* (1985) investigated the long-term motion of all periodic comets with well-known orbits and found several that had either been in temporary Jupiter orbit in the past, or would enter temporary orbit in the relatively near future. Tancredi *et al.* (1990) investigated the temporary capture of comet P/Helin-Roman-Crockett by Jupiter during intervals centered on close approaches to Jupiter in 1976 and 2075. Using more recent orbit solutions, with nongravitational effects included when appropriate, we studied the motions of seven comets other than SL9 that either have been, or will be, temporary satellites of Jupiter (Yeomans and Chodas 1994b).

By early May, the span of astrometric observations was sufficiently long to begin to reveal the true collision trajectory of the comet. Amateur observers had contributed a large number of valuable measurements, and as more and more of these were used, orbital computations by S. Nakano and Marsden began to indicate the possibility of impact in July 1994. Now, this was truly unprecedented. Marsden alerted us of this exciting development on May 21, and provided the key set of recent astrometric measurements. We immediately confirmed Nakano and Marsden's computations, and computed that the probability of impact was about 50%. Our software had just recently been augmented with the capability to estimate probability of impact, in preparation for a study of the hazards of near-Earth objects (Chodas 1993; Yeomans and Chodas 1995). The dramatic announcement of the impending collision was issued the next day (Marsden 1993c), along with Nakano's orbit solution (Nakano 1993). One of our initial orbit solutions appeared in the Minor Planet Circulars shortly after (Yeomans and Chodas 1993b). Within a few days of the impact announcement, as more astrometric data became available, the probability of impact rose to 64% (Yeomans and Chodas 1993a), and it reached 95% only a week later.

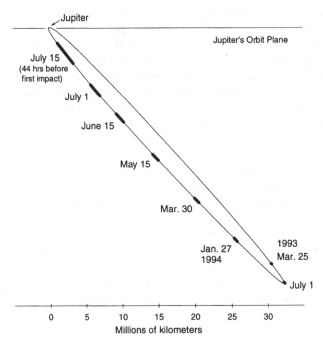

FIGURE 1. The orbit of comet Shoemaker-Levy 9 about Jupiter, as viewed from the direction of Earth on May 15, 1994. The length and orientation of the train of fragments are shown to scale on eight dates. The train is nearly aligned along the velocity vector, except near apojove. The orbit is somewhat foreshortened in this view; the major axis rises out of the plane of the diagram towards the viewer at an angle of about 20 degrees.

To determine the basic characteristics of SL9's orbit and its impending impact, we quickly modified our software to provide jovicentric information, including position, velocity, and orbital elements as a function of time. It became clear that the comet was approaching the apojove of an extremely eccentric orbit about Jupiter, with eccentricity ~ 0.99 and apojove distance ~ 0.33 AU (see Fig. 1). By June 1, we had determined that the impact would occur in the mid-southern latitudes of Jupiter, and, unfortunately, on the side of the planet facing away from Earth. We defined impact to occur when the comet reached the one bar pressure level in Jupiter's atmosphere, which we modeled as an oblate spheroid with radius and flattening given by Davies *et al.* (1992). Finding the moment of impact in these early solutions required searching through tables of numbers, but we soon automated this important function. We also wrote software to compute and plot the motion of the comet in various jovicentric frames, which helped in visualizing its trajectory (Yeomans and Chodas 1993c). With it, we determined that the Galileo spacecraft would likely have a direct view of the impact, although this was far from certain because the predicted impact was right on the limb.

There was much more to study with this dynamically fascinating object. For one thing, it had a whole train of nuclei to consider. As SL9 passed through apojove around July 13, the question arose as to whether the fragments would reverse their order on the sky as viewed from the Earth. After all, they had 'turned the corner' and were heading back to Jupiter. But in fact, the appearance of the train did not change because the fragments were not all on the same orbit. A useful analogy is to think of throwing a handful of pebbles upwards, each receiving a slightly different vertical velocity. The slowest pebble trails the others going upwards, but it reaches its apex first and is the first to hit the ground; furthermore, the separations between the pebbles increase monotonically during their flight. With SL9, the eastern-most fragment trailed the others on the outbound leg, reached apojove first, and became the leading fragment inbound. The fragments passed through apojoves in sequence over a period of a few days, while the train continued its expansion and remained approximately pointed at Jupiter throughout.

As SL9 headed into solar conjunction in late July, other important questions were being raised. Would all the fragments collide with Jupiter? Would any of them impact on the side of Jupiter visible from Earth? Unfortunately, no astrometric measurements of individual nuclei were available for computing orbit solutions. In anticipation of their availability, however, we wrote new software to investigate how slight variations in an object's orbit at one epoch would affect its position on the plane of sky at later epochs. We would soon use this software to study the tidal splitting of the comet and explain the appearance of the train. Scotti and Melosh (1993) used a similar model when, armed with Scotti's measurements of the train length and orientation, they determined that the progenitor nucleus needed to be only 1 km in radius to explain the observed train dimensions, assuming disruption occurred at perijove. They also determined that the entire train would impact Jupiter over a period of 5.6 days. With Scotti's train measurements in hand, and using their assumptions, we confirmed these results and furthermore determined that all the fragments would impact Jupiter on the far side as viewed from Earth.

The first astrometric measurements of individual nuclei became available in October 1993. These were positions of 21 fragments obtained by Jewitt *et al.* (1993) on four nights from March through July, 1993. The positions were offsets from the brightest nucleus, designated 7 in Jewitt's numbering system. In collaboration with Z. Sekanina, we determined the effective time of tidal disruption and the impulse each fragment must have received in order to appear at the observed relative positions. This approach provided the first orbital solutions and predicted impact times for individual fragments, which we denoted A through W (Chodas and Yeomans 1993). Because the relative times were known much better than the absolute times, these first impact time predictions were given relative to the impact time of the center of the train. The relative times all turned out to be within 40 minutes of the actual impact times relative to the center time, remarkable precision considering the prediction for each fragment was based on only 4 measurements taken over a year before impact. This accuracy attests to the precision of Jewitt *et al.*'s measurements, and indicates the great utility of the tidal disruption approach for computing orbit solutions. Our orbit solutions indicated that, to match the observed position angle history of the train at the 0.1 degree level, the effective time of tidal breakup of the progenitor comet had to be ~ 2.2 hr after perijove passage in 1992. From this and other evidence, we concluded that the radius of the progenitor comet was probably ~ 5 km. (Sekanina *et al.* 1994).

Other important predictions required at this time included the expected uncertainties of the predicted impact times in the last weeks and days before impact. The impact time accuracy was required to plan impact observations, especially those to be made by spacecraft, which had to programmed well in advance of the event. The rate of decrease of the impact time uncertainty depended upon the number of the astrometric measurements used in the orbit solutions, as well as their quality. Assuming a conservative 9 measurements per month and a 1-arcsec measurement accuracy, we estimated the 1-sigma impact time uncertainty a month before impact would be ~ 13 minutes, and a week before impact, ~ 7 minutes. If only two more measurements could be made on the two days before impact, the uncertainty would drop to ~ 3 minutes. Clearly, the most powerful observations for determining impact times were those closest to impact. But how close to bright Jupiter could the faint cometary fragments be observed?

In November, the first batches of absolute astrometric measurements for individual fragments became available: J. V. Scotti and T. Metcalfe provided 250 measurements obtained from Kitt Peak over the period March through July, and A. Whipple and P. Shelus provided 54 measurements obtained from McDonald Observatory taken in April and May. Marsden (1993d) used the Kitt Peak data to compute the first independent orbit solutions for individual fragments (*i.e.*, solutions which made no assumptions about the tidal disruption process), and we used both observation sets in similar solutions a few days later. Only the nine brightest fragments (E, G, H, K, L, Q, R, S, and W) had enough astrometric data to yield well-determined solutions. The impact times derived from these solutions were about 18 hours earlier than those based on the center-of-train solution and relative astrometry. This jump was most likely due to errors in locating the center of the train: the east end of the train may have been too faint to be seen by many of the observers. The new orbit solutions superceded the center-of-train solutions, which were now abandoned; astrometry and orbit computations from this time onwards referred only to individual fragments. The impact time uncertainties actually increased slightly with the new solutions, because the fragments had fewer measurements than the center-of-train orbits, but at least the solutions were now tied to well-defined points.

The emergence of SL9 from solar conjunction was greatly anticipated. Although attention focused on possible changes in the appearance of the comet, we were anxious because new astrometric data would dramatically improve the orbit solutions. On December 9, sooner than expected, Scotti recovered the comet. Although the train had lengthened, the fragments appeared much the same as before conjunction. Marsden (1993e) computed new orbit solutions for the nine brightest fragments and found that the impact times were almost a day earlier than in previous solutions. We confirmed Marsden's computations, and found an exciting new result: the predicted impact locations, though still on the far side of Jupiter, were now much closer to the morning terminator, only 5–10 degrees behind the limb as seen from the Earth, with the later impacts closest to the limb (Yeomans and Chodas 1993d, Chodas and Yeomans 1994a). The impact sites had also moved well onto the hemisphere visible to Galileo. Would the predicted impact locations continue to move towards the limb, and possibly even onto the near side? Unfortunately not. These would be the last large changes in the predictions, because the orbit solutions had become relatively well determined. Based on Monte Carlo analyses which used actual orbit uncertainties and correlations, we concluded that there was little chance that any of the fragments would impact the side of Jupiter visible to the Earth.

In mid-December, our impact predictions, together with orbital elements and ephemerides for the nine brightest fragments, were posted on the special SL9 electronic bulletin board operated at the Planetary Data Systems' Small Bodies Node at the University of Maryland (UMD). Over the remaining seven months before impact, we posted over a dozen more sets of predictions, and associated data. The predicted parameters in our tables included impact time, jovicentric latitude, meridian angle, and the Earth-Jupiter-fragment (E-J-F) angle at impact. This latter angle indicated how far behind the limb the impact would occur. The meridian angle was defined as the jovicentric longitude of impact relative to the midnight meridian, measured towards the morning terminator. This relative longitude was known much more accurately than the Jupiter-fixed longitude, because of the large uncertainty in the impact times and Jupiter's fast rotation. Basically, the approach trajectory of each fragment was known much more accurately than the fragment's location on that trajectory at any given time. Predictions of absolute jovicentric longitude were not included in our tables until later, when the impact times were better known. Also added later were predictions of satellite longitudes at impact for four of the inner jovian satellites, Amalthea, Io, Europa, and Ganymede, kindly supplied by P. D. Nicholson.

Keeping track of all the fragments was a continual challenge. Not only were there a lot of them to consider, but each seemed to have its own personality. In January 1994, only seven fragments (G, H, K, L, Q, S, and W) had well-established orbit solutions and consistent impact predictions, while solutions for E and R remained a little erratic, as they were based on fewer measurements. The rest of the fragments had too little astrometric data to determine reliable independent orbit solutions, so we applied our tidal disruption model as we had done earlier, although now we varied the orbit of fragment Q instead of the orbit for the center of the train. By the end of the month, as more observations became available, the solutions for fragments E and R became consistent with the rest, and fragment F graduated to the group with independent solutions. The image of SL9 taken by the Hubble Space Telescope (HST) in late January revealed changes in the SL9 menagerie: fragments J and M had disappeared completely, and fragments in the P–Q region had clearly split. For a time, there was confusion in identifying fragments in this region, with N identified as P and the P sub-fragments identified as Q3 and Q4, but by mid-February the P1/P2 and Q1/Q2 nomenclature was established (Marsden 1994a). Correctly identifying the fainter fragments in ground-based observations was a recurring problem, as the fragments were often near the limits of detectability. We checked observer's identifications by comparing observations against positions predicted from orbit solutions, but this was an imperfect process because the orbit solutions for these faint fragments were not well-determined either.

By late February, independent orbit solutions had been computed for 19 fragments by both Marsden (1994b) and ourselves, although only 12 fragments (the original nine, plus F, N, and P2) had solutions reliable enough to be used in our impact predictions (Chodas *et al.* 1994). By early June we had adopted independent orbit solutions for all fragments but Q2, although the impact predictions for the extremely faint fragments T and U continued to be erratic for several more weeks. Fragment Q2 was especially difficult, as it had very few measurements. Using HST measurements of the offset of Q2 from Q1, Sekanina (1995) applied our disruption model to determine that Q2 likely broke away from Q1 in the March–April 1993 period, right around the time of discovery of SL9. Not until July did the separation between Q1 and Q2 increase to the point that many ground-based observers could resolve the two fragments; we finally adopted an independent solution for Q2 in the last set of predictions before its impact.

In April, we upgraded the dynamical models used in our orbit determinations and impact predictions. Up until this time, we had used only point mass perturbations by the sun and planets, with planetary positions and masses taken from JPL planetary ephemeris DE200 (Standish 1990). But now, we switched to the more accurate planetary ephemeris DE245, and refined our models to include perturbations due to the Galilean satellites and the J2 and J4 zonal harmonic terms of Jupiter's gravity field. The positions of the Galilean satellites were computed using the analytic theory by Lieske (1977, 1994), while the parameters for Jupiter's gravity field were obtained from Campbell and Synnott (1985). Since the SL9 fragments approached Jupiter from the south, and impacted in the southern hemisphere, they did not come near the Galilean satellites on their final approach, and, as a result, the inclusion of the satellite perturbations had only a minor effect on the impact times. Similarly, the inclusion of the Jupiter oblateness perturbations made only a small difference in the predicted impact times. Both perturbations, however, were important in the long term backward integration of the comet's motion, discussed in section 6.

As SL9 passed through opposition in the April–May 1994 period, the number of astrometric observations increased dramatically, and the measurements themselves became more powerful in reducing orbital uncertainties, simply because the Earth was closer to the comet. During this time, the predictions drifted towards later impact times for most fragments, until, at the end of May, they were about an hour later than they had been in March. Meanwhile, the formal impact time uncertainties fell from about 30 minutes to 18 minutes (1-sigma), for the brightest fragments. The drift in impact times reversed itself in June and early July, with times sliding earlier by 30–50 minutes on average, while the impact time uncertainties fell to less than 10 minutes. The relatively large shifts in predicted impact times were a concern in the period from mid-June to early July, as final predictions had to be made for use in the Galileo impact observation sequences. The spacecraft was programmed to observe during a window of only 20–60 minutes around each of the predicted times. As it turned out, of the 16 impacts observed by Galileo instruments, only one was missed because the event shifted out of the observing window.

The most likely explanation for the large shifts in the predicted impact times was the presence of systematic errors in the reference star catalogs used by observers in reducing their measurements. Star catalog errors can be a major error source for precision orbit determination of comets and asteroids. Since background stars in an astrometric image are used as reference points in determining the position of an object of interest, systematic errors in the tabulated coordinates and proper motions of the reference stars lead to systematic errors in the deduced positions for the object. Most of the astrometric data for SL9 were reduced with respect to versions 1.1 or 1.2 of the *Hubble Space Telescope Guide Star Catalog* (GSC), which contains systematic errors of ∼ 0.5 arcsec for some regions of the sky. These errors are significantly larger than the typical errors incurred in actually measuring the position of the nucleus in the image, which could be as small as ∼ 0.2 arcsec in the best ground-based observations. In our orbit solutions, we modeled measurement errors simply as zero-mean Gaussian noise, and used a standard deviation, or noise value, of 1 arcsec for most observations, to account for the star catalog errors.

Since the most powerful astrometric data for reducing uncertainties in the predicted impact times would be those data taken closest to impact, it was especially important to try to reduce systematic star catalog errors in the region occupied by the comet near impact. To this end, J. V. Scotti generated and distributed a special reference star catalog for the region traversed by the comet in the last week before impact. Scotti made offset corrections to GSC reference stars by differencing the positions of stars common to the GSC and the more accurate PPM catalog (Roeser and Bastian 1989). Observations reduced with respect to Scotii's special catalog were assigned a noise value of 0.6 arcsec in our solutions.

A pre-publication version of the Hipparcos star catalog, kindly provided by M. Perryman and C. Turon of the Hipparcos project, was used by R. West and O. Hainaut in the reduction of a number of observations from the European Southern Observatory (ESO) taken between May 1 and July 14, 1994. Because the Hipparcos catalog is known to be highly accurate, systematic star catalog errors should be largely absent from these measurements, and we therefore assigned them noise values of 0.3 arcsec in our solutions. The post-fit root-mean-square (rms) of the ESO observation residuals was about one third the size of the rms of all the residuals.

The ESO group was able to obtain astrometric images close to Jupiter, with enough sensitivity to see even the fainter fragments. Theirs were the last astrometric observations taken of eight of the faintest fragments, ranging from 2.3 to 7 days before impact. Several other groups attempted to observe the SL9 fragments even closer to impact by using coronographs to block out the light from Jupiter, but this proved to be a very difficult task. Only two groups succeeded in obtaining astrometric data using this approach: D. Rabinowitz and H. Butner at Las Campanas obtained the last astrometric observations of fragments E, G, and L, with E seen only 1.45 days before its impact, and D. Jewitt and D. Tholen on Mauna Kea obtained the last astrometry for fragments P2, Q2, Q1, R, and S, with P2 caught only 1.33 days before its impact. Observations of several fragments even closer to impact were made by the Hubble Space Telescope, but these did not provide astrometry, except for a measurement of Q2 relative to Q1 within 10 hours of the Q2 impact.

Our final set of predicted impact parameters was issued on the UMD e-mail exploder only 4 hours before impact A. The impact time uncertainties were down in the 3–5 minute range (1-sigma) for most of the fragments, not much different from our original projection of 3 minutes made nine months earlier. Although for most of the fragments, observers had not been able to obtain astrometry as close to impact as we had hoped, they had contributed many more measurements than we had anticipated—about 3200 in total, spread over 20 fragments. Extensive and accurate astrometric data had been received from several observatories, including Catalina Station, Kavalur, Klet, Kuma Kogen, La Palma, La Silla, Mauna Kea, McDonald, Siding Spring, Steward, and the U.S. Naval Observatory at Flagstaff. The highly-accurate Hipparcos-based astrometry was also unanticipated, and it contributed greatly to the accuracy of the orbital solutions.

3. Estimates of impact times from observed phenomena

During and after impact week, our attention turned to the problem of determining the actual impact times, based on the timing of observed impact phenomena. This was especially important for maximizing the data return from the Galileo spacecraft, which had viewed the impacts directly. Because of difficulties with its main antenna, the spacecraft had recorded most of its impact observations on tape, and could replay only a small fraction of the data back to Earth. Accurate impact time estimates would help to quickly locate the portions of data obtained around the times of the impacts. Fortunately, observers using Earth-based telescopes and the HST had detected a variety of impact phenomena, and promptly made the times available on the e-mail exploder.

After the first few impact events, it became clear that our predicted impact times were systematically early by 5–10 minutes. This conclusion was based on the assumption that the impacts occurred around the times of the earliest phenomena for each event. Although various types of impact observations were reported, the most robust and consistent set were the phenomena seen in the near-infrared and mid-infrared wavelengths. These light curves followed a consistent pattern, starting with a *precursor* flash, and sometimes even two, followed ∼ 6 minutes later by the start of a dramatic brightening which later became known as the *main event*. This surprisingly bright feature peaked about 10 minutes after the precursor (see the chapter by Nicholson for details).

The interpretation of the IR light curve features was initially puzzling, with the unusual viewing geometry complicating an already poorly-understood process. The limb of Jupiter just barely occulted the impact sites, and Jupiter's rotation brought them into full view anywhere from 20 minutes later for impact A, to 10 minutes later for W. The precursor was generally believed to be associated with the impact itself, but whether it was the meteor phase being observed directly, or an indirect view of the impact explosion reflected off incoming cometary debris, was not clear. Based on our predictions of how far behind the limb the impacts occurred, the meteors would have to be very high in Jupiter's atmosphere to be visible from Earth, especially for the earlier impacts. The interpretation of the main event was also uncertain. It could not be the plume rising above the limb of the planet, or the plume emerging into sunlight, because models suggested this would occur only a minute or two after impact (Boslough *et al.* 1994). Another possibility was that the main event was simply the impact site rotating into view, but then there should have been a variation in the time between precursor and main event, according to how far behind the limb the impact occurred.

Confirmation that the IR precursors occurred near the time of impact came from photometric observations obtained by the Photopolarimeter Radiometer (PPR) instrument on board Galileo. Transmitted to Earth within a day of the events, the PPR light curves of the H and L impacts displayed a 2-second rise to peak, followed by a plateau and slow decrease, lasting a total of 25–35 s (Martin *et al.* 1995). The sharp rise was interpreted as corresponding to the final moments of the bolide's trajectory, while the plateau and decay were due to the subsequent expanding and cooling fireball. The times of the initial PPR detection of the H and L impacts matched the times of precursors flashes to within a minute or so, although most of the reported flashes followed the PPR start times by about 1 minute. The PPR times also provided the first accurate calibration of our predicted impact times. The predictions for H and L were an average of 7 minutes early, an effect we subsequently concluded was due to systematic errors in the star catalogs.

Shortly after the impacts ended, we compiled our best estimates of the actual impact times, based on the reported times of various observed phenomena (Yeomans and Chodas 1994a). For impacts H and L, we simply adopted the times of initial detection in the PPR data. For the majority of the other impacts, which had consistent reports of precursor flashes and main events starting 5–6 minutes later, we generally took the impact time to be one minute before the flash time, or $\sim$ 6 minutes before the main event start. We also considered a set of impact times determined from measurements of the longitudes of impact spots seen in HST images (Hammel *et al.* 1995). The measured longitudes were differenced with predicted longitudes, converted to time differences by dividing by the rotation rate of Jupiter, and added onto the predicted impact times. These times could only be used as guides, however, as they seemed to be uncertain by 3–4 minutes. Finally, for fragments with no observed impact phenomena, we simply added an empirical correction of 7 minutes to the predicted impact times (Chodas and Yeomans 1994b).

The estimates of the actual impact times were used to position the Galileo tape for playback of selected portions of the data during the period from August 1994 through February 1995. Images of impacts K, N, and W taken by the Solid State Imager (SSI) were successfully returned, as were time series of spectra for impacts G and R taken by the Near-Infrared Mapping Spectrometer (NIMS) and Ultraviolet Spectrometer (UVS), as well as a PPR light curve for impact G. The Galileo data yielded accurate impact times for a total of 8 impacts: G, H, K, L, N, Q1, R, and W. The new impact time data confirmed our conclusion that the impact predictions were $\sim$ 7 minutes early.

The NIMS light curves for both G and R showed two phases—a fireball phase, due to the hot, expanding plume formed from the impact explosion, and a splash phase attributed to plume material falling back onto the atmosphere, heating it and producing thermal emission. For both the G and R events, the splash phase started $\sim$ 360 seconds after impact, and continued increasing for several minutes, through the end of the data sets (Carlson *et al.* 1995b). The delay between impact and onset of the splash phase seemed to be an intrinsic property of the impacts. Furthermore, it matched the 6 minute delay between first precursor and main event start seen in ground-based IR light curves of all the well-observed impacts, suggesting that the onset of the main event was not controlled by observing geometry, and the region of atmospheric heating was directly observable from Earth for most, if not all, the impacts (Zahnle and MacLow 1995).

With this new piece of the puzzle in place, a convincing explanation of the IR light curves was proposed by a number of authors, including Boslough *et al.* (1995), Zahnle and MacLow (1995), Hamilton *et al.* (1995), and Nicholson *et al.* (1995). The scenario, described in detail in the chapters by Nicholson and Sekanina, is summarized as follows. The first precursor (PC1) is due to thermal emission of the meteor trail in the Jupiter's upper atmosphere; its flux peaks as the bolide passes behind the limb, ranging from ~ 15 s before impact for fragment A to ~ 5 s before impact for W. The impact itself occurs at the initial peak of the PPR and SSI light curves, and is not visible from the Earth. At ~ 100 s after impact for A, decreasing to ~ 30 s after impact for W, a self-luminous fireball rises above the limb into Earth view, giving rise to the start of the second precursor (PC2). As the fireball rises and expands, the IR flux increases, but the plume rapidly cools, causing the signal to decay. Still rising, the plume emerges into sunlight, and reaches a maximum height of ~ 3000 km above the 100-mbar pressure level about 8 minutes after impact (Hammel *et al.* 1995). Meanwhile, the main event (ME) starts ~ 360 s after impact, as plume material begins splashing down onto the top of jovian atmosphere.

Table 1 summarizes, in chronological order, the times of key events in the observed impact phenomena, from which we can infer the actual impact times. The list is not meant to be exhaustive: it includes only the Earth-based infrared observations, events observed from Galileo, and relevant images from HST. The data were obtained from published reports, private communications, and a survey of the participants at IAU Colloquium 156. The listed times are generally mid-exposure times, while the uncertainties generally reflect the sampling times of the observations.

We estimated the actual impact times by fitting the times of the observed phenomena to the generic interpretation of light curves described above. These estimates are included in Table 5, along with a host of other results which are discussed later. As an aid in interpreting the impact phenomena, we have included in Table 1 the times of observed events relative to our estimated impact times, T_*. For some of the impact events, the interpretation of phenomena is uncertain, as outlined in the following paragraphs. The orbital solutions referred to in these notes are discussed in the next section.

• Impact A: Hammel *et al.* (1995) suggest that the HST image centered at 20:13:23 UT shows the bolide, since the next frame, centered at 20:15:18 UT shows nothing. Herbst *et al.* (1995) argue that the bright pixels in the first HST frame are due to the plume, since a precursor was seen from Calar Alto over two minutes earlier. Why then does the HST frame at 20:15:18 show nothing? Possibly because it was a short exposure, and possibly because the plume had cooled and had not yet emerged into sunlight. Although Herbst *et al.* could not identify which precursor they saw, due to a data outage, it seems likely that it was PC2. None of the impacts earlier than G produced first precursors, as they were simply too far behind the limb. If the precursor really was PC2, however, it occurs somewhat too soon after our estimated impact time, which was derived from the ME start time. It is possible that the impact occurred ~ 1 minute earlier, and the main event start was delayed because the splash area was entirely beyond the limb. The orbital solutions certainly favor an earlier impact.

• Impact B: The 17-minute duration of the faint event observed from Keck suggests that it was a faint main event, and our impact time estimate is based on this interpretation. However, the orbital solution clearly favors a later impact time, indicating the Keck observation may be a long second precursor.

• Impact M: We assume the faint brightening seen from Keck was a very faint main event. The orbital solution for this lost fragment is so poorly determined that the predicted impact time cannot assist the interpretation.

Imp	Date	Time (UT)	± (s)	Event[†]	$T - T_*$ (s)	Reference
A	16	20:11:29	5	PC2? start (Calar Alto 2.3 μm)	+ 49	Herbst et al. 1995
		20:13:23	7	HST—Bright pixels (888 nm)	+163	Hammel et al. 1995
		20:15:18	2	HST—Nothing (888 nm)	+278	Hammel et al. 1995
		20:16:56	5	ME start (Calar Alto 2.3 μm)	+376	Herbst et al. 1995
		20:18:24	8	HST—Plume in sunlight (953 nm)	+464	Hammel et al. 1995
B	17	02:56		ME? start (Keck, 3.3 μm)	+360	de Pater et al. 1994
C	17	07:11:57	15	PC2 start (AAT)	+ 67	Meadows, priv. comm.
		07:12:00	30	PC2 start (ANU)	+ 70	McGregor, priv. comm.
		07:12:07	5	PC2 start (Okayama, 2.3 μm)	+ 77	Takeuchi et al. 1995
		07:12:42	15	PC2 peak (AAT)	+112	Meadows priv. comm.
		07:13:20	30	PC2 peak (ANU)	+150	McGregor, priv. comm.
		07:16:30	30	ME start (Okayama, 2.3 μm)	+340	Takeuchi et al. 1995
		07:16:45	15	ME start (AAT)	+355	Meadows et al. 1995
		07:17:00	30	ME start (ANU)	+370	McGregor, priv. comm.
		07:17		ME start (IRTF)	+370	Orton et al. 1995b
D	17	11:54:30	30	PC2 peak (ANU)	+120	McGregor, priv. comm.
		11:54:46	180	PC2 peak (AAT)	+136	Meadows et al. 1995
		11:58:30	30	ME start (ANU)	+360	McGregor, priv. comm.
		12:00:30	180	ME start (AAT)	+480	Meadows et al. 1995
		12:00:48	45	ME start (SPIREX)	+498	Severson, priv. comm.
E	17	15:16:54	30	ME start (SAAO)	+314	Sekiguchi, priv. comm.
		15:17:49	0	ME start (SPIREX)	+369	Severson, priv. comm.
		15:17:56	5	ME start (Calar Alto 2.3 μm)	+376	Herbst et al. 1995
		15:19:31	15	HST—Plume in sunlight (888 nm)	+471	Hammel et al. 1995
F	18			No impact observations reported		

TABLE 1. Compilation of times of selected impact phenomena. The column entitled $T - T_*$ gives the event times relative to the accepted impact times, which appear in Table 5.

Imp	Date	Time (UT)	± (s)	Event†	$T - T_*$ (s)	Reference
G	18	07:32:20	30	PC1 start (ANU)	− 73	McGregor, priv. comm.
		07:32:58	180	PC1 start (AAT)	− 35	Meadows et al. 1995
		07:33:31	15	HST—Bright pixels (888 nm)	− 2	Hammel et al. 1995
		07:33:33	5	Galileo PPR peak (945 nm)	0	Martin et al. 1995
		07:33:37	5	Galileo NIMS start (1–5 μm)	+ 4	Carlson et al. 1995a
		07:34:32	30	PC2 start (SPIREX)	+ 59	Severson, priv. comm.
		07:34:43	30	PC2 start (ANU)	+ 70	McGregor, priv. comm.
		07:35:24	8	HST—Emission in shadow (888 nm)	+111	Hammel et al. 1995
		07:35:30	30	PC2 peak (ANU)	+117	McGregor, priv. comm.
		07:35:50	180	PC2 (AAT)	+137	Meadows priv. comm.
		07:38:24	8	HST—Plume in sunlight (953 nm)	+291	Hammel et al. 1995
		07:39:30	30	ME start (ANU)	+357	McGregor, priv. comm.
		07:39:37	40	ME start (SPIREX)	+364	Severson, priv. comm.
		07:39:41	3	Galileo splash start (NIMS)	+368	Carlson et al. 1995b
		07:41:15	180	ME start (AAT)	+462	Meadows priv. comm.
H	18	19:31:38	3	PC1 start (Pic du Midi 2.1 μm)	− 21	Colas et al. 1995
		19:31:45	20	PC1 peak (Calar Alto 2.3 μm)	− 14	Hamilton et al. 1995
		19:31:46	4	PC1 peak (Pic du Midi 2.1 μm)	− 13	Colas et al. 1995
		19:31:59	1	Galileo PPR peak (945 nm)	0	Martin et al. 1995
		19:32:30	1	PC2 start (Calar Alto 3.1 μm)	+ 31	Tozzi et al. 1995
		19:32:41	2	PC2 start (Pic du Midi 2.1 μm)	+ 42	Colas et al. 1995
		19:32:47	12	PC2 start (Calar Alto 2.3 μm)	+ 48	Hamilton et al. 1995
		19:32:57	3	PC2 start (La Silla)	+ 58	Email exploder report
		19:33:06	2	PC2 peak (Pic du Midi 2.1 μm)	+ 67	Colas et al. 1995
		19:33:10	1	PC2 peak (Calar Alto 3.1 μm)	+ 71	Tozzi et al. 1995
		19:37:25	2	ME start (Pic du Midi 2.1 μm)	+326	Colas et al. 1995
		19:37:27	15	ME start (Calar Alto 2.3 μm)	+328	Hamilton et al. 1995
		19:37:30	1	ME start (Calar Alto 3.1 μm)	+331	Tozzi et al. 1995
J	19			No impact observations reported		

TABLE 1. *Continued*

Imp	Date	Time (UT)	± (s)	Event[†]	T − T* (s)	Reference
K	19	10:20:41	30	Leader start (ANU)	−216	McGregor, priv. comm.
		10:22:42	40	Leader start (AAT)	− 95	Meadows et al. 1995
		10:23:03	30	PC1 start (ANU)	− 74	McGregor, priv. comm.
		10:23:19	40	PC1 start (AAT)	− 58	Meadows priv. comm.
		10:23:57	40	PC1 peak (AAT)	− 20	Meadows priv. comm.
		10:24:03	5	PC1 peak (Okayama, 2.3 μm)	− 14	Watanabe et al. 1995
		10:24:17	2	Galileo SSI peak (890 nm)	0	Chapman et al. 1995
		10:25:03	5	PC2 start (Okayama, 2.3 μm)	+ 46	Watanabe et al. 1995
		10:25:24	30	PC2 start (ANU)	+ 67	McGregor, priv. comm.
		10:25:26	5	PC2 peak (Okayama, 2.3 μm)	+ 69	Watanabe et al. 1995
		10:25:30	30	PC2 peak (ANU)	+ 73	McGregor, priv. comm.
		10:30:23	5	ME start (Okayama, 2.3 μm)	+366	Watanabe et al. 1995
		10:30:30	30	ME start (ANU)	+373	McGregor, priv. comm.
L	19	22:16:18	3	PC1 start (Calar Alto 2.3 μm)	− 31	Hamilton et al. 1995
		22:16:41	3	PC1 peak (Calar Alto 2.3 μm)	− 8	Hamilton et al. 1995
		22:16:49	1	Galileo PPR peak (945 nm)	0	Martin et al. 1995
		22:17:27	3	PC2 start (Calar Alto 2.3 μm)	+ 38	Hamilton et al. 1995
		22:17:29	5	PC2 start (Pic du Midi 2.1 μm)	+ 40	Colas et al. 1995
		22:17:40	2	PC2 start (La Palma 12 μm)	+ 51	Lagage et al. 1995
		22:17:58	5	PC2 peak (Pic du Midi 2.1 μm)	+ 69	Colas et al. 1995
		22:22	60	ME start (Calar Alto 2.3 μm)	+311	Hamilton et al. 1995
		22:22:55	18	ME start (Pic du Midi 2.1 μm)	+366	Colas et al. 1995
M	20	06:08	60	ME? start (Keck)	+360	de Pater, priv. comm.
N	20	10:29:20	2	Galileo SSI peak (890 nm)	0	Chapman et al. 1995
		10:35:40	30	ME start (ANU)	+380	McGregor, priv. comm.
P2	20			No impact observations reported		
P1	20			No impact observations reported		

TABLE 1. *Continued*

Imp	Date	Time (UT)	± (s)	Event†	T − T* (s)	Reference
Q2	20	19:44		PC2? start (Pic du Midi 2.1 μm)	0	Email exploder report
		19:44:10	1	PC2? start (Calar Alto 3.1 μm)	+10	Tozzi et al. 1995
		19:44:40	1	PC2? peak (Calar Alto 3.1 μm)	+40	Tozzi et al. 1995
		19:44:47	3	PC2? start (Calar Alto 2.3 μm)	+47	Herbst et al. 1995
		19:52:10	1	ME start (Calar Alto 3.1 μm)	+490	Tozzi et al. 1995
		19:52:24	15	ME start (Calar Alto 2.3 μm)	+504	Herbst et al. 1995
Q1	20	20:09:50	1	PC0 start (Calar Alto 3.1 μm)	−243	Tozzi et al. 1995
		20:10:30	1	PC0 peak (Calar Alto 3.1 μm)	−203	Tozzi et al. 1995
		20:13		PC1 (Pic du Midi 2.1 μm)	−53	Email exploder report
		20:13:15	1	PC1 start (Calar Alto 3.1 μm)	−38	Tozzi et al. 1995
		20:13:15	3	PC1 start (Calar Alto 2.3 μm)	−38	Herbst et al. 1995
		20:13:40	1	PC1 peak (Calar Alto 3.1 μm)	−13	Tozzi et al. 1995
		20:13:53	1	Galileo PPR peak (945 nm)	0	Martin et al. 1995
		20:18:10	1	PC3 start (Calar Alto 3.1 μm)	+257	Tozzi et al. 1995
		20:18:40	1	PC3 peak (Calar Alto 3.1 μm)	+287	Tozzi et al. 1995
		20:19:15	1	ME start (Calar Alto 3.1 μm)	+322	Tozzi et al. 1995
		20:19:47	3	ME start (Calar Alto 2.3 μm)	+354	Herbst et al. 1995
R	21	05:34:32	180	PC1 start (AAT)	−25	Meadows priv. comm.
		05:34:44	8	PC1 start (Keck 2.3 μm)	−13	Graham et al. 1995
		05:34:52	10	PC1 peak (Palomar 4.5 μm)	−5	Nicholson et al. 1995
		05:34:52	8	PC1 peak (Keck 2.3 μm)	−5	Graham et al. 1995
		05:35:08	22	Galileo NIMS start (1–5 μm)	+11	Carlson et al. 1995b
		05:35:27	10	PC2 start (Palomar 4.5 μm)	+30	Nicholson et al. 1995
		05:35:46	8	PC2 start (Keck 2.3 μm)	+49	Graham et al. 1995
		05:35:48	10	PC2 peak (Palomar 4.5 μm)	+51	Nicholson et al. 1995
		05:35:54	8	PC2 peak (Keck 2.3 μm)	+57	Graham et al. 1995
		05:38:34	30	ME start (Palomar, 3.2 μm)	+217	Nicholson et al. 1995
		05:39:49	10	ME start (Palomar, 4.5 μm)	+292	Nicholson et al. 1995
		05:40:00	30	ME start (ANU)	+303	McGregor, priv. comm.
		05:40:57	8	ME start (Keck 2.3 μm)	+360	Graham et al. 1995
		05:41:00	11	Galileo splash start (NIMS)	+363	Carlson et al. 1995b
		05:42:22	180	ME start (AAT)	+445	Meadows et al. 1995

TABLE 1. *Continued*

Imp	Date	Time (UT)	± (s)	Event†	$T - T_*$ (s)	Reference
S	21	15:17:36	30	PC2 peak (SAAO)	+ 66	Sekiguchi, priv. comm.
		15:19:20		PC2 start (Vainu Bappu, 1.65 μm)	+170	Bhatt 1994
		15:22	120	ME start (Calar Alto 2.3 μm)	+330	Herbst et al. 1995
		15:22:40	30	ME start (SAAO)	+370	Sekiguchi, priv. comm.
T	21			No impact observations reported		
U	21	22:00:37		ME? start (McDonald)	+ 35	Cochran, priv. comm.
V	22	04:23:09	10	PC1? start (Palomar)	− 11	Nicholson, this vol
		04:23:13	60	PC1? (AAT)	− 7	Meadows et al. 1995
W	22	08:06:16	2	Galileo SSI peak (560 nm)	0	Chapman et al. 1995
		08:06:16	1	HST—Emission in shadow (540 nm)	0	Hammel et al. 1995
		08:06:24	30	PC2 start (ANU)	+ 8	McGregor, priv. comm.
		08:06:54	180	PC2 start (AAT)	+ 38	Meadows, priv. comm.
		08:07:10	30	PC2 peak (ANU)	+ 54	McGregor, priv. comm.
		08:09:21	5	HST—Plume in sunlight (409 nm)	+185	Hammel et al. 1995
		08:12:00	30	ME start (ANU)	+344	McGregor, priv. comm.
		08:12:20	180	ME start (AAT)	+364	Meadows et al. 1995

TABLE 1. *Concluded*

† Table 1 abbreviations: AAT: Anglo-Australian Telescope, Mount Stromlo and Siding Spring Observatories, ANU: Australian National University, Mount Stromlo and Siding Spring Observatories, HST: Hubble Space Telescope, IRTF: NASA Infrared Telescope Facility, Mauna Kea, NIMS: Near Infrared Mapping Spectrometer, PPR: Photopolarimeter Radiometer, SAAO: South African Astrophysical Observatory, SPIREX: South Pole Infrared Experiment, SSI: Solid State Imager.

• Impact Q2: We have assumed the precursor observed at Calar Alto was PC2. If it was PC1, the estimated impact time would be $\sim$ 1 minute later. Either way, the main event starts later than expected, possibly because it was very faint, and its real onset was below the limits of detectability.

• Impact Q1: A total of three precursors were seen from Calar Alto in the 3.1 μm band (Tozzi *et al.* 1995). Only one of these (PC1) fits the expected pattern relative to the main event start time; the others have been labeled PC0 and PC3 in the table, and remain unexplained.

• Impact R: Although Galileo NIMS data are available for this impact, they cannot be used to constrain the impact time very precisely, because the sampling time was large ($\sim$ 11 s), and the sample nearest the impact time was missed (Carlson *et al.* 1995b). We adopt the impact time derived by Sekanina (this volume) using the ground-based IR light curves.

• Impact U: A possible detection is listed for this impact. It is not clear why larger telescopes, observing at the same time under excellent conditions, did not see the event. The reported time is consistent with the impact time derived from our final orbital solution, which is the time we adopt.

• Impact V: Light curves displayed only a short flash, which had the appearance of a faint first precursor; no main event was seen. The V fragment may have been too small to produce a plume or main event.

4. Post-impact improvements to the orbit solutions

In order to obtain the most accurate possible final estimates of the impact parameters, it was necessary to refine the orbit solutions from which they were computed. The most important improvement needed was to make the orbit solutions consistent with the observed impact times. If the solutions could be updated to "predict" the correct impact times, estimates of other parameters such as the impact locations would also become more accurate. Although this update could have been accomplished by making the impact time a constraint and forcing the orbit solution to satisfy it exactly, a better approach was simply to use the impact time as an additional measurement in the solution process. Accordingly, we augmented our orbit determination software to handle an impact time as a new measurement type. The new measurements were assigned conservative uncertainties—typically 5 s (1-sigma) for the impacts observed by Galileo, and 60 s or larger for fragments with impact times inferred from ground-based observations. We also modified our definition of impact slightly, raising it up to the 100-mbar level of Jupiter's atmosphere as defined by Lindal *et al.* (1981), but this change made little difference in the final solutions.

Using the impact times as measurements now, we computed new orbital solutions for the 16 fragments with observed impact phenomena. As an additional refinement, the planetary ephemeris was updated to the more accurate DE403 (Standish 1995, private communication). As a check, impact times were "predicted" from the new solutions; as expected, they matched the accepted times to within the assigned uncertainties. The systematic 7-minute error had been eliminated, at least for the 16 fragments whose impacts were observed. Significantly, the inclusion of the impact time in the orbit solutions did not adversely affect the residuals for the remaining observations. Typically, they increased by less than 0.1 arcsec over the entire observation span, although the differences ranged as high as 0.3 arcsec for some fragments. The largest changes in residuals were nearest impact. Clearly, systematic star catalog errors did not have to be very large to cause the observed 7-minute error in our predicted impact times. Assuming the catalog

Fragment	Data Interval	Number of Obs.	Impact Obs.	Weighted r.m.s. (″)	Orbit Ref.
A	93 March 27–94 July 12.99	54	Y	0.22	A38
B	93 March 27–94 July 12.99	75	Y	0.29	B34
C	93 March 27–94 July 14.00	82	Y	0.19	C28
D	93 March 27–94 July 12.99	53	Y	0.21	D29
E	93 March 27–94 July 14.97	193	Y	0.20	E50
F	93 March 27–94 July 14.97	120	N	0.19	F33
G	93 March 27–94 July 17.18	268	Y	0.18	G52
H	93 March 27–94 July 15.17	272	Y	0.18	H43
K	93 March 27–94 July 14.98	281	Y	0.16	K45
L	93 March 27–94 July 16.16	249	Y	0.20	L49
N	93 March 27–94 July 11.09	63	Y	0.19	N36
P2	93 March 27–94 July 19.31	107	N	0.19	P37
P1	93 July 1–94 June 17.56	35	N	0.21	PA6
Q2	93 March 30–94 July 20.42	19	Y	0.10	QB13
Q1	93 March 27–94 July 19.31	273	Y	0.21	Q63
R	93 March 27–94 July 19.31	185	Y	0.22	R58
S	93 March 27–94 July 19.34	239	Y	0.23	S62
T	93 March 27–94 July 7.98	35	N	0.25	T22
U	93 March 27–94 July 7.98	26	N	0.29	U24
V	93 March 27–94 July 14.98	55	Y	0.23	V27
W	93 March 27–94 July 16.16	180	Y	0.26	W52

TABLE 2. Summary of orbit solutions. The data interval indicates the dates of the first and last observations used in the solutions, excluding the impact observation. The number of observations similarly excludes the impact observation. The following column indicates whether the impact time was used as an observation in the solution. The final two columns give the weighted rms residual and the orbit reference identifier.

errors were the culprit, the new residuals were now a better representation of the actual measurement errors. In other words, the inclusion of the impact time had moved at least a portion of the star catalog errors out of the orbit solution into the residuals, where they belong. A small effect on the predicted 1992 perijove distances was also noted—the new solutions lowered them by ∼ 500 km.

The inclusion of impact times was a powerful method for improving orbit solutions, but it was applicable only to fragments whose impacts were observed. How could the orbit solutions for the other fragments be improved as well? One possible technique was simply to add an empirical 7-minute correction to the predicted impact times for those fragments, and use these as pseudo-impact times when computing new solutions. But this was rather *ad hoc*. The approach we adopted was to improve orbit solutions by improving the measurements upon which the solutions were based.

Observers typically captured several fragments in each of their astrometric images, and reduced the positions of all the fragments using the same stars. Our technique took advantage of the fact that errors in the star positions produced the same measurement bias for all fragments in a given observation set. (An 'observation set' is the set of individual fragment measurements made from a single astrometric image and reduced together, presumably relative to a single set of reference stars.) The measurement bias can be seen clearly in a plot of the fragment residuals in a given set. The residuals typically cluster around a point offset from the origin by a few tenths of an arc second,

	e	q (AU)	ω (deg)	Ω (deg)	i (deg)	T_p (1994 TDB)
A	0.21620917	5.38056310	354.89352	220.537655	6.003294	Mar. 24.10320
B	0.21561980	5.38065243	354.90065	220.565286	5.990216	Mar. 24.52991
C	0.21516872	5.38041144	354.90826	220.581389	5.981965	Mar. 24.81127
D	0.21472534	5.38036971	354.91311	220.600306	5.972968	Mar. 25.10038
E	0.21441065	5.38031828	354.91615	220.613781	5.966629	Mar. 25.30070
F	0.21358484	5.38036243	354.93310	220.651606	5.948463	Mar. 25.96840
G	0.21288148	5.38011243	354.93416	220.680082	5.935514	Mar. 26.32445
H	0.21177932	5.37997339	354.94843	220.728309	5.912868	Mar. 27.08466
K	0.21042545	5.37977493	354.96756	220.788975	5.885070	Mar. 28.05310
L	0.20936108	5.37963171	354.98453	220.837514	5.863131	Mar. 28.84854
N	0.20827689	5.37953629	354.99998	220.886270	5.840458	Mar. 29.63448
P2	0.20788730	5.37960765	355.01194	220.904724	5.831257	Mar. 30.00933
P1	0.20774507	5.37968923	354.99938	220.909078	5.829288	Mar. 29.92297
Q2	0.20745337	5.37940298	355.01107	220.924560	5.823607	Mar. 30.23482
Q1	0.20740426	5.37934828	355.01099	220.927985	5.822819	Mar. 30.27142
R	0.20658149	5.37923176	355.02229	220.966459	5.805868	Mar. 30.87835
S	0.20573678	5.37912914	355.03337	221.006655	5.788483	Mar. 31.50454
T	0.20550407	5.37931862	355.03349	221.016204	5.783002	Mar. 31.63843
U	0.20516743	5.37914092	355.04151	221.034174	5.776310	Mar. 31.94418
V	0.20461625	5.37904612	355.05009	221.060787	5.764890	Apr. 1.37239
W	0.20428226	5.37890776	355.05614	221.077213	5.758370	Apr. 1.63988

TABLE 3. Osculating **heliocentric** orbital elements for the fragments of comet Shoemaker-Levy 9 at epoch 1994 May 8.0 TDB = JD 2449480.5 TDB. The elements are eccentricity (e), perihelion distance (q), argument of perihelion (ω), longitude of the ascending node (Ω), inclination (i), and time of perihelion passage (T_p). The angular orbital elements are referred to the ecliptic plane and equinox of J2000.

with a scatter much smaller than the bias. We concluded that the bias was mostly due to star catalog errors, while the scatter was mostly due to the actual errors of measurement.

When looking at residuals, we concentrated on the six fragments with the most accurate orbit solutions, G, H, K, L, Q, and W, which we called the *primary* fragments. These had the largest astrometric data sets and impact times known to within a few seconds from Galileo observations. Almost all of the 370 observation sets contained at least one primary fragment, and most contained several. Residuals for the primary fragments typically clustered around the bias point with a scatter smaller than that of the other residuals. Our estimate of the measurement bias of each observation set was obtained by averaging the residuals of the primary fragments. We then subtracted this bias from all measurements in the set to obtain corrected 'synthetic' observations. The measurements in each set were assigned a single noise value according to the scatter of the residuals. Because biases had been removed, most noise values were much smaller than in previous solutions, typically 0.2 arcsec for high-quality observations.

To test this method, we applied it to individual primary fragments to see whether we could correctly predict the impact times. For example, to test the method on fragment G, we computed observation biases by averaging the residuals of the other primary fragments, adjusted the G observations by subtracting off the biases, and computed the synthetic solution *without using the G impact time as an observation*. The impact times predicted by these test solutions were very close to the accepted times, within 30 s in most cases, giving us confidence that our approach could predict accurate impact times even for those fragments whose impacts were not observed.

	e	q (km)	ω (deg)	Ω (deg)	i (deg)	T_p (1994 TDB)
A	0.99860178	37359.28	43.224821	284.754106	88.510529	July 16.98246
B	0.99859179	37635.83	43.245621	284.022794	87.742186	July 17.26248
C	0.99860684	37238.92	43.218524	285.032690	88.799176	July 17.44353
D	0.99860765	37223.04	43.219369	285.010446	88.776404	July 17.64044
E	0.99860837	37208.32	43.217578	285.104456	88.875632	July 17.77938
F	0.99860544	37300.48	43.229297	284.546587	88.280449	July 18.17450
G	0.99861241	37121.95	43.213977	285.353708	89.136959	July 18.46555
H	0.99861664	37024.28	43.212378	285.468492	89.256703	July 18.96760
K	0.99862089	36930.22	43.209596	285.784875	89.585060	July 19.59097
L	0.99862481	36841.23	43.209208	285.978211	89.783022	July 20.08892
N	0.99863152	36676.73	43.209216	285.807826	89.604233	July 20.60125
P2	0.99863451	36604.26	43.213545	285.252492	89.013508	July 20.80606
P1	0.99863423	36611.50	43.233503	284.468538	88.212721	July 20.85685
Q2	0.99863402	36621.70	43.208505	286.026438	89.832856	July 20.98913
Q1	0.99862993	36731.86	43.208967	286.330548	90.150681	July 21.00939
R	0.99863325	36655.41	43.209414	286.439518	90.264716	July 21.40164
S	0.99863588	36598.27	43.209765	286.452323	90.278231	July 21.80819
T	0.99864005	36489.40	43.216762	285.166900	88.935962	July 21.93115
U	0.99863734	36567.67	43.208609	286.028497	89.832581	July 22.09104
V	0.99863940	36521.02	43.208271	286.195726	90.006578	July 22.35899
W	0.99863886	36538.68	43.212640	286.774798	90.611613	July 22.51406

TABLE 4. Osculating **jovicentric** orbital elements for the fragments of comet Shoemaker-Levy 9 at epoch 1994 May 8.0 TDB = JD 2449480.5 TDB. The elements are eccentricity (e), perijove distance (q), argument of perijove (ω), longitude of the ascending node (Ω), inclination (i), and time of perijove passage (T_p). The angular orbital elements are referred to the ecliptic plane and equinox of J2000.

Our final set of orbit solutions were computed using the synthetic method just described, with impact times used as observations when available. Table 2 summarizes these solutions, giving for each fragment the data interval, number of observations, the weighted rms residual, an indication of whether we used the impact time as an observation, and our orbit reference identifier. The final orbital elements themselves are given in heliocentric form in Table 3, and jovicentric form in Table 4. An independent orbit solution for fragment P1 is included for the first time. The extremely small weighted rms residuals for these solutions, less than 0.2 arcsec for half of the solutions, and less than 0.3 arcsec for the rest, is due to the removal of the measurement biases. The attentive reader may note that our data interval for fragment Q2 begins on March 30, 1993, well before Q2 was seen on its own. We have used the Q1 position on this date as a pseudo-measurement to constrain the Q2 solution because this was approximately the time Q2 split away from Q1 (Sekanina 1995).

It is interesting to integrate the orbit solutions backward to the 1992 perijove to see how closely the fragments come together. Figure 2 shows the clustering of the 1992 perijove times and perijove distances. Only on-train fragments are included, as off-train fragments presumably split well after perijove. The perijove times all fall within a 45-minute period, and the perijove distances within a 500-km range. The inter-fragment distances themselves are quite large, however, because of the dispersion in the perijove times.

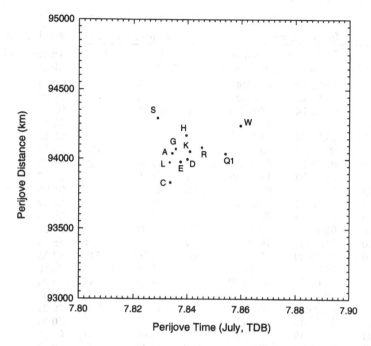

FIGURE 2. Plot of 1992 perijove distances vs. perijove times of our final orbit solutions for 12 on-train fragments. Even though the solutions have not been constrained to come together at this perijove, they do cluster fairly well. The perijove times, shown as day and fraction of day in July 1992, all fall within a 45-minute period.

5. Summary of impact times, impact locations, and impact geometries

Our final estimates of the impact times and locations of the fragments of SL9 are given in Table 5. The estimated impact times in this table are slight revisions to earlier estimates compiled at IAU Colloquium 156. Impact was defined to occur at the 100-mbar level of Jupiter's atmosphere. The impact estimates for all fragments except J and M are based on the independent orbit solutions discussed in the previous section. The estimates for the 'lost' fragments J and M were obtained by applying our tidal disruption model to the orbit for fragment Q1 and matching the astrometry of these two fragments relative to Q1. The third column of Table 5 contains our final pre-impact prediction for each of the fragments, taken from the sets of predictions we distributed electronically on the UMD e-mail exploder. The fourth column lists our final best estimates, which were inferred directly from impact phenomena for 16 fragments, as described in section 3, and computed from the orbit solutions for the rest. All times are as viewed from the Earth, and therefore include the light travel time. The impact time uncertainties are rough estimates which indicate our confidence level in the accepted time; they are not formal 1-sigma uncertainties. The impact latitude is jovicentric, while the longitude is System III, measured westwards on the planet. The meridian angle is the jovicentric longitude of the impact point measured from the midnight meridian towards the morning terminator. At the latitude of the impacts, the limb as viewed from the Earth is at meridian angle 76 deg, and the terminator is at meridian angle 87 deg.

The final column of Table 5 gives the angular distance of the impacts behind the limb, a more useful parameter than the Earth-Jupiter-fragment (E-J-F) angle we gave in our

Event	Impact Time (UTC)			Impact Location		Merid.	Ang. Dist.
	Date (July)	Predicted h m s	Accepted h m s	± (s)	Lat. (deg)	Lon. (deg)	Angle (deg)	Behind Limb (deg)
A	16	19:59:40	20:10:40	60	−43.35	184	65.40	7.7
B	17	02:54:13	02:50:00	180	−43.22	67	63.92	8.8
C	17	07:02:14	07:10:50	60	−43.47	222	66.14	7.1
D	17	11:47:00	11:52:30	60	−43.53	33	66.16	7.1
E	17	15:05:31	15:11:40	120	−43.54	153	66.40	6.9
F	18	00:29:21	00:35:45	300	−43.68	135	65.30	7.7
G	18	07:28:32	07:33:33	3	−43.66	26	67.09	6.4
H	18	19:25:53	19:31:59	1	−43.79	99	67.47	6.1
J	19	02:40	01:35	3600	−43.75	∼ 316	68.05	∼6
K	19	10:18:32	10:24:17	2	−43.86	278	68.32	5.5
L	19	22:08:53	22:16:49	1	−43.96	348	68.86	5.1
M	20	05:45	06:00	600	−43.93	∼ 264	69.25	∼5
N	20	10:20:02	10:29:20	2	−44.31	71	68.68	5.1
P2	20	15:16:20	15:21:11	300	−44.69	249	67.58	5.8
P1	20	16:30	16:32:35	800	−45.02	∼ 293	65.96	6.9
Q2	20	19:47:11	19:44:00	60	−44.32	46	69.26	4.7
Q1	20	20:04:09	20:13:53	1	−44.00	63	69.85	4.3
R	21	05:28:50	05:34:57	10	−44.10	42	70.21	4.1
S	21	15:12:49	15:16:30	60	−44.22	33	70.34	4.0
T	21	18:03:45	18:09:56	300	−45.01	141	67.73	5.7
U	21	21:48:30	22:00:02	300	−44.48	278	69.54	4.5
V	22	04:16:53	04:23:20	60	−44.47	149	69.96	4.2
W	22	17:59:45	08:06:16	1	−44.13	283	71.19	3.4

TABLE 5. Summary of impact times and locations

earlier sets of predictions. The use of the E-J-F angle has led to a small error in computing the precise distance of the impact behind the limb. Because of Jupiter's oblateness, the limb of Jupiter cannot be assumed to be located at an E-J-F angle of 90 deg. In fact, at the latitude of the impacts, the limb was at an E-J-F angle of ∼ 90.3 deg, moving the impacts a little closer to the limb than previously thought. Our final estimates put impact W less than 3.5 deg behind the limb.

Table 6 summarizes the impact velocities and directions as computed from our final orbit solutions. These parameters are all related to the velocity of the fragment relative to the impact point in a frame rotating with Jupiter at the System III rotation rate. Thus, the relative velocity includes a small component due to Jupiter's rotation. The incidence angle is measured from the local vertical, while the azimuth angle is measured from north towards the west.

6. Pre-breakup orbital history

Backward numerical integrations of SL9's orbital motion can provide clues as to the nature and origin of the object. Accurate knowledge of the comet's pre-breakup motion is essential in searches for the progenitor comet in existing image libraries. A pre-breakup detection would enable limits to be set on the size of the progenitor nucleus, and even a non-detection is useful, if we could be sure of the ephemeris. Tancredi *et al.* (1993) reported that they did not see the comet in a 90-min exposure of the Jupiter region taken in March 1992, which had a limiting magnitude of 21.3. The investigation of SL9's pre-breakup motion also helps determine when the comet was likely captured by Jupiter,

Event	Velocity (km s^{-1})	Incidence Angle (deg)	Azimuth Angle (deg)
A	61.23	43.30	14.37
B	61.12	43.27	13.34
C	61.29	43.29	14.89
D	61.29	43.25	14.91
E	61.31	43.26	15.08
F	61.23	43.12	14.33
G	61.36	43.24	15.59
H	61.39	43.20	15.86
K	61.46	43.21	16.48
L	61.50	43.18	16.86
N	61.48	43.00	16.73
P2	61.40	42.75	15.93
P1	61.28	42.49	14.77
Q2	61.53	43.02	17.15
Q1	61.57	43.21	17.59
R	61.60	43.17	17.8
S	61.61	43.12	17.96
T	61.41	42.59	16.06
U	61.55	42.95	17.38
V	61.58	42.98	17.69
W	61.68	43.20	18.57

TABLE 6. Summary of impact velocities and directions

and provides insight into the object's pre-capture heliocentric orbit. Unfortunately, SL9's orbit about Jupiter was among the most chaotic of any known solar system body, with an effective Lyapunov time on the order of 10 years (Benner and McKinnon 1995). As a result, a single backward numerical integration does not provide definitive answers on the orbital history of this object. A better approach is to account for the uncertainties in the initial conditions of the backward integrations, and to investigate the motion in a statistical manner using a Monte Carlo analysis (Chodas and Yeomans 1995).

The first difficulty encountered when investigating SL9's pre-breakup orbital history is how to solve the Humpty-Dumpty problem, *i.e.*, how to obtain the orbit for the progenitor nucleus from the orbits of the fragments. Our solution to this problem was simply to assume that fragment K was near the center of mass of the original nucleus, and that its motion was unaffected by the breakup. As initial conditions for the progenitor nucleus, we used our orbit solution for fragment K, along with the actual orbit uncertainties and their correlations. Fragment K was a natural choice, since it was closest to the mid-point of the train. We repeated our analyses with fragment L, which was also close to the mid-point, and obtained essentially the same results.

Our approach was to create a random ensemble of 1000 initial conditions whose statistics matched the actual orbital element uncertainties and correlations. Effectively, a six-dimensional uncertainty ellipsoid in orbital element space was populated with 1000 random points to obtain an ensemble of initial conditions consistent with the actual 6×6 covariance matrix of the orbital solution. As before, our dynamic model included solar and planetary perturbations, as well as perturbations from the Galilean satellites and Jupiter's oblateness (J2 and J4 terms).

Each sample point was integrated backward in time until it escaped from Jupiter, at which point its heliocentric orbital elements were tabulated. Orbits which had encounters

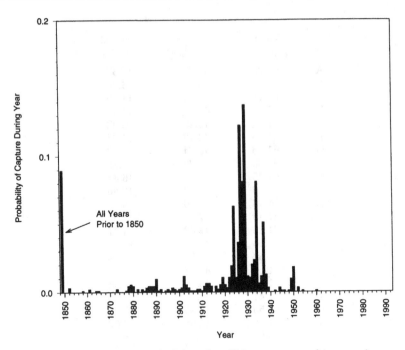

FIGURE 3. Histogram showing the probability that SL9 was captured in any given year back to 1850, based on a Monte Carlo analysis of 938 sample points. The most likely time of capture is 1929 ± 9 years (72% probability).

with Jupiter closer than that in 1992 were discarded. Escape was defined to occur when the jovicentric eccentricity exceeded unity and the distance from Jupiter exceeded 0.7 AU. Of course, the moment of escape in the backward integration is really the moment of capture when viewed in the forward direction.

Figure 3 shows a histogram of the number of samples which escaped from Jupiter each year back to 1850, when our integrations stopped. Nine percent of the samples were still in Jupiter orbit at the end of integrations. The most likely time of capture, with a probability of 72%, was 1929 ± 9 years. During its several-decade residency as a captured comet, SL9 orbited Jupiter with a period of 2–3 years and a semi-major axis of ~ 0.2 AU. Its orbit was highly inclined to Jupiter's equator, and oscillated between periods of near-circularity and periods of high eccentricity. Throughout this time, the comet remained within four degrees of Jupiter, as viewed from the Earth. Its pre-discovery ephemeris is fairly well-determined, at least as far back as 1979, and the ephemeris uncertainties grow to no more than 0.25 degree (1-sigma) during this time. Figure 4 shows a representative trajectory for the captured comet in a rotating jovicentric frame, following the comet from capture in 1928 to the comet's final orbit in 1992–1994. Although in this example the comet was captured from the direction of the Sun, other cases show the comet being pulled in from the anti-solar direction. Comets are typically captured as they pass near the libration points on the Jupiter-Sun line.

The pre-capture heliocentric orbits of our samples were all of low inclination ($i < 6$ deg) and moderately low eccentricity ($e < 0.4$). As shown in Fig. 5, the pre-capture orbits fell into two groups—those orbits well inside Jupiter's orbit, and those well outside. On the orbits interior to Jupiter's orbit, capture occurred at aphelion, while on those exterior, capture occurred at perihelion. None of the pre-capture orbits crossed Jupiter's orbit.

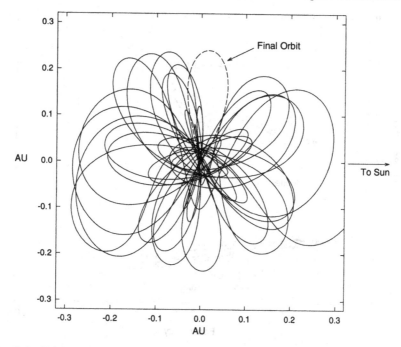

FIGURE 4. Orbital behavior of a representative trajectory for SL9, shown in a jovicentric rotating frame, and projected into the orbital plane of Jupiter. The comet enters the diagram from the solar direction in 1930, and completes 25 orbits about Jupiter before impact.

These findings are consistent with the general results of Carusi and Valsecchi (1979), and Kary and Dones (1995), who have shown that captures occur when minor bodies approach Jupiter along nearly-tangent orbits. In our analysis for SL9, capture from orbits interior to Jupiter's orbit was three times more likely than capture from orbits exterior to Jupiter's orbit. Benner and McKinnon (1995) obtained a similar result in their integrations for SL9, and noted that the preference for capture from interior orbits is really just a measure of the comparative ease with which captures (or escapes) occur at Jupiter's two libration points.

An important parameter used in classifying orbits of comets and asteroids is the Tisserand invariant with respect to Jupiter, T, which is approximately constant during encounters with the planet (Kresák 1979). The critical value of $T_o = 3$, where T_o is the value of T for $i = 0$, can be used to distinguish between cometary type orbits ($T_o < 3$) and asteroidal type orbits ($T_o > 3$). The Tisserand parameters for the samples in our analysis for SL9 straddled this boundary, with T_o values ranging from 2.99 to 3.04, and a mean of ~ 3.02, indicating that SL9's pre-capture orbit was probably asteroid-like. However, as noted by Benner and McKinnon (1995) the inner distribution of possible pre-capture orbits for SL9 overlaps a group of known comets, referred to as quasi-Hildas by Kresák. With Tisserand parameter values ranging from 3.00 to 3.04, these comets also have asteroid-like orbits. In fact, they occupy the same region in a/e phase space as the Hilda asteroids, although they are not in the same stable 3:2 resonances as the Hildas. Three members of the quasi-Hilda group, P/Gehrels 3, P/Smirnova-Chernykh, and P/Helin-Roman-Crockett, and one former member, P/Oterma, are plotted in Fig. 5. They reside comfortably inside the inner distribution of possible SL9 orbits. P/Oterma made a close approach to Jupiter in 1963, moved to an orbit exterior to that of Jupiter,

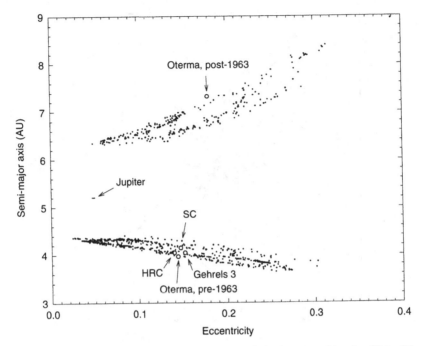

FIGURE 5. Scatter plot of ~ 850 possible pre-capture heliocentric orbits for SL9. The orbits of four known comets are shown as open circles; SC denotes P/Smirnova-Chernykh, and HRC denotes P/Helin-Roman-Crockett. P/Oterma's orbit is shown both before and after its 1963 close approach to Jupiter.

and now resides in the outer distribution of possible SL9 orbits. All four of these comets have either been temporarily captured by Jupiter in the past, or will be temporarily captured in the future (Yeomans and Chodas 1994b). Before its final capture, SL9 was probably also a member of this quasi-Hilda cometary group.

7. Summary and conclusions

This paper has reviewed the early orbit computation efforts for SL9, including the surprising discoveries that the comet was in orbit about Jupiter, and that it would impact the planet. We confirmed these results, and computed the probability of impact, which rose from 50% to near unity during the two-week period after the impact announcement, as more astrometric measurements were added to the orbit solutions. We also determined that the impact of the center of the train would occur on the far side of the planet. After solar conjunction, and in the months leading up to the impacts, we computed increasingly accurate orbit solutions for the individual fragments, using the growing set of astrometric observations. Our predictions of the times and locations of the impacts were regularly made available to the astronomical community via the electronic bulletin board and e-mail exploder operated by the University of Maryland.

After the impacts, we estimated the actual impact times from the times of observed impact phenomena, which we have compiled in Table 1. Our final predicted impact times were systematically early by ~ 7 minutes, probably due to systematic errors in the reference star catalogs used in the reduction of the fragments' astrometric positions. We refined our orbit solutions by using the observed impact times as additional data for 16 of the fragments, and by estimating and removing star catalog errors from all the astrometric observations. Our final orbit solutions for 21 fragments are summarized in Table 2, and the heliocentric and jovicentric orbital elements are presented in Tables 3 and 4, respectively. Our best estimates of the impact times and impact locations are given in Table 5. Sixteen of the impact times were derived from the times of observed impact phenomena, while the remaining times were computed directly from the orbit solutions. The new estimates for the impact locations are 0.5–1 deg closer to the limb than in previous estimates.

We investigated the pre-breakup orbital history of SL9 by performing a Monte Carlo analysis of backward integrations, using an ensemble of orbits whose mean and covariance were consistent with our orbit solution for fragment K. We assumed that this fragment originated near the center of mass of the progenitor nucleus, and that its motion was un-affected by the breakup process. Our analysis showed that SL9 had been orbiting Jupiter for decades before its discovery, and that it was most likely captured from heliocentric orbit in 1929 ± 9 years. Prior to capture, SL9 was in a low-inclination, low-eccentricity heliocentric orbit, entirely inside Jupiter's orbit, or, less likely, entirely outside. Its pre-capture orbit is consistent with a group of known comets called the quasi-Hildas.

As a part of our investigation of SL9, we developed a number of new techniques with regard to cometary orbit determination. We successfully determined the probability of collision of a comet and a planet. We accurately predicted the times and locations of the impacts of the cometary fragments on Jupiter. Our orbit computations used not only planetary and solar perturbations, but also perturbations due to the Galilean satellites and Jupiter's oblateness. We included the observed Jupiter impact times as data in our post-impact orbit solutions, and successfully removed star catalog biases from the sets of astrometric data. To our knowledge, the dynamical modeling of this comet's motion is the most complex cometary orbit determination problem yet undertaken, and our resultant orbit solutions for the 21 fragments of comet Shoemaker-Levy 9 have the smallest rms residuals of any comet to date.

This work would not have been possible without the selfless contributions of the many observers who supplied astrometric data for SL9. We would also like to thank the many observers who provided us with the precise times of impact-related phenomena. Finally, we wish to thank Z. Sekanina and P. Nicholson for helpful comments and suggestions. This work was supported by the NASA Planetary Astronomy Program. The research was performed at the Jet Propulsion Laboratory, California Institute of Technology, under contract with the National Aeronautics and Space Administration. Support for this work was also provided by NASA through grant number GO-5624.03-93A from the Space Telescope Science Institute, which is operated by the Association of Universities for Research in Astronomy, Incorporated, under NASA contract NAS5-26555.

REFERENCES

BENNER, L. A. M. & MCKINNON, W. B. 1995 On the orbital evolution and origin of comet Shoemaker-Levy 9. *Icarus* **118**, 155–168.

BHATT, H. C. 1994 *IAU Circular 6039*, July 27, 1994.

BOSLOUGH, M., CRAWFORD, D., ROBINSON, A., & TRUCANO, T. 1994 Mass and penetration depth of Shoemaker-Levy 9 fragments from time-resolved photometry. *Geophys. Res. Lett.* **21**, 1555.

BOSLOUGH, M., CRAWFORD, D., TRUCANO, T., & ROBINSON, A. 1995 Numerical modeling of Shoemaker-Levy 9 impacts as a framework for interpreting observations. *Geophys. Res. Lett.* **22**, 1821.

CAMPBELL, J. K. & SYNNOTT, S. P. 1985 Gravity field of the jovian system from Pioneer and Voyager tracking data. *Astron. J.* **90**, 364–372.

CARLSON, R. W., WEISSMAN, P. R., SEGURA, M., HUI, J., SMYTHE, W. D., JOHNSON, T. V., BAINES, K. H., DROSSART, P., ENCRENAZ, TH., LEADER, F. E. 1995a Galileo infrared observations of the Shoemaker-Levy 9 G impact fireball: A preliminary report. *Geophys. Res. Lett.* **22**, 1557.

CARLSON, R. W., WEISSMAN, P. R., HUI, J., SMYTHE, W. D., BAINES, K. H., JOHNSON, T. V., DROSSART, P., ENCRENAZ, T., LEADER, F., & MEHLMAN, R. 1995b Some timing and spectral aspects of the G and R collision events as observed by the Galileo near-infrared mapping spectrometer. In Proceedings of the European SL-9/Jupiter Workshop (eds. R. West and H. Bohnhardt) pp. 69–73. ESO.

CARUSI, A., & VALSECCHI, G. B. 1979 Numerical simulations of close encounters between Jupiter and minor planets. In *Asteroids,* (ed. T. Gehrels), pp. 391–416. University of Arizona Press.

CARUSI, A., KRESAK, L., PEROZZI, E., VALSECCHI, G. B. 1985 Long-term evolution of short-period comets. Adam Hilger Ltd.

CHAPMAN, C. R., MERLINE, W. J., KLAASEN, K., JOHNSON, T. V., HEFFERNAN, C., BELTON, M. J. S., INGERSOLL, A. P., & THE GALILEO IMAGING TEAM 1995 Preliminary results of Galileo direct imaging of S-L 9 impacts. *Geophys. Res. Lett.* **22**, 1561.

CHODAS, P. W. 1993 Estimating the impact probability of a minor planet with the earth. *Bull. Am. Astron. Soc.* **25**, 1236.

CHODAS, P. W., & YEOMANS, D. K. 1993 The upcoming collision of comet Shoemaker-Levy 9 with Jupiter. *Bull. Am. Astron. Soc.* **25**, 1042.

CHODAS, P. W., & YEOMANS, D. K. 1994a The impact of comet Shoemaker-Levy 9 with Jupiter. *Bull. Am. Astron. Soc.* **26**, 1022.

CHODAS, P. W., & YEOMANS, D. K. 1994b Comet Shoemaker-Levy 9 impact times and impact geometries. *Bull. Am. Astron. Soc.* **26**, 1569.

CHODAS, P. W. & YEOMANS, D. K. 1995 The pre-breakup orbital history of comet Shoemaker-Levy 9. *Bull. Am. Astron. Soc.* **27**, 1111.

CHODAS, P. W., YEOMANS, D. K., & SEKANINA, Z. 1994 *IAU Circular 5941*, February 24, 1994.

COLAS, F., TIPHENE, D., LECACHEUX, J., DROSSART, P., DE BATZ, B., PAU, S., ROUAN, D., & SEVRE, F. 1995 Near infrared imaging of SL9 impacts on Jupiter from Pic-du-Midi observatory. *Geophys. Res. Lett.* **22**, 1765–1768.

DAVIES, M. E., ABALAKIN, V. K., BRAHIC, A., BURSA, M., CHOVITZ, B. H., LIESKE, J. H., SEIDELMANN, P. K., SINCLAIR, A. T., & TJUFLIN, Y. S. 1992 Report of the IAU/IAG/COSPAR working group on cartographic coordinates and rotational elements of the planets and satellites: 1991. *Cel. Mech.* **53** 377–397.

DE PATER, I., GRAHAM, J., & JERNIGAN, G. 1994 *IAU Circular 6024*, July 17, 1994.

GRAHAM, J. R., DE PATER, I., JERNIGAN, J. G., LIU, M. C., BROWN, M. E. 1995 The fragment R collision: W. M. Keck Telescope Observations of SL9. *Science* **267**, 1320–1223.

HAMILTON, D. P., HERBST, T. M., RICHICHI, A., BOHNHARDT, H., & ORTIZ, J. L. 1995 Calar Alto observations of Shoemaker-Levy 9: Characteristics of the H and L impacts. *Geophys. Res. Lett.* **22**, 2417–2420.

HAMMEL, H. B., BEEBE, R. F., INGERSOLL, A. P., ORTON, G. S., MILLS, J. R., SIMON, A. A., CHODAS, P. W., CLARKE, J. T., DEJONG, E., DOWLING, T. E., HARRINGTON, J., HUBER, L. F., KARKOSCHKA, E., SANTORI, C. M., TOIGO, A., YEOMANS, D. K., &

WEST, R. A. 1995 Hubble Space Telescope imaging of Jupiter: atmospheric phenomena created by the impact of comet Shoemaker-Levy 9. *Science* **267**, 1288–1296.

HERBST, T. M., HAMILTON, D. P., BOHNHARDT, H., & ORTIZ-MORENO, J. L. 1995 Near infrared imaging and spectroscopy of the SL-9 impacts from Calar Alto. *Geophys. Res. Lett.* **22**, 2413–2416.

JEWITT, D., LUU, J., & CHEN, J. 1993 Physical properties of split comet Shoemaker-Levy 9. *Bull. Am. Astron. Soc.* **25**, 1042.

KARY, D. M., & DONES, L. 1995 Capture statistics of short-period comets: Implications for comet Shoemaker-Levy 9. *Icarus*, (submitted).

KRESÁK, L. 1979 Dynamical interrelations among comets and asteroids. In *Asteroids*, (ed. T. Gehrels). pp. 289–309. University of Arizona Press.

LAGAGE, P. O., ALDEMARD, P. H., PANTIN, E., JOUAN, R., MASSE, P., SAUVAGE, M., OLOFSSON, G., HULDTGREN, M., NORDH, L., BELMONTE, J. A., REGULO, C., RODRIGUEZ ESPINOSA, J. M., VIDAL, L., MOSSER, B., ULLA, A., & GAUTIER, D. 1995 Collision of Shoemaker-Levy 9 fragments A, E, H, L, Q1 with Jupiter: Mid-Infrared light curves. *Geophys. Res. Lett.* **22**, 1773–1776.

LIESKE, J. H. 1977 Theory of motion of Jupiter's Galilean satellites. *Astron. Astrophys.* **56**, 333–352.

LIESKE, J. H. 1994 Galilean satellite ephemerides E4. *JPL Engineering Memorandum 314-545*, June 19, 1994.

LINDAL, G. F., WOOD, G. E., LEVY, G. S., ANDERSON, J. D., SWEETNAM, D. N., HOTZ, H. B., BUCKLES, B. J., HOLMES, D. P., DOMS, P. E., ESHLEMAN, V. R., TYLER, G. L., & CROFT, T. A. 1981 The atmosphere of Jupiter: An analysis of the Voyager radio science occultation measurements. *J.G.R.* **86**, 8721–8727.

MARSDEN, B. G. 1993a *IAU Circular 5726*, March 27, 1993.

MARSDEN, B. G. 1993b *IAU Circulars 5744-5745*, April 3, 1993.

MARSDEN, B. G. 1993c *IAU Circular 5800-5801*, May 22, 1993.

MARSDEN, B. G. 1993d *IAU Circular 5892*, November 22, 1993.

MARSDEN, B. G. 1993e *IAU Circular 5906*, December 14, 1993.

MARSDEN, B. G. 1994a *IAU Circulars 5936-5937*, February 18–21, 1994.

MARSDEN, B. G. 1994b *Minor Planet Circulars 23105-23107*, February 26, 1994.

MARTIN, T. Z., ORTON, G. S., TRAVIS, L. D., TAMPPARI, L. K., & CLAYPOOL, I. 1995 Observation of Shoemaker-Levy impacts by the Galileo Photopolarimeter Radiometer. *Science* **268**, 1875–1879.

MEADOWS, V., CRISP, D., ORTON, G., BROOKE, T., & SPENCER, J. 1995 AAT IRIS observations for the SL-9 impacts and initial fireball evolution. In *Proceedings of the European SL-9/Jupiter Workshop*, (eds. R. West and H. Bohnhardt) pp. 129–134. ESO.

NAKANO, S. 1993 *IAU Circular 5800*, May 22, 1993.

NICHOLSON, P. D., GIERASCH, P. J., HAYWARD, T. L., MCGHEE, C. A., MOERSCH, J. E., SQUYRES, S. W., VAN CLEVE, J., MATTHEWS, K., NEUGEBAUER, G., SHUPE, D., WEINBERGER, A., MILES, J. W., & CONRATH, B. J. 1995 Palomar observations of the R impact of comet Shoemaker-Levy 9: I. Light curves. *Geophys. Res. Lett.* **22**, 1613–1616.

ORTON, G. S., WITH HELP FROM THE NIMS, PPR, SSI, AND UVS EXPERIMENT TEAMS 1995a Comparison of Galileo SL-9 impact observations. In *Proceedings of the European SL-9/Jupiter Workshop*, (eds. R. West and H. Bohnhardt), pp. 75–80. ESO.

ORTON, G., A'HEARN, M., BAINES, K., DEMING, D., DOWLING, T., GOGUEN, J., GRIFFITH, C., HAMMEL, H., HOFFMANN, W., HUNTEN, D., JEWITT, D., KOSTIUK, T. et al. 1995b Collision of comet Shoemaker-Levy 9 with Jupiter observed by the NASA Infrared Telescope Facility. *Science* **267**, 1277–1282.

ROESER, S. & BASTIAN, U. 1989 *PPM: Positions and proper motions of 181731 stars north of −2.5 degrees declination for equinox and epoch J2000*. Astronomisches Rechen-Institut, Heidelberg.

SCOTTI, J. V., & MELOSH, H. J. 1993 Estimate of the size of comet Shoemaker-Levy 9 from a tidal breakup model. *Nature* **365**, 733–735.

SEKANINA, Z. 1995 The splitting of the nucleus of comet Shoemaker-Levy 9. In *Proceedings of the European SL-9/Jupiter Workshop*, (eds. R. West and H. Bohnhardt), pp. 43–55. ESO.

SEKANINA, Z., & YEOMANS, D. K. 1985 Orbital motion, nucleus precession, and splitting of period comet Brooks 2. *Astron. J.* **90**, 2335–2352.

SEKANINA, Z., CHODAS, P. W., & YEOMANS, D.K. 1994 Tidal disruption and the appearance of periodic comet Shoemaker-Levy 9. *Astron. Astrophys.* **289**, 607–636.

SHOEMAKER, C. S., SHOEMAKER, E. M., & LEVY, D. 1993 *IAU Circular 5725*, March 26, 1993.

STANDISH, E. M. 1990 The observational basis for JPL's DE200, the planetary ephemerides of the Astronomical Almanac. *Astron. Astrophys.* **233**, 252–271.

TAKEUCHI, S., HASEGAWA, H., WATANABE, J., YAMASHITA, T., ABE, M., HIROTA, Y., NISHI-HARA, E., OKUMURA, S., & MORI, A. 1995 Near-IR imaging observations of the cometary impact into Jupiter: Time variation of radiation from impacts of fragments C, D, and K. *Geophys. Res. Lett.* **22**, 1581.

TANCREDI, G., LINDGREN, M., & LAGERKVIST, C.-I. 1993 *IAU Circular 5892*, November 22, 1993.

TANCREDI, G., LINDGREN, M., & RICKMAN, H. 1990 Temporary satellite capture and orbital evolution of comet P/Helin-Roman-Crockett. *Astron. Astrophys.* **239**, 375–380.

TOZZI, G. P., RICHICHI, A., & FERRARA, A. 1995 High temporal resolution near-IR observations of impacts H and Q from Calar Alto and interpretation. In *IAU Colloquium 156: The Collision of Comet P/Shoemaker-Levy 9 and Jupiter*, p. 111.

WATANABE, J., YAMASHITA, T., HASEGAWA, H., TAKEUCHI, S., ABE, M., HIROTA, Y., NISHI-HARA, E., OKUMURA, S., & MORI, A. 1995 Near-IR observation of cometary impacts to Jupiter: Brightness variation of the impact plume of fragment K. *Publ. Astron. Soc. Japan* **47**, L21.

YEOMANS, D. K., & CHODAS, P. W. 1993a *IAU Circular 5807*, May 28, 1993.

YEOMANS, D. K., & CHODAS, P. W. 1993b Minor Planet Circular 22197, June 4, 1993.

YEOMANS, D. K., & CHODAS, P. W. 1993c The collision of comet Shoemaker-Levy 9 with Jupiter in July 1994. *JPL Interoffice Memorandum 314.10-50*, July 21,1993.

YEOMANS, D. K., & CHODAS, P. W. 1993d *IAU Circular 5909*, December 17, 1993.

YEOMANS, D. K., & CHODAS, P. W. 1994a Comet Shoemaker-Levy 9 impact times. *JPL Interoffice Memorandum 314.10-87*, August 2, 1994.

YEOMANS, D. K., & CHODAS, P. W. 1994b Comet Shoemaker-Levy 9 in orbit about Jupiter. *Bull. Am. Astron. Soc.* **26**, p. 3.

YEOMANS, D. K., & CHODAS, P. W. 1995 Predicting close earth approaches of asteroids and comets. In *Hazards Due to Comets and Asteroids*, (ed. T. Gehrels), pp. 241–258. University of Arizona Press.

ZAHNLE, K., & MAC LOW, M.-M. 1995 A simple model for the light curve generated by Shoemaker-Levy 9 impact. *J. Geophys. Res.* **100**, 16885.

Observational constraints on the composition and nature of comet D/Shoemaker-Levy 9

By JACQUES CROVISIER

Observatoire de Paris-Meudon, CNRS URA 1757, F-92195 Meudon, France

What did the break-up of comet Shoemaker-Levy 9 (SL9) and its subsequent impact on Jupiter teach us about the nature and constitution of this comet? The break-up of the comet apparently triggered activity of the fragments. Although a dust coma was continuously present around the fragments that orbited Jupiter, spectroscopic observations did not reveal any sign of gas. The impact itself was so energetic that most molecules of the impactor were dissociated and that any chemical memory was lost. Ultraviolet and visible spectroscopy of the impact sites revealed emission lines from several atoms, giving potential information on elemental abundances. However, the fact that both neutral and ionized atoms are emitting, and that both fundamental and inter-system lines are present, suggest that the medium is out-of-equilibrium and that emitting mechanisms other than simple resonance fluorescence are at work. Ultraviolet, infrared, and radio spectroscopy revealed lines of several molecular species, in emission and/or absorption, that are not normally present in Jupiter's upper atmosphere. In the visible, dark spots due to aerosols developed at the impact sites. It is not clear at the present time which part of this material is coming from preserved impactor material, from the recombination of the dissociated impactor material, from reactions between the impactor's and Jupiter's material, or from material coming from the lower layers of Jupiter's atmosphere. Realistic modelling of the impacts and of the following chemical reactions will be necessary to address all these issues. In this chapter, we will review the observational clues to the composition and structure of SL9 in the context of our knowledge of the composition and structure of comets and asteroids.

---

1. Introduction

The usual way to study the chemical nature of comets is by remote sensing, through the spectroscopic study of the volatile constituents that sublime under solar heating, and of the small dust particles that are shed during this sublimation process. Such observations, however, are biased towards the study of active regions of the nucleus surface.

We now have a global idea of the composition of cometary volatiles using spectroscopy ranging from radio to UV wavelengths (Crovisier 1994). This knowledge, however, is biased towards comets observed around 1 AU, where their activity is governed by water sublimation. Due to sublimation fractionation, the composition of released cometary gases depends upon the heliocentric distance and does not reflect directly the composition of cometary ices.

On special occasions, it was possible to observe *sungrazing* comets that passed so near the Sun that part of their refractory component was vaporized. In the past century, the spectrum of the Great Comet of 1882 was observed visually by Copeland & Lohse (1882), revealing the emission lines of several metals. More recently, comet Ikeya-Seki (1965 VIII) was the subject of extensive spectroscopic observations (*e.g.,* Preston 1967) when it passed at 0.0078 AU from the Sun.

The *in situ* exploration of comet Halley allowed one to investigate the elemental composition of its dust by mass spectroscopy and to evaluate the elemental relative abundances from carbon to nickel (Jessberger & Kissel 1988); they were found close to the solar values. This study was restricted to small-size dust particles. If cometary nuclei contain a population of large refractory particles, their composition is still unknown.

The impact of comet Shoemaker-Levy 9 (D/1993 F2 in the new-style designation of comets; hereafter SL9) on Jupiter gave us the unique occasion to witness the complete disruption of a comet into its elemental constituents and, potentially, to study the bulk composition of its nucleus.

The present chapter reviews the observations of SL9 and of its impact on Jupiter and discusses their implications for the composition and structure of the impactors. To set the stage, Section 2 summarizes our present knowledge of the composition and structure of comets and primitive asteroids. Section 3 is devoted to the pre-impact observations: imaging, photometry, spectroscopy. The study of the disruption of SL9 itself, which gives crucial clues to the structure of this body, is not treated here, since it is the subject of the chapter by Sekanina. Section 4 discusses impact and post-impact observations. The impact itself and its energetics inform us on the mass and density of the impactors; this topic will be only briefly evoked (Section 4.1), since it is the subject of several observational and theoretical reviews in this book. Spectroscopic observations revealed atomic lines (Section 4.2) and molecular signatures (Section 4.3); the observations of these latter are only summarized, since they are presented in detail in the chapter by Lellouch. Section 4.3 draws the consequences of the observations of aerosols on the impact sites, which is also the topic of the chapter by West. In Section 4, we attempt to draw the consequences of the observed molecular and atomic abundances on the composition of SL9 and its nature. Section 5 concludes this review and advocates future work.

2. The composition of "conventional" comets and asteroids

The composition of comets is known from the analysis of the output of their activity. For about a century, our investigations were limited to remote sensing by visible spectroscopy. This allowed us to identify several of the dissociation products—radicals, ions, atoms—of the volatile molecules sublimed from the nucleus. We then had to guess the identity of their parent molecule. Only recently was it possible to directly observe these parent species, benefitting from the development of spectroscopy at UV, infrared and radio wavelengths. Our present knowledge of the composition of cometary volatiles is summarized in Fig. 1. More details on the derivation of the composition of cometary ices can be found in Festou *et al.* (1993), Arpigny (1994) and Crovisier (1994).

An important fraction of cometary material is in the form of refractories, which are partly released as dust particles. An indication of its nature is revealed by the existence of the spectral signature of silicates in the infrared. The space probes to comet Halley had the unique opportunity—up to now—to measure by mass spectroscopy the elemental composition of cometary dust, as reported in Table 1. Another unique event was the observation of the sungrazing comet Ikeya-Seki (1965 VIII), which passed so close to the Sun (0.0078 UA) that part of its refractory component was vaporized and dissociated into atoms. The fluorescence lines of many metals and heavy atoms were then observed in its spectrum (Na, K, Ca, Ca^+, V, Cr, Mn, Fe, Co, Ni, Cu).

There are many issues still pending concerning the composition of comets. The relative abundances of *volatiles* are usually observed at heliocentric distances of the order of 1 AU, where cometary activity is controlled by water sublimation. They may differ from the relative abundances of cometary *ice* within the nucleus because of fractionation during the sublimation process and because the most volatile species may be already depleted from sublimation at larger heliocentric distances. Indeed, recent radio observations of P/Schwassmann-Wachmann 1 (hereafter P/SW1), an active comet with a nearly circular

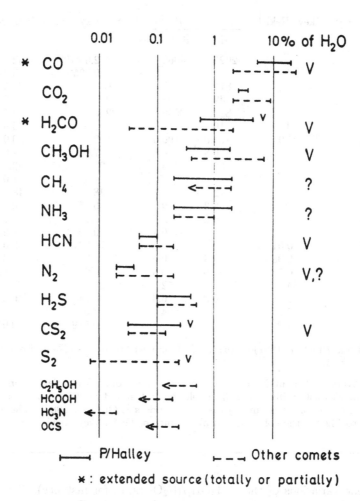

FIGURE 1. Relative abundances of the cometary volatiles expressed in percentage of the water content, by number. "v" indicates temporal variability in a given comet, and "V" in the last column refers to variation from comet to comet. From Arpigny (1994).

orbit at $r_h \simeq 6$ AU (*i.e.*, beyond Jupiter), revealed that its activity may be governed by CO sublimation. The cometary activity which is observed far from the Sun—now revealed in more and more comets with the increasing sensitivity of modern techniques— is presumably due to the sublimation of such very volatile species.

Refractories have only been partially sampled in the sungrazing comet Ikeya-Seki and more thoroughly in P/Halley. In fact, only a very limited fraction of the cometary material (volatiles or refractories) has been sampled either by remote sensing or by *in situ* analysis. We thus have no direct idea of the *bulk* composition of comet nuclei. In particular, large size dust particles (*rocks*)—the existence of which is attested by radar and radio continuum observations—practically escaped all investigations.

To further complicate the situation, there may be a high degree of diversity of composition among comets resulting from their formation in different regions of the solar nebula. The direct measurements of parent volatiles are still sparse (although they in-

Element	Ikeya-Seki [†]	P/Halley		Solar system	CI-chondrite
		dust	dust + ice		
H		2025.	4062.	2.6×10^6	492.
Li	†			0.0053	0.0053
C		814.	1010.	940.	70.5
N		42.	95.	291.	5.6
O		890.	2040.	2216.	712.
Na	†	10.	10.	5.3	5.3
Mg	(=100.0)	=100.0	=100.0	=100.0	=100.0
Al		6.8	6.8	7.9	7.9
Si		185.	185.	93.	93.
P				1.0	1.0
S		72.	72.	48.	48.
K	†	0.2	0.2	0.35	0.35
Ca		6.3	6.3	5.7	5.7
Ti	<0.02	0.4	0.4	0.22	0.22
V	0.01			0.027	0.027
Cr	0.08	0.9	0.9	1.3	1.3
Mn	0.5	0.5	0.5	0.89	0.89
Fe	84.	52.	52.	84.	84.
Co	0.4	0.3	0.3	0.21	0.21
Ni	7.2	4.1	4.1	4.6	4.6
Cu	0.2			0.049	0.049

[†] In addition, Preston (1967) determined abundances relative to Na of 1.6×10^{-3} for K and $< 2.5 \times 10^{-5}$ for Li.

TABLE 1. Average elemental abundances measured in comet Ikeya-Seki (from Arpigny *et al.* 1995), in Halley's dust grains and in the whole comet dust and volatiles (from Jessberger & Kissel 1991), and in other solar system objects (from Anders & Grevesse 1989). The abundances are normalized to Mg (to the solar system abundance of Fe for Ikeya-Seki, in which Mg was not observed).

dicate strong variations of the $[CH_3OH]/[H_2O]$ ratio, for instance). The investigation of the radical abundances in significant samples reveals relative abundance variations of OH, CN, C_2, C_3 and NH. Thus A'Hearn *et al.* (1996) could separate two classes of comets from the observations of their radicals: "typical" and "C_2-depleted"; these latter being related to Jupiter-family comets. Also suspected is the heterogeneity of cometary nuclei, resulting from the possible aggregation of planetesimals of different origins.

The density of cometary nuclei is not well known. The nucleus size is known directly from imaging only for P/Halley. It can, in principle, be determined from photometry, but practically it is very difficult to disentangle the respective contributions of the nucleus and of the dust coma for an active comet. On the other hand, the only way to have access to the nucleus mass at the present time is by relating the evaluation of non-gravitational forces—due to anisotropic outgassing—to their perturbating effect on the cometary orbits. This method is still highly model-dependent. Available evaluations lead to densities in the range $0.2 < \rho < 0.9$ g cm^{-3} for P/Halley and very few other comets (Festou *et al.* 1993, Rickman 1994).

Since there is no release of material from *bona fide* asteroids, information on their composition can only come from spectroscopy of the solid minerals at their surface. This is of little help, since this material is probably weathered and has little to do with the internal composition of these differentiated bodies. However, it is usually believed that most meteorites recovered on Earth are of asteroidal origin, thus providing models for asteroidal matter. But the precise links between meteorites and asteroids are not easy to establish.

Although asteroid sizes are known for a large sample of objects, there is little information on their masses which could lead to density determinations. Evaluations from mutual perturbations are rare and inaccurate (Hoffmann 1989). Recently, the *Galileo* probe, when imaging (243) Ida, discovered a second component, Dactyl. Unfortunately, the orbit of the binary system could not be accurately determined and the density of Ida is poorly constrained: $2.1 < \rho < 3.1$ g cm^{-3} (Belton *et al.* 1995). This indicates a chondritic composition and/or a highly porous structure.

Table 1 shows that there are little differences between different objects with regard to the abundances of heavy elements; all abundances are close to "cosmic". This is no longer the case for light elements; these elements are likely to be deposited in highly volatile components; their abundances are thus highly dependent on the physical conditions of the formation and evolution of these various objects.

3. Pre-impact observations

3.1. *Imaging and photometry*

The pre-impact images of SL9 were obtained at several observatories: *e.g.*, Chernova *et al.* (1995) with the ESO 1-m; Cochran *et al.* (1995) at McDonald; Colas *et al.* (1995) at Pic-du-Midi; Jewitt (1995a) at the Mauna Kea telescopes; Scotti *et al.* (1994) at the Spacewatch station; Stüwe *et al.* (1995) with the ESO-NTT; Trilling *et al.* (1995) at Calar-Alto; Weaver *et al.* (1994, 1995) with the HST; R. M. West *et al.* (1995) with several ESO telescopes.

Tancredi & Lindgren (1994) have reported a negative search for comets in the vicinity of Jupiter undertaken at ESO in March-April 1992, *i.e.*, one year before the discovery of SL9 and several months before its disruption by Jupiter. From their limiting magnitude $B = 21.5$, they infer an upper limit of 7.2 km on the radius of SL9 nucleus before its disruption.

3.1.1. *Sizes of nuclei from the images*

Retrieving the nucleus sizes from the images is a difficult problem. The nuclei are not resolved. One could hope to separate the relative contributions of the star-like nucleus and of the dust distribution from high-resolution images. Assuming an albedo similar to that observed for other comet nuclei, it would then be possible to derive the nucleus cross-section. The problem is that the dust distribution does not follow the typical $1/r$ law observed in other comets; it is steeper and curved, precluding an easy extrapolation to the unresolved nuclear region. To further complicate the problem, we may have to deal with multiple nuclei (or even with swarms of nuclei, as postulated by Xie *et al.* 1995).

From ground-based observations, Jewitt (1995a) gave an upper limit of 1 km to the fragment sizes. From the HST images, Weaver *et al.* (1994, 1995) derived sizes ranging from 0.6 to 4.1 km in diameter for the various fragments (assuming an albedo of 0.04), which are in fact conservative upper limits to the true values. Sekanina (1995b) independently tried to deconvolve the HST images and argued that the fragments could have, in fact, multiple secondary nuclei in addition to a large main component.

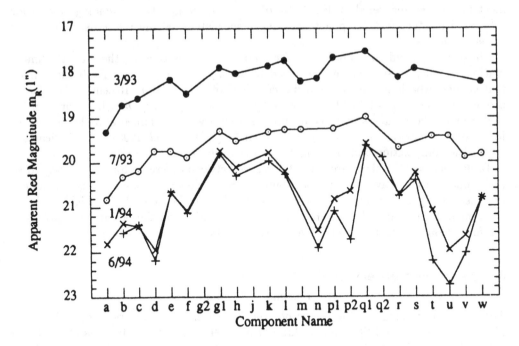

FIGURE 2. The evolution of the fragment brightness: R-filter photometry of the SL9 components with a 1″ radius aperture, at three different epochs, from Jewitt (1995a). All components fade as dust dissipates; however, different components fade at different rates.

3.1.2. *Activity*

It is of paramount importance to determine whether the comae observed around every fragment are the remnants of the dust outburst which followed the 1992 break-up, or are due to continuous activity. This is still a controversial point which has been discussed by several authors (Sekanina *et al.* 1994; Weaver *et al.* 1994, 1995; Jewitt 1995a; Sekanina 1995a; R. A. West, this volume).

The evolution of the fragment brightnesses, as observed by Jewitt (1995a), is shown in Fig. 2. There is undoubtedly an important fading of all components, revealing a decrease of activity, or even the dissipation of dust in the absence of any activity. It has been noted that in the absence of activity, the persistence of the central coma can only be explained by the presence of large-size (centimetre-size) particles. From ground-based observations, Jewitt (1995a) estimated that the observed dust distribution could be consistent with a dust production rate of no more than 1 kg s^{-1} for the largest components, whereas the analysis of the HST observations by Weaver *et al.* (1995) yielded a production rate of at most 5 kg s^{-1} (the difference between these two evaluations could be ascribed to differences in the dust models rather than to the observations themselves).

Each component being accompanied by its own tail, the tail morphology and evolution are important clues to the dust properties and production. Most of the time, the tail's orientation was close to the opposite Sun direction, as are all well-behaved cometary tails due to a continuous dust production under the influence of solar radiation pressure. Just a few days before the impacts, the tails and dust distributions were deformed and elongated along the comet-Jupiter direction (R. M. West *et al.* 1995), presumably under the influence of Jovian tidal forces. At the moment of opposition at the end of April

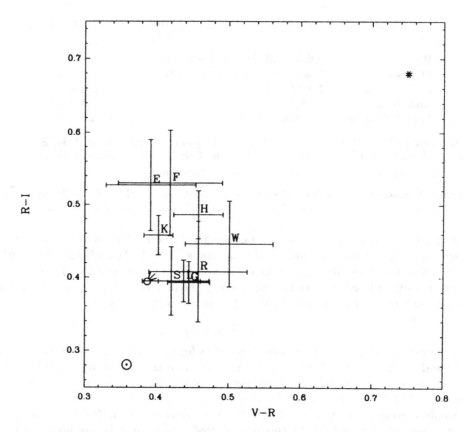

FIGURE 3. The colours of the fragments of SL9, plotted from the Mauna Kea data of mid-January 1994 reported by Meech and Weaver (1995); only the fragments with the best data are plotted, with $\pm 1\sigma$ error bars; symbols indicate the colours of the Sun ($\odot$), of a typical comet ($\odot\!\!\prec$) and of Pholus (*).

1994, the phase angle was very close to 0. One would then expect a rapid rotation of the tail's position angle, following the anti-Sun direction. It was noted by several observers (Jewitt 1995a, Scotti et al. 1994) that this was not exactly the case. Detailed modelling of this process, in order to retrieve dust properties and the evolution of dust production from available images, is highly desirable (see also Sekanina, this volume).

3.1.3. Were all fragments alike?

In addition to their difference in sizes, in activity (production of dust, fading, further splitting), the SL9 fragments might have differences in composition. Such an effect might result from the formation of the initial SL9 nucleus itself, like from the accretion of planetesimals of different origins. A test of this could come from broadband photometric or spectrophotometric observations, which could reveal differences in the colours of the fragments. Such observations were performed by several teams: Chernova et al. (1995); Cochran et al. (1994), Rauer and Osterloh (1995); Meech and Weaver (1995); Trilling et al. (1995). No significant differences between the fragments were reported.

For example, Fig. 3 shows the colours of the fragments as observed by Meech and Weaver (1995) at Mauna Kea. The results are dispersed and some fragments have colours

Instrument[†]	Date	$Q[\mathrm{OH}]$ $[\mathrm{s}^{-1}]$	$Q[\mathrm{CN}]$ $[\mathrm{s}^{-1}]$
HST 2.2-m	Jul. 93–Jul. 94	$< 1 - 2 \times 10^{27}$	
McDonald 2.7-m[‡]	Feb. + Mar. 94	$< 8.4 \times 10^{26}$	$< 1.7 \times 10^{24}$
CFHT 3.6-m	5–6 Mar. 94		$< 3 \times 10^{24}$
Keck 10-m	10 Apr. 94		$< 3 \times 10^{23}$
ESO/NTT 3.5-m	1–2 Jul. 94		$< 1.8 - 2.8 \times 10^{23}$

[†] References. HST: Weaver *et al.* 1994, 1995; McDonald: Cochran *et al.* 1994; ESO/NTT: Stüwe *et al.* 1995; CFHT and Keck: Jewitt, Luu and Owen, *private communication*.

[‡] These limits pertain to a combination of several fragments.

TABLE 2. Spectroscopic limits (3σ) on gas production rates in SL9 prior to the impacts.

differing by more than the $\pm 1\sigma$ errors, but not much more. This does not seem to be a convincing clue for diversity among the fragments. If real, these effects would rather reflect different size distributions of the dust particles. The figure shows, however, that the dust is definitely redder than the Sun and marginally redder than the average comet. It is not, however, as red as the enigmatic distant asteroid Pholus.

3.2. *Spectroscopy*

Several sensitive pre-impact spectroscopic searches for signs of gas in the visible (CN and CO^+) and in the near-UV (OH) were undertaken from the ground and with the HST. They are summarized in Table 2. They were negative. Upper limits on the production rates were $Q[\mathrm{OH}] < 1 - 2 \times 10^{27}$ s^{-1} and $Q[\mathrm{CN}] < 2 - 3 \times 10^{23}$ s^{-1}. The upper limit on CN would correspond to $Q[\mathrm{OH}] < 10^{26}$ s^{-1} if one assumes $Q[\mathrm{OH}]/Q[\mathrm{CN}] \simeq 350$, as for comets observed at $r_h \simeq 1$ AU (A'Hearn *et al.* 1995), which may be invalid at 5 AU.

Searches for CO at radio wavelengths were also negative at the CSO (Jewitt & Senay 1994, *private communication*), at the SEST (Rickman 1994, *private communication*), as well as at IRAM where observations were conducted at the very moment preceding the impacts (Festou & Lellouch 1994, *private communication*). No limits on the corresponding CO production rate were given up to now; they are presumably on the order of several 10^{28} s^{-1}).

A comparison can be made with the activity of other *distant* comets (see Huebner *et al.* [1993] for a review): spectroscopic detections are very rare for comets at $r_h > 5$ AU. CN was detected in Chiron at 10 AU, P/SW1 at 6 AU and P/Halley when it was at 4.8 AU pre-perihelion; CO^+ was only found in P/SW1. Recently, the radio rotational lines of CO were observed in P/SW1 (Senay & Jewitt 1994; Crovisier *et al.* 1995) and in the long-period comet C/1995 Q1 (Hale-Bopp) (Matthews *et al.* 1995; Rauer *et al.* 1995); the corresponding CO production rate in these comets is as large as $2 - 5 \times 10^{28}$ s^{-1}.

At 5.4 AU the equilibrium temperature of a rapidly rotating nucleus is 120 K, which is not sufficient to allow the sublimation of water: the sublimation rate, which is of the order of 10^{17} water molecules cm^{-2} s^{-1} at 1 AU, drops to 3×10^9 cm^{-2} s^{-1}—more than seven orders of magnitudes lower—at 5.4 AU. Other effects which have been invoked to explain distant activity are: (*i*) a slowly rotating nucleus or (*ii*) small icy grains which could still be heated up to 150 K at these distances and have significant water ice sublimation rates ($\sim 10^{14}$ cm^{-2} s^{-1}); (*iii*) an internal source of energy such as the exothermic conversion of amorphous ice into crystalline ice; or (*iv*) the sublimation of more volatiles species, which scales at r_h^{-2} up to much larger values of r_h. The recent detection of a large CO production rate in distant comets shows that the last process is indeed at work.

As stated in Section 3.1.2, it is not yet clear from the analyses of the SL9 images whether the release of dust was continuous or restricted to the period following the July 1992 disruption. If continuous, the dust production rate was very low (no more than 1–5 kg s^{-1} per fragment). When scaled to P/SW1 according to its dust production, the expected production rate of CO from the fragments of SL9 is too low to permit detection (P/SW1 has a dust production rate of the order of 10–100 kg s^{-1} and Q[CO]$\simeq$ 5×10^{28} s^{-1}). At 4.8 AU from the Sun, Q[CN] was 3×10^{24} s^{-1} for P/Halley (Wyckoff *et al.* 1985) whereas its dust production was ~ 3 kg s^{-1} as estimated by Weaver *et al.* (1995), but this was at r_h somewhat smaller than SL9. Thus, the non-detection of any gas signature in the spectrum of SL9 is not really surprising.

Another peculiar example of distant activity is that developed by the dynamically new comet Bowell (1982 I). This comet showed a significant production of OH (7.5 $\times$ 10^{28} s^{-1}) pre-perihelion as far as 5.25 AU from the Sun, which did not increase much when the comet reached perihelion at 3.26 AU (A'Hearn *et al.* 1984). This unusually high water production rate at this heliocentric distance was interpreted as coming from the sublimation of water from icy grains; the very large surface of the grains compensates for the very low sublimation rate at this distance, ensuring an efficient water production. At similar heliocentric distances, the dust production parameter $Af\rho$† was $\simeq 7000$ cm for comet Bowell (A'Hearn *et al.* 1984) and $\simeq 200-500$ cm for the strongest fragments of SL9 in early 1994, from photometric measurements of Cochran *et al.* (1994) and Jewitt (1995a and Fig. 2). If Q[OH]/$Af\rho$ were the same for both objects, one would expect Q[OH] $\simeq 5 \times 10^{27}$ s^{-1} for the strongest fragments of SL9, which is significantly more than the observed upper limits (table 2). It can be inferred that the grains of SL9 had a different nature from those of comet Bowell, or that they lost their icy component in the first months following the disruption of the comet by Jupiter in July 1992.

A puzzling flash of Mg$^+$ emission (doublet near 280 nm) was fortuitously observed with the HST/FOS on fragment G on 14 July 1994, four days before impact (Weaver *et al.* 1995; Feldman *et al.* 1996). This emission, which decayed with a time constant of one minute, was followed by an increase of the dust scattered emission. This transient event may be associated with the entry of the fragment in Jupiter's magnetosphere, a possible source mechanism being ion sputtering of the dust surrounding the fragment, as suggested by Feldman *et al.* (1995).

4. Impact and post-impact observations

Cometary molecules impacting at a velocity of 60 km s^{-1} have a kinetic energy about two orders of magnitude higher than their chemical binding energy. Therefore, not only will they be dissociated, but the excess energy of their fragments will be high enough to initiate several chemical reactions (dissociations) within Jupiter's atmosphere. In the shock (or hot) chemistry which will follow, all memory of the initial chemical composition of the impactor will be lost (*e.g.*, Borunov *et al.* 1995).

However, we cannot exclude the possibility that part of the initial cometary molecules may be preserved (for instance, after successive fragmentations, small particles might

---

† The parameter $Af\rho$, first introduced by A'Hearn *et al.* (1984), is the product of dust albedo A, dust filling factor f and field of view diameter ρ. It is homogeneous to a distance. It can be straightforwardly derived from photometry and it is a convenient measure of the dust production rate.

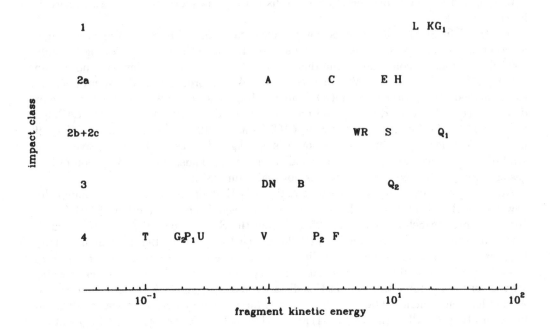

FIGURE 4. The *classes* of the impacts, as defined by Hammel *et al.* (1995), as a function of the fragment relative kinetic energies, evaluated by Weaver *et al.* (1995) from the brightnesses measured by the HST.

be decelerated without being destroyed). The proportion of such preserved material is unknown, but it is presumably small compared to the amount of chemically processed material.

Thus, clues to the chemical composition of the impactor must be searched for in the elemental abundances. The upper atmosphere of Jupiter is devoid of heavy elements, of oxygen and sulphur compounds, and is depleted in nitrogen. The detection of such elements after the impacts could betray the presence of SL9 material. However, such clues might be masked by material coming from the deep atmosphere of Jupiter (and especially from the NH_3, NH_4SH and water clouds) if the fragments went so far.

4.1. *Clues from the energetics of the impacts*

The manifestations of the impact phenomena, such as their light curves, the fireball brightnesses and sizes, and the depths to which the impactors penetrated into Jupiter's atmosphere, critically depend upon the sizes and density of the fragments. This is discussed in the theoretical chapters on the physics of the impacts (chapters by Crawford, Mac Low, Zahnle, this volume). For instance, Zahnle & Mac Low (1995) estimated from the light curve of its impact that fragment R had diameter 450–500 m and mass $\sim 2 - 3 \times 10^{13}$ g.

It was noted by Weaver *et al.* (1995) and by several other observers that the brightest fragments did not yield the strongest impacts. This is illustrated in Fig. 4 which shows the *classes* of the impacts, as defined by Hammel *et al.* (1995) on the basis of the aspects of the impact sites, as a function of the pre-impact brightnesses of the fragments as measured by the HST. Although the figure shows a clear correlation between these two

species	line or band	frequency wavelength	instrument[†]	impact
radio:		[GHz]		
H_2O	$6_{16} - 5_{23}$	22	Medicina	E
HCN	1-0	89	IRAM-inter	A
CO	1-0	115	IRAM-30m	*several*
OCS	18-17	219	IRAM-30m	K W
CO	2-1	230	IRAM-30m, SEST	*several*
CS	5-4	245	IRAM-30m	*several*
HCN	3-2	266	JCMT	*several*
HCN	4-3	345	JCMT	*several*
infrared:		[μm]		
H_2O	rot.	23.9, 22.6	KAO/KEGS	G K
HCN	ν_2	13.4	NASA/IRTF	*several*
C_2H_4	ν_7	11.0	NASA/IRTF	K
silicates		$\simeq 10.0$	Palomar	R
PH_3	ν_2	10.1	Palomar	L
H_2O	ν_2	6.6	KAO/HIFOGS	R W
H_2O	ν_2	7.7	KAO/KEGS	G K
CO	$\Delta v = 1$	4.7	*several*	*several*
H_2O	ν_3	2.7	Galileo/NIMS	G R
CO	$\Delta v = 2$	2.3	*several*	*several*
H_2O	$\nu_3, \nu_2 + \nu_3$	2.38, 2.0	AAT	C K
NH_3	*several*		*several*	*several*
CH_4	*several*		*several*	*several*
H_2	*several*		*several*	*several*
H_3^+	*several*		*several*	*several*

[†] References to the Table. Medicina: Montebugnoli *et al.* 1995; IRAM-inter: Wink *et al.* 1994; IRAM-30m: Lellouch *et al.* 1995; SEST: Bockelée-Morvan *et al.* 1995; JCMT: Marten *et al.* 1995; KAO/KEGS: Bjoraker *et al.* 1995; KAO/HIFOGS: Sprague *et al.* 1995; NASA/IRTF: Bézard *et al.* 1995, Griffith *et al.* 1995, Orton *et al.* 1995; AAT: Crisp & Meadows 1995; Galileo/NIMS: Carlson *et al.* 1995; Palomar: Conrath *et al.* 1995; Nicholson *et al.* 1995.

TABLE 3. Detected radio and infrared spectral lines and bands during the SL9/Jupiter impact event.

parameters, some fragments depart from this law: A and C gave relatively stronger impacts, whereas the impacts of Q1 and Q2 were relatively fainter. This could betray a diversity among the fragments; if the impact class corresponds to the fragment mass, and the fragment brightness to the fragment size, this would suggest a diversity of densities. But it could also correspond to a pre-fragmentation of some of the fragments, or to diversity in their emission of dust.

4.2. *Clues from atomic lines*

Emissions of atomic lines were reported in the UV from the HST (Noll *et al.* 1995) and the IUE (Prangé *et al.* 1994), and in the visible from several ground-based observations (Catalano *et al.* 1995; Costa *et al.* 1994; Fitzsimmons *et al.* 1995; Roos-Serote *et al.* 1995a, b). They are listed in Table 4. It should be noted that several unidentified lines were also observed that might be due to atoms or ions.

The atomic lines observed in the visible from the ground (Figs. 5, 6 and 7) were emitted in the splashback phase of the collision, as could be determined from the time-resolved

species	line or band	wavelength [nm]	instrument[†]	impact
visible:				
Fe	$a\,^5F - z\,^7D^0$	804.8	OMP	L
K	$4s\,^2S - 4p\,^2P^0$	766.5	OMP	L Q$_1$
Li	$2s\,^2S - 2p\,^2P^0$	670.8	OMP, Brazil	L
Ca	$4s^2\,^1S - 4p\,^3P^0$	657.3	OMP, Brazil, INT	L
H	H α	656.3	OMP	L
Fe	$a\,^5F - z\,^7F^0$	635.9	OMP, Brazil	L
Na	$3s\,^2S - 3p\,^2P^0$	589.6 & 589.0	OMP, Brazil, INT, Catania	H L Q$_1$ Q$_2$
Mn	$a\,^6S - z\,^8P^0$	543.2	INT	L
Fe	$a\,^5D - z\,^7D^0$	511.0	INT	L
Mg	$3s^2\,^1S - 3p\,^3P^0$	457.2	INT	L
Fe	$a\,^5D - z\,^7F^0$	437.7	INT	L
Cr	$a\,^7S - z\,^7P^0$	425.4	INT	L
ultraviolet:				
Na	$3s\,^2S - 4p\,^2P^0$	330.2	IUE	A
?		309.6	HST *unidentified*	S
S$_2$	$B^3\Sigma_u^- - X^3\Sigma_g^-$	293-255	HST	G S
?		288.4	HST *unidentified*	S
Mg	$3s^2\,^1S - 3p\,^1P^0$	285.3	HST	S
Mg$^+$	$3s\,^2S - 3p\,^2P^0$	279.6 & 280.5	HST	pre-impact, S
Fe$^+$	$4s\,^6D - z\,^6D^0$	262.	HST	S
CS	$A^1\Pi - X^1\Sigma$	257.8	HST	S
Si	$3p_2\,^3P - 4s\,^3P^0$	251.5 & 252.5	HST	S
Fe	$4s^2\,^5D - x\,^5F^0$	248.4	HST	S
NH$_3$	$\tilde{A}^1A"_2 - \tilde{X}^1A_1$	230-180	HST	G
CS$_2$	$\tilde{A}^1B_2 - \tilde{X}^1\Sigma_g^+$	206.4-187.3	HST	S
H$_2$S	$\tilde{A} - \tilde{X}^1A_1$	~ 200	HST *tentative*	G
S	$3p^4\,^4S^0 - 4s\,^3S^0$	180.8-182.0	IUE (*or* Si$^+$?)	A
Si$^+$	$3p\,^2P^0 - 3p^2\,^2D$	180.7-181.7	IUE (*or* S ?)	A
?		176 & 178	IUE *unidentified*	A
Al$^+$	$3s^2\,^1S - 3p\,^1P^0$	167.1	IUE *tentative*	K
C	$2p^2\,^3P - 3s\,^3P^0$	165.7	IUE *tentative*	K
H$_2$	$B^1\Sigma_u^+ - X^1\Sigma_g^+$ (Lyman)	155.-162.	IUE	K S
S$^+$	$3p^3\,^4S^0 - 3p^4\,^4S$	125.	HST	G
H$_2$	$C^1\Pi_u - X^1\Sigma_g^+$ (Werner)	123.-130.	IUE	K S
H	Ly α	121.5	IUE	K S
He	$1s\,^1S - 2p\,^1P^0$	58.4	EUVE	Q$_1$

[†] References to the Table. OMP: Roos-Serote *et al.* 1995a; Brazil: Costa *et al.* 1994; INT: Fitzsimmons *et al.* 1995; Catania: Catalano *et al.* 1995; IUE: Prangé *et al.* 1994, Ballester *et al.* 1995; HST: Noll *et al.* 1995, Weaver *et al.* 1995, Yelle & McGrath 1995; EUVE: Gladstone *et al.* 1995.

TABLE 4. Detected visible and ultraviolet spectral lines and bands during the SL9/Jupiter impact event.

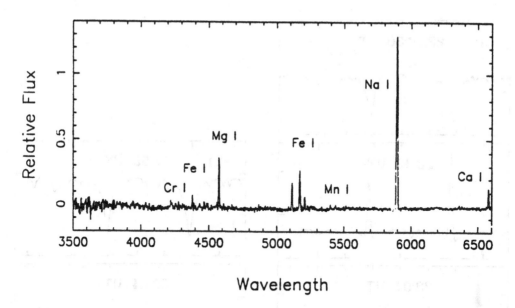

FIGURE 5. Atomic line emissions observed in the visible with the 2.5-m Isaac Newton Telescope at La Palma observatory in the L-plume on July 19 at 22:35 UT, just after the impact. The wavelength is given in ångstrøms. From Fitzsimmons *et al.* (1995).

observations of impact L at La Palma (Fitzsimmons *et al.* 1995). No emission could be detected more than one hour after the impact. The strongest lines of Na, Ca, Fe and even Li were visible in the raw spectra, whereas weaker lines of Fe and Ca multiplets and the Hα emission could only be spotted in spectra corrected for the solar reflected spectrum (Roos-Serote *et al.* 1995b). The IUE observations (Ballester *et al.* 1995) revealed emission lines and bands of H Lyman α and H_2 in spectra of impact sites K and S observed for $\simeq$ 20 min around impact times. The far-UV emission lines of carbon and Al^+ were tentatively identified in impact K. With the exception of the Mg^+ flash pre-impact observation (see Section 3.2), the metallic emission lines of Mg, Mg^+, Fe, Fe^+ and Si detected in the UV by the HST (Fig. 8) were observed in a field encompassing the sites of three-day old impact G and of the fresh impact S just 45 min after its formation. It is thus likely that these lines are coming from new impact S rather than old impact G.

All these observations remind us of the observed spectra of sungrazing comets which showed the lines of several metals (Na, K, Ca, V, Cr, Mn, Fe, Co, Ni, Cu; see also Table 1). Lithium is a relatively rare element which was not detected before in any comet. An upper limit was reported for comet Ikeya-Seki (Table 1). It was observed, however, in the spectra of some meteoroids (*e.g.*, in the Benešov bolide; Borovička & Spurný 1995).

Since, due to sedimentation, metallic compounds are only expected to be present at very deep levels in Jupiter's atmosphere, there is little doubt that the atomic lines in the SL9 plumes are coming from the impactor itself. Translating the observed line intensities into elemental abundances requires the knowledge of the emission mechanism.

Fluorescence is the usual mechanism invoked to explain cometary atomic and molecular emissions, as well as the sodium emission lines observed in Io's torus. As far as the

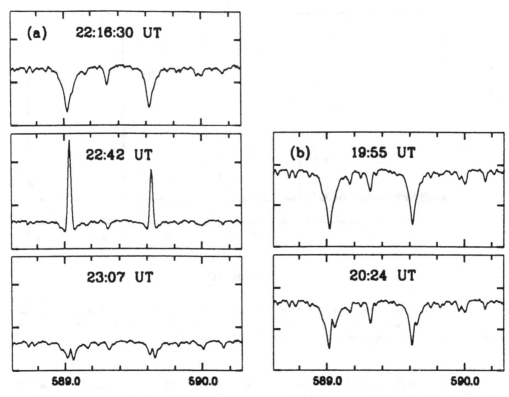

FIGURE 6. A time sequence showing the evolution of the Na emission line doublet during the L impact (left) and the Q impacts (right), observed at the 2-m telescope of the Pic-du-Midi observatory. The raw spectra are shown here, uncorrected for the reflected solar spectrum. The wavelength is given in nanometres. From Roos-Serote *et al.* (1995b).

lines are not saturated, the line intensities I are proportional to the column densities $< N >$ through the simple law $I = g < N >$, where g is the usual g-factor for resonant fluorescence. Saturated lines cannot exceed $I = \Phi \Delta \nu$, where Φ is the solar flux density at Jupiter and $\Delta \nu$ is the line width. This intensity was exceeded for the lines observed by the IUE (the H Lyman α, C and possibly Al$^+$ lines, as reported by Ballester *et al.* 1995) as well as for most lines observed in the visible. For the HST observations, Noll *et al.* (1995) initially proposed emission due to resonant fluorescence. The observed line intensities, at least for the Mg and Mg$^+$ lines, are however large and could only be explained by saturated lines of width of 5 km s^{-1}. This is significantly broader than the thermal line width (about 1.0 km s^{-1} at 1000 K) or the width due to bulk motions within the impact sites (presumably < 1 km s^{-1}). This suggests that the lines are broadened by heavy saturation (optical depths of the order of 10^6 are necessary to broaden the lines by a factor of 4). Noll *et al.* derived column abundances and masses based upon the law $I = g < N >$, valid for the optically thin case. The derived masses could thus be underestimated by many orders of magnitudes. They could still be consistent, however, with the elemental masses contained in a $10^{14} - 10^{15}$ g fragment with cometary atomic abundances. It is unfortunate that in this heavily saturated situation, atomic abundances cannot be retrieved with any degree of accuracy.

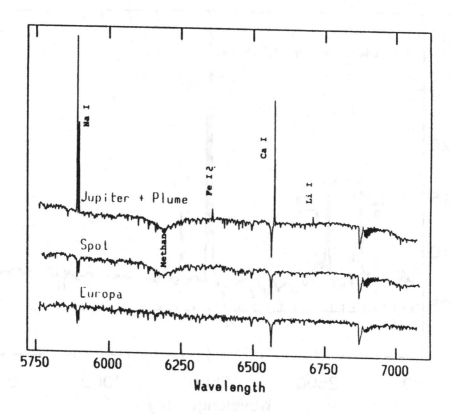

FIGURE 7. Spectrum in the 575–700 nm region of the plume of the L-impact observed with the 1.6-m telescope of the Laboratório Nacional de Astrofísica at Brazópolis. Emission lines of Na, Fe, Ca and Li are present. Below are shown, for comparison, the spectrum of the same region one hour later and that of Europa. The wavelength is given in ångstrøms. From Costa *et al.* (1994).

There are several pieces of evidence suggesting that fluorescence from atoms in the ground state is not the mechanism (or not the only one) responsible for these line emissions. As noted by Carpenter *et al.* (1995), several of the lines observed in the UV are terminating at excited levels significantly above ground. In the visible, a strong inter-system Ca line at 657.3 nm is observed, whereas the fundamental Ca line at 422.8 nm is not detected in the La Palma spectra (Fig. 5).

Thermal excitation by collisions is a plausible mechanism for the transient emission of atomic lines. It can explain the spectrum of meteors in the Earth atmosphere, and it may be noted that the velocity of the splashback impacts on Jupiter was $\simeq 10$ km s^{-1}, of the same order as the entry velocities of meteors in the Earth atmosphere. Other possible mechanisms are electronic recombinations and prompt emissions following chemical reactions. The modelling of these time-dependent excitation phenomena requires the knowledge of the time evolution of the plume temperature, of the plume and impact site geometry and its evolution (to account for *beam dilution*), and of course of the line saturation.

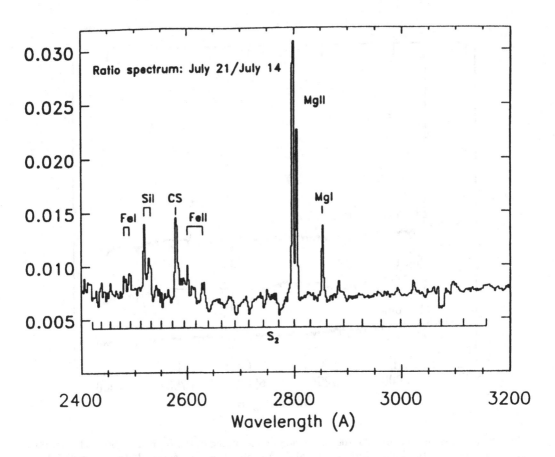

FIGURE 8. Atomic lines observed in the 240–320 nm region with the HST on the S impact site, 45 min after the impact. The spectrum has been ratioed by a pre-impact Jupiter spectrum. Weak unidentified emission lines are present around 300 nm. From Noll *et al.* (1995).

4.3. *Clues from molecular lines*

Molecular lines and bands detected following the SL9 impacts on Jupiter are reported in Tables 3 and 4. These observations are discussed in detail in the chapter by Lellouch (this volume). Some of these "detections" should be taken with caution, because they rely on a single instrument observation with a poor signal-to-noise ratio (like the radio observation of H_2O), or on a broad spectral feature which cannot provide an unambiguous identification (like H_2O in the near-infrared or H_2S in the UV).

During the SL9 impact observing campaign, special attention was paid to the searches for water signatures, since it was believed that the presence or absence of this species in the impact sites would betray the nature of the impactor, *comet nucleus* or *asteroid*. Such signatures, which could not be found in the quick-look inspections, were only revealed after careful analysis of the data. Infrared spectroscopy revealed transient emissions of water lines just after the impacts (Table 3). The KAO observed emission lines of water at 7.67–7.72 μm (ν_2 ro-vibrational lines) and at 22.6–23.9 μm (rotational lines) just after the impacts of fragments G and K (Bjoraker *et al.* 1995) and at 6.6 μm after impacts of R and W (Sprague *et al.* 1995). These lines correspond to high-energy rotational levels. They rapidly decreased in intensity and disappeared, which can be attributed to cooling of the gas rather than to the disappearance of water. Sprague *et al.* estimated the water masses corresponding to these observations to be 1.4 and 2.8×10^{12} g for G and K (Bjoraker *et al.*), 6 and 10×10^{10} g for R and W (Sprague *et al.*), equivalent to pure ice spheres of 140, 180, 50 and 60 m in diameter, respectively. Observations of several impacts at the AAT revealed a rapidly varying infrared spectrum which could be fitted with transient emission of the ν_3 and $\nu_2 + \nu_3$ water bands (Crisp & Meadows 1995). The observers of Galileo/NIMS also claim to have observed the 2.7 μm water band in the splashbacks of G and R, despite very poor spectral resolution (Carlson *et al.* 1995). A possible detection of the water line at 22 GHz in the weeks following the impacts has also been reported (Montebugnoli *et al.* 1995).

The various observers of molecules detected at UV, infrared, and radio wavelengths, following the impacts have reported masses (or masses upper limits) that may strongly disagree (by orders of magnitudes!) from one technique to another. This should not be a surprise, since we are still in a preliminary state of analysis and:

- we are sometimes comparing different impacts;
- we are dealing with either transient effects observed in the plume or the rapidly cooling impact site (UV, IR), or medium- or long-term effects (radio);
- we have to take care of the different geometries (radio observations were affected by beam-dilution and encompassed several impact sites; UV observations may give lower limits due to limited sampling);
- line saturation is a problem which has to be considered (*e.g.*, for fluorescence emission; see Section 4.2);
- the different observations were sounding different levels and the vertical distributions of the molecular species are poorly known; and
- assumptions must be made on the vertical temperature profiles (for the radio observations) and on the temperature evolution (for the UV and IR observation of the rapidly cooling impact sites).

Table 5 lists the masses of the various molecules observed in the major impact sites, adopted from the "baseline" values reported in the chapter by Lellouch (this volume); the reader is referred to this chapter for all details on their determination and their reliability. For the sake of discussion, Table 5 also lists the corresponding masses for several key elements contained in these molecules, as well as those contained in a comet or a CI chondrite of 10^{15} g mass, for the typical abundances of Table 1. The mass of 10^{15} g roughly corresponds to a kilometre-size fragment of density 1, and is probably the upper bound of the plausible mass of the largest SL9 fragments.

Little can be learned from the mass of carbon, because most of this element was presumably provided by Jovian methane. In contrast, the mass of oxygen gives perhaps the most stringent constraint on the impactor: the upper atmosphere of Jupiter is devoid of oxygen compounds and it seems improbable that the impactors hit the hypothetical deep water clouds. Oxygen from CO alone amounts to 1.5×10^{14} g, corresponding to a fragment mass of at least 3×10^{14} g, if having either a cometary or a CI chondritic composition. To this, one should add the mass of oxygen contained in water: this latter is difficult to estimate since only "hot" water was detected in the plumes or the splashbacks. Cold water or condensed water ice (which might be an important component of the haze, unfortunately without specific spectral signature) eluded detection. Assuming that the mass of the initial nucleus was about ten times that of the big fragments, we obtain a lower limit of 3×10^{15} g for the mass of the nucleus, corresponding to a diameter of 2 km (for a density of 0.5 g cm^{-3}).

The sulphur mass is poorly known, mainly due to the still uncertain S_2 mass and to that of OCS, a molecule only marginally detected in the radio. Given these uncertainties,

molecule	observation[†]	molecular mass[‡] in a SL9 impact	elemental mass in a SL9 impact	elemental mass in a 10^{15} g fragment	
				comet	CI-chondrite
H_2O	**IR, radio**	$>2 \times 10^{12}$	$> 1.5 \times 10^{14}$ [O]	5×10^{14} [O]	5×10^{14} [O]
CO	IR, **radio**	2.5×10^{14}			
NH_3	UV, **IR**	1×10^{13}	1×10^{13} [N]	2×10^{13} [N]	3×10^{11} [N]
HCN	IR, **radio**	6×10^{11}			
S_2	**UV**	1.5×10^{12} ?			
CS_2	**UV**	1.5×10^{11}			
CS	**UV, radio**	5×10^{11}	5×10^{12} [S]	4×10^{13} [S]	6×10^{13} [S]
H_2S	**UV**	?			
OCS	**radio**	3×10^{12}			
			1×10^{14} [C]	2×10^{14} [C]	3×10^{13} [C]
PH_3	**IR**	?	3×10^{11} [P]	1×10^{12} [P]	
silicates	**IR**	6×10^{12}	8×10^{13} [Si]	1×10^{14} [Si]	
				4×10^{13} [Mg]	1×10^{14} [Mg]

[†] In bold are the observations from which the abundances are taken.

[‡] Some of these values are preliminary and subject to revision. See Tables 3 and 4 for references and Lellouch (this volume) for a discussion.

TABLE 5. Amounts (in g) of detected molecules and corresponding elemental masses following the impacts of SL9 on Jupiter, with a comparison with elemental masses expected from a comet nucleus or a carbonaceous chondrite.

the [S]/[O] ratio is compatible with cometary or asteroidal abundances, and there is no need to invoke a contribution from the hypothetical NH_4SH Jovian cloud (as it was by Noll *et al.* [1995], on the basis of a much higher preliminary estimation of the S_2 mass).

Is it possible to discriminate between an impactor of cometary or asteroidal origin from these elemental abundances? The oxygen abundances are nearly the same in a cometary nucleus or in a primitive asteroid with composition supposed to be similar to that of carbonaceous chondrites; most of the oxygen is in water ice and dust silicates in the comet, in crystallization water and in silicates in the asteroid. Thus, the observed oxygen mass can inform us on the mass of the fragments, but not on their origin.

Light elements, however, such as H, C, N, are significantly depleted in primitive asteroids (they could be still more depleted in more evolved asteroids). Nothing could be learned from C or H, which could come from Jovian H_2 and CH_4. The nitrogen mass is important and corresponds to [N]/[O] $\simeq 0.07$; this is compatible with a cometary composition ([N]/[O] $\simeq 0.04$), but is much larger than a chondritic composition ([N]/[O] $\simeq 6 \times 10^{-4}$). However, most of the observed nitrogen is coming from ammonia: this argument would be ruined if this ammonia were coming from the NH_3 cloud hit by the impactor.

It would also be interesting to know the D/H ratio for the molecules observed in the impact sites. This could allow us to distinguish cometary material from processed

Jovian material, since deuterium is believed to be heavily enriched in comets. D/H was measured to be 3×10^{-4} for the water of P/Halley (Balsiger *et al.* 1995; Eberhardt *et al.* 1995) whereas it is $\simeq 2 \times 10^{-5}$ in Jupiter (Lécluse *et al.* 1995), similar to the protosolar value. Unfortunately, no information on isotopic ratios is available from spectroscopy of the SL9 impacts.

4.4. *Clues from aerosols*

The impacts were followed by the apparition of spots, or blemishes, that persisted for months for the strongest impacts. These spots were attributed to the formation of dark aerosols (see the chapters by Hammel and West).

Several authors attempted to derive the impactor's masses from the observations of the aerosols in the impact sites (Field *et al.* 1994; Knacke & A'Hearn 1994; Tejfel *et al.* 1995; Vanysek 1995; R. A. West *et al.* 1995). The observations give directly the spot sizes (5000 to 15000 km diameter) and the filling factors (equal to the optical depth, 0.01 to 0.3) of the aerosol particles. With reasonable assumptions of the density (0.5 to 3 g cm^{-3}) and the size (0.5 to 10 μm) of the particles, one can derive the total mass of the visible aerosols: 3×10^{10} to 10^{15} g in a single impact site. To further obtain the masses of the impactors, one has then to estimate the fraction of the refractories which effectively recondense into aerosols, and the fraction of the impactor which is composed of refractories. Due to the large uncertainties on all parameters, we must admit that these masses are not seriously constrained.

The nature of the aerosols is still the subject of debates. Multi-wavelength observations permitted measurement of the imaginary part of the refractive index of the aerosol material, and thus a constraint on its nature (R. A. West *et al.* 1995). The dark brown colour suggests graphite, dirty silicates, organics, sulphur and nitrogen compounds.

A key observation was the identification of the silicate emission feature at 10 μm in a spectrum obtained immediately after impact R at Palomar observatory (Nicholson *et al.* 1995). It corresponded to a mass of silicates of 6×10^{12} g, assuming grains of radius ≤ 1 μm and density 3.3 g cm^{-3}.

Recondensed refractories are thus highly plausible candidates. Models of recondensation of silicates have been achieved by Friedson (1995) and Hasegawa *et al.* (1995). Poly-HCN has also been proposed because its imaginary refractive index closely reproduces that observed (R. A. West, this volume, R. A. West *et al.* 1995); it can be noted that HCN is an identified component of both cometary nuclei and the impact sites (Marten *et al.* 1995). A chemical model by Lyons (1995), however, showed that poly-HCN is unlikely to form in the post-impact reactions. Wilson & Sagan (1995) remarked that the observed imaginary refractive index is also close to that of the organic residue of the Murchison carbonaceous chondrite, and they proposed that the aerosols could come from chemically preserved debris of the impactor.

It is probable that the aerosol is of a complex composition and that its grains are heterogeneous due to the variety of the input material. It should also be stressed that the aerosols may consist, in a large part, of ices of water, ammonia and carbon dioxide, which escaped detection because of their small absorption coefficients.

5. Conclusion

Retrieving the chemical composition of SL9 from the spectroscopic observations of the SL9 event is a difficult problem. A careful modelling of the physical conditions of the impacts (geometry, temperature, time evolution) is required, as well as of the chemical processes. There is no doubt that these challenging issues could only be solved by the close collaboration of all observers and modellers, in order to put all the pieces of the puzzle together.

At the present stage of the interpretation of the observations, SL9 appears to be a very brittle body for which we have only soft constraints on the composition. These constraints agree with our current knowledge of the cometary and asteroidal compositions. Trying to classify SL9 as a *comet* or an *asteroid* is perhaps a meaningless point: we now know that these two kinds of small bodies are closely related and that they may evolve from one class to another.

Therefore, it seems appropriate to conclude by the following quotations:

There is no compelling need to invoke a special class of object for SL9. The classification of SL9 as a presumably normal Jupiter-family short-period comet is not contradicted by any observation so far. (R. M. West 1995.)

Most of the objects classified as asteroids at and beyond Jupiter's orbit are likely to conceal buried volatiles, and thus are more usefully considered as comets. (Jewitt 1995b.)

I would like to thank all the observers of comet SL9 and of its impacts on Jupiter who communicated their results prior to publication, and the editors for their comments and their help in improving this paper.

REFERENCES

A'HEARN, M. F., MILLIS, R. L., SCHLEICHER, D. G., OSIP, D. J. & BIRCH, P. V. 1995 The ensemble properties of comets: results from narrowband photometry of 85 comets, 1976–1992. *Icarus*, **118**, 223–270.

A'HEARN, M. F., SCHLEICHER, D. G., FELDMAN, P. D., MILLIS, R. L. & THOMPSON, D. T. 1984 Comet Bowell 1980b. *Astron. J.* **89**, 579–591.

ANDERS, R. & GREVESSE, E. 1989 Abundances of the elements: meteoritic and solar. *Geochim. Cosmochim. Acta* **53**, 197–214.

ARPIGNY, C. 1979 Relative abundances of the heavy elements in comet Ikeya-Seki (1965 VIII). *Mem. Soc. Roy. Liège* **22**, 189–197.

ARPIGNY, C. 1994 Physical chemistry of comets: models, uncertainties, data needs. In *Molecules and Grains in Space*, (ed. I. Nenner), AIP Conference Proceedings **312**, pp. 205–238.

ARPIGNY, C., DOSSIN, F., WOSZCZYK, A. *et al.* 1995 Atlas of cometary spectra. Kluwer (in press).

ATREYA, S. K., EDGINGTON, S. G., TRAFTON, L. M. *et al.* 1995 Abundance of ammonia and carbon disulfide in the Jovian stratosphere following the impact of comet Shoemaker-Levy 9. *Geophys Res. Let.* **22**, 1625–1628.

BALLESTER, G. E., HARRIS, W. M., GLADSTONE, G. R. *et al.* 1995 Far-UV emissions from the SL9 impacts with Jupiter. *Geophys. Res. Let.* **22**, 2425–2429.

BALSIGER, H., ALTWEGG, K. & GEISS, J. 1995 D/H and $^{18}O/^{16}O$-ratio in the hydronium ion and in neutral water from in situ ion measurements in comet Halley. *J. Geophys. R.* **100**, 5827–5834.

BELTON, M. J. S., CHAPMAN, C. R. *et al.* 1995 Bulk density of asteroid 243 Ida from the orbit of its satellite Dactyl. *Nature* **374**, 785–788.

BÉZARD, B., GRIFFITH, C. A., KELLY, D. *et al.* 1995 Mid-IR high-resolution spectroscopy of the SL9 impact sites: temperature and HCN retrievals. *IAU Coll. No. 156* (poster).

BJORAKER, G. L., STOLOVY, S. R., HERTER, T. L., GULL, G. E. & PIRGER, B. E. 1995 Detection of water after the collision of fragments G and K of comet Shoemaker-Levy 9 with Jupiter. *Icarus* (submitted).

BOCKELÉE-MORVAN, D., COLOM, P., DESPOIS, D. *et al.* 1995 Observations of the Shoemaker-Levy 9 impacts on Jupiter at the Swedish-ESO submillimetre telescope. *The Messenger* **79**, 29–31.

BOROVIČKA J. & SPURNÝ P. 1995 Radiation study of two very bright terrestrial bolides. *Icarus* (submitted).

BORUNOV, S., DROSSART, P., ENCRENAZ, T. & DOROFEEVA, V. 1995 High temperature chemistry in the fireballs formed by the impacts of comet P/Shoemaker-Levy 9 in Jupiter. *Icarus* (submitted).

CARLSON, R. W., WEISSMAN, P. R., SEGURA, M. *et al.* 1995 Galileo infrared observations of the Shoemaker-Levy 9 G impact fireball: a preliminary report. *Geophys Res. Let.* **22**, 1557–1560.

CARPENTER, K. G., MCGRATH, M. A., YELLE, R. V., NOLL, K. S. & WEAVER, H. A. 1995 Formation of atomic emission lines in the atmosphere of Jupiter after the comet Shoemaker-Levy 9 S impact. *Bull. Amer. Astron. Soc.*, **27**, 64.

CATALANO, S., RODONO, M., VENTURA, R. & CACCIANI, A. 1995 Observations of the SL-9 impacts with a sodium magneto-optical filter. *European SL-9/Jupiter Workshop*, (eds. R. M. West and H. Boehnhardt), ESO Conference and Workshop Proceedings No. 52, pp. 209–214.

CHERNOVA, G. P., JOCKERS, K. & KISELEV, N. N. 1995 Imaging photometry and color of comet Shoemaker-Levy 9. *Icarus* (submitted).

COCHRAN, A. L., WHIPPLE, A. L., MACQUEEN, P. J. *et al.* 1994 Preimpact characterization of comet P/Shoemaker-Levy 9. *Icarus* **112**, 528–532.

COLAS, F., JORDA, L., LECACHEUX, J. *et al.* 1995 Pre-impact observations of Shoemaker-Levy 9 at Pic du Midi and Observatoire de Haute Provence. *European SL-9/Jupiter Workshop*, (eds. R. M. West and H. Boehnhardt), ESO Conference and Workshop Proceedings No. 52, pp. 23–28.

CONRATH, B. J., GIERASCH, P. J., HAYWARD, T. *et al.* 1995 Palomar mid-infrared observations of comet Shoemaker-Levy 9 impact sites. *IAU Coll. No. 156* (poster).

COPELAND, R. & LOHSE, J. G. 1882 Spectroscopic observations of comets III and IV, 1881, comet I, 1882, and the Great Comet of 1882. *Copernicus* **2**, 225–244.

COSTA, R. D. D., DE FREITAS PACHECO, J. A., SINGH, P. D., DE ALMEDA, A. A. & CODINA, S. J. 1994 Detection of lithium in the plume of the L-fragment impact of comet Shoemaker-Levy 9 with Jupiter. *Preprint*.

CRISP, D. & MEADOWS, V. 1995 Near-infrared imaging spectroscopy of the impacts of SL9 fragments C, D, G, K, N, R, V, and W with Jupiter. *IAU Coll. No. 156* (poster).

CROVISIER, J. 1994. Molecular abundances in comets. In *Asteroids, Comets, Meteors 1993*, (eds. A. Milani *et al.*). pp. 313–326. Kluwer.

CROVISIER, J., BIVER, N., BOCKELÉE-MORVAN, D. *et al.* 1995 Carbon monoxide outgassing from comet P/Schwassmann-Wachmann 1. *Icarus* **115**, 213–216.

EBERHARDT, P., REGBER, M., KRANKOWSKY, D. & HODGES, R. R. 1995 The D/H and $^{18}O/^{16}O$ ratios in water from comet P/Halley. *Astron. Astrophys.*

FELDMAN, P. D., WEAVER, H. A., BOICE, D. C. & STERN, S. A. 1995 HST observations of Mg^+ in outburst from comet D/Shoemaker-Levy 9. *Icarus* (in press).

FESTOU, M. C., RICKMAN, H. & WEST, R. M. 1993 Comets. *Astron. Astrophys. Reviews* **3**, 363–447 and **5**, 37–163.

FIELD, G. B., TOZZI, G. P. & STANGA, R. M. 1994. Dust as the cause of spots on Jupiter. *Astron. Astrophys.* **294**, L53–L55.

FITZSIMMONS, A., LITTLE, J. E., ANDREWS, P. J. *et al.* 1995a Optical imaging and spectroscopy of the impact plumes on Jupiter. *European SL-9/Jupiter Workshop*, (eds. R. M. West and H. Boehnhardt), ESO Conference and Workshop Proceedings No. 52, pp. 197–201.

FRIEDSON, A. J. 1995 Refractory grain formation in Shoemaker-Levy fireballs. *Icarus* (submitted).

GLADSTONE, G. R., HALL, D. T. & WAITE, JR., J. H. 1995 EUVE observations of Jupiter during the impact of comet Shoemaker-Levy 9. *Science* **268**, 1595–1597.

GRIFFITH, C. A., BÉZARD, B., KELLY, D. *et al.* 1995. Mid-IR spectroscopy and NH₃ and HCN images of K impact site. *IAU Coll. No. 156* (poster).

HAMMEL, H. B., BEEBE, R. F., INGERSOLL, A. P. *et al.* 1995 HST imaging of atmospheric phenomena created by the impact of comet Shoemaker-Levy 9. *Science* **267**, 1288–1296.

HASEGAWA, H., TAKEUCHI, S., YAMASHITA, T. & WATANABE, J. 1995 Reflectivities of the cometary impact sites. *IAU Coll. No. 156* (poster).

HOFFMANN, M. 1989. Asteroid mass determination: present situation and perspectives. In *Asteroids II*, (eds. R. P. Binzel *et al.*), pp. 228–239. University of Arizona Press.

HUEBNER, W. F., KELLER, H. U., JEWITT, D., KLINGER, J. & WEST, R. (eds.) 1993 Workshop on the Activity of Distant Comets. SWRI.

JESSBERGER, E. K. & KISSEL, J. 1991 Chemical properties of cometary dust and a note on carbon isotope. In *Comets in the Post-Halley Era*, (eds. R. L. Newburn *et al.*), pp. 1075–1092. Kluwer.

JEWITT, D. 1995a Pre-impact observations of SL-9. *European SL-9/Jupiter Workshop*, (eds. R. M. West and H. Boehnhardt), ESO Conference and Workshop Proceedings No. 52, pp. 1–4.

JEWITT, D. 1995b From comets to asteroids: when hairy stars go bald. In *Small Bodies in the Solar System and their Interactions with the Planets*, (eds. H. Rickman and M. Dahlgren). Kluwer (in press).

KNACKE, R. F. & A'HEARN, M. F. 1994 An estimate of the comet Shoemaker-Levy 9 fragment sizes from the debris fields. *Earth Moon Planets* **66**, 11–12.

LÉCLUSE, C., ROBERT, F., GAUTIER, D. & GUIRAUD, M. 1995 Laboratory determinations of deuterium exchange kinetics: application to the determination of the D/H ratio in giant planets, *Icarus* (in press).

LELLOUCH, E., PAUBERT, G., MORENO, R. *et al.* 1995 Chemical and thermal response of Jupiter's atmosphere following the impact of comet Shoemaker-Levy 9. *Nature* **373**, 592–595.

LYONS, J. R. 1995. Thermochemical and kinetic modeling of the SL9 impact debris. *IAU Coll. No. 156* (poster).

MARTEN, A., GAUTIER, D., OWEN, T. *et al.* 1995 The collision of the comet Shoemaker-Levy 9 with Jupiter: detection and evolution of HCN in the stratosphere of the planet. *Geophys. Res. Lett.* **22**, 1589–1592.

MATTHEWS, H. E., JEWITT, D. & SENAY, M. C. 1995 *IAU Circ. No. 6234.*

MEECH, K. J. & WEAVER, H. A. 1995 Unusual comets(?) as observed from the Hubble Space Telescope. In *Small Bodies in the Solar System and their Interactions with the Planets*, (eds. H. Rickman and M. Dahlgren). Kluwer (in press).

MONTEBUGNOLI, S., BORTOLOTTI, C., CATTANI, A. *et al.* 1995 Detection of the 22-GHz line of water during and after the SL-9/Jupiter impact. *European SL-9/Jupiter Workshop*, (eds. R. M. West and H. Boehnhardt), ESO Conference and Workshop Proceedings No. 52, pp. 261–266.

NICHOLSON, P. D., GIERASCH, P. J., HAYWARD, T. L. *et al.* 1995 Palomar observations of the R impact of comet Shoemaker-Levy 9: II. Spectra. *Geophys. Res. Lett.* **22**, 1617–1620.

NOLL, K. S., MCGRATH, M. A., TRAFTON, L. M. *et al.* 1995.. HST spectroscopic observations of Jupiter after the collision of comet P/Shoemaker-Levy 9. *Science* **267**, 1307–1313.

ORTON, G., A'HEARN, M., BAINES, D. *et al.* 1995 Collision of comet Shoemaker-Levy 9 with Jupiter observed by the NASA Infrared Telescope Facility. *Science* **267**, 1277–1282.

PRANGÉ, R., EMERICH, C., TALAVERA, A. *et al.* 1994 *IAU Circ.* No. 6041.

PRESTON, G. W. 1967 The spectrum of comet Ikeya-Seki (1965f). *Astron. J.* **147**, 718–742.

RAUER, H. & OSTERLOH, M. 1995. A search for differences in colour of the fragments of comet Shoemaker-Levy 9. *Astron. Astrophys.* **295**, L31–L34.

RAUER, H., DESPOIS, D., MORENO, R. *et al.* 1995 *IAU Circ. No. 6236.*

RICKMAN, H. 1994 Cometary nuclei. In *Asteroids, Comets, Meteors 1993*, (eds. A. Milani *et al.*), pp. 297–312. Kluwer.

ROOS-SEROTE, M., BARUCCI, A., CROVISIER, J. *et al.* 1995a Metallic emission lines during the impacts L and Q₁ of SL-9 on Jupiter. *Geophys. Res. Let.* **22**, 1621–1624.

ROOS-SEROTE, M., BARUCCI, A., CROVISIER, J. *et al.* 1995b Pic-du-Midi observations of atomic lines following impacts L and Q_1 of comet SL-9 with Jupiter. *European SL-9/Jupiter Workshop*, (eds. R. M. West and H. Boehnhardt), ESO Conference and Workshop Proceedings No. 52, pp. 203–208.

SCOTTI, J. V., GEHRELS, T. & METCALFE, T. S. 1994 Spacewatch observations of the surface brightness of comet P/Shoemaker-Levy 9. *Preprint*.

SEKANINA, Z. 1995a The splitting of the nucleus of comet Shoemaker-Levy 9. *European SL-9/Jupiter Workshop*, (eds. R. M. West and H. Boehnhardt), ESO Conference and Workshop Proceedings No. 52, pp. 43–55.

SEKANINA Z. 1995b Nuclei of SL-9 on images taken with the Hubble Space Telescope. *European SL-9/Jupiter Workshop*, (eds. R. M. West and H. Boehnhardt), ESO Conference and Workshop Proceedings No. 52, pp. 29–35.

SEKANINA, Z., CHODAS, P. W. & YEOMANS, D. K. 1994 Tidal disruption and the appearance of periodic comet Shoemaker-Levy 9. *Astron. Astrophys.* **289**, 607–636.

SENAY, M. C. & JEWITT, D. 1994 Coma formation driven by carbon monoxide release from comet Schwassmann-Wachmann 1. *Nature* **371**, 229–231.

SPRAGUE, A. L., BJORAKER, G. L., HUNTEN, D. M. *et al.* 1995 Observations of H₂O following the R and W impacts of comet SL-9 into Jupiter's atmosphere. *Science* (submitted).

STÜWE, J. A., SCHULZ, R. & A'HEARN, M. F. 1995 NTT observations of SL-9: imaging and spectroscopy. *European SL-9/Jupiter Workshop*, (eds. R. M. West and H. Boehnhardt), ESO Conference and Workshop Proceedings No. 52, pp. 17–20.

TANCREDI, G. & LINDGREN, M. 1994 Searching for comets encountering Jupiter: first campaign. *Icarus* **107**, 311–321.

TEJFEL, V. G., KHARITONOVA, G. A., SINYAEVA, N. V. *et al.* 1995 The reflectivity of the aerosol material in the G, L and E impact-regions. *IAU Coll. No. 156* (poster).

TRILLING, D. E., KELLER, H. U., RAUER, H., SCHULZ, R. & THOMAS, N. 1995 Observations of P/Shoemaker-Levy 9 in Johnson B, V and R filters from Calar Alto Observatory on 2/3 June, 1994. *European SL-9/Jupiter Workshop*, (eds. R. M. West and H. Boehnhardt), ESO Conference and Workshop Proceedings No. 52, pp. 37–41.

VANYSEK, V. 1995 A note on the fragments size of SL-9 and debris field. *European SL-9/Jupiter Workshop*, (eds. R. M. West and H. Boehnhardt), ESO Conference and Workshop Proceedings No. 52, pp. 297–298.

WEAVER, H. A., A'HEARN, M. F., ARPIGNY, C. *et al.* 1995 The Hubble Space Telescope (HST) observing campaign on comet P/Shoemaker-Levy 9. *Science* **267**, 1282–1288.

WEAVER, H. A., FELDMAN, P. D., A'HEARN, M. F. *et al.* 1994 Hubble Space Telescope observations of comet P/Shoemaker-Levy 9 (1993e), *Science* **263**, 787–791.

WEST, R. A., KARKOSHKA, E., FRIEDSON, A. J. *et al.* 1995 Impact debris particles in Jupiter's atmosphere. *Science* **267**, 1296–1301.

WEST, R. M. 1995 Nature and Structure of the impacting objects. *European SL-9/Jupiter Workshop*, (eds. R. M. West and H. Boehnhardt), ESO Conference and Workshop Proceedings No. 52, pp. 407–410.

WEST, R. M., HOOK, R. & HAINAUT, O. 1995. A morphological study of SL-9 CCD images obtained at La Silla (July 1–15, 1994). *European SL-9/Jupiter Workshop*, (eds. R. M. West and H. Boehnhardt), ESO Conference and Workshop Proceedings No. 52, pp. 5–10.

WILSON, P. D. & SAGAN, C. 1995 Chemistry of the Shoemaker-Levy 9 jovian impact blemishes: indigenous cometary vs. shock-synthesized organic matter. *IAU Coll. No. 156* (poster).

WINK, J., LUCAS, R., GUILLOTEAU, S. & DUTREY, A. 1994 Interferometric observations of HCN J=1-0 during the SL-9 crash on Jupiter. *IRAM Newsletter* **18**, 5.

WYCKOFF, S., WAGNER, M., WEHINGER, P., SCHLEICHER, D. & FESTOU, M. 1984 Onset of sublimation in comet P/Halley. *Nature* **316**, 241–242.

XIE, X., MUMMA, M. J. & OLSON, K. M. 1995 Physical properties of swarmed SL9 fragments at impact. *IAU Coll. No. 156* (poster).

YELLE, R. V. & MCGRATH, M. A. 1995 Ultraviolet spectroscopy of the SL9 impact sites, I: the 175–230 nm region. *Icarus* (submitted).

ZAHNLE, K. & MAC LOW, M.-M. 1995 A simple model for the light curve generated by a Shoemaker-Levy 9 impact. *J. Geophys. Res.* **100**, 16885–16894.

Tidal breakup of the nucleus of Comet Shoemaker–Levy 9

By ZDENEK SEKANINA

Jet Propulsion Laboratory, California Institute of Technology, Pasadena, CA 91109, USA

The breakup of Comet Shoemaker-Levy 9 is discussed both in the context of splitting as a cometary phenomenon, comparing this object with other split comets, and as an event with its own idiosyncrasies. The physical appearance of the comet is described, features diagnostic of the nature of tidal splitting are identified, and the implications for modelling the event are spelled out. Among the emphasized issues is the problem of secondary fragmentation, which documents the comet's continuing disintegration during 1992–94 and implies that in July 1992 the parent object split tidally near Jupiter into 10–12, not 21, major fragments. Also addressed are the controversies involving models of a strengthless agglomerate versus a discrete cohesive mass and estimates for the sizes of the progenitor and its fragments.

1. Introduction

Splitting is a relatively common phenomenon among comets, even though its detection is observationally difficult because companions are almost invariably very diffuse objects with considerable short-term brightness variations. Comet Shoemaker-Levy 9's behavior was generally less erratic than that of an average split comet, which may have in part been due to a major role of large-sized dust. The breakup products that contributed most significantly to the comet's total brightness are referred to below as *components*, or, because of their diffuse appearance, as *condensations*, both common terms of cometary phenomenology. The terms *nuclei* and *fragments* are instead reserved for genuine solid bodies of substantial dimensions ($\gtrsim 1$ km across) that were "hidden" in the condensations. A condensation may contain many fragments or nuclei, besides large amounts of material of subkilometer-sized and smaller particulates, the entire population of which is characterized by a certain size distribution function.

A total of 21 split comets had been documented in the literature by 1980 (for a review, see Sekanina 1982) and eleven additional ones have been reported since. Of these recent entries, fully eight are or were short-period comets (79P/du Toit-Hartley, 108P/Ciffréo, 101P/Chernykh, D/Shoemaker-Levy 9, P/Machholz 2, 51P/Harrington, 73P/Schwassmann-Wachmann 3, and the parent of 42P/Neujmin 3 and 53P/Van Biesbroeck); two are (or were) "old" comets (of which Takamizawa-Levy 1994 XIII = C/1994 G1 is one, while the breakup products of the other were discovered as two separate objects, Levy 1987 XXX = C/1988 F1 and Shoemaker-Holt 1988 III = C/1988 J1, orbiting the Sun in virtually identical paths with a period of ~14,000 years and passing through perihelion $2\frac{1}{2}$ months apart); and one is a "new" comet from the Oort cloud (Wilson 1987 VII = C/1986 P1). Shoemaker-Levy 9 was unique among all the split comets in that the maximum number of condensations observed *at the same time* was by far the largest. It is shown below, however, that without the italicized qualification the statement would not be valid.

Comet Shoemaker-Levy 9 is one among only a few multiple comets that are known to have fragmented due primarily—if not entirely—to the action of tidal forces during their extremely close encounters with Jupiter or the Sun. Besides Shoemaker-Levy 9, direct evidence exists for 16P/Brooks 2, which missed Jupiter by one planet's radius above the cloud tops in 1886; and for two or three members of the sungrazing comet group (1882 II = C/1882 R1, Ikeya-Seki 1965 VIII = C/1965 S1, possibly also Pereyra 1963 V = C/1963 R1), whose perihelia were located within $\frac{2}{3}$ the Sun's radius above the photosphere. For other split comets, the nature of the disruption mechanism is not fully understood, although jettisoning of pancake-shaped fragments of an insulating mantle from the nuclear surface by stresses, built up unevenly beneath it, is consistent with the evidence suggesting that these comets "peel off" rather than break up (Sekanina 1982).

2. The number of condensations

The detected number of Shoemaker-Levy 9's condensations depended not only on the imaging circumstances and the instrument used, but also on the time of observation, because some of the condensations disappeared with time while others began to develop companions of their own. Accounts of high-resolution observations indicate that no more than 22 condensations were detected *at a time*. On the other hand, the combined number of condensations reported on visually inspected images obtained *at various times* appears, collectively, to total 25. This is short of the record held by the progenitor of the group of sungrazing comets, from which all the observed members were found by Marsden (1989) to derive. The known sungrazers represent at least three generations of fragmentation products and include: (i) four members discovered between 1843 and 1887, one of which (1882 II) was observed after perihelion to have at least five components and another (1887 I = C/1887 B1) always appeared as a headless object (Kreutz 1888); (ii) four members discovered between 1945 and 1970; (iii) six members detected with a coronagraph onboard the SOLWIND spacecraft between 1979 and 1984; (iv) 10 members detected with a coronagraph onboard the Solar Maximum Mission spacecraft between 1987 and 1989; and (v) any possible precursor objects, of which an uncertain orbit exists for one (the comet of 1668 = C/1668 E1) and very little information on two more (the comets of 1106 and 371 BC). Excluding the dubious companion to 1963 V, a few unlikely candidates in the 17th through 19th centuries (for an overview, see Marsden 1967), and the comets of 1106 and 371 BC, one still finds a total of 29 comets and companions—a number that moderately exceeds the 25 condensations of Shoemaker-Levy 9—observed over a period of nearly 150 years. However, if this system of comet parentage is accepted, the number of catalogued split comets should be decreased by one, because 1882 II and 1965 VIII would not then be listed as separate entries. On the other hand, if only first-generation products of an object should be counted, the number of sungrazers would drop, but so would the number of condensations of comet Shoemaker-Levy 9, as discussed in some detail in Sec. 4.3.

3. Appearance of comet Shoemaker-Levy 9

The comet's condensations were all aligned in an essentially rectilinear configuration, which extended almost perfectly along a great circle of the projected orbit and whose appearance has often been fittingly compared to a *string of pearls*. In the technical literature, the collection of the condensations is usually referred to as the *nuclear train* or just the *train*. Even though the condensations were the most prominent features contributing substantially to the total brightness, significant amounts of material were also situated in between them, along the train's entire length.

The comet further exhibited three other kinds of morphological features. Extending from the train on either side were *trails* or *wings*, of which the east-northeastern one appeared to be slightly inclined relative to the train. Subtending a relatively small angle with the train and pointing generally to the west was a set of straight, narrow *tails*, whose roots coincided with the distinct condensations in the train. These parallel tails were immersed in, and on low-resolution images gradually blended with, an enormous, completely structureless *sector of material*, which was stretching to the north of its sharp boundary delineated by the nuclear train and the two trails.

3.1. *The nuclear train*

To describe the train's structure in detail, two notations were proposed to identify the condensations. The system introduced by Sekanina *et al.* (1994; hereafter referred to as Paper 1) has been employed, especially after the impacts, almost universally: the eastern-most condensation, the first to crash, was named A; the westernmost, which crashed the last, W. The letters I and O were excluded to avoid any confusion with the symbols used for the respective digits. The relationship between this notation and Jewitt *et al.*'s (1993) system, which numbers the condensations, is: $A \equiv 21$, $B \equiv 20$, ..., $W \equiv 1$.

The train's length, defined by the projected distance between the condensations A and W, continuously grew with time, from ~50 arcsec shortly after discovery in late March to almost 70 arcsec by mid-July of 1993, to more than 2 arcmin by the beginning of 1994, to about 5.5 arcmin by early May, and to some 10 arcmin, equivalent to a projected distance of more than 2 million km, by early July 1994. The train's enormous extent was reflected in the time span of 5.5 days between the first and the last impacts: July 16.84 UT for A and July 22.34 UT for W (Chodas & Yeomans 1994).

The train's orientation varied relatively insignificantly during the period of more than one year between discovery and collision. The position angle, measured in the direction from A to W, was within 1° of 256° between late March and mid-July 1993, decreasing to 245° by the beginning of 1994, reaching a minimum of 241° in early March, a maximum of 244° at the beginning of June, and decreasing again, at an accelerating rate, to ~240° by mid-July 1994.

A detailed analysis of the alignment of the individual condensations showed that five of them—B, J, M, P (later resolved into P1 and P2; see Sec. 3.2), and T—exhibited small but detectable *off-train* deviations on high-resolution images as early as March–July 1993, during the first four months after discovery (Table 8 of Paper 1). From a comparison of the 1993 pre- and post-conjunction observations, it became apparent that this group of "anomalous" condensations also included F. And more recently, with the use of a large number of 1994 observations, this category of condensations grew further, now also encompassing G2, N, Q2 (see Sec. 3.2), U, and V. Thus, eleven (or more) of the condensations were found to deviate noticeably from the nearly perfectly aligned *on-train* condensations. The status of some of them, in particular C and/or D, is not entirely clear to this time.

3.2. *Physical evolution of the train*

Two of the condensations, J and M, had only been detected by Jewitt *et al.* (1993) on four occasions between late March and mid-July 1993. They are not apparent on the images taken with the Hubble Space Telescope (HST) on July 1, 1993 (Weaver *et al.* 1994) and were not reported at any time during 1994. An unpublished account of a possible detection of J on the comet's image taken at Mauna Kea on December 14, 1993 does not appear to be correct.

The first signs of impending dramatic changes in the appearance of the train became evident on the July 1993 images obtained with the HST (Weaver *et al.* 1994). While the components J and M vanished, the condensation Q, the brightest at the time, appeared to have a faint, diffuse companion some 0.3 arcsec away at a position angle of ~ 30°. At the location of the condensation P two very diffuse nebulosities can be seen, less than

1 arcsec apart and aligned approximately with the train. The condensation L may also have exhibited a companion nebulosity to the north. The spatial brightness distribution in the condensations was reported by Weaver *et al.* to be significantly flatter than the inverse first power of distance from the center, with no molecular emissions in the spectral region between 2220 and 3280 Å.

By late January 1994, the time of the next HST observation (Weaver *et al.* 1995), the P–Q region had developed considerably. The condensation Q was manifestly broken into two, a brighter Q1 and a fainter Q2, 1.2 arcsec apart and the fainter to the north-northwest of the brighter. The condensation P also consisted of two widely separated nebulosities. The fainter and poorly condensed component, P1, was 4.2 arcsec from Q1 and almost exactly to the north, while the brighter, P2, was 5.0 arcsec from Q1 in the north-northeastern direction. The separation of P1 from P2 was 2.2 arcsec, the former nearly to the west of the latter. Both components looked elongated: P1 very distinctly to the west-northwest, in the direction of the tail; P2 less noticeably to the southwest. Another significant development was the issuance, in the southward direction, of a bright "spur" from the condensation S (Weaver 1994). The spatial distribution in the condensations became steeper than the inverse first power of distance. The appearance of the nuclear train, except for the condensation W, is shown in Fig. 1.

Further morphological changes were noticed on the HST images of the comet taken on March 29–30, 1994. The condensation P2 had become double, the fainter component appearing rather diffuse, and the spur of S had grown fainter (Weaver 1994; Weaver *et al.* 1995). While most condensations were still sharply defined, the components P1 and T were barely discernible as virtually uncondensed masses of material. Again, no molecular emissions were detected in the 2220–3280 Å region.

The HST monitoring of the comet continued throughout the months of May–July 1994, one of the highlights of this period being the spectral detection of a strong outburst of Mg^+ on July 14 (Weaver *et al.* 1995, Feldman *et al.* 1996; the continuum outburst observed ~18 mins after the Mg^+ outburst was probably due to the passage of a faint star through the spectrograph aperture, according to the latter reference). The absence of any neutral molecular emissions in the spectrum between 2220 and 3280 Å was again confirmed. The images show that the central regions of the condensations, a few seconds of arc across, remained spherically symmetric until one week or so prior to impact, at which time they began to grow strikingly elongated along the direction of the train.

In the meantime, ground-based observations—interrupted in late July 1993 because of the comet's approaching conjunction with the Sun—resumed in early December 1993 and continued during 1994 until impact. The condensation P2 was observed extensively and as late as July 14 at the European Southern Observatory at La Silla and July 19 at the Mauna Kea Observatory. Ground-based observations of P1 apparently terminated in late March. The condensation Q2 may have been sighted on only a few occasions and was measured perhaps just once. On May 7, 1994 Jewitt & Trentham (1994) detected a companion to the condensation G, 5.1 arcsec to the northeast of it. The existence of this companion G2 was confirmed on the HST images taken on May 17, by which time its separation distance increased to 5.9 arcsec (Noll & Smith 1994). On the other hand, its identification with a faint object located 4.1 arcsec to the north-northeast of G on the HST images from March 29 is doubtful.

3.3. *The dust trails*

Observations of the two trails are very limited, compared with the extensive amount of information available on the nuclear train. Almost all the published data refer to the early post-discovery period, late March through late May 1993. Their nearly complete

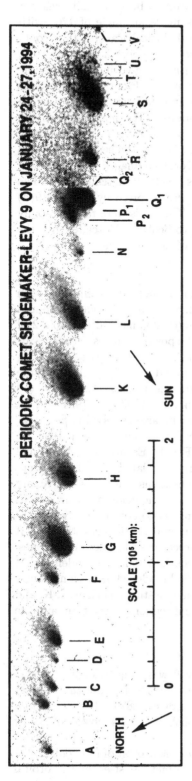

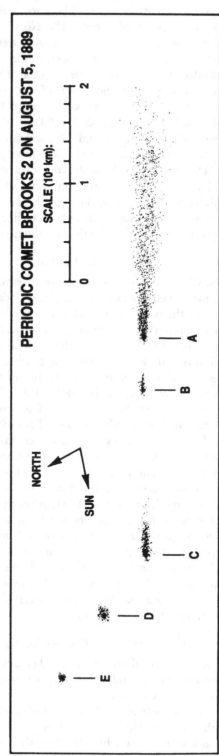

FIGURE 1. Comparison of the appearance of periodic comets Shoemaker-Levy 9 and Brooks 2. *Top:* Mosaic image of the nuclear train (except for the condensation W) and the tails of Comet Shoemaker-Levy 9 taken with the Wide Field Planetary Camera 2 of the Hubble Space Telescope on January 24–27, 1994. The 20 visible condensations are identified in the commonly used notation introduced by Sekanina *et al.* (1994) and later expanded to accommodate the additional condensations. The projected linear scale at the comet and the directions of the north and the Sun are shown. (Adapted from an image provided by courtesy of H. A. Weaver and T. E. Smith, Space Telescope Science Institute.) *Bottom:* Drawing of P/Brooks 2 made by E. E. Barnard (1889) and based on his visual observation with the 91-cm equatorial of the Lick Observatory on August 5, 1889. The notation used for the five condensations is that of Barnard. The scale and the orientation of the drawing are also shown.

list is presented in Table 2 of Paper 1. The west-southwestern trail was perfectly aligned with the nuclear train and its southern boundary appeared to be marginally sharper than that of the east-northeastern trail. The two trails made an angle of 176° with each other, with an uncertainty of about ±2°. No morphological features that could possibly suggest the presence of faint condensations were ever reported in either trail, but it appears that not enough effort has been made with this goal in mind. To the north, the trails blended into the structureless sector of material. From the descriptions based on 10 low-resolution images, the length of the east-northeastern branch was, on the average, 0.63 the length of the west-southwestern branch. The maximum lengths were reported by Scotti (1993) on March 30, 1993: ∼10.4 arcmin at a position angle of 260° for the west-southwestern branch and ∼6.2 arcmin at 75° for the east-northeastern branch, measured from the train's midpoint. Computer processed images show, however, no clear difference between the lengths of the two branches (Scotti & Metcalfe 1995). The only 1994 observations of the trails that I am aware of were reported by Lehký (1994) and by Scotti & Metcalfe (1995), both during February. The trails were fading rapidly with time and their expansion was consistent, according to Scotti & Metcalfe, with the assumption of no evaporation and no further production of dust material.

3.4. *The tails*

The tails associated with the condensations were observed virtually at all times between discovery and collision. Their orientations and lengths derived from the comet's ground-based images taken in the early period after discovery were summarized in Table 2 of Paper 1. The tails were reported to point at this time at the position angles between 280° and 300°, making an angle of ∼ 30° with the nuclear train. Their lengths, depending on observing conditions and on the intrumental resolution and sensitivity, the exposure time, and the spectral window, were found to be up to ∼80 arcsec. Additional tail observations were made by Scotti & Metcalfe (1995) in both 1993 and 1994 and by R. M. West *et al.* (1995) shortly before impact. These results are discussed in Sec. 4.4.

The tails are displayed prominently on most of the HST images obtained in 1994, an example of which, from late January, is reproduced in Fig. 1. Each condensation had its own tail, whose length and degree of prominence clearly correlated with the "parent" condensation's brightness. The tails of the major condensations, such as G or K, are seen to have extended at this time all the way to the edge of the field and their lengths must have greatly exceeded 25 arcsec, or 100,000 km in projection onto the sky plane.

The apparent breadth a few seconds of arc from the condensation is estimated at ∼ 6–7 arcsec for the brightest tails, but only at ∼ 2 arcsec or so for the fainter ones, corresponding to projected linear widths of 8000 to 25,000 km. There is evidence that the tails of some of the condensations (such as E, H, and S) broadened more significantly with distance from the train than did the tails of other condensations (such as C, K, and L). There also is an indication that the angle between the directions of the train and the tails might have been getting smaller with time.

3.5. *Comparison with other tidally split comets*

Even though the appearance of comet Shoemaker-Levy 9 was unquestionably unique among observed comets, certain similarities, however remote, can be found with two other tidally disrupted comets, P/Brooks 2 (1889 V) and the sungrazer 1882 II. The recognition of such similarities is of the essence in the context of establishing the diagnostics of tidally split objects and their observable characteristics.

During its approach to Jupiter on July 21, 1886, the jovicentric orbit of P/Brooks 2 was slightly hyperbolic (Sekanina & Yeomans 1985) and the comet's post-encounter orbital evolution was very different from that of Shoemaker-Levy 9. After passing two planet's radii from the center of Jupiter, P/Brooks 2 settled in a new heliocentric orbit, whose perihelion distance was 1.95 AU. The comet was discovered about three years later and was 1.16 AU from the Earth and 2.01 AU from the Sun and approaching the perihelion, when Barnard (1889) made the object's drawing on August 5, 1889, shown in Fig. 1. Two of the morphological features recognized in comet Shoemaker-Levy 9 are apparent on this drawing: (i) the nuclear train, consisting of the condensations A, B, and C and (ii) the tails, which in this case were aligned with the train. The condensations D and E, either of which was seen on only two occasions, represent a pair of "anomalous", off-train components analogous to Shoemaker-Levy 9's condensations B, F, N, etc. In addition, on one night Barnard (1889, 1890) detected four other faint companions to the south of the train and more distant from A than was E, but all of them remained unconfirmed, as did a companion reported by Renz (1889) on another day. Barnard (1890) suspected that at least some of these objects may have in fact been faint, unrecorded nebulae.

There was no evidence for wings extending from the condensations in either direction or for a sector of diffuse material spreading to either side of the train. However, Weiss (1889) reported the detection of a nebulous sheath encompassing the condensations A and B, while Barnard (1890) remarked on the absence of any such nebulosity during his observations on the same dates.

The relatively straightforward, empirically inferred parallelism between some of the morphological properties of comets Shoemaker-Levy 9 and Brooks 2 is found to be rather encouraging and appealing. Yet, the differences between the two objects and between their dynamical evolutions following the close encounter with Jupiter are significant enough that caution should be exercised not to overinterpret the similarities.

The nuclear region of the sungrazing comet 1882 II was observed to consist of at least five condensations after perihelion. They lined up in the orbital plane in a direction that, after correcting for effects of foreshortening, was lagging the Sun-comet line by more than 20° in the early post-perihelion period, but gradually less with time, until the two directions coincided some 5–6 months later, at the time of the final observations (Kreutz 1888). The brightness of the individual components varied with time, but the second and the third condensations from the train's sunward end were consistently the most prominent ones. An elongated nebulous sheath of material was enclosing the entire train. When last measured, about 150 days after perihelion, the sheath was almost 3 arcmin long, which is equivalent to a projected extent of more than 300,000 km. It is conceivable that the far regions of the sheath would have eventually evolved into wings similar to those displayed by comet Shoemaker-Levy 9, if they were sufficiently bright to have remained under observation still longer.

4. Implications for modelling comet Shoemaker-Levy 9

Before turning to the discussion of the various models proposed for comet Shoemaker-Levy 9, I list a few critical issues that represent excellent test criteria and are therefore to be addressed first. The merit of a paradigm depends primarily on the degree of its conformity with observations used in the course of its formulation, on the plausibility of the assumptions employed, and obviously very much on the success of any verifiable predictions that it might offer.

4.1. Morphological test criteria

From the descriptions of the comet's appearance in Sec. 3, two criteria are identified for testing a model: (i) it should explain quantitatively the four classes of morphological features detected (the nuclear train, trails, tails, and the sector of material) and their evolution and properties; and (ii) it should be conceptually consistent with unambiguous, model-independent conclusions based on direct observational evidence. Each model should also be judged in a broader context, in such terms as the object's implied long-term dynamical stability, the plausibility of postulated properties, and the paradigm's compatibility with current views on the physical behavior of comets.

The aspects of the problem that are considered the most significant are individually discussed below. The first point to make is that *each* of the four classes of morphological features presents an important piece of observational evidence. As such, these diverse features should either be consistently interpreted as different manifestations of the same process and as individual products of the parent object in the context of the tidal breakup and its consequences; or else be explained by a credible, independent hypothesis that should account for their existence, appearance, and temporal evolution. A model that leaves some of this morphology unexplained, is inevitably incomplete and tentative and its merits and the conclusions to which it leads should be viewed with caution and some skepticism.

The second point to make is that although the nuclear train should not be singled out as the *only* target of modelling efforts, its morphology offers more information than do the other features and its analysis should be of top priority. Some of the train's properties are diagnostically so critical that they deserve special attention.

4.2. *The train's orientation*

Temporal variations in the train's orientation are among the dynamical characteristics of comet Shoemaker-Levy 9 that have been determined with very high accuracy and are of fundamental importance from the standpoint of modelling. The dependence of the position angle of the nuclear train on the conditions at the time of tidal disruption was investigated in Paper 1. The orbital calculations showed conclusively that for a given trajectory of the comet and a fixed time of breakup, no measurable effect on the train's orientation could be generated by variations, within physically meaningful ranges, in the initial radial distances of the fragments from Jupiter (which represent the dimensions of the parent nucleus) or in their orbital velocities (which simulate any effects in the rotational and/or translational momentum), or some combination thereof. The only parameter that demonstrably affects the train orientation in a systematic manner is the model-independent *time of dynamical separation*, also called the *effective time of breakup*.

In practice, the train's position angles derived from astrometric observations were found to depend slightly, but measurably, on the selection of the condensations. For the sake of uniformity, it was necessary to select a "standard" set of condensations. Since the origin of the group of anomalous condensations (B, F, J, M, N, etc.) was suspect, its members were the first to be excluded from any such standard set. The selection was essentially dictated by a balance between two somewhat contradictory requirements: (i) by the need to employ the largest possible number of observations and (ii) by a condition that the position angles calculated from the standard set be representative of the "intrinsic" orientation of the train as a whole. Since most observers measured only brighter condensations, the first point implies the need to employ as few condensations as possible. By contrast, the second point requires a large number. After much experimentation, the standard set was defined in Paper 1 by eight condensations— E, G, H, K, L, Q (later Q1), S, and W. Comparisons with other sets showed that, except when the anomalous condensations were much involved, the position angles differed at most by a few hundredths of a degree. This is only a small fraction of the typical uncertainty involved in the position-angle determinations, which was about $\pm 0°\!.1$ for

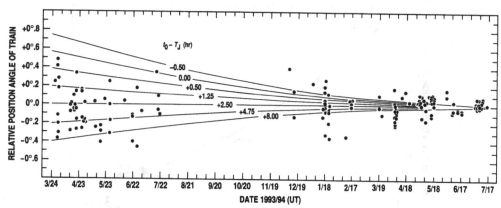

FIGURE 2. The temporal distribution of residuals of the nuclear train's position angle as a function of the assumed effective time of tidal disruption, t_0. The dots represent the 144 data points, determined by fitting astrometric observations of the standard set's eight condensations E through W. The curves show the variations in the calculated position angle for seven different effective times of breakup measured relative to the time of perijove, $T_J = 1992$ July 7.860364 TDB. For the optimized solution, $t_0 - T_J = +2.5$ hours, the calculated variations are represented by a straight line. (After Sekanina *et al.* 1995.)

high-resolution images (telescopes of ~ 1.5-meter aperture or larger) and some $\pm 0°.3$ or so for images of lower resolution (telescopes of ~ 1-meter aperture or smaller).

Although relatively subtle, the deviations of the nuclear train orientation from the values predicted for perijove were found in Paper 1 to be sufficiently pronounced that the *effective time of breakup*, t_0, could be determined, with an estimated error of ± 0.5 hour or so, from observations made primarily in the early post-discovery period of time. The time t_0, which describes the completion of a post-breakup collisional redistribution of the debris rather than the initiation of tidal fracture, was calculated in Paper 1 to equal 2.2 hours after perijove, based on a total of 42 astrometric positions from 1993. A new result based on the recently expanded database, including all 144 relevant observations from 1993–1994 (Sekanina *et al.* 1995), essentially confirms the solution derived from the smaller sample; the statistically best estimate is now 2.5 hours, again with an estimated error of ± 0.5 hour. The distribution of the residuals in Fig. 2 shows that the 1994 observations, while entirely consistent with the 1993 ones, fail to improve the solution significantly because of a dramatic decrease in the sensitivity. The detailed interpretation of the effective breakup time is an issue that each model is to settle within its own framework. I will return to this problem in Sec. 5.

4.3. *Secondary fragmentation*

The next fundamental issue is that of the origin and the nature of the off-train, anomalous condensations. This point, too, was addressed in Paper 1 on the example of the motion of the component P relative to Q. The 12 relative positions between late March 1993 and mid-January 1994 could be satisfied with a mean residual of ± 0.28 arcsec, if P separated from Q in mid-November 1992 (with an uncertainty of a few weeks) at a relative velocity of 0.9 m s^{-1}. Evidence for similar events of *secondary fragmentation* became ubiquitous during 1994. Two examples are reproduced in Figs. 3 and 4, both based on the results of a work in progress (Sekanina *et al.* 1995). Figure 3 displays an updated solution to the motion of P2 relative to Q, which satisfies 74 observations between March 27, 1993 and July 19, 1994 with a mean residual of ± 0.29 arcsec. The breakup was found to have occurred on 1992 Dec. 29 ± 9 days with a separation velocity of 1.14 ± 0.08 m s^{-1}.

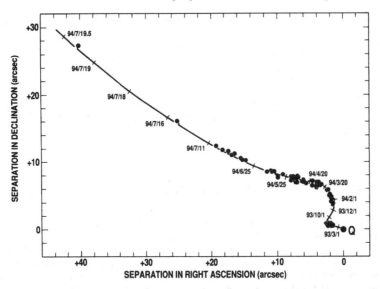

FIGURE 3. Projected motion of the component P2 relative to Q, based on 74 observations between March 27, 1993 and July 19, 1994 and interpreted as due to a splitting of the two condensations on Dec. 29, 1992 with a relative velocity of 1.14 m s^{-1}. The dots are the observations and the curve shows the optimized dynamical solution. (After Sekanina *et al.* 1995.)

Figure 4 displays the motion of Q2 relative to Q1, based on very accurate measurements from the HST images. The eight positions cover the period of time from July 1, 1993, when the two components were just 0.3 arcsec apart, to July 20, 1994. The solution, leaving a mean residual of ±0.029 arcsec (!), yields 1993 Apr. 12 ± 8 days for the date of splitting and 0.32 ± 0.02 m s^{-1} for the separation velocity.

The fact that the comet's fragmentation continued as a sequence of discrete events for a considerable time after the 1992 grazing encounter with Jupiter has enormous ramifications. First of all, secondary fragmentation was positively nontidal in nature. Next, for all practical purposes it is certain that *all* of the off-train condensations were products of this process, so that the number of major components split off from the parent comet by tidal fracture was not 21, but most probably 10–12. The preliminary results also indicate no need to introduce differential accelerations in the motions of the off-train condensations, although much work still remains to be completed. The apparent absence of such accelerations lends support to *tentative* conclusions that neither effects of solar radiation pressure nor nongravitational effects have been detected, and that these condensations were neither loose assemblages of *small-sized* particulate material nor did they display any activity.

Secondary fragmentation will unquestionably stay in the forefront of attention for some time to come. A plausible physical model for Shoemaker-Levy 9 must account for this sequence of discrete events, most of which appear to have taken place between early July 1992 and the beginning of 1993, when the nuclear train was receding from Jupiter, and also explain the observed relative velocities involved, of up to at least 1 m s^{-1}.

4.4. *Tail orientation and morphology: Relation to the problem of activity*

The problem of Shoemaker-Levy 9's outgassing activity, especially after discovery, has been a hotly debated issue. With the exception of the brief appearance of Mg$^+$ shortly before impact (as already mentioned in Sec. 3.2), no emission was ever detected spectroscopically. The unsuccessful search for the hydroxyl radical with the HST's Faint Object

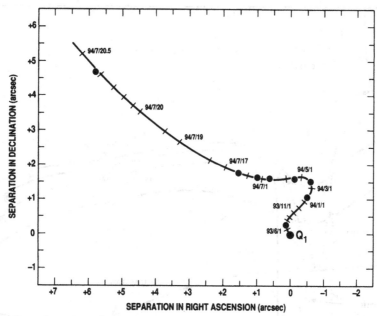

FIGURE 4. Projected motion of the component Q2 relative to Q1, based on eight HST observations made between July 1, 1993 and July 20, 1994 and interpreted to result from a splitting of the two condensations on Apr. 12, 1993 with a relative velocity of 0.32 m s^{-1}. The dots are the observations and the curve shows the optimized dynamical solution. (After Sekanina *et al.* 1995.)

Spectrograph and from the ground yielded a 3σ upper limit on the water production rate of 30 to 60 kg s^{-1} (Weaver *et al.* 1995; quoted here is the range of corrected values).

Indirect information on the object's possible activity is provided by the orientation and morphology of the dust tails. It was stated in Paper 1 that the orientation of the tails observed in March 1993, shortly after the comet's discovery, essentially supports the conclusion that they consisted of particulate material that had been released during the tidal breakup in early July 1992 and subsequently subjected to effects of solar radiation pressure. The results of R. M. West *et al.*'s (1995) careful study of the tails of the fragments G and K in the period of July 1–15, 1994 are also generally consistent with this conclusion. The new preliminary results of our work (Sekanina *et al.* 1995), which is still continuing, indicate that, using an updated set of orbital elements for the fragments, we have apparently detected small but *systematic* deviations of the tails' reported position angles from the tidal-breakup synchrone. The dependence of the tail orientation on the time of particle release is plotted in Fig. 5. The observed tail orientations are from four sources: (i) most of the data in March 1993 are from Table 2 of Paper 1; (ii) the majority of the points before July 1994 are preliminary values for the fragment G by Scotti & Metcalfe (1995), with an estimated uncertainty of several degrees; (iii) two points are the author's estimates on available HST prints (for six fragments in each case); and (iv) the points in July 1994 are averages of R. M. West *et al.*'s (1995) results for the fragments G and K at 15,000 km from the condensation. It is evident from Fig. 5 that the synchrones corresponding to times of release in September–November 1992 are formally more consistent with the observed orientations than the synchrone of early July 1992. Because of uncertainties involved in the measured position angles and the poor resolution due to the "crowding" of the synchrones, all one can conclude at this time is that the

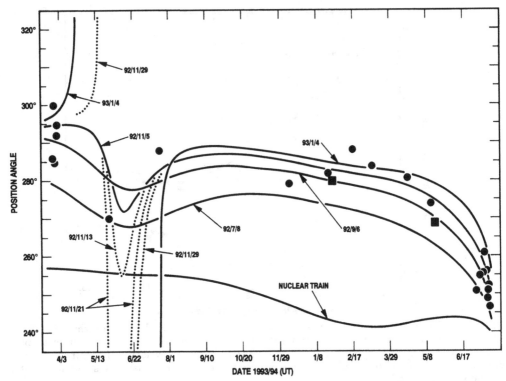

FIGURE 5. The tail orientation as a function of the assumed time of release of particulate material. The position angles measured on ground-based images are plotted as circles; those estimated by the author from two HST prints, as squares. Most of the March 1993 data are from Table 2 of Paper 1; the majority of the points before July 1994 are preliminary values for the fragment G by Scotti & Metcalfe (1995); and those from July 1994 are averages of R. M. West *et al.* (1995) results for the fragments G and K at 15,000 km from the condensation. The synchrones for particles released in mid- through late November 1992 became nearly aligned with the comet-Earth line during May and June 1993, resulting in a rapid variation in the position angle (dotted curves). The nuclear-train orientation is plotted for comparison. (After Sekanina *et al.* 1995.)

observed tails consisted of dust particles released most probably during the period of time between early July and the end of 1992. Available evidence does not make it possible to decide (i) whether the release of the particulates was continuous or proceeded in a sequence of discrete events and (ii) whether or not it was outgassing-driven.

From the temporal coincidence it is tempting to associate the tail formation with both the tidal breakup and the events of secondary fragmentation. However, unless more is learnt about Shoemaker-Levy 9's tails in the future, the proposed relationship with the secondary-fragmentation events will necessarily remain speculative.

It was argued in Paper 1 that if the fragments were active, they would have displayed tails to the east-southeast from the condensations during certain periods of time (spanning many months), contrary to the observations. In order to address a counterargument (Weaver *et al.* 1995) that such tails may not have been detectable, if the dust production rates were very low, it clearly is desirable to obtain at least crude quantitative estimates for critical dust production rates. Although this issue is not investigated here systematically, two examples shown below are sufficient to illustrate approximate limits that such considerations imply. In both examples the particle size distribution function is assumed to vary inversely as the fourth power of the particle size—a reasonable approximation supported by results on a number of comets, the visual geometric albedo of the dust is taken to be equal to 0.04, and the total dust production rate is understood to refer to particles whose sizes range from 0.1 μm to 1 cm. For the assumed size distribution, the results are relatively insensitive to the choice of the size limits; increasing, for example, the upper limit from 1 cm to 1 m would increase the production rate by a factor of only 1.4. In the first of the two cases, I assume dust particles ejected in early January 1993. In late May 1993 some of these ejecta would be located in a tail about 15 arcsec away from the parent fragment at a position angle of $\sim$ 110°. A conservative assumption of a tail detection threshold of, say, $\sim$23 mag/arcsec^2 in the visual passband on ground-based images implies an upper limit on the dust production rate of $\sim$ 5 kg s^{-1}, equal to the steady-state rate derived by Weaver *et al.* (1995) from the observed spatial profile of the brighter condensations. In the second case, I assume ejections near March 20, 1994. On the comet's image taken with the HST on May 17, 1994 (Fig. 1 of Weaver *et al.* 1995), a tail consisting of these ejecta should be at a position angle of $\sim$ 103°. At $\sim$ 5 arcsec from the parent fragment, a surface brightness of 2 ADU per WF pixel^2 in a red passband, which should have easily been detectable (Weaver 1995), implies a conservative upper limit of merely 0.5 kg s^{-1}, one order of magnitude lower than the limits quoted above.

The conclusion that the tails consisted of dust released exclusively (or nearly so) in the second half of 1992, sets tight constraints on particle sizes and velocities. First of all, solar radiation pressure accelerations on dust grains in the tails are found to have been extremely low. For example, on the July 1994 images taken by R. M. West *et al.* (1995) a typical acceleration ratio of radiation pressure to solar gravity on particulates located 15,000 km from the parent fragment amounted to 0.00005, if they were released in early July 1992, or to 0.00020, if in late 1992. These are pebble-sized objects, with typical diameters between 1 and 10 cm, depending on their bulk density. The width of the tails is determined by limits to the particle velocity distribution in the plane normal to the orbital motion; from the projected linear widths in late January 1994 (Sec. 3.4) and the age of the particulates at the time ($\sim$ 13–19 months), one finds characteristic velocities in the range from 0.1 to 0.4 m s^{-1}, comparable with the separation velocities of the products of secondary fragmentation (Sec. 4.3). And since each tail is an obvious outgrowth of its parent condensation, particles in the latter must have had even larger sizes and lower velocities.

An independent argument against any meaningful level of activity of Shoemaker-Levy 9 is the absence of detection of any diagnostic morphology in the condensations. If a comet ejects dust—especially as far from the Sun as 5–6 AU—it does so not from the entire nuclear surface, not even from its entire sunlit side, but from a discrete, suddenly activated spot. As the ejected material propagates through the comet's atmosphere, it forms some kind of a gradually expanding feature, whose coma-residence time scale is usually a few weeks at heliocentric distances comparable with that of Jupiter. One such example is 29P/Schwassmann-Wachmann 1. Another, appearing quite recently, is Hale-Bopp (1995 O1). *Absolutely no such feature* was ever observed in Shoemaker-Levy 9. Even the bright "spur" (Weaver 1994) in the condensation S would not do, as it has an explanation that requires no activity (Sekanina 1995a,b).

5. Models for the splitting of comet Shoemaker-Levy 9

All published models for comet Shoemaker-Levy 9 are based on the assumption that the Jovian tidal forces were either entirely, or primarily, responsible for the comet's breakup.

This notion is supported by the orbital calculations, which show that the spatial positions of the various condensations, extrapolated from their absolute astrometric observations in 1993–1994 back in time, essentially coincided with one another at the time of closest approach to Jupiter on 7 July 1992 (Yeomans & Chodas 1994), the most satisfactory correspondence being indicated by the on-train condensations.

5.1. *Classification of the models and a historical perspective*

There are basically two critical discriminants for the various models: (i) the gross nucleus structure and morphology of the progenitor comet and (ii) the role of particle-particle collisions. The first discriminant divides the existing models of the parent nucleus into two groups: either a strengthless "rubble pile," an agglomerate of smaller building blocks of material held together *entirely* by self-gravity; or a discrete mass of solids—whether or not of aggregate structure—possessing a definite cohesive strength. The second discriminant separates the models that account for effects of collisions accompanying the breakup from the models that treat the problem as that of a collisionless process.

Diverse morphology of the proposed models for Shoemaker-Levy 9 is reminiscent of an old controversy between the *icy conglomerate* nucleus of Whipple (1950, 1951) and the *sand bank* model, which went on for decades and was thought—prematurely, as it now appears—to have been settled in favor of Whipple's concept by the images of Halley's nucleus returned by the Giotto and Vega spacecraft. As pointed out by Whipple (1961, 1963), the sand bank paradigm has been around, in one form or another, for a century or so. Its extreme variation—a diffuse, loose swarm of particles—was advocated by Lyttleton (1953), but in other versions the cometary nucleus was envisaged as a much more compact agglomerate. For example, Vorontsov-Velyaminov (1946) maintained that the nucleus of Halley's comet is $\sim$ 30 km in diameter and consists of a cluster of meteoric blocks, each $\sim$ 150 meters across, which are nearly in contact. Schatzman (1953) concluded that a compact sand bank assemblage could collapse under certain circumstances, if protected from dispersive forces, but he cautioned that in relevant scenarios the process might be too slow. Whipple (1961, 1963) showed that the compact sand bank model encounters most of the difficulties of the diffuse model and pointed out that gravitational coherence alone is unlikely to keep the nucleus intact over extended periods of time. While one may argue that comet Shoemaker-Levy 9 is an exception to the rule, an *ad hoc* postulate of this kind is questionable and should be viewed with skepticism.

5.2. *Strengthless agglomerate models: The pros and cons*

At least two models describe the nucleus of comet Shoemaker-Levy 9 as a strengthless agglomerate of subkilometer-sized cometesimals (Asphaug & Benz 1994, 1996, Solem 1994, 1995). Both are concerned only with the nuclear train and the main difference between them is that Solem's model accounts for effects of particle-particle collisions. A third model, briefly described by Rettig *et al.* (1994), was at the time of this writing still in the process of development; in some of its versions a nonzero mechanical strength among the assumed $\sim$ 50-meter-sized cometesimals is considered (Mumma 1995). All distances in the strengthless models scale with simple similarity and the results depend on the bulk density of the assemblage, as can readily be illustrated on a spherical body, whose radius is R, bulk density ρ, and gravitational pressure at its center P_c. The gravitational attraction between its two hemispheres is $\frac{1}{2}\pi R^2 P_c$, while the net tidal force from Jupiter, of radius R_0 and density ρ_0, amounts to $\pi R^2 P_c (\rho_0/\rho)(R_0/\Delta)^3$ at a distance Δ from the planet. The necessary condition for a hemispheric separation isindependent of the radius R and is given by the following expression (Aggarwal & Oberbeck 1974, Dobrovolskis 1990)

$$\frac{\rho}{\rho_0} \left(\frac{\Delta}{R_0} \right)^3 < 2. \qquad (5.1)$$

For Shoemaker-Levy 9 one finds $\rho < 1.1$ g cm^3, probably a soft limit, since the breakup is likely to have begun before perijove (cf. also the investigations by Boss 1994 and by Greenberg *et al.* 1995). A more meaningful constraint is offered by P/Brooks 2, which approached Jupiter to 2 Jovian radii, so $\rho < 0.34$ g cm^3 (Sekanina & Yeomans 1985).

An attractive attribute of the self-gravitating strengthless compact agglomerate models is their apparently successful simulation of the progenitor's tidal disruption into a fairly small number (between a few and two dozen or so) of discrete clumps of debris, if the bulk density of the cometesimals is confined to a narrow interval of plausible values. The reason for this unexpected result is that shortly after passing through perijove the cometesimals begin to reassemble gravitationally, but because of the considerable tidal elongation of the cloud at that time, the coagulation proceeds only locally, if the bulk density ρ is, in the particular case of Shoemaker-Levy 9, close to 0.5 g cm^3.

Unfortunately, the Asphaug-Benz-Solem paradigm of a strengthless nucleus has a number of grave conceptual weaknesses. Besides the fact that the complete absence of material cohesion is unphysical (Greenberg *et al.* 1995), two of the most serious shortcomings of these proposed strengthless models are (i) the assumption of *equal-sized* cometesimals that make up the nucleus and (ii) their "typical" chosen size of more than 100 meters across. Attempts to match the comet's brightness, reported by visual observers to have attained a total magnitude of 12–13 shortly after discovery, and thereby to accommodate the object's enormous cross-sectional area, run into unsurmountable problems. For example, with the parent nucleus 1.5 km and each cometesimal 0.15 km in diameter, typical of the Asphaug-Benz-Solem model, one infers from it—even in the absence of inevitable mutual particle shadowing—that the comet should have had an apparent visual magnitude of 22–23 for a plausible geometric albedo of <10 percent, tacitly implying that the comet *should not have been discovered* with the Palomar 0.46-cm Schmidt telescope! Even with an entirely unrealistic albedo of unity, the comet still could only have been of magnitude 19. On the other hand, a *constant* size of the cometesimals that would satisfy the conditions for the parent comet's diameter and the observed brightness is found to be a small fraction of 1 cm, or more than four orders of magnitude smaller than the size used in the models. The only possible way to avoid the contradiction is to allow the cometesimals to possess a *broad size (and mass) distribution*, as considered by Sekanina *et al.* (1994). Since the self-gravity field of a cloud of particles depends on their masses, the introduction of a broad mass distribution has to have a dramatic effect on the cloud's stability, as well as on the tidal disruption and the subsequent clumping of the cometesimals. This mass distribution effect can in no way be predicted or estimated from the results of the proposed equal-size strengthless models.

The strengthless agglomerate models also experience difficulties with explanations of two well documented phenomena. One is secondary fragmentation (Sec. 4.3), the other is the sharp contrast, shortly before crash, between Jovian-gravity driven, progressive *stretching* of each condensation's dust coma and the complete absence of this effect in its innermost, brightest part, which retained the appearance of an unresolved dot (cf. the last HST frame of Q1 and Q2 in Weaver *et al.*'s 1995 Fig. 2). A supporter of the strengthless agglomerate model now has a choice: he can assume that, following the comet's 1992 tidal disruption and subsequent clumping, the central fragments in the condensations became gravitationally either stable or quasi-stable. If they were stable, they could not disrupt again months later and secondary fragmentation remains unexplained. If they were quasi-stable, secondary fragmentation is possible, but then the central fragments should have stretched during their 1994 return to Jupiter just as did the surrounding dust clouds, because gravitational stretching is mass independent, except in stable configurations, which are explicitly ruled out. The lack of progressive stretching of the central objects in the condensations now remains unexplained.

While the Asphaug-Benz-Solem strengthless assemblages are not defined in sufficient detail to choose between the stable and quasi-stable scenarios, the above example shows that either scenario can selectively be manipulated to explain *one* phenomenon at a time, but not to offer conceptually consistent interpretations of *two or more* phenomena. From Asphaug & Benz's (1994, 1996) analysis it appears that the clumps must have been relatively stable, if self-gravity began to prevail over the Jovian tides as early as several hours after perijove. However, the perception of the Asphaug-Benz paradigm in the comet community is clearly exemplified by Weissman's (1994) grossly incorrect predictions for the impact phenomena, based on this model. The motivation for such misleading conclusions is obvious: loose assemblages of hundred-meter-sized cometesimals should indeed have disintegrated into debris at high Jovian altitudes and we should have witnessed a cosmic fizzle. Whereas observations do provide substantial evidence (*e.g.,* Meadows *et al.* 1995) that *some*, and probably *most*, of each condensation's mass disintegrated high in the Jovian atmosphere, large enough residual masses of 11 of the components penetrated into the lower stratosphere, or perhaps still deeper, where they exploded and generated huge ejecta clouds, some of which were imaged with the HST (Hammel *et al.* 1995). This was hardly what one would expect of a strengthless assemblage of hundred-meter-sized cometesimals!

Also of concern is the sensitivity of the strengthless agglomerate models to the comet's rotation. Both Asphaug & Benz and Solem found that retrograde progenitor rotation states yield train configurations containing a dominant central clump, instead of the observed quasi-uniform strand of clumps. The validity of their conclusion that the parent comet could not have been a retrograde rotator is limited by their model's assumptions and does not apply as a general rotation constraint.

The sensitive dependence of the strengthless agglomerate models on the choice of bulk density is even more worrisome. While Shoemaker-Levy 9 yields a plausible value for the bulk density, the application to other tidally split comets leads to far less satisfactory results. For P/Brooks 2, which could not have split at a jovicentric distance of less than two planet's radii (cf. Sec. 3.5), these models would unquestionably predict a significantly lower density. Worse yet, the strengthless assemblage paradigm offers particularly discouraging results for the sungrazing group of comets. Two of them, 1882 II and Ikeya-Seki 1965 VIII (both mentioned in Sec. 1), had a nearly identical perihelion distance of 1.67 solar radii, but 1882 II, the much brighter (and almost certainly much larger) of the two was observed after perihelion to have split into six discrete condensations (Kreutz 1888), while 1965 VIII into only two (Sekanina 1977). Furthermore, only *one possible* secondary condensation was reported (Roemer 1965) for another well-observed member of the sungrazing group, Pereyra (1963 V), and none for 1843 I (= 1843 D1) and 1880 I (= 1880 C1), even though their perihelia were still closer to the Sun, between 1.09 and 1.19 solar radii. Since all the sungrazers have a single common parent (Marsden 1967, 1989), major variations in their effective bulk density are unlikely and the examples provide strong evidence against the validity of the strengthless agglomerate models.

One can point out additional problems with the strengthless models: doubts whether they can explain the observed nuclear train orientation more than eight months after the breakup (Sec. 4.2), an issue not addressed by the Asphaug-Benz-Solem papers; a contradiction in the tacit assumption that while the dynamical behavior of the assemblage is governed purely by self-gravity, the cometesimals themselves are structurally so cohesive that their strength is not even questioned; and a highly problematic long-term dynamical stability of strengthless agglomerates, a major problem for all sand bank models. Aggregate structures dominated by water ice were recently shown to possess a tensile strength on the order of 10^{-3} bar (Greenberg *et al.* 1995). All these arguments support the notion expressed by Whipple a long time ago (Sec. 5.1) that gravitational coherence *alone* cannot provide the basis for a realistic model of cometary nuclei.

5.3. *Models for a discrete mass of limited mechanical strength*

Tidal splitting of a discrete nucleus that possesses some, however limited, strength is governed by different conditions. As mentioned in Secs. 5.1 and 5.2, cohesion in such nuclei of aggregate structure should vary due to unevenly strong mechanical bonds among its building blocks or due to uneven cementing of the interiors of the individual blocks, or both. Considering a self-gravitating, incompressible elastic sphere, Aggarwal & Oberbeck (1974) showed that fracture starts either at the body's center or on its surface. In general, their result can be written in a form analogous to (5.1),

$$\frac{\rho}{\rho_0}\left(\frac{\Delta}{R_0}\right)^3 < k\frac{P_c}{U}, \tag{5.2}$$

where k is a constant on the order of unity. When fracture starts at the surface, U equals T, the body's tensile strength. When it starts at the center, $U = T + P_c$. Aggarwal & Oberbeck argued that fracture is completed when the tensile strength is exceeded by the greatest principal stress both at the center and on the surface of the body. Dobrovolskis (1990) pointed out, however, that Aggarwal & Oberbeck's approach underestimated the extent of fracture because their calculation of the stress field did not account for its changes as the fissure propagates. In any case, the critical tensile strength at which the body would begin to break apart varies as the central pressure and is therefore proportional to the square of the body's size. In scenarios with fixed values of ρ, ρ_0, R_0, and Δ, *the larger a comet's nucleus is the easier it is to split it tidally.* The size dependence represents a fundamental difference between the behavior of strengthless agglomerates and nuclei that are at least weakly cemented. It is noted that the proposed size dependence is consistent with the uneven numbers of condensations observed in the sungrazers 1882 II and 1965 VIII (Sec. 5.2).

Secondary fragmentation is one of the observed phenomena that makes the concept of discrete cometary nuclei of limited and variable strength very attractive. The fact that the products of a secondary fragmentation event appeared as two distinct condensations, rather than an elongated cloud of dust, testifies to the presence of a dominant mass in each daughter condensation. Their diffuse appearance suggests that much, and possibly all, of the involved dust population was formed during the discrete event. Otherwise one would have to postulate a bimodal velocity distribution in the parent condensation's dust cloud, a premise with no physical or dynamical rationale and difficult to defend.

Events of secondary fragmentation can readily be understood in the framework of a discrete nucleus as a result of a gradual fissure propagation. Asphaug & Benz (1994, 1996) argued that a body of any realistic density could not have been broken up into ~ 21 pieces by the tidal forces, regardless of its strength. They proposed that if the comet was not a strengthless "rubble pile" to begin with, it would have to be a structurally weak aggregate shattered by impact during its inbound passage through the Jovian ring plane. This scenario fails to explain the breakup of P/Brooks 2, which never approached the planet close enough to cross the ring. Since the *initial* number of major fragments of comet

Shoemaker-Levy 9 was just 10 to 12 (Sec. 4.3), Asphaug & Benz's objection becomes irrelevant. Also, their argument applies neither to scenarios involving additional forces, such as rotational stresses, that can assist the tides in breaking the nucleus up, nor to irregular bodies and/or to bodies of nonuniform strength. In fact, Greenberg *et al.* (1995) show that sequential tidal splitting of the parent nucleus of a nonzero strength can result in still more than ~ 20 pieces, depending on the comet's original nuclear size.

However, Asphaug & Benz's suggestion that growing cracks in a nucleus of some intrinsic strength may stall is of interest, paralleling somewhat our argument in Paper 1. The implications are that a body of *uniform* strength is pure fiction and any arguments against its expected behavior are meaningless. A crack could stall whenever it encounters a mass of greater strength during its propagation through the nucleus. As the comet rotates, the centrifugal force would assist the tides in some parts of the nucleus, depending on the orientation of the spin vector. Even for a nonrotating comet, the configuration of regions of variable strength in the nucleus interior should be changing rapidly relative to Jupiter (and its tidal field) near perijove because of the sharply curved orbit. An obvious inference is that there should have been *large fragments that had survived the Jovian encounter cracked but not completely broken* and that *some of the cracks would have been extended to the point of fracture at later times* in those among the fragments that happened to have been spun up as a result of the collisional angular momentum redistribution in the cloud of debris.

The plausibility of the concept of discrete nuclei of limited and variable strength is also illustrated by other idiosyncrasies of the condensations of Shoemaker-Levy 9, both in interplanetary space and upon entering the Jovian atmosphere. One of these peculiarities is the gradual disappearance of a condensation, which is particularly well documented by the HST observations of P1 (Fig. 2 of Weaver *et al.* 1995), but was obviously also experienced by J, M, and P2b. Common to these condensations was apparently their extremely poor cementing throughout their interiors on such scales that no fragment of a size detectable by the HST could survive even in interplanetary space. This critical size at the comet's distance is about 1 km in diameter, so the gradual diappearance of these condensations does not provide a very strict limit on the maximum fragment size and on the density of lines of extreme structural weakness in these objects.

The next group includes other off-train condensations, such as B, F, etc., which did not disintegrate during the months before impact but generated no detectable ejecta. Since stresses (such as rotational) acting on comets and their fragments in interplanetary space are generally lower than the tidal forces very close to Jupiter, it appears that lines of high structural weakness were less densely distributed in the interiors of these fragments, but that areas of some structural weakness were still sufficiently extensive in these objects for their tidal breakup shortly before entering the atmosphere (Sekanina 1993).

Finally, the on-train condensations apparently contained kilometer-sized fragments in which areas of high structural weakness were still less common, so that these fragments survived the tidal action and began to fragment precipitously only under the effects of aerodynamic pressure, which for the impact velocity of Shoemaker-Levy 9 began to exceed the tidal force at altitudes of about 300 km or so above the 1-bar level.

These groups of fragments clearly correlate with the classes introduced by Hammel *et al.* (1995) and suggest that the distribution of lines of extreme structural weakness is this classification's criterion. However, one should not think in terms of discrete categories; instead, each fragment is likely to have its own position in the hierarchy of structural strength. One observational implication is that the off-train condensations of structurally weaker fragments should have generally appeared brighter than the on-train condensations, because a greater mass fraction of the off-train condensations was contained in the debris near the lower end of the size spectrum, so they had a higher apparent cross-sectional area per unit mass. This effect was particularly well illustrated by the first two condensations to crash, A and B, as A was in fact fainter, yet apparently much more massive. Of the on-train condensations, one example of an excessively fragile fragment was Q, which over a time gave birth to perhaps as many as five secondary fragments. This condensation had long been the brightest component of the train, yet the impact of Q1 was a relatively disappointing event.

The high susceptibility of cometary objects to fragmentation during flight through the Earth's atmosphere has long been known from observational evidence accumulated in the field of meteor physics (*e.g.*, McCrosky & Ceplecha 1970). In fact, fragmentation—especially discrete fragmentation events, which trigger flares or outbursts along the atmospheric path—is clearly the dominant ablation process for massive cometary impactors. Borovička & Spurný (1995) recently analyzed a photographed cometary bolide, which morphologically represents a good analogue for the fragments of comet Shoemaker-Levy 9. The bolide's maximum brightness normalized to a distance of 100 km reached panchromatic magnitude -21.5, its bulk density was ~ 0.1 g cm^3, and its initial (preatmospheric) mass ~ 5 tons. Borovička & Spurný also found that along the entire luminous path the bolide's flight was only marginally decelerated by the atmosphere. The bolide became visible at an altitude of 99 km above sea level, where the dynamic pressure reached ~ 2 mbar and the atmospheric pressure was only 0.4 μbar. An equivalent altitude in the Jovian atmosphere is ~ 380 km above 1 bar. The bolide disintegrated *entirely* by the time it reached an altitude of 59 km (an equivalent Jovian altitude of ~ 190 km above 1 bar), at a dynamic pressure of ~ 1 bar and an atmospheric pressure of 0.25 mbar. Borovička & Spurný's modelling of the object's light curve shows that almost 50 percent of the initial mass was lost in the brightest flare alone, whose FWHM duration was ~ 0.05 second, and that the residual mass after this event amounted to less than 1 percent of the initial mass. This outburst was observed at an altitude of 67 km, where the dynamic pressure reached ~ 0.4 bar and the atmospheric pressure was 0.08 mbar. The fragments of Shoemaker-Levy 9 were subjected to the same dynamic pressure at an altitude of ~ 200 km above 1 bar, where the atmospheric pressure was ~ 0.15 mbar.

To summarize, the paradigm of a cometary nucleus that possesses a limited but variable strength avoids conceptual pitfalls of the strengthless agglomerate models. The results for the tidal breakup are no longer critically sensitive to the bulk density, for which values significantly lower than 0.5 g cm^3 are preferred. This limit is based not on the results obtained from the strengthless agglomerate models, but from the simple application of Eq. (5.1) to P/Brooks 2. The Shoemaker-Levy 9 progenitor could have begun to break apart perhaps as early as one hour or more before perijove, especially if its nuclear dimensions were relatively large (Sec. 5.4). The interpretation of secondary fragmentation and the explanation of the dramatic differences in the behavior among the condensations both before and upon their atmospheric entry are thus logical outgrowths of the fundamental conclusion on low and highly variable strength of tidally generated fragments and their products.

5.4. *Sizes of the progenitor nucleus and its fragments*

The dimensions of the progenitor nucleus and its major fragments have been a subject of continuing controversy. A relatively soft upper limit on the progenitor's photometric cross section results from a failure to find the comet on prediscovery exposures taken with the 100-cm Schmidt telescope at the European Southern Observatory in March 1992 (Tancredi *et al.* 1993, Tancredi & Lindgren 1994), on which the limiting magnitude for an object having the comet's motion was $B = 21.3$. Even the largest estimates for the nuclear size indicate that the comet, if inactive, would have been at least 0.5 magnitude fainter, assuming appropriate values for the albedo and the phase coefficient.

In the papers by Scotti & Melosh (1993), by Asphaug & Benz (1994, 1996), by Solem (1994, 1995), and by Chernetenko & Medvedev (1994) the effective diameter of the original nucleus was estimated at $\lesssim 2$ km. Scotti & Melosh found 2.3 km, but they used an early orbit whose 1992 perijove distance was too large. Refined orbits yielded a smaller miss distance (Yeomans & Chodas 1994), requiring a revision of Scotti & Melosh's value to 1.8 km. Asphaug & Benz, employing a recent orbit, derived an effective diameter of 1.5 km, while Solem obtained 1.8 km, and Chernetenko & Medvedev, 1.1 km.

The agreement among these results is not surprising, because they all were determined from the same observed quantity—the length of the nuclear train, always interpreted as a product of radial differential perturbations by Jupiter. Only in two of the four studies was the issue of the trails addressed at all, very briefly in either case. Asphaug & Benz's discussion was limited to an obvious remark that trails of debris could be expected on either side of the major clumps, but they offered no quantitative information. Scotti & Melosh concluded that the trails were either made up of remnants of a dust coma that the comet had possessed before its close encounter with Jupiter, or consisted of dust liberated during the breakup and subsequently perturbed by various forces. The extent of the trails was noted by these authors to correspond to a diameter of ~ 20–35 times the diameter of the parent comet but they did not elaborate on the significance of this finding. The other two papers ignored the existence of the trails altogether and none of the four studies paid any attention to the system of tails or to the sector of material to the north of the train and the trails.

In the meantime, the comet's first observations by the HST, made on July 1, 1993, were analyzed by Weaver *et al.* (1994). After subtracting the light of the surrounding comae, the magnitudes of the central nuclei in the 11 brightest condensations were calculated to imply effective diameters in the range from 2.5 to 4.3 km, at an assumed geometric albedo of 0.04. Even though Weaver *et al.* remarked that the derived nuclear magnitudes may not have been entirely free from a contamination by residual dust in the employed 3×3 pixel box centered on the brightest pixel, the effect could not possibly have amounted to > 3 magnitudes, nor could the albedo have been underestimated by a factor of > 20 to make the results compatible with the progenitor's diameter of $\lesssim 2$ km.

Weaver *et al.* (1995) subsequently applied the same technique to the HST observations from January and March 1994, finding that the spatial brightness distribution in the condensations could not be fitted by a simple model and that there was no reliable way of deconvolving the contributions by any unresolved sources from the surrounding dust clouds. On the other hand, Sekanina's (1995a,b) application of his independent deconvolution technique to the HST images taken in January, March, and early July 1994 resulted in positive detections of unresolved sources in nearly all condensations under investigation. The dimensions calculated for the major fragments were virtually independent of the law employed to approximate the brightness distribution of the extended source (the surrounding dust cloud) and agreed closely with those derived by Weaver *et al.* (1994) from the July 1993 data.

Other, independent lines of evidence also suggest that the original nucleus could not possibly be $\lesssim 2$ km in diameter. Using the formalism of sequential tidal fragmentation introduced by Dobrovolskis (1990), Greenberg *et al.* (1995) showed that with a derived tensile strength of 0.0027 bar and a plausible bulk density of ~ 0.3 g cm^3, the nucleus of Shoemaker-Levy 9 could not break up *at all* unless its diameter was at least ~ 5 km. In addition, their numbers imply that the nucleus could not break up into 10–12 pieces (Sec. 4.3) unless its diameter was at least 7–8 km. Greenberg *et al.* noted an encouraging agreement between their results and the nuclear sizes inferred by Weaver *et al.* (1994) and by Sekanina *et al.* (1994). For a bulk density of 0.2 g cm^3, assumed in this paper, the comet's nucleus could not split unless it was at least 6 km in diameter and could not split into the observed number of major fragments unless at least 9–10 km in diameter. Also of interest are the implications for 1882 II, 1965 VIII, and the parent comet of P/Brooks 2, whose minimum diameters are found to be equal, respectively, to 12 km, 8 km, and 11 km. On the other hand, if 1963 V did not break up, its nucleus could not be more than 4 km across.

Further evidence that the parent nucleus of Shoemaker-Levy 9 was $\gg$ 2 km in diameter is provided by the studies of various aspects of the impact phenomena.

From their analysis of the optical depth distribution of the dark impact debris on Jupiter imaged with the HST between 1 day and 1 month after the last impact, R. A. West *et al.* (1995) concluded that the mean particle radius was between 0.2 and 0.3 μm and their total volume was equal to that of a sphere 1.0 km in diameter. Since these aerosol particles are believed to have represented condensates of supersaturated vapor originating from the hot gas in the raising plumes of debris, they consisted primarily—and perhaps exclusively—of the impactor's mass and, unlike in the original cometary environment, had densities close to the mineralogical densities of the involved materials. R. A. West *et al.* adopted a density of 2 g cm^3, which yields a total of 1.0×10^{15} g for this *optically recovered* mass of the refractory material, which *alone* corresponds to an effective diameter of 2.1 km for the comet's plausible bulk density of 0.2 g cm^3 (cf. Secs. 5.2 and 5.3). There is no doubt whatsoever that this optically recovered aerosol mass represents only a fraction of the total recondensed refractory mass of the fragments, which, in turn, represents only a small fraction of the total mass delivered to Jupiter by the fragments and by the dust clouds in which the fragments were immersed before impact. Even this total delivered mass obviously does not represent the entire mass of the original comet, although the two may be comparable in magnitude.

The incompleteness of the optically recovered aerosol mass is plainly illustrated by the fact that the contributions to the dark debris from most off-train condensations remained undetected. Besides, R. A. West *et al.* (1995) emphasize that significant nucleation requires supersaturation and that cooling below the saturation point is not a sufficient condition for the aerosol formation. They also find that, in a dense plume of debris, silicates and similar refractory materials could condense into grains larger than 10 μm in radius, whose sedimentation times in the Jovian atmosphere are only a fraction of 1 day. Such large particles obviously could not survive in a debris imaged days or weeks after impact. When R. A. West *et al.*'s results are combined with Vanýsek's (1995) estimate that only $\sim$ 1 percent of the delivered mass should have contributed to the recondensation process, one finds a total delivered cometary mass of $\sim 10^{17}$ g. And since the optically recovered aerosol mass was probably derived from the recondensed residual mass involved in the explosions in the lower stratosphere and/or the troposphere, it would not include the fraction of the original mass of the impactors that was lost by their precipitous fragmentation in the upper atmosphere prior to the explosions and may have been responsible for, or contributed to, the detected heating of the stratosphere (*e.g.*, Lellouch *et al.* 1995, Bézard *et al.* 1995). This missing mass has remained unaccounted for, even though it may have represented a significant fraction of the initial mass of the large fragments. Finally, from observations at millimeter wavelengths, additional substantial amounts of the delivered mass were identified by Lellouch *et al.* (1995) in the form of volatile compounds, concentrated near the 0.5 mbar pressure level and probably involved in the shock chemistry. For the fragment K, for example, a mass of 2.5×10^{14} g of carbon monoxide was detected several hours after impact (Lellouch, this volume), which alone represents ~ 30 percent of the *total* mass of the Asphaug-Benz *parent* comet. Recovery of all these impressively large amounts of mass from an original nucleus of less than 2 km in diameter would surely represent a *humpty-dumpty* feat of unrivalled proportions.

5.5. *Rotation model for the progenitor nucleus and collisional evolution of the debris*

One of the first results of the numerical experiments conducted in Paper 1 was a finding that equivalent values could be established for an initial radial separation of the fragments or for their orbital velocity increment or for various combinations of these quantities so that they yielded identical temporal variations in the nuclear train's apparent evolution (its length and orientation), a fact that can also be derived from the virial theorem. If the breakup is assumed to have occurred at closest approach, the relevant values are 1.26 km for the radial separation (that is, the nucleus diameter) and 0.17 m s^{-1} for the orbital velocity increment. Thus, the breakup of the progenitor 1.26 km in diameter represents only one of an infinite number of possible solutions, based on a number of assumptions regarding the time of the event (*exactly* at perijove) and the comet's rotation vector (no rotation or the axis aligned with the orbital velocity vector). This is such an exceptionally special case that the probability of its having actually taken place is virtually nil. The equivalence of effects due to Jovian perturbations and due to an orbital velocity impulse signifies a basic dynamical indeterminacy of the problem, which is also reflected in the major role of the comet's rotation recognized by Asphaug & Benz (1994, 1996) and by Solem (1994, 1995) and which makes the tight constraints on the parent comet's size and bulk density in these models vulnerable and suspect.

A Monte Carlo simulation of ubiquitous low-velocity particle-particle collisions, carried out in Paper 1, showed that the initial rotational velocities were rearranged into a rapidly "thermalized" distribution, characterized by a long tail of fairly high velocities (up to ~ 7 m s^{-1}) for the debris that eventually populated far regions of the trails. The period of intense particle-particle collisions was estimated to have continued for at least a few hours, at which time the systematic forces began to dominate. The particle mass distribution of the fragments appears to have been relatively flat near the upper end of the size spectrum but steeper for pebble-sized and smaller debris. Fine dust effectively provided a temporary viscous medium for the major fragments. Dimensions of fragments populating the west-southwestern trail probably ranged from several hundred meters down to a few centimeters, the latter constraint being dictated by the absence of measurable solar radiation pressure effects. The debris in the east-northeastern trail was mostly submillimeter- and millimeter-sized. All the debris to the north of the train-trail boundary was affected by solar radiation pressure and made up of particles that were microns to several millimeters across, the size being the largest near the boundary.

To constrain the comet's bulk properties, a rotational model was formulated by us in Paper 1 and a search was initiated for solutions consistent with evidence on the nuclear train and the trails, while also accommodating limited information on the tails and the sector of material. The maximum dynamically plausible train and trail lengths, searched for as functions of a location on the nucleus, depend on the nuclear dimensions and the rotation vector of the parent comet, on the effective breakup time (Sec. 4.2), on the particle-mass distribution of the debris, and on the collisional-velocity enhancement factors. Although no unique solution could be derived, models for the parent comet that fitted the constraints best implied a nuclear diameter of ~ 10 km, a spin axis nearly in the jovicentric orbital plane, and a short rotation period, perhaps 7–8 hours. For a bulk density of 0.2 g cm^3, the net tidal stress is calculated to have been 0.0038 bar at perijove, 0.0008 bar 1 hour earlier, and 0.0002 bar 2 hours earlier, comparable with the central gravitational pressure and the centrifugal stress due to rotation. It thus appears that the comet's spin assisted the tidal forces in splitting the nucleus apart.

6. Summary and conclusions

The events experienced by comet Shoemaker-Levy 9 near Jupiter in early July 1992 began with fissures propagating throughout its nucleus, about 10 km in diameter or 10^{17} g in mass at an assumed density of 0.2 g cm^3. The cracks were caused by tidal stresses exerted by the planet, with some assistance from the comet's rotation. Probably even before reaching perijove, the inflicted structural failures resulted in the body's gradual breakup, first into a couple of large fragments accompanied by immense amounts of small-sized debris. Because of a distribution of rotation velocities, collisions became inevitable and, together with the continuing tidal forces, contributed to further fragmentation. The collisional velocity distribution rapidly "thermalized" and developed a long tail, populated by particulates with relative velocities of up to ~ 7 m s^{-1}. Intensive collisions did not terminate until after perijove, defining the effective time of breakup (dynamical separation). The 10–12 largest fragments contained apparently close to 90 percent of the total mass of the progenitor. The largest fragment was estimated to have been at least 4 km in diameter. A mean fragment size gradually decreased from the train to the two trails, the tails, and the sector of material that contained microscopic debris. Definite evidence for discrete events of secondary fragmentation indicates that the comet's disintegration continued long after its 1992 encounter with Jupiter. Observed effects on the unevenly susceptible fragments provide intriguing information on the complex morphology of the comet's nucleus interior. The orbital calculations offer an independent insight by showing that fragments that ended up nearer the planet at the end of the collisional period remained so throughout the orbit until collision, while fragments with greater velocity increments in the direction of the comet's motion had larger orbital dimensions and impacted Jupiter later.

The comet's fragments of the estimated mass will have delivered a total energy of tens of millions of megatons of TNT upon impact. Much of this energy was rapidly dissipated over huge volumes of the Jovian atmosphere in the early phase of each fragment's entry and only a fraction was apparently transformed into more persisting, readily detectable effects. Whereas the impact phenomena provide critical information on the nature of the fragments' interaction with the Jovian atmosphere, the comet's tidal disruption deserves attention in a broader context, including the role of nuclear splitting in the evolution of comets. A particularly diagnostic property concerns systematic differences in fragment configurations of tidally and nontidally split objects. Whatever the mechanism(s) of nontidal breakup may be (Sec. 1), it is well known (Sekanina 1977, 1982) that the configuration of fragments is, in these cases, controlled primarily by differential forces acting along the direction of the radius vector. Relative to the *principal* (parent) nucleus, which is usually (but not necessarily) the brightest component, the companions are lined up approximately along the antisolar direction shortly after their separation, but rotate their positions gradually with time and end up eventually—if they are still observable—in the direction of the reverse orbital-velocity vector, that is, they follow the parent object in its heliocentric orbit. Hence, the characteristic attribute of such configurations is that the parent nucleus is always situated at the *leading end* of the fragment lineup. This dynamical evolution is of course readily predictable from considerations of the orbital angular momentum and is indeed consistent with observations of fragments of most split comets. For only three among the split comets with more than two components ever observed was the brightest condensation situated at a "wrong" location, after each of them had broken tidally in the immediate proximity of the Sun or Jupiter: the sungrazer 1882 II, P/Brooks 2, and Shoemaker-Levy 9. This evidence suggests that the fragment configurations of tidally split comets are determined primarily by the conditions at breakup and not by the differential forces that the fragments are subjected to following their separation.

I thank H. A. Weaver for providing his measurements of the offsets of the component Q2 from Q1 on the HST images. I also thank J. V. Scotti for communicating his and T. S. Metcalfe's results on the dust trails and tails before publication. This research was carried out by the Jet Propulsion Laboratory, California Institute of Technology, under contract with the National Aeronautics and Space Administration and was supported in part through Grants GO-5021 and GO-5624 from the Space Telescope Science Institute, operated by the Association of Universities for Research in Astronomy, Inc., under contract with the National Aeronautics and Space Administration.

REFERENCES

AGGARWAL, H. R. & OBERBECK, V. R. 1974 Roche limit of a solid body. *Astrophys. J.* **191**, 577–588.

ASPHAUG, E. & BENZ, W. 1994 Density of comet Shoemaker-Levy 9 deduced by modelling breakup of the parent "rubble pile." *Nature* **370**, 120–124.

ASPHAUG, E. & BENZ, W. 1996 The tidal disruption of strengthless bodies: lessons from comet Shoemaker-Levy 9. *Icarus*, in press.

BARNARD, E. E. 1889 Discovery and observations of companions to comet 1889...(Brooks July 6). *Astron. Nachr.* **122**, 267–268.

BARNARD, E. E. 1890 Physical and micrometrical observations of the companions to comet 1889 V (Brooks). *Astron. Nachr.* **125**, 177–196.

BÉZARD, B., GRIFFITH, C. A., KELLY, D., LACY, J., GREATHOUSE, T. & ORTON, G. 1995 Mid-IR high-resolution spectroscopy of the SL9 impact sites: Temperature and HCN retrievals. Poster paper presented at IAU Colloq. No. 156 *The Collision of Comet P/Shoemaker-Levy 9 and Jupiter*, Baltimore, Maryland, May 1995.

BOROVIČKA, J. & SPURNÝ, P. 1995 Radiation study of two very bright terrestrial bolides. Poster paper presented at IAU Colloq. No. 156 *The Collision of Comet P/Shoemaker-Levy 9 and Jupiter*. ST ScI.

BOSS, A. P. 1994 Tidal disruption of periodic comet Shoemaker-Levy 9 and a constraint on its mean density. *Icarus* **107**, 422–426.

CHERNETENKO, Y. A. & MEDVEDEV, Y. D. 1994 Estimate of the Shoemaker-Levy 9 nucleus size from position observations. *Planet. Space Sci.* **42**, 95–96.

CHODAS, P. W. & YEOMANS, D. K. 1994 Comet Shoemaker-Levy 9 impact times and impact geometries. *Bull. Amer. Astron. Soc.* **26**, 1569.

DOBROVOLSKIS, A. R. 1990 Tidal disruption of solid bodies. *Icarus* **88**, 24–38.

FELDMAN, P. D., WEAVER, H. A., BOICE, D. C. & STERN, S. A. 1996 HST observations of Mg^+ in outburst from comet D/Shoemaker-Levy 9. *Icarus*, in press.

GREENBERG, J. M., MIZUTANI, H. & YAMAMOTO, T. 1995 A new derivation of the tensile strength of cometary nuclei: application to comet Shoemaker-Levy 9. *Astron. Astrophys.* **295**, L35–L38.

HAMMEL, H. B., BEEBE, R. F., INGERSOLL, A. P., ORTON, G. S., MILLS, J. R., SIMON, A. A., CHODAS, P., CLARKE, J. T., DE JONG, E., DOWLING, T. E., HARRINGTON, J., HUBER, L. F., KARKOSCHKA, E., SANTORI, C. M., TOIGO, A., YEOMANS, D. & WEST, R. A. 1995 HST imaging of atmospheric phenomena created by the impact of comet Shoemaker-Levy 9. *Science* **267**, 1288–1296.

JEWITT, D., LUU, J. & CHEN, J. 1993 Physical properties of split comet Shoemaker-Levy 9. *Bull. Amer. Astron. Soc.* **25**, 1042.

JEWITT, D. & TRENTHAM, N. 1994 Periodic comet Shoemaker-Levy 9 (1993e). *IAU Circ.* No. 5999.

KREUTZ, H. 1888 Untersuchungen über des Cometensystem 1843 I, 1880 I und 1882 II. *Publ. Sternw. Kiel* No. 3, pp. 1–111.

LEHKÝ, M. 1994 Periodic comet Shoemaker-Levy 9 (1993e). *Int. Comet Q.* **16**, 35.

LELLOUCH, E., PAUBERT, G., MORENO, R., FESTOU, M. C., BÉZARD, B., BOCKELÉE-MORVAN, D., COLOM, P., CROVISIER, J., ENCRENAZ, T., GAUTIER, D., MARTEN, A., DESPOIS, D., STROBEL, D. F. & SIEVERS, A. 1995 Chemical and thermal response of Jupiter's atmosphere following the impact of comet Shoemaker-Levy 9. *Nature* **373**, 592–595.

LYTTLETON, R. A. 1953 *The Comets and Their Origin.* Cambridge University.

MARSDEN, B. G. 1967 The sungrazing comet group. *Astron. J.* **72**, 1170–1183.

MARSDEN, B. G. 1989 The sungrazing comet group. II. *Astron. J.* **98**, 2306–2321.

McCROSKY, R. E. & CEPLECHA, Z. 1970 Fireballs and physical theory of meteors. *Bull. Astron. Inst. Czech.* **21**, 271–296.

MEADOWS, V., CRISP, D., ORTON, G., BROOKE, T. & SPENCER, J. 1995 AAT IRIS observations of the SL-9 impacts and initial fireball evolution. In *European Shoemaker-Levy 9/Jupiter Workshop* (eds. R. West & H. Böhnhardt), pp. 129–134. European Southern Observatory.

MUMMA, M. J. 1995 Personal communication.

NOLL, K. S. & SMITH, T. E. 1994 Periodic comet Shoemaker-Levy 9 (1993e). *IAU Circ.* No. 6010.

RENZ, F. 1889 Ueber die Begleiter des Cometen 1889...(Brooks Juli 6). *Astron. Nachr.* **122**, 413–416.

RETTIG, T. W., MUMMA, M. J., TEGLER, S. C. & HAHN, J. 1994 Are the fragments of comet Shoemaker-Levy 9 swarms of meter-sized planetesimals? *Bull. Amer. Astron. Soc.* **26**, 862.

ROEMER, E. 1965 Observations of comets and minor planets. *Astron. J.* **70**, 397–402.

SCHATZMAN, E. 1953 La structure et l'evolution des noyaux cometaires. *Mém. 8° Soc. Roy. Sci. Liège* (Sér. 4) **13** (Fasc. 1–2), 313–323.

SCOTTI, J. V. 1993 Periodic comet Shoemaker-Levy 9. *Minor Planet Circ.* Nos. 21988–21989.

SCOTTI, J. V. & MELOSH, H. J. 1993 Estimate of the size of comet Shoemaker-Levy 9 from a tidal breakup model. *Nature* **365**, 733–735.

SCOTTI, J. V. & METCALFE, T. S. 1995 Personal communication.

SEKANINA, Z. 1977 Relative motions of fragments of the split comets. I. A new approach. *Icarus* **30**, 574–594.

SEKANINA, Z. 1982 The problem of split comets in review. In *Comets* (ed. L. L. Wilkening), pp. 251–287. University of Arizona, Tucson.

SEKANINA, Z. 1993 Disintegration phenomena expected during collision of comet Shoemaker-Levy 9 with Jupiter. *Science* **262**, 382–387.

SEKANINA, Z. 1995a Nuclei of comet Shoemaker-Levy 9 on images taken with the Hubble Space Telescope. In *European Shoemaker-Levy 9/Jupiter Workshop* (ed. R. West & H. Böhnhardt), pp. 29–35. European Southern Observatory.

SEKANINA, Z. 1995b Evidence on sizes and fragmentation of the nuclei of comet Shoemaker-Levy 9 from Hubble Space Telescope images. *Astron. Astrophys.* **304**, 296–316.

SEKANINA, Z., CHODAS, P. W. & YEOMANS, D. K. 1994 Tidal disruption and the appearance of periodic comet Shoemaker-Levy 9. *Astron. Astrophys.* **289**, 607–636. (Paper 1.)

SEKANINA, Z., CHODAS, P. W. & YEOMANS, D. K. 1995 In preparation.

SEKANINA, Z. & YEOMANS, D. K. 1985 Orbital motion, nucleus precession, and splitting of periodic comet Brooks 2. *Astron. J.* **90**, 2335–2352.

SOLEM, J. C. 1994 Density and size of comet Shoemaker-Levy 9 deduced from a tidal breakup model. *Nature* **370**, 349–351.

SOLEM, J. C. 1995 Cometary breakup calculations based on a gravitationally-bound agglomeration model: the density and size of Shoemaker-Levy 9. *Astron. Astrophys.* **302**, 596–608.

TANCREDI, G. & LINDGREN, M. 1994 Searching for comets encountering Jupiter: First campaign. *Icarus* **107**, 311–321.

TANCREDI, G., LINDGREN, M. & LAGERKVIST, C.-I. 1993 Periodic comet Shoemaker-Levy 9 (1993e). *IAU Circ.* No. 5892.

VANÝSEK, V. 1995 A note on the fragment size of SL-9 and debris field. In *European Shoemaker-Levy 9/Jupiter Workshop* (ed. R. West & H. Böhnhardt), pp. 297–298. European Southern Observatory.

VORONTSOV-VELYAMINOV, B. 1946 Structure and mass of cometary nuclei. *Astrophys. J.* **104**, 226–233.

WEAVER, H. A. 1994 Periodic comet Shoemaker-Levy 9 (1993e). *IAU Circ.* Nos. 5947 & 5973.

WEAVER, H. A. 1995 Personal communication.

WEAVER, H. A., FELDMAN, P. D., A'HEARN, M. F., ARPIGNY, C., BROWN, R. A., HELIN, E. F., LEVY, D. H., MARSDEN, B. G., MEECH, K. J., LARSON, S. M., NOLL, K. S., SCOTTI, J. V., SEKANINA, Z., SHOEMAKER, C. S., SHOEMAKER, E. M., SMITH, T. E., STORRS, A. D., YEOMANS, D. K. & ZELLNER, B. 1994a Hubble Space Telescope observations of comet P/Shoemaker-Levy 9 (1993e). *Science* **263**, 787–791.

WEAVER, H. A., A'HEARN, M. F., ARPIGNY, C., BOICE, D. C., FELDMAN, P. D., LARSON, S. M., LAMY, P., LEVY, D. H., MARSDEN, B. G., MEECH, K. J., NOLL, K. S., SCOTTI, J. V., SEKANINA, Z., SHOEMAKER, C. S., SHOEMAKER, E. M., SMITH, T. E., STERN, S. A., STORRS, A. D., TRAUGER, J. T., YEOMANS, D. K. & ZELLNER, B. 1995 The Hubble Space Telescope (HST) observing campaign on comet Shoemaker-Levy 9. *Science* **267**, 1282–1288.

WEISS, E. 1889 Ueber die Erscheinungen am Cometen 1889... (Brooks Juli 6). *Astron. Nachr.* **122**, 313–316.

WEISSMAN, P. 1994 The big fizzle is coming. *Nature* **370**, 94–95.

WEST, R. A., KARKOSCHKA, E., FRIEDSON, A. J., SEYMOUR, M., BAINES, K. H. & HAMMEL, H. B. 1995 Impact debris particles in Jupiter's stratosphere. *Science* **267**, 1296–1301.

WEST, R. M., HOOK, R. N. & HAINAUT, O. 1995 A morphological study of SL-9 CCD images obtained at La Silla (July 1–15, 1994). In *European Shoemaker-Levy 9/Jupiter Workshop* (ed. R. West & H. Böhnhardt), pp. 5–10. European Southern Observatory.

WHIPPLE, F. L. 1950 A comet model. I. The acceleration of comet Encke. *Astrophys. J.* **111**, 375–394.

WHIPPLE, F. L. 1951 A comet model. II. Physical relations for comets and meteors. *Astrophys. J.* **113**, 464–474.

WHIPPLE, F. L. 1961 Problems of the cometary nucleus. *Astron. J.* **66**, 375–380.

WHIPPLE, F. L. 1963 On the structure of the cometary nucleus. In *The Moon, Meteorites, and Comets* (ed. B. M. Middlehurst & G. P. Kuiper), pp. 639–664. University of Chicago.

YEOMANS, D. K. & CHODAS, P. W. 1994 Comet Shoemaker-Levy 9 in orbit about Jupiter. *Bull. Amer. Astron. Soc.* **26**, 1566.

Earth-based observations of impact phenomena

By PHILIP D. NICHOLSON

Department of Astronomy, Cornell University, Ithaca, NY, 14853, USA

Earth-based observations at near- and mid-infrared wavelengths were obtained for at least 15 of the SL9 impacts, ranging from the spectacular G, K and L events to the barely-detected N and V impacts. Although there were a few exceptions, most of the IR lightcurves fit a common pattern of one or two relatively faint precursor flashes, followed several minutes later by the main infrared event as the explosively-ejected plume crashed down onto the jovian atmosphere. Correlations with the impact times recorded by the Galileo spacecraft and plumes imaged by the Hubble Space Telescope lead to an interpretation of the twin precursors in terms of (i) the entry of the bolide into the upper atmosphere, and (ii) the re-appearance of the rising fireball above Jupiter's limb. Positive correlations are observed between the peak IR flux observed during the splashback phase and both pre-impact size estimates for the individual SL9 fragments and the scale of the resulting ejecta deposits. None of the fragments observed to have moved off the main train of the comet by May 1994 produced a significant impact signature. Earth-based fireball temperature estimates are on the order of 750 K, 30–60 sec after impact. For the larger impacts, the unexpectedly protracted fireball emission at 2.3 μm remains unexplained. A wide range of temperatures has been inferred for the splashback phase, where shocks are expected to have heated the re-entering plume material at least briefly to several thousand K, and further modelling is required to reconcile these data.

1. Introduction

The impacts of the 20 or so fragments of comet Shoemaker-Levy 9 (henceforth SL9 for simplicity) with Jupiter in July 1994 were observed from a wide range of Earth-based telescopes, including instruments in Europe, South and North America, Hawaii, Australia, Japan, South Africa and even Antarctica. In this review we will concentrate on Earth-based near-infrared and mid-infrared observations of prompt phenomena, defined as those occurring within approximately one hour after the impacts. The extensive sets of post-impact imaging and spectroscopic studies of the impact sites, including compositional analyses of the impact debris, are reviewed in the chapters by West, Lellouch, Moses, and Conrath. Magnetospheric effects are reviewed in the chapter by Ip.

The most extensive sets of ground-based data were obtained in the near-infrared, especially at wavelengths of 2.3 and 3.5 μm where methane absorption bands in the jovian spectrum greatly reduce the planetary background and enhance the contrast of the impact features. By a fortunate coincidence, this spectral region was also near the peak of thermal emission from the impact plumes and their remnants. A new generation of infrared cameras and spectrometers provided the bulk of the data, and we may reflect upon our good fortune that the impact of SL9 did not occur a decade earlier.

Although Jupiter rose in the early afternoon and set before midnight at most sites (the notable exception being the South Pole station where almost continuous monitoring was possible), the wide geographical distribution of observers ensured that attempts were made to observe every predicted impact. Of the 24 predicted impacts (Sekanina *et al.* 1994; Chodas & Yeomans 1994 [email predictions]), convincing detections of 15 have been reported to date, and another three (B, M and U) may have been marginally detected. Six fragments (F, G2, J, P1, P2 and T) appear either to have disintegrated completely before encountering Jupiter, or to have disappeared into the planet's atmosphere without a trace.

After a discussion of a typical SL9 lightcurve, and what has become the standard interpretation of its main features, we begin in § 2 by summarizing the observed phenomena and classifying the impacts into five categories. In § 3 we present a detailed description of the successive phases in a generic lightcurve, drawing on individual examples for illustrations. An attempt is made to relate these observations to those made simultaneously by the Galileo spececraft (see chapter by Chapman in this volume) and by the Hubble Space Telescope (see chapter by Hammel). A simple ballistic model of the impact plumes which is consistent with the ensemble of observed lightcurves is used as common framework for interpretation of the observations. This model is based loosely on the numerical models developed in the chapters by Crawford, Zahnle and MacLow. In § 4 we discuss the relatively small number of spectroscopic observations made during the actual impact events, and their interpretation in terms of plume and fallback temperatures. Sections 5 and 6 cover other prompt phenomena associated with the impacts, while in § 7 we summarize our conclusions.

1.1. *A typical example: the R impact*

Most, if not all, of the features seen in the near-infrared SL9 impact lightcurves are illustrated in the data obtained for the medium-sized R impact and shown in Figure 1. The 2.3 μm observations were made at Mauna Kea with the Keck 10m telescope and the near-IR camera (Graham *et al.* 1995), while the 3.2 and 4.5 μm observations were made at the 5m Hale telescope at Palomar, using simultaneously-mounted near-IR and mid-IR cameras (Nicholson *et al.* 1995a). Sampling times were 8–10 sec at 2.3 and 4.5 μm and 30 sec at 3.2 μm. The 4.5 μm fluxes have been corrected for background light from Jupiter, but the steady pre-impact slope in the 2.3 μm data is due to the rotation of the older G impact site onto the jovian limb. Black bars under the lightcurves indicate the times (corrected for light travel time) during which the Galileo NIMS instrument observed detectable flux from the impact (Carlson *et al.* 1995b).

Shortly before the initial Galileo flash, a brief flash was observed at all three wavelengths with a duration of ~ 30 sec. About 60 sec after the first signal, a second, brighter flash began abruptly, only to decay over the next minute or so (within 30 sec at 4.5 μm, but extending over 3 min at 2.3 μm). These two events have come to be known as the *first* and *second precursors*, respectively. The precursor events are shown at an expanded scale in Fig. 1(b). Approximately 6 min after the initial flash a dramatic brightening commenced, eventually reaching a peak flux about one hundred times greater than that of the second precursor, 10 min after the impact. This *main event* was observed simultaneously by earth-based telescopes and by Galileo NIMS. Following the peak of the main event the infrared flux decayed with a time constant of ~ 3 min. This fading was interrupted by a *shoulder*, or secondary maximum, occuring ~ 9 min after the peak. At 3.2 μm there is an indication of a second, weaker shoulder ~ 18 min after the main peak.

The generic sequence of events represented by this lightcurve is illustrated schematically in Fig. 1 in the chapter by Zahnle, and in Fig. 2 of Boslough *et al.* (1995). Figure 2 shows the particular geometry for earth-based observations of the R impact. As seen from Earth, the impact itself occurred $\sim 5.7°$ in longitude behind the planet's dawn limb; the events were observed directly only by Galileo. The first precursor apparently corresponded to thermal emission from the trail left by the passage of the fragment through Jupiter's upper atmosphere, and was too faint to be detected by the less-sensitive Galileo instruments. About 30 sec after Galileo observed the intensely bright terminal phase of the impact, the expanding, incandescent fireball rose above the limb into our

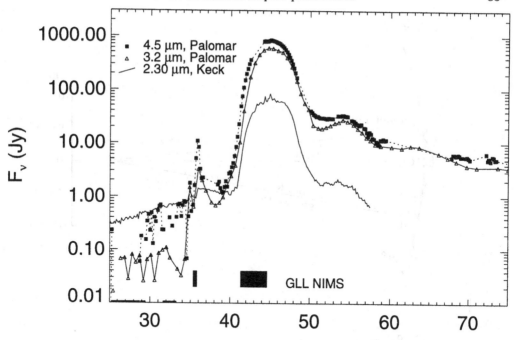

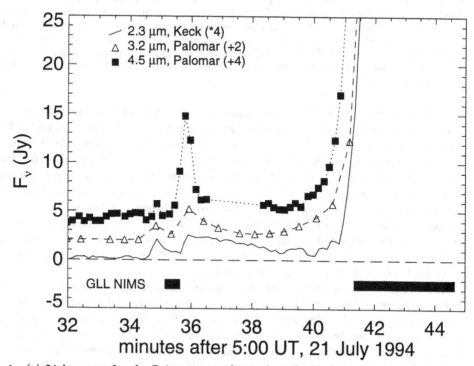

FIGURE 1. (*a*) Lightcurves for the R impact, as observed at the Keck 10m and Palomar 5m telescopes, on a logarithmic scale. Gaps in the 4.5 μm data correspond to periods during which spectroscopic measurements were obtained. Solid black bars indicate the periods of emission observed by the Galileo NIMS instrument. (*b*) Expanded plot of the R precursor events on a linear scale. Note the scale factor applied to the 2.3 μm data.

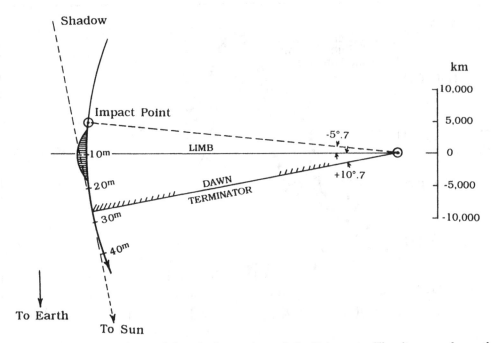

FIGURE 2. Geometry for earth-based observations of the R impact. The diagram shows the view from above Jupiter's north pole, with the Earth towards the bottom of the page and the planet's center towards the right edge. The R impact site, indicated by the ⊙ symbol, was behind the limb and out of direct sight. Ten min after the impact, and near the time of peak IR flux, the site reached the limb. The plume was also near its maximum height, as indicated schematically by the hatched region. Another 17 min elapsed before the impact site reached the dawn terminator and direct sunlight. The geometry for the other impacts was similar, with earlier impacts generally occurring further behind the limb and later impacts even closer (Sekanina *et al.* 1994).

line of sight, although still in the pre-dawn shadow. This is the second precursor, whose decay represents the rapidly cooling fireball. Now quite cold, but emerging after 1–2 min into sunlight and visibility by the Hubble Space Telescope (henceforth HST), the debris plume reached a maximum altitude of ∼ 3200 km above Jupiter's cloud tops (Hammel *et al.* 1995), before falling back into the planet's atmosphere. (This is based on observations of other impacts; HST did not observe the R impact.) For R, the maximum height was reached just as the impact site itself rotated into view, but this was not the case for earlier impacts, which occurred further behind the limb. The infrared main event corresponds to this extended period of fallback, which lasted from 6 to ∼ 15 min after the impact, and during which the descending debris was shock-heated again to temperatures of at least 500–1000 K, and perhaps much higher.

The origin of the secondary shoulders is less certain, but they may be due to re-entering material 'bouncing' off the top of the atmosphere and re-entering a second (or even a third) time (Deming *et al.* 1995). Not until ∼ 27 min after the impact did the R impact site cross the dawn terminator and emerge into direct sunlight. In practice, it is impossible to distinguish in near-IR images of the impacts between the dying thermal stages of the main event and the appearance of the fresh impact site in reflected sunlight on the terminator.

With minor modifications, this scenario seems to fit the observations of almost all of the impacts, while also being consistent with numerical models of the process of fireball formation and evolution (Boslough *et al.* 1994; Crawford *et al.* 1994; Zahnle & MacLow 1994; Takata *et al.* 1994). We will use it as a working model in describing some of the detailed features exhibited by other lightcurves in § 3 below.

2. Lightcurve classification

2.1. *Lightcurve comparisons*

Comparison of lightcurves for different events is often complicated by differences in telescope aperture, wavelength of observation, sampling time and weather conditions. Two extensive series of homogeneous observations which avoid these pitfalls are shown in Figs. 3 and 4. Figure 3, from McGregor *et al.* (1995), presents near-IR lightcurves obtained at the 2.3 m telescope at Siding Spring Observatory, Australia for events C, D, G, K, R and W. Impact N was detected only weakly in 3 frames, while V was not detected at all. The data were obtained with the CASPIR near-IR camera, at exposure intervals of ~ 50 sec, using a narrow-band 2.34 μm filter. These data include examples of large, medium-size and small impacts, and in addition to precursors for events C, D, G, K and W, the lightcurves show prominent shoulders for C, D and G. (The lightcurves for the later impacts were followed only through the main peak.)

Figure 4, from Lagage *et al.* (1995), presents a series of mid-IR lightcurves for the A, E, H, L and Q1 impacts obtained with the 10 μm CAMIRAS camera at the 2.6 m Nordic Optical Telescope at La Palma, in the Canary Islands. Impacts F, P2, Q2, T and U were not detected. The filter passband was 10.5–13 μm, and the original 1.1 sec samples have been binned at intervals of ~ 60 sec. The background thermal emission from Jupiter has been subtracted from the images. The lightcurve for the extremely bright L impact also shows a brief precursor ~ 60 sec after the Galileo flash, with a peak flux of 300 Jy. A similar precursor was seen at 10 μm for the H impact by Livengood *et al.* (1995), using the TIMMI camera at the ESO 3.6 m telescope at La Silla, Chile. Although not prominent in this presentation, each of the lightcurves shows a definite shoulder ~ 20 min after the impact. The shoulders are most prominent for H and L.

2.2. *Summary of observations*

Table 1, compiled by P. Chodas from published reports and a survey of participants at IAU Colloquium 156, and with some additions by the author, summarizes the observed phenomena associated with each impact. The adopted impact times are based on Galileo observations of the initial flash where available, on Earth-based precursor times, or—where necessary—on predictions or extrapolation from the onset of the of the main event. For further details the chapter by Chodas & Yeomans should be consulted. PC1 and PC2 refer to detections of the first and second precursors, respectively, and ME to observations of the infrared main event. The columns labelled 'spot' and 'ejecta' refer to the presence of these features in HST images (Hammel *et al.* 1995). The last two columns give the morphological class assigned by Hammel *et al.*, which reflects a combination of characteristics such as central spot size, scale of ejecta blanket, and prominence of waves (if any), and the Galileo instruments which observed the event, where applicable.

In Fig. 5 we plot the main event peak flux at 2.3 μm, a quantity which is available for most of the observed impacts from published lightcurves, vs. the HST class. On the basis of this diagram, and on the observability of the post-impact features in ground-based images of Jupiter, the impacts can be fairly cleanly (if somewhat arbitrarily) divided into five categories.

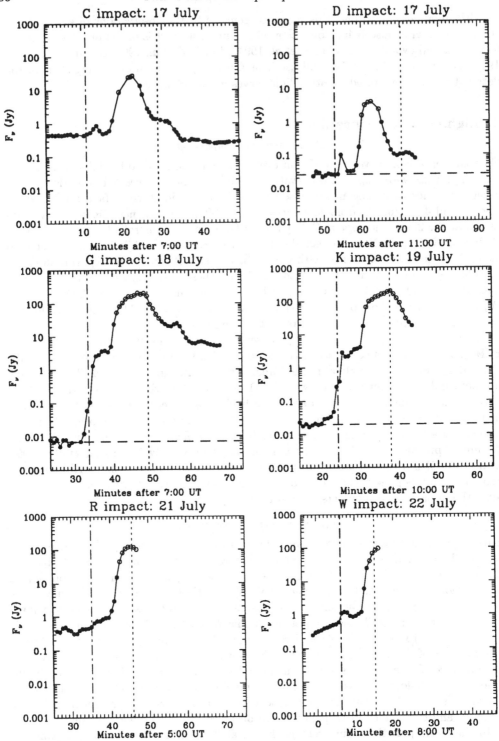

FIGURE 3. Lightcurves at 2.34 μm obtained at Siding Spring Observatory, Australia. Open symbols indicate frames in which the brightest part of the image exceeded the linear range of the detector, and the total flux is likely to be underestimated. Vertical dot-dashed lines indicate the adopted impact times from Table 1, while vertical dashed lines indicate the time at which the impact site reached the planet's limb (see Fig. 2). From McGregor *et al.* (1995).

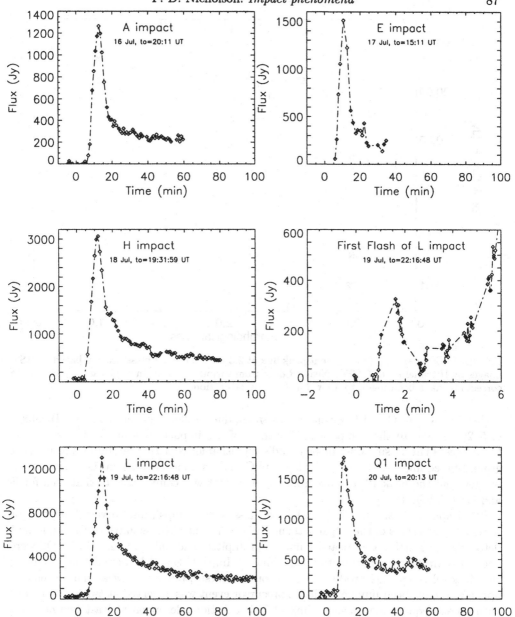

FIGURE 4. Lightcurves at 12 μm obtained at La Palma in the Canary Islands. The abscissa is time from the nominal impact time, as indicated in each panel. From Lagage *et al.* (1995).

(i) Large impacts G, K and L produced peak IR fluxes of 200 Jy or greater, and were associated with prominent post-impact sites visible in both the HST images and in Earth-based near-IR images.

(ii) Medium-sized impacts (peak fluxes in the range 50–200 Jy) include E, H, Q1, R, S and W. All except S and W (which landed very close to the already complex G and K sites and were difficult to distinguish as a result) produced post-impact features of class 2 in the HST images, which were also readily detectable in Earth-based images.

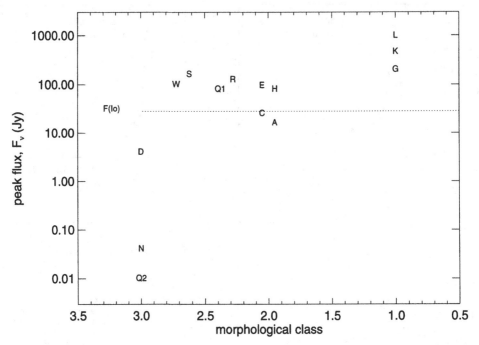

FIGURE 5. Comparison of main-event peak flux at 2.3 μm vs. impact classification based on HST images by Hammel *et al.* (1995). Near-IR data from various sources. In some cases approximate corrections have been made to allow for detector saturation.

(iii) Impacts A, C and D we classify as small, due to their smaller peak near-IR fluxes of 5–20 Jy, and to the comparative faintness of the impact sites in Earth-based near-IR images. The D site was, in fact, indistinguishable from the nearby G site in most ground-based images. However, both A and C were rated as class 2a on the basis of HST images, and the peak 12 μm flux from the A impact was comparable to that seen for E and Q1 (cf. Fig. 4).

(iv) Impacts B, M, N, Q2, U and V we classify as 'wimps', due to very faint and/or uncertain detections of the impacts themselves. For none of these impacts was a remnant detectable in Earth-based near-IR images of Jupiter, and only sites B, N and Q2 were detectable in the higher-resolution HST images. Impacts B and M were detected only by the Keck telescope. Q2 produced an obvious precursor, but a very weak main event. V appears to have been unique in that a precursor event was observed at two stations, but with no subsequent main event. Impact U is questionable, given at least two negative reports from larger telescopes. See § 5 for further details on these events.

(v) No credible reports of impact signatures exist for fragments F, G2, J, P1, P2 and T, and there are no identifiable impact sites corresponding to the predicted impact locations in the HST images. Fragments J and M had in fact disappeared at least six months prior to July 1994, while fragments P1 and P2 had been observed to be further disintegrating (Weaver *et al.* 1995).

It is also of interest to compare the luminosities of the impact events with the relative sizes of the fragments inferred from their brightness in pre-impact HST images (Weaver *et al.* 1995). In Fig. 6 we make such a comparison, again using the peak 2.3 μm flux as a yardstick. (A'Hearn *et al.* (1995) present a similar plot using peak and integrated Galileo fluxes at 0.945 μm.) On average, the peak luminosity is roughly proportional to the cube of the estimated pre-impact diameter, *i.e.*, to the putative mass and ki-

Fragment	Date/time (UT)	PC1	PC2	ME	Spot	Ejecta	HST	Galileo
A	July 16, 20:12	–	Y	Y	Y	Y	2a	–
B	July 17, 02:50	–	–	?	Y	–	3	–
C	July 17, 07:11	–	Y	Y	Y	Y	2a	–
D	July 17, 11:53	–	Y	Y	Y	–	3	–
E	July 17, 15:12	–	–	Y	Y	Y	2a	–
F	July 18, 00:37	–	–	–	–	–	–	–
G	July 18, 07:33:32	Y	Y	Y	Y	Y	1	UVS/PPR/NIMS
H	July 18, 19:31:59	Y	Y	Y	Y	Y	2a	PPR
J	July 19, 01:35	–	–	–	–	–	–	–
K	July 19, 10:24:13	Y	Y	Y	Y	Y	1	SSI
L	July 19, 22:16:48	Y	Y	Y	Y	Y	1	PPR
M	July 20, 06:02	–	–	?	–	–	–	–
N	July 20, 10:29:17	–	–	Y	Y	–	3	SSI
P2	July 20, 15:23	–	–	–	–	–	–	–
P1	July 20, 16:37	–	–	–	–	–	–	–
Q2	July 20, 19:44	–	Y	Y	Y	–	3	–
Q1	July 20, 20:13:52	–	Y	Y	Y	Y	2b	PPR
R	July 21, 05:35:03	Y	Y	Y	Y	Y	2b	NIMS
S	July 21, 15:16	–	Y	Y	Y	?	2c	–
T	July 21, 18:11	–	–	–	–	–	–	–
U	July 21, 21:56	–	–	?	–	–	–	–
V	July 22, 04:23:10	Y?	–	–	–	–	–	–
W	July 22, 08:06:14	–	Y	Y	Y	?	2c	SSI

TABLE 1. Summary of impact times, observed phenomena, HST impact site classifications and availability of Galileo data for the individual SL9 fragments. PC1 and PC2 refer to Earth-based detections of the first and second precursors, ME to observations of the infrared main event, and 'spot' and 'ejecta' to the visibility of central spots and crescent-shaped ejecta blankets in the Hubble Space Telescope (HST) images. Question marks refer to uncertain observations, or to the uncertain interpretation of the V event. Galileo instruments are indicated by their standard abbreviations.

netic energy of the fragment, as might be expected. The correlation between fragment size (inferred from their relative brightnesses) and impact flux is, however, by no means a one-to-one relationship. Although the brightest IR signals were associated with the large G, K and L fragments, the impact of the equally-bright Q1 produced an unexpectedly faint infrared signature—and a correspondingly average-sized impact scar on the planet—while the slightly fainter Q2 fragment yielded one of the weakest main events detected. Fragments F, P2, and B might have been expected to produce signatures at least comparable to those readily detected from the smaller A and D fragments.

More significant perhaps is the observation by Weaver *et al.* (1995) that the fragments located off the line of the main SL9 train produced smaller ejecta patterns than their brightnesses would have predicted. These include fragments B, F, G2, N, P2, P1, Q2, T, U and V. If the vanished J and M fragments are included, this list is identical to that of our categories (iv) and (v) above. It seems likely that these fragments were of lower density and much less cohesive than their siblings, perhaps being no more than weakly-bound clumps of debris which entered the jovian atmosphere as meteor showers rather than as coherent bolides.

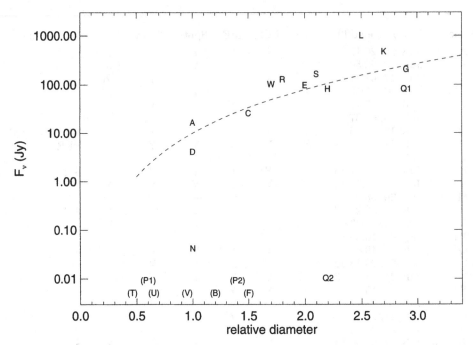

FIGURE 6. Comparison of main-event peak flux at 2.3 μm vs. pre-impact relative fragment sizes derived from HST images by Weaver *et al.* (1995). Near-IR data as in Fig. 5. Labels in parentheses indicate either non-detections or very weak detections. The dashed curve is a simple cubic fit.

3. Anatomy of a lightcurve

The various phases in the impact lightcurves are seen most clearly in the R lightcurves in Fig. 1 and in the data from the well-observed G, H, K and L impacts. These lightcurves show very similar characteristics, and we are fortunate in having Galileo observations for all five events, as well as a series of HST images for the G impact. Figures 7 and 8 show the early phases of the G and K impacts, using 2.3 μm lightcurves from Siding Spring (McGregor *et al.* 1995) and the Okayama Astrophysical Observatory, Japan (Watanabe *et al.* 1995; Takeuchi *et al.* 1995). In Fig. 7 the periods during which the Galileo PPR, UVS and NIMS instruments observed emission are indicated (Martin *et al.* 1995; Hord *et al.* 1995; Carlson *et al.* 1995a), as are the exposure times of individual HST images (Hammel *et al.* 1995). The period during which the Galileo SSI instrument observed the flash from the K impact (Chapman *et al.* 1995) is shown in Fig. 8.

The early phases of the H and L impacts—the latter being arguably the most energetic of all the SL9 events—are shown in Fig. 9, with lightcurves from the 2.2 and 3.5 m telescopes at Calar Alto, Spain (Hamilton *et al.* 1995) and the 1 m telescope at Pic du Midi, France (Colas *et al.* 1995; Drossart *et al.* 1995). For L, Fig. 9(a) shows a very abrupt first precursor at 22:16:41 UT, followed by a decay over the next 50 sec until the onset of the second precursor at 22:17:30 UT saturated the detector. The Pic du Midi observations in Fig. 9(b) carry the L lightcurve through the second precursor phase, until saturation occurred at the beginning of the main event at about 22:23 UT.

3.1. *A simple ballistic model*

Although a full understanding of the process whereby the kinetic energy of the incoming fragments of SL9 produced the spectacular events seen by Galileo, HST and Earth-based

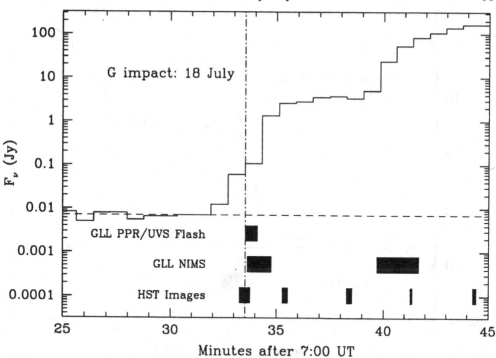

FIGURE 7. Lightcurve for the G impact obtained at Siding Spring Observatory, Australia at 2.34 μm. Steps indicate individual 30 sec exposures at $\sim$ 50 sec intervals. A vertical dot-dashed line indicates the adopted impact time, as set by the onset of the Galileo PPR/UVS flash. Blocks indicate periods of emission observed by Galileo, and exposure times for HST images. From McGregor *et al.* (1995).

telescopes depends on detailed numerical modelling of the 3-D hydrodynamics of the impact (cf. the chapter by Crawford in this volume), the subsequent evolution of the plumes themselves was largely governed by simple ballistics (see chapters by Zahnle and MacLow). Guided by the observations of maximum plume height and debris distribution by HST (Hammel *et al.* 1995), and the conceptual models of Boslough *et al.* (1995) and Zahnle and MacLow (1995), we can construct a purely ballistic model which accounts for the timing of most of the features seen in the SL9 lightcurves (cf. also Takeuchi *et al.* 1995; Drossart *et al.* 1995).

A maximum vertical ejection velocity for the impact debris cloud of $v_z \simeq 12.5$ km s^{-1} is set by the observed maximum plume height of $v_z^2/2g = 3200$ km (Hammel *et al.* 1995). The 12,500 km outer radius of the 'crescent' of debris seen around the G impact site in HST images (West *et al.* 1995) suggests a maximum ejection velocity of $v_e = \sqrt{gr} \simeq$ 18 km s^{-1}, although Zahnle has argued that the crescent may have expanded significantly *after* re-entry of the plume material. Both results are consistent with a relatively flat plume trajectory, with a maximum elevation angle $\theta_{max} \simeq 45°$. An ejection angle of 45° would also match the incoming trajectory of the fragment relative to Jupiter's cloud tops, back along which the bulk of ejecta is predicted to have been ejected (Boslough *et al.* 1995). In reality, of course, the ejecta emerged with a distribution of initial velocities and/or directions, as demonstrated by the broad crescent-shaped ejecta patterns, although the asymmetry of the ejecta blankets seen in the HST images confirms that the plume material was primarily ejected back along the fragments' inbound trajec-

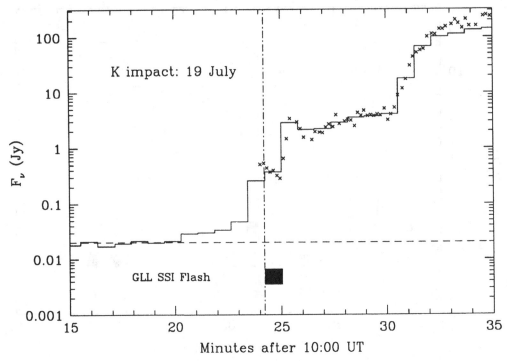

FIGURE 8. Lightcurves for the K impact obtained at Siding Spring Observatory, Australia at 2.34 μm (histogram) and the Okayama Astrophysical Observatory, Japan at 2.35 μm (crosses). The OAO data are 1 sec exposures taken with an IR camera and narrowband 2.35 μm filter, at 10 sec intervals. Neutral density filters were used during the period of peak brightness to avoid detector saturation. A vertical dot-dashed line indicates the adopted impact time, as set by the onset of the Galileo SSI flash. From McGregor *et al.* (1995).

tory, or towards the south-east once allowance is made for Coriolis deflection (Hammel *et al.* 1995).

Figure 10 shows sample trajectories for plume tracers ejected at a common velocity of 18 km s^{-1} and at elevation angles of 15°, 30° and 45° from the horizontal. Superimposed on the particle trajectories are a set of curves indicating the time and altitude at which the ejecta rises above the planet's limb into the view of Earth-based observers, for an ejection azimuth 45° east of south. (These curves also depend weakly on the assumed elevation angle of the ejecta, as they are controlled by both the planet's rotation and the horizontal velocity of the plume: the plotted curves are for $\theta = 30°$.) The different curves for different impacts reflect the varying location of the impact points behind the limb; later impacts generally falling closer to the limb (Chodas & Yeomans, this volume). The G plume, for example, would have become directly visible 32 sec after impact for $\theta = 45°$, whereas the W plume emerged into view after only 12 sec.

The dashed curves in Fig. 10 indicate the time and altitude at which the ejecta rose above the shadow of the jovian limb, and into direct sunlight. Again, the plumes from the later impacts, starting closer to the terminator, rise into sunlight earlier and at a lower altitude. For G, the plume remains in shadow until 130 sec after impact and 1500 km altitude, while for W the interval is only 90 sec and the corresponding altitude ~ 1000 km, all for $\theta = 45°$. These times are consistent with the sequences of images obtained for these two impacts by HST (Hammel *et al.* 1995): the G plume is seen in

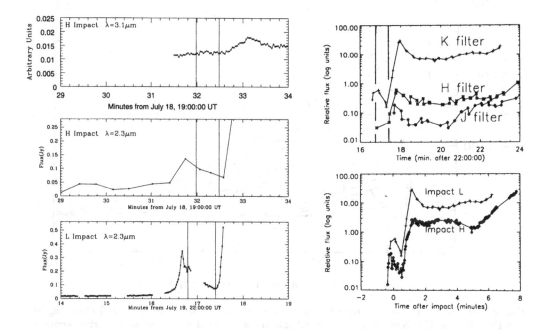

FIGURE 9. (a) Precursor lightcurves for the H and L impacts obtained at the 2.2 and 3.5 m telescopes at Calar Alto Observatory, Spain. The 2.30 μm data were obtained with an IR camera and a narrowband methane filter, while the 3.1 μm data were obtained with a high-speed photometer. For L the time resolution was 1.2 sec. From Hamilton *et al.* (1995). (b) Lightcurves for the H and L precursors obtained at the 1 m telescope at Pic du Midi Observatory, France. The data were obtained with a multi-channel IR camera and standard J, H and K' (1.25, 1.65 and 2.12 μm) filters, at $\sim$ 18 sec intervals. The lower panel compares the 2.12 μm data on a common scale. From Colas *et al.* (1995).

thermal emission at $T_i + 2.0$ min and in sunlight at $T_i + 5.0$ min; the W plume is seen emerging into sunlight at $T_i + 3.0$ min.

We now discuss the successive phases in a typical SL9 lightcurve, with this highly simplified but illustrative model as a guide to their interpretation.

3.2. Leader emission

For some minutes prior to the actual impact time, faint emission is seen in the G, K and L lightcurves. This *leader* emission brightens gradually, and then more steeply, until it merges into the first precursor flash. Its duration ranges from $\sim$ 30 sec for L (Fig. 9) to 3.5 min for K (Fig. 8), and is far too long to be accounted for by the passage of the main fragment through Jupiter's atmosphere at the entry velocity of 60 $\mathrm{km\,s^{-1}}$. More likely, this emission is due to an extended stream of dust preceding the main fragment, which arrived as a 'meteor shower' in Jupiter's atmosphere (Meadows *et al.* 1995; McGregor *et al.* 1995). The durations of the leaders are consistent with the elongations of the cometary comae over several tens of thousands of km observed prior to impact (Weaver *et al.* 1995). For the L event, a fainter feature was visible on the limb for at least 20 min prior to the impact, but it is not clear if this also represents 'leader' emission or

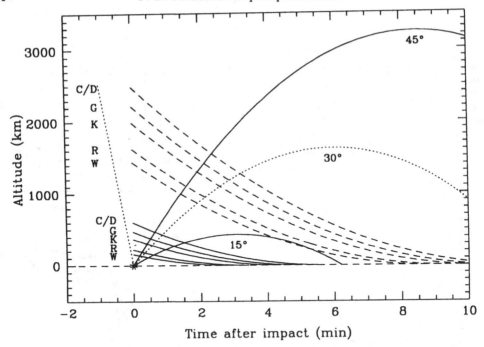

FIGURE 10. Plume altitude vs. time for an ejection velocity $v_e = 18\ \mathrm{km\,s^{-1}}$ and ejection angles $\theta = 15°$, $30°$ and $45°$. Solid sloping lines indicate the altitude at which plume material becomes directly visible to an earth-based observer, for representative impacts, while the dashed lines indicate where the plume enters sunlight for the same impacts. The trajectory of the incoming fragments is shown by the dotted line at negative time. From McGregor *et al.* (1995).

the remnant of an earlier impact by an uncatalogued or lost (J?) fragment (Hamilton *et al.* 1995).

3.3. *First precursor*

First precursors have been identified convincingly only in the lightcurves for the G, H, K, L and R impacts, with the highest resolution data being available for the R, K and L events (Figs. 1, 8 & 9(a)). The first L precursor peaked sharply 7 sec prior to the initial Galileo PPR detection, declined to a plateau for about 8 sec, and then faded gradually over the next 40 sec (Hamilton *et al.* 1995). A similar pattern was observed for K, where the first precursor appeared abruptly and peaked 11 sec prior to the beginning of the flash observed by the Galileo SSI instrument (Watanabe *et al.* 1995), before decaying slowly over the next 60 sec. The first R precursor shows a similar development at 2.3 μm, beginning abruptly $\sim$ 24 sec prior to the Galileo NIMS flash, and then decaying slowly (Graham *et al.* 1995). The Siding Spring lightcurve for G shows but does not fully resolve the first precursor, which appears only as a 2-frame pause before it is overtaken by the second precursor (McGregor *et al.* 1995). In Calar Alto observations of the H impact, the first precursor appears in only four frames (Hamilton *et al.* 1995). For both G and H the initial abrupt flux increase occurs in the image obtained immediately prior to the onset of the corresponding Galileo flash, consistent with the higher-resolution K, L and R data.

The timing of the first precursor, as well as the steeply-brightening leader seen for L, strongly suggests that it was due to thermal emission from the main fragment as it entered Jupiter's upper atmosphere, 400–650 km above the effective impact level. Inspection of

Fig. 10 shows that the K and L meteors were indeed last directly visible above the limb at about this altitude. The ~ 10 sec gap between the first precursor peak and the onset of the Galileo flash represents the flight time of the bolide behind the jovian limb. The slow decay of the first precursor has been attributed to cooling of the heated trail left by the entering body (Hamilton *et al.* 1995; McGregor *et al.* 1995), while it has been suggested (Chapman 1995, private communication) that the plateau in the L data may be due to reflection of the initial fireball from infalling dust high in the atmosphere.

Finally, we note that a 30 sec exposure of the G impact taken by HST centered on the start time of the Galileo PPR/UVS flash (cf. Fig. 7 and Hammel *et al.* 1995) shows a point-like source of emission in the planet's shadow. Since the rising plume cannot have been visible at the Earth within 15 sec after the impact (see Fig. 10), this frame must also have captured the meteor during its passage through the upper atmosphere ~ 10 sec prior to impact.

3.4. *Second precursor*

Approximately 55–60 sec after the onset of the first precursor, a much brighter second precursor was seen for each of the larger impacts (cf. Figs. 1(b), 7, 8 & 9). In the case of the less energetic impacts such as A, C, D, Q2, Q1, S and W, only a single precursor was detected; we assume that this corresponds to the second precursor, although for the late W impact both precursors may have merged into a single event (McGregor *et al.* 1995). The rise time of the second precursor was 15–20 sec, and the duration (FWHM) was typically 30–60 sec. For the weaker impacts (*e.g.*, C, D, Q2, Q1 and W), the flux soon thereafter returned to the pre-impact level (cf. Fig. 3 and Herbst *et al.* 1995, Fig. 1).

For the bright G, H, K and L impacts, on the other hand, the flux level after the second precursor dipped only slightly and then remained relatively constant—or even increased slowly—for ~ 4 min until the main event began. The R impact presents a particularly interesting case, with the fluxes at 3.2 and 4.5 μm fading quite rapidly after the second precursor peaked, while the 2.3 μm flux declines only slowly (cf. Fig. 1(b)). A similar wavelength-dependence is shown by the H impact, where the flux was observed to drop more sharply after the second precursor at 3.1 μm than at 2.3 μm (Hamilton *et al.* 1995). Mid-IR observations of the H and L impacts show second precursors for these events at 10–12 μm, which also faded after 30–60 sec (Livengood *et al.* 1994; Lagage *et al.* 1995) in contrast to their persistence at 2.3 μm.

There is some indication that the time of peak flux for the second precursor is also wavelength-dependent, although results are inconsistent. Hamilton *et al.* (1995) found that the H precursor peaked 10–20 sec earlier at 3.1 μm than at 2.3 μm. Nicholson *et al.* (1995a) similarly noted that the R precursor began ~ 20 sec earlier at 4.5 μm than at 2.3 μm. However, Colas *et al.* (1995) find that the L precursor peaked ~ 15 sec later at 2.12 μm than at 1.25 or 1.65 μm. Drossart *et al.* (1995) and Lagage *et al.* (1995) note a 30 sec delay between the onset of the second L precursor at 2.3 μm and at 12 μm, although the second H precursor seems to have been seen essentially simultaneously at 2.3 μm and at 10 μm (Livengood *et al.* 1994). No obvious pattern emerges from these reports.

The second precursor has generally been interpreted as the first appearance of the fireball as seen from the Earth, rising above the jovian limb (*e.g.*, Graham *et al.* 1995). Comparison of the reported times for the second precursor with those of the initial Galileo flashes show that the 2.3 μm second precursors began 70 ± 18 sec (G), 48 ± 12 sec (H), 50 ± 7 sec (K), 39 ± 4 sec (L), 38 ± 11 sec (R) and 10 ± 16 sec (W) after the actual impact times (see chapter by Chodas & Yeomans). This pattern of decreasing intervals is consistent with the progressively shorter times taken by the rising fireball to reach an altitude at which it is directly visible from earth, as illustrated in Fig. 10. The actual intervals between the impact flash and the arrival of the fireball at the limb also depend on jovian atmospheric extinction. Estimates of the pressure level at which the line-of-sight optical depth is of order unity range from ~ 15 mb at 2.3 or 3.5 μm (Meadows *et al.* 1995) to 100 mb at 10 μm (Livengood *et al.* 1995). These levels are, respectively, 90 km and 45 km above the 1 bar level, the approximate depth at which the explosion is believed to have been initiated (Boslough *et al.* 1995; Zahnle and MacLow 1995). Assuming a maximum vertical velocity of 12.5 km s^{-1}, and neglecting the initial acceleration phase of the explosion, we find that the G and H fireballs should have become visible at 2.3 μm no sooner than 40 sec after impact, while the R and W fireballs should have appeared after about 24 and 20 sec, respectively. These predictions are in reasonable agreement with the observed intervals quoted above.

This interpretation is strengthened by Galileo NIMS observations of the G impact, which clearly show a rising, cooling fireball at the same time as the second precursor was beginning (cf. Fig. 7). An HST image taken immediately after the Galileo flash had faded (at 7:35:16 UT) shows thermal emission in the planet's shadow from the rising plume. Near-IR spectra taken during the second precursor phase of the K impact show a blue continuum during the first minute of the second precursor, which then rapidly cooled over the succeeding few minutes (Meadows & Crisp 1995)—again consistent with a rising fireball.

The absence of any noticeable increase in brightness in the second precursors ~ 2 min after impact as the plumes crossed the terminator into sunlight (cf. Fig. 10) argues against any appreciable fraction of the near-IR flux at this time being due to reflected sunlight from condensates which may have formed in the plume.

In the context of a rising, cooling plume, the flat or slowly increasing flux level exhibited by the G, H, K and L precursors at 2.3 μm is puzzling. McGregor *et al.* (1995) have suggested that this may be due to the continued ejection of hot material from the entry site after the initial fireball for the most energetic (and deepest penetrating?) impacts. Some support for this explanation is provided by an HST image of the G plume taken at 7:38:16 UT, almost 5 min after the impact. This image appears to show the plume in sunlight, but with a faint 'tail' of thermal emission in the shadow region (Hammel *et al.* 1995). This tail could be interpreted in terms of continued ejection along the entry corridor. It is not clear, however, how the more rapid decay of the second precursors at 3.1 μm and longer wavelengths is to be interpreted in this context, as an increase in ejecta temperature over time seems rather unlikely. A similar problem arises in connection with the slow decay of the second R precursor at 2.3 μm.

3.5. *Main event*

Approximately 6 min after each recorded impact, a steady and in some cases truly awesome brightening was observed at near-IR and mid-IR wavelengths. In the case of some of the smaller impacts, or when observing conditions were below par, this *main event* was the only feature observed. An initially rapid rise over 1–2 min was followed by a more gradual increase, the intensity reaching a peak 10–15 min after the moment of impact. The interval between impact and peak luminosity was correlated with the peak flux, the brighter events taking longer to reach maximum (Meadows *et al.* 1995); the average interval was 12 min. The initial decline in intensity following the peak had an e-folding time of ~ 3 min at 2.3 μm, but was somewhat slower at 10–12 μm (cf. Figs. 3 and 4). In many cases the decay was interrupted by a plateau or *shoulder* at ~ 20 min after the impact.

At their peaks, the G and K fluxes at 2.3 μm probably both reached $\sim$ 450 Jy, or 15 times the brightness of Io, although the G lightcurve is partly saturated. The peak brightness of the L impact seems to have been even greater, and this event strongly saturated the array detectors on larger telescopes. Observing at the 0.4 m telescope at Whately Observatory, Massachusetts, with an IR array and narrowband 2.30 μm filter, Skrutskie & Aas (1995) were able to follow the main L event through only modest saturation to a peak brightness of 35 times that of Io, or $\sim$1000 Jy. Very similar main events were observed for impacts at 2.3, 3.1–3.2, 4.5, 10 and 12 μm, suggesting that the temperature of the emission remained relatively constant during the entire duration of this phase. At 12 μm the peak flux for the L impact reached 13,000 Jy, over 4 times the brightness of the H event and 8 times that of E and Q1 (cf. Fig. 4).

Although initial discussions of the main event for R (cf. Graham *et al.* 1995; Nicholson *et al.* 1995a) emphasized the rough coincidence between the time of peak flux and the time at which the impact site rotated onto the planet's limb, once a larger suite of observations became available it became apparent that this purely geometric interpretation of the lightcurves was not tenable. The extensive series of 2.3 μm lightcurves obtained at Siding Spring (McGregor *et al.* 1995; see Fig. 3), and the similar series of 12 μm lightcurves obtained at La Palma (Lagage *et al.* 1995; see Fig. 4), demonstrate conclusively that the main event began $\sim$ 6 min after the impact time in *all* cases, irrespective of the distance of the impact point behind the limb. In fact, for the early C and D impacts, the main event was almost over by the time the impact site had rotated onto the limb (cf. dashed lines in Fig. 3). Moreover, the Galileo NIMS instrument recorded the beginning of the main events for G and R 370 sec after the initial flashes (Carlson *et al.* 1995a; 1995b), at almost exactly the same time as did earth-based observers at 2.3 μm (cf. Figs. 1 and 7). The average interval between the adopted impact time and the onset of the main event observed in 12 lightcurves (mostly at 2.3 μm) was 5.9 ± 0.3 min.

It is thus apparent that the origin of the main peak must be sought in the impact phenomenon itself, rather than in the observational geometry. The rapid adiabatic cooling of the rising plume makes it highly unlikely that the plume itself contributes significantly to thermal emission in the near-IR during the main event (Graham *et al.* 1995; Nicholson *et al.* 1995a), although this view has been disputed by Meadows *et al.* (1995) on the basis of spectroscopic evidence. Although several alternative suggestions have been made (*e.g.*, dust formation in the cooling plume—Hasegawa *et al.* 1995), the most likely explanation is that the tremendous burst of infrared flux is associated with the re-entry of the plume into the jovian atmosphere, where its kinetic energy is converted back to thermal energy via shock-heating (Zahnle & MacLow 1995; Boslough *et al.* 1995)—picturesquely described as the 'splashback' or 'hypervelocity splat'. Note that, because of the range of vertical ejection velocities represented in the plume, there is no contradiction in the lower velocity material beginning its re-entry at about the same time as the highest velocity material is nearing its peak altitude, as seen in the HST plume images (Boslough *et al.* 1995). Simple models for the infrared luminosity of the re-entering R impact plume have been generated by Zahnle and MacLow (1995), and are reviewed in the chapter by Zahnle in this volume. These models are quite successful in reproducing both the amplitude and the temporal evolution of the R main event at 2.3–4.5 μm, although it should be noted that the opacity source for the shock-heated debris is at present unknown and the model opacity was simply adjusted to fit the observations.

The common interval between impact and onset of the main event can be interpreted as requiring a minimum vertical re-entry velocity in order to generate significant shock-heating and/or thermal radiation. A 6 min flight time corresponds to a vertical velocity component $v_z = gt/2 = 4.5\,\mathrm{km\,s^{-1}}$. Zahnle (this volume) has in fact suggested that a minimum vertical re-entry velocity of 4–5 $\mathrm{km\,s^{-1}}$ is needed to heat jovian air sufficiently for shock synthesis to produce abundant organic solids as opacity sources. At lower shock velocities, the heated gas remains transparent, and little IR radiation or organic aerosol production would be expected. As well as neatly explaining the 6 min delay of the main event, this hypothesis can also account for the hollow, crescent-shaped ejecta blankets observed by HST around the larger impact sites: material falling closer to the impact site does so at lower velocities, in general.

Given that the impacts occurred behind the planet's limb, and that the time for the impact site itself to rotate onto the limb ranged from 19 min for A to 9 min for W (see chapter by Chodas & Yeomans), it must be asked how the re-entering plume material could be seen by earth-based observers only 6 min after impact. The answer lies in the fact that most of the debris was apparently ejected not vertically upward, as in the simple 'toy plume' model of Zahnle & MacLow (1995), but in a south-easterly direction back along the inbound trajectory of the cometary fragments relative to the surface of Jupiter, and thus towards the limb. Although the simple ballistic model illustrated in Fig. 10 shows that material ejected at 45° elevation and our nominal ejection velocity of 18 $\mathrm{km\,s^{-1}}$ takes ~ 17 min to land, material ejected at lower velocities and on flatter trajectories lands sooner. Specifically, debris ejected at 18 $\mathrm{km\,s^{-1}}$ and $\theta = 15°$ re-enters after only $2v_e \sin\theta/g = 6$ min, but remains in direct view of the Earth, *i.e.*, above the solid curves in Fig. 10. This is due in roughly equal parts to jovian rotation and downrange motion of the plume.

The overall duration of the typical main event of $\sim$10 min fits well with the period during which plume material ejected at 18 $\mathrm{km\,s^{-1}}$ and elevations ranging from 15° to 45° would have landed. Some material may have been ejected at steeper elevations, but the maximum observed plume height of 3200 km (Hammel *et al.* 1995) sets an upper limit of $\sim 12.5\,\mathrm{km\,s^{-1}}$ on the vertical velocity, and a corresponding upper limit of $2v_z/g = 17$ min on the ballistic flight time.

The relation between time-of-flight and longitudinal distance travelled by the plume material is shown more clearly in Fig. 11. For simplicity, the azimuth of ejection is assumed to be a constant 45° east of south, and the ejection velocity $v_e = 17.9\,\mathrm{km\,s^{-1}}$. Points along the solid curve correspond to differing ejection angles θ, starting with $\theta = 0$ at the origin and increasing to $\theta = 45°$ at the maximum range of 12,800 km (10.5° in longitude on Jupiter at latitude 44°S). (In reality, the full 3-dimensional plume involved a range of ejection velocities and azimuths as well, but these parameters are somewhat less important, within reasonable limits, in determining the maximum distance travelled downrange by the plume at any given time.) Horizontal dashed lines indicate significant events in the evolution of a typical impact plume: the start of the main event, the time at which the plume reaches maximum altitude, the time at which CO emission at 2.3 μm appears in the spectrum (Meadows & Crisp 1995), and the time of the shoulder in many lightcurves. Squares at $\theta = 15°$ and $\theta = 30°$ correspond approximately to the onset and peak of the main event, at which times the vertical velocity of the re-entering ejecta was 4.5 and 9 $\mathrm{km\,s^{-1}}$, respectively.

Superimposed on the time-of-flight curve are a set of straight diagonal lines showing the distances of several representative impact sites behind the limb as functions of time from impact. At the intersection of the appropriate diagonal line with the time-of-flight curve, the expanding curtain of re-entering ejecta crosses the limb onto the visible face of Jupiter. Ejecta from the G impact, for example, began to land on the visible disk 6.3 min after impact, while that from the R impact crossed the limb within only 4.5 min.

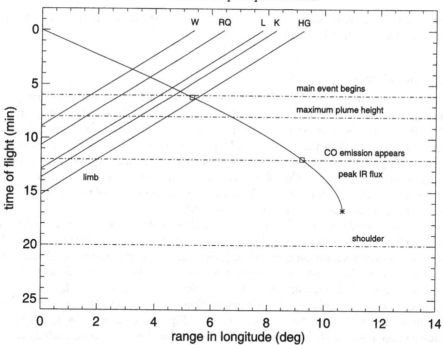

FIGURE 11. Time of flight vs. downrange distance (in longitude) for plume material ejected at $v_e = 17.9$ km s^{-1} and azimuth 45° east of south, for a range of elevation angles θ. An asterisk indicates the maximum range of 12,800 km (or 10.5° in longitude) reached by ejecta at $\theta = 45°$, while the squares correspond to $\theta = 15°$ and $\theta = 30°$. Diagonal lines indicate where the falling ejecta crosses the limb onto the visible hemisphere of Jupiter.

The G and R impact sites themselves did not cross the limb until 15.5 min and 10.6 min, respectively, as indicated by the intercept of the diagonal lines on the vertical axis.

3.6. *Shoulders*

The declining phases of many of the SL9 lightcurves show shoulders, in the form of either changes in slope or even secondary maxima, 19–20 min after impact (see examples in Figs. 1, 3 and 4). These shoulders are seen both at 2.3 μm and at longer wavelengths. The R lightcurve also shows some evidence for subsequent, but more subdued shoulders at ~ 28 and 38 min after the impact; similar features are seen in the C and G lightcurves in Fig. 3.

The origin of these features is unclear. Suggestions that they are due to the impact ejecta crossing the terminator, and thus becoming visible in reflected light, run afoul of two facts: (i) the visibility of the shoulders at 12 μm as well as the near-IR, and (ii) the observation that they appear at a more-or-less fixed time interval after impact. Most other explanations involve a dynamical response of the atmosphere to the re-entering debris, either in the form of a vertically propagating sound wave reflected from the tropopause and/or the upper stratosphere (Livengood *et al.* 1995; Nicholson *et al.* 1995a), or an actual re-launching of some of the ejecta (MacLow 1994, private communication). A 1-dimensional 'bounce' model, powered by the adiabatic compression of the atmosphere under the weight of the collapsing plume, was explored numerically by Deming *et al.* (1995) who obtained an oscillation period of ~ 10 min, in good agreement with the observations. Their model, however, ignored radiative damping and may thus have overestimated the amplitude of the bounce. Two-dimensional hydrodynamical models of the re-entering plume, including radiative losses, also show such a bounce (see chapter by MacLow in this volume).

4. Temperatures

Although a significant number of spectroscopic observations were made during several of the larger impacts, most of these data have yet to be published and few quantitative results are available at the time of writing. Compositional inferences from these measurements are reviewed in the chapter by Lellouch in this volume, while implications for the perturbed post-impact thermal structure of the jovian atmosphere are reviewed in the chapter by Conrath. We therefore restrict our discussion here to estimates of temperatures during the fireball and splashback phases, as these are of potential importance in constraining models of the impact process itself.

Temperature estimates may be divided into three categories, in order of decreasing probable reliability. First are physically-motivated spectroscopic estimates based, for example, on rotation-vibration line ratios. Even these estimates require a physical model for the emission region, and make assumptions of LTE, both of which may introduce significant errors (cf. discussion by Lellouch). Second are blackbody fits to regions of apparent continuum emission, perhaps representing the temperature of hot dust produced in the initial impact, or in the fallback shock. Figure 12 shows a series of medium-resolution grism spectra of the K impact obtained at the AAT which beautifully illustrate the evolution of the 2.0–2.4 μm spectrum during the fireball and splashback phases of this event (Meadows & Crisp 1995). Third are simple color temperatures derived from photometric flux ratios, which may suffer large systematic errors if the measured flux in one or more of the bands used is in fact dominated by line emission. We will group the available estimates by impact phase.

4.1. *First precursor (the bolide?)*

Very few multi-wavelength observations, and little spectroscopic data, have been reported for this short-lived phase in the lightcurves. Peak fluxes for the R impact first precursor at 2.3, 3.2 and 4.5 μm (cf. Fig. 1(b)) are consistent with a color temperature of 1000 ± 120 K (Nicholson *et al.* 1995b). A comparison of the flux measured by HST at 0.89 μm for what is interpreted as the meteor phase of the G impact (Hammel *et al.* 1995) with the simultaneously-measured first precursor brightness at 2.3 μm (cf. Fig. 7) yields a color temperature of 1800 K (McGregor *et al.* 1995). However, in both cases the measurements themselves were not made simultaneously, and it is likely that the fluxes changed appreciably during the integration times of 5–30 sec, casting considerable doubt on the reliability of the derived temperatures. The K impact spectrum at this time shows elevated emission longward of 2.05 μm, which matches a 6700 K blackbody (Fig. 12, panel 3).

4.2. *Second precursor (rising fireball)*

Thermal emission from the rising fireball was detectable for several minutes in the case of the larger impacts, permitting some spectroscopic observations. During most of the second precursor phase (Fig. 12, panels 4–6), the K impact spectra show primarily continuum emission with a blackbody temperature which declined from 720 K at $T_i + 3.5$ min to 420 K at $T_i + 6$ min. This presumably reflects the temperature of the plume after it had risen into view from the Earth; Galileo NIMS observations for the G and R impacts show considerably higher temperatures in the first 30 sec of the plume's development, when it was still hidden behind the limb (Carlson *et al.* 1995a).

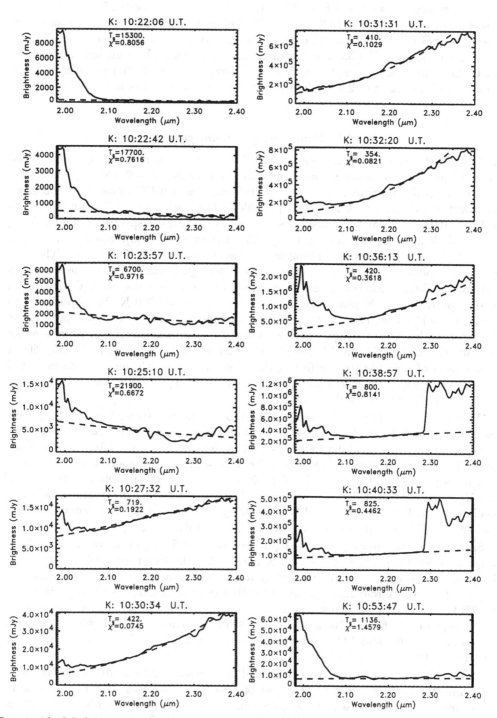

FIGURE 12. 2.0–2.4 μm spectra obtained at the Anglo-Australian Telescope during the K impact. Dashed curves indicate blackbody fits to the continuum regions of the spectra. The Galileo impact time was T_i=10:24:13 UT, while the main event began at ~10:31. Pre-impact leader emission is seen in panels 1 & 2, and the first precursor in panel 3. The second precursor begins in panel 4 and continues through panels 5 and 6. Panels 7–11 cover the main event, with panel 9 at T_i+12 min falling near the peak of the 2.3 μm lightcurve. Absolute calibration is provisional. Adapted from Meadows & Crisp (1995).

Color temperatures derived from 2.3/3.1 μm flux ratios during the latter part of the H second precursor are 900–1300 K (Hamilton *et al.* 1995), while during the equivalent phase of the R impact the 2.3/4.5 um color temperature slowly increased from 650 K at T_i+1 min to 850 K at $T_i+3.5$ min. Spanning a wider wavelength range, measurements of the H and L precursor peaks at 10 and 12 μm (Livengood *et al.* 1995; Lagage *et al.* 1995) combined with data at at 2.3 μm imply similar color temperatures of 600–625 K.

4.3. *Main event (collapsing plume)*

In the early stages of the main event, the 2.0–2.4 μm spectrum of the K impact continued to be dominated by continuum emission, with temperatures of ∼ 400 K. Beginning near the time of peak flux, and for ∼ 10 min thereafter, prominent CO emission was observed longward of 2.30 μm, with NH_3 and possible H_2O emission below 2.1 μm (Meadows & Crisp 1995). Similar CO emission was observed in near-IR spectra of the C, D, G, H, R and W impacts (Meadows & Crisp 1995; Herbst *et al.* 1995). Excitation of these $\Delta\nu = 2$ vibrational bands usually implies a temperature of 2000 K or higher. The NH_3 emission implies a similarly high temperature, although Lellouch cautions against such a simple LTE-based interpretation. In contrast, fits to the emission by H_2O at 2.04 μm in the spectra at the K peak suggest a temperature of 600–700 K (Meadows & Crisp 1995), while Encrenaz *et al.* (1995) estimate a temperature of 750 K at the peak of the H lightcurve, based on CH_4 spectra at 3.5 μm and an assumed effective pressure level for the radiating region of 1 mb. Maillard *et al.* (1995) derived a temperature of 750–1500 K at the peak of the C lightcurve from CH_4 emission lines at 3.3 μm. The onset of CO emission may signal the increasing vertical re-entry velocity of the plume, $v_z = gt/2$: by $T_i + 12$ min this had reached 9 km s^{-1}.

Bjoraker *et al.* (1995), using the Kuiper Airborne Observatory with an echelle grating spectrometer at 7.7 μm, observed the main events for the G and K impacts. Emission due to hot methane, water and dust was observed to commence ∼ 7 min after the K impact, peak at 13–14 min post-impact, and then decay by ∼ 25 min. A temperature of ∼ 1000 K and a base pressure of 3 μb at the K peak was derived from the relative strengths of hot H_2O and CH_4 lines. Lower-resolution 5–10 μm spectroscopic observations were also made with the KAO for the R and W impacts by Sprague *et al.* (1995), revealing strong emission by CH_4 and H_2O at a temperature of ∼ 500 K.

Simultaneous monitoring of the R lightcurve at 3.2 and 4.5 μm showed a surprisingly constant color temperature of ∼ 1000 K during the fallback phase, although this may be strongly affected by CH_4 line emission at 3.2 μm. In fact, the best overall fit to the R impact spectrum from 2.3 to 13 μm at the time of peak flux implies either a blackbody at 700 K, or an opacity which scales as λ^{-1}, as expected for emission by sub-micron dust particles, and $T \sim 600$ K (Nicholson *et al.* 1995b). Simultaneous narrowband measurements at 7.85, 10.3 and 12.2 μm by Friedson *et al.* (1995) are also consistent with this fit (cf. Fig. 13).

Color temperatures derived from published lightcurves at 2.3 and 10–12 μm for the E, H, L and R impacts, evaluated at the time of peak flux, also fall in the range 550–700 K. Of course, these estimates may be systematically skewed by CO emission at 2.3 μm and/or silicate emission at 10 μm (Nicholson *et al.* 1995b). Perhaps none of the above blackbody temperature estimates should be taken too seriously until a physical model of the emission from the shocked mixture of cometary debris and jovian atmosphere is developed which takes into account both molecular and dust emission, as well as the near-limb viewing geometry.

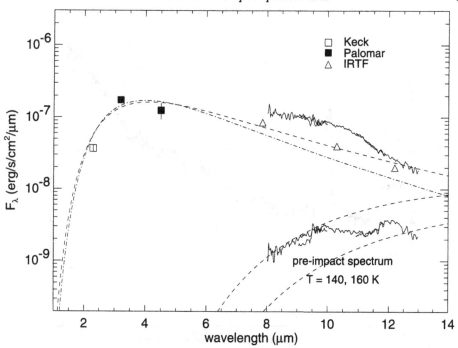

FIGURE 13. Composite spectrum for the R impact, at the peak of the main event, based on narrowband imaging at 2.3, 3.2, 7.85, 10.3 and 12.2 μm, broadband imaging at 4.5 μm, and low-resolution spectroscopy at 8–13 μm. Data from Nicholson *et al.* (1995a), Graham *et al.* (1995) and Friedson *et al.* (1995). The dashed curve shows the best-fitting blackbody spectrum (700 K, with an optical depth-solid angle product $\Omega\tau = 0.01$ sq arcsec), while the dot-dashed curve is the best-fitting dust-opacity ($\tau \sim \lambda^{-1}$) model, with $T = 600$ K and $\Omega\tau = 0.01$ sq arcsec at 10 μm. The spectrum at lower right was obtained immediately prior to the impact, and reflects normal jovian tropospheric and lower stratospheric temperatures of 140–160 K. Adapted from Nicholson *et al.* (1995b).

5. Anomalous lightcurves

A few apparently aberrant lightcurves do not appear to fit comfortably into the scenario described in §3. The strong Q1 precursor, similar in shape to the second L precursor, is reported to have started 37 sec *before* the inital Galileo PPR detection (Herbst *et al.* 1995), an anomaly with no obvious explanation.

The Q2 impact, which occurred 29 min before Q1, showed a single precursor followed about 8 min later by a very subdued 'main event', which was fainter than the precursor (Herbst *et al.* 1995).

The V impact was observed only as a brief flash, $\sim$ 30 sec in duration, with no subsequent main event at all (Weinberger 1994; Meadows 1994, private communications). Based on its duration and its similarity to the first R precursor at 2.3 μm (Fig. 14), we tentatively identify this as a 'first precursor', perhaps due to the entry flash of a particularly small and/or weak fragment of SL9.

For each of the B, M and N impacts, there is a single report of a brief (1–2 min in duration) period of weak 2.3 μm emission (de Pater 1994, private comunication; McGregor *et al.* 1995), which most likely reflects a very faint main event with no discernable precursor. The N event was first seen at 10:36:30 UT, 7 min after the impact flash was detected by the Galileo SSI instrument, consistent with a barely-detectable main event.

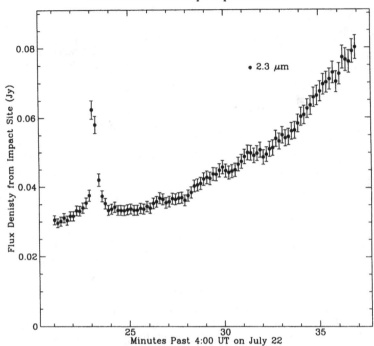

FIGURE 14. Palomar lightcurve showing the very brief V impact flash at 2.3 μm, at 10 sec resolution. The increasing background signal is due to the old E impact site rotating across the terminator. A simultaneous flash was recorded at the AAT. Possible faint leader emission is seen before the main flash, but there appears to have been no detectable 'main event'. Unpublished data from A. Weinberger.

A similar report of a faint U event was made by Cochran *et al.* (1995), but simultaneous observations with larger telescopes under clear skies failed to detect any signature of this impact (Hamilton *et al.* 1995). Likewise, early reports of an F impact have not been confirmed by other observers, and were likely due to confusion with the re-appearance of the old E impact site on the limb.

6. Other phenomena

6.1. *Satellite flashes*

Considerable attention before and during the SL9 impacts was devoted to predicting and searching for impact flashes reflected from the Galilean satellites. Despite several efforts to record a reflected flash with photometers and CCD cameras, no unambiguous reflected signals were detected. Consolmagno & Menard (1995) monitored Europa during the A impact and Europa during the H, Q1 and Q2 impacts. Possible 2-σ detections of flashes from A, Q2 and Q1 were reported, but none was considered to be convincing. A'Hearn *et al.* (1995) and Woodney *et al.* (1995) report observations of Io and Europa during the D, E and K impacts, and derive upper limits on the peak flash luminosity at 0.42 and 0.62 μm. However, the D and E limits far exceed the peak luminosities measured by the Galileo PPR and SSI instruments for several brighter impacts. It is apparent from these results that the flashes—particularly when reflected from sunlit satellites—were far too faint for detection at visual wavelengths.

A more favourable opportunity was provided by the K impact, which occurred while Europa was in eclipse but visible to earth-based observers if illuminated by the flash.

However, no detectable visual signal was observed, and the flux at 2.34 μm did not exceed 6 mJy (McGregor *et al.* 1995). Both visible and near-IR upper limits on the K flash exceed by a factor of $\sim$ 10 those expected based on the peak luminosity measured by the Galileo SSI experiment, assuming isotropic emission and temperatures of several thousand K.

6.2. *Post-impact features on Jupiter*

Due to the strong methane absorption in the underlying atmosphere, the ejecta blankets deposited by the larger impacts were readily visible in reflected sunlight as bright features in images taken at wavelengths of 2.3 or 3.5 μm. The structure of the ejecta blankets is best seen in the HST images (cf. chapter by Hammel in this volume), but some information concerning the impact process itself may be gleaned from the near-IR images. Although the impact site was generally unresolved in images taken near the peak of the main event, implying a source size of order 4000 km or less, later images frequently showed an extended region of emission, elongated along the planet's limb (Graham *et al.* 1995; Nicholson *et al.* 1995a; Lagage *et al.* 1995). Dimensions ranged from $\sim$ 8000 km for the R impact at 2.3 and 4.5 μm to 30,000 km for L at 12 μm. The smaller measurement is compatible with the size of the ejecta blankets imaged in the visible by HST (Hammel *et al.* 1995).

Near-IR images of several sites taken during the impact week show a distinct central spot and a crescent-shaped feature to the south-east, reminiscent of the celebrated HST image of the G site (West *et al.* 1995). Although only a few quantitative analyses have yet been performed using these data (Banfield *et al.* 1995; Chanover *et al.* 1995; Ortiz *et al.* 1995; Moreno *et al.* 1995), it is clear that the ejecta was deposited fairly high in the stratosphere, at pressure levels of 10 mb or less. For more details, see the chapter by West in this volume.

6.3. *Three-micron rings*

More unexpected was the discovery of very large circular rings surrounding the C, G and K impact sites and visible only between 3.1 and 4.0 μm (McGregor *et al.* 1995). Over this range, their spectra are extremely blue. Measurements of re-projected images show that the G and K rings expanded at a rate of $\sim$ 1400 m s^{-1} to radii of $\sim$ 18,500 km 2 hr after the impacts, before slowly fading from view 1.5 hrs later. No sign of these features is seen in many images taken at shorter wavelengths. The rings are centered $\sim$ 3500 km to the south-east of the impact sites, in the same direction as the displacement of the HST ejecta patterns, and appear to have expanded from an initial radius of $\sim$ 8000 km, within the optical crescent. It is not clear if the rings are seen in emission or reflected sunlight, although the former seems more consistent with the rapid fading. Their spectral signature is consistent with emission by poly-HCN (P. Wilson 1995, private communication). These rings are quite distinct from the expanding waves seen in the HST images of several sites; the latter are restricted to radii of 5000 km or less and expanded at a much slower rate of 450 m s^{-1} (Hammel *et al.* 1995).

Although no good model exists for these features, it has been suggested (Zahnle 1995, private communication) that they could be due to impact debris 'sliding' horizontally across the top of the atmosphere, from an initial location in the crescent. Alternative explanations include mildy supersonic shock waves propagating in the stratosphere, where the sound speed is $\sim$ 980 m s^{-1}.

7. Summary

Despite some predictions to the contrary, almost all phases of the impact process—with the exception of the terminal phase of the bolide and the initial seconds of the fireball—were apparently recorded by ground-based instruments. Although we are still far from having quantitative models which successfully account for all the features observed in the SL9 impact lightcurves, a viable qualitative picture of the various impact phenomena does seem to have emerged, capable of explaining the principal observations. A remarkably consistent set of characteristics emerges for the best-observed events, beginning with the first and second precursor flashes as the bolide entered the jovian atmosphere and the hot fireball emerged from behind the limb, and culminating in the infrared main event as the ejecta plume re-entered the jovian atmosphere. Only a handful of weaker events failed to follow this pattern. A simple ballistic plume model is successful in accounting for the timing of the major features in the lightcurves; in the future more realistic 3-D hydrodynamic simulations will be required to model quantitatively the observed fluxes and extract fragment energies, penetration depths, re-entry altitudes, etc. Particular problems include the persistence of the second precursor for the larger impacts, and the multiple shoulders exhibited by several lightcurves.

A general correlation is observed between the peak brightness of the infrared signal and the estimated sizes of the pre-impact fragments, and with the morphology of the post-impact sites as seen in the HST images. The faintest (or undetected) impact signatures were associated with fragments displaced from the main SL9 train, strongly suggesting that some 'fragments' were little more than fluffy aggregates, or perhaps simply agglomerations of dust.

Temperature estimates derived from near-IR and mid-IR photometry and spectroscopy range from 1000 K or higher for the entry trail of the bolide, through ~ 750 K for the fireball after it had cleared the limb. Temperatures derived during the splashback phase are less concordant, varying from 500 K to 2000 K. This dispersion may reflect a combination of non-LTE effects, non-blackbody emission, and real temperature variations, both spatial and temporal. Development of more realistic radiative models for the re-entering plume is clearly a high priority if the wealth of available spectroscopic and color data is to be fully interpreted.

Over the year since the events of July 1994 the author has greatly benefitted from discussions with many individuals, and it is impossible at this date to attribute correctly all original ideas. Much of the impact scenario presented here was developed in discussions at the AAS Division for Planetary Sciences conference in October 1994 and the European SL9 meeting in February 1995, and refined during IAU Colloquium 156 in May 1995. Particular thanks go to Don Banfield, Clark Chapman, Paul Chodas, Imke de Pater, Doug Hamilton, Tom Hayward, Peter McGregor, Keith Matthews, Vikki Meadows, Gerry Neugebauer, Glenn Orton, Alycia Weinberger, Kevin Zahnle and, last but not least, Audrey and Alicia. Preparation of this article was supported by the NSF and the NASA Planetary Astronomy program.

REFERENCES

A'HEARN, M. F., MEIER, R., WELLNITZ, D., AND WOODNEY, L. 1995 Were Any Impact Flashes Seen in Reflection from the Satellites? In *Proceedings of the European SL-9/Jupiter Workshop* (eds. R. West and H. Bohnhardt) pp. 113–118. ESO.

BANFIELD, D., GIERASCH, P., SQUYRES, S. W., NICHOLSON, P. D., CONRATH, B. AND MATTHEWS, K. 1995 2 μm Spectrophotometry of Jovian Stratospheric Aerosols— Scattering Opacities, Vertical Distributions and Wind Speeds. *Icarus*, (submitted).

BJORAKER, G. L., STOLOVY, S. R., HERTER, T. L., GULL, G. E. AND PIRGER, B. E. 1995 Detection of Water After the Collision of Fragments G and K of Comet Shoemaker-Levy 9 with Jupiter. *Science* (submitted).

BOSLOUGH, M., CRAWFORD, D., ROBINSON, A., AND TRUCANO, T. 1994 Mass and penetration depth of Shoemaker-Levy 9 fragments from time-resolved photometry. *Geophys. Res. Lett.* **21**, 1555.

BOSLOUGH, M., CRAWFORD, D., TRUCANO, T., AND ROBINSON, A. 1995 Numerical modeling of Shoemaker-Levy 9 impacts as a framework for interpreting observations. *Geophys. Res. Lett.* **22**, 1821.

CARLSON, R. W., WEISSMAN, P. R., SEGURA, M., HUI, J., SMYTHE, W. D., JOHNSON, T., BAINES, K. H., DROSSART, P., ENCRENAZ, TH., AND LEADER, F. E. 1995a Galileo infrared observations of the Shoemaker-Levy 9 G impact fireball: A preliminary report. *Geophys. Res. Lett.* **22**, 1557.

CARLSON, R. W., WEISSMAN, P. R., HUI, J., SMYTHE, W. D., BAINES, K. H., JOHNSON, T. V., DROSSART, P., ENCRENAZ, T., LEADER, F. E. AND MEHLMAN, R. 1995b Some timing and spectral aspects of the G and R collision events as observed by the Galileo near-infrared mapping spectrometer. In *Proceedings of the European SL-9/Jupiter Workshop* (eds. R. West and H. Bohnhardt) pp. 69–73. ESO.

CHANOVER, N. J., BEEBE, R. F., MURRELL, A. S., AND SIMON, A. A. 1995 Absolute Reflectivity Spectra of Jupiter: 0.25–3.5 Microns. *Icarus*, (submitted).

CHAPMAN, C. R., MERLINE, W. J., KLAASEN, K., JOHNSON, T. V., HEFFERNAN, C., BELTON, M. J. S., INGERSOLL, A. P., AND THE GALILEO IMAGING TEAM 1995 Preliminary results of Galileo direct imaging of S-L 9 Impacts. *Geophys. Res. Lett.* **22**, 1561.

COCHRAN, A. L., ARMOSKY, B. J., PULLIAM, C. E., CLARK, B. E., COCHRAN, W. E., FRUEH, M., LESTER, D. F., TRAFTON, L., KIM, Y., NA, C., AND PRYOR, W. 1995 An Update on Imaging Observations from McDonald Observatory. In *IAU Colloquium 156: The Collision of Comet P/Shoemaker-Levy 9 and Jupiter*, p. 21.

COLAS, F., TIPHENE, D., LECACHEUX, J., DROSSART, P., deBATZ, D., PAU, S., ROUAN, D., AND SEVRE, F. 1995 Near-infrared imaging of SL9 impacts on Jupiter from Pic-du-Midi Observatory. *Geophys. Res. Lett.* **22**, 1765.

CONSOLMAGNO, G. J. AND MENARD, G. 1995 A search for variations in the light curves of Io and Europa during the impact of Comet SL9: A, H, and Q events. *Geophys. Res. Lett.* **22**, 1633.

CRAWFORD, D., BOSLOUGH, M., TRUCANO, T., AND ROBINSON, A. 1994. The impact of comet Shoemaker-Levy 9 on Jupiter. *Shock Waves*, **4**, 47.

DEMING, D AND HARRINGTON, J. 1995. no title; poster paper at IAU Colloquium 156, Baltimore MD.

DROSSART, P., ENCRENAZ, T., LECACHEUX, J., COLAS, F., AND LAGAGE, P.O. 1995 The time sequence of SL9/impacts H and L from infrared observations. *Geophys. Res. Lett.* **22**, 1769.

ENCRENAZ, T., SCHULZ, R., STÜWE, J. A., WIEDEMANN, G., DROSSART, P., AND CROVISIER, J. 1995 Near-IR spectroscopy of Jupiter at the time of SL9 impact: Emissions of CH_4, H_3^+, and H_2. *Geophys. Res. Lett.* **22**, 1577.

FRIEDSON, A. J., HOFFMANN, W. F., GOGUEN, J. D., DEUTSCH, L. K., ORTON, G. S., HORA, J. L., DAYAL, A., SPITALE, J. N., WELLS, W. K., AND FAZIO, G. G. 1995 Thermal infrared lightcurves of the impact of comet Shoemaker-Levy 9 Fragment R. *Geophys. Res. Lett.* **22**, 1569.

GRAHAM, J., DE PATER, I., JERNIGAN, J., LIU, M., AND BROWN, M. 1995. W. M. Keck telescope observations of the comet P/Shoemaker-Levy 9 fragment R Jupiter collision. *Science* **267**, 1320.

HAMILTON, D. P., HERBST, T. M., RICHICHI, A., BÖHNHARDT, H., ORTIZ, J. L. 1995 Calar
    Alto Observations of Shoemaker-Levy 9: Characteristics of the H and L Impacts. *Geophys.
    Res. Lett.* **22**, 2417.

HAMMEL, H. B., BEEBE, R. F., INGERSOLL, A. P., ORTON, G. S., MILLS, J. R., SIMON,
    A. A., CHODAS, P., CLARKE, J. T., DE JONG, E., DOWLING, T. E., *et al.* 1995 HST
    Imaging of Atmospheric Phenomena Created by the Impact of Comet Shoemaker-Levy 9.
    Science **267**, 1288.

HASEGAWA, H., TAKEUCHI, S., WATANABE, J. 1995 Grain Formation in Cometary Impact
    Plume. In *Proceedings of the European SL-9/Jupiter Workshop* (eds. R. West and H. Bohn-
    hardt) pp. 279–286. ESO.

HERBST, T. M., HAMILTON, D. P., BÖHNHARDT, H., AND ORTIZ-MORENO, J. L. 1995 Near
    Infrared Imaging and Spectroscopy of the SL-9 Impacts from Calar Alto. *Geophys. Res.
    Lett.* **22**, 2413.

HORD, C. W., PRYOR, W. R., STEWART, A. I. F., SIMMONS, K. E., GEBBEN, J. J., BARTH,
    C. A., McCLINTOCK, W. E., ESPOSITO, L. W., TOBISKA, W. K., WEST, R. A., ED-
    BERG, S. J., AJELLO, J. M. AND NAVIAUX, K. L. 1995 Direct observations of the comet
    Shoemaker-Levy 9 fragment G impact by Galileo UVS. *Geophys. Res. Lett.* **22**, 1565.

LAGAGE, P. O., GALDEMARD, PH., PANTIN, E., JOUAN, R., MASSE, P., SAUVAGE, M., OLOF-
    SSON, G., HULDTGREN. M., NORDH, L., BELMONTE, J. A., REGULO, C., RODRIGUEZ
    ESPINOSA, J. M., VIDAL, L., MOSSER, B., ULLA, A., AND GAUTIER, D. 1995 SL-9 frag-
    ments A, E, H, L, Q1 collision on to Jupiter: Mid-infrared light curves. *Geophys. Res. Lett.*
    22, 1773.

LIVENGOOD, T. A., KÄUFL, H. U., KOSTIUK, T., BJORAKER, G. L., ROMANI, P. N., WIEDE-
    MANN, G., MOSSER, B., AND SAUVAGE, M. 1995 Multi-Wavelength Thermal-Infrared
    Imaging of SL9 Impact Phenomena. In *Proceedings of the European SL-9/Jupiter Workshop*
    (eds. R. West and H. Bohnhardt) pp. 1437–146. ESO.

MAILLARD, J. P., DROSSART, P., BÉZARD, B., DE BERGH, C., LELLOUCH, E., MARTEN,
    A., CALDWELL, J., HILICO, J. C. 1995 Methane and carbon monoxide infrared emissions
    observed at the CFH telescope during the collision of comet SL-9 on Jupiter. *Geophys. Res.
    Lett* **22**, 1573.

MARTIN, T. Z., ORTON, G. S., TRAVIS, L. D., TAMPPARI, L. K., AND CLAYPOOL, I. 1995 Ob-
    servation of Shoemaker-Levy Impacts by the Galileo Photopolarimeter radiometer. *Science*
    268, 1875.

McGREGOR, P., NICHOLSON, P. D., AND ALLEN, M. 1995. CASPIR observations of the collision
    of Comet Shoemaker-Levy 9 with Jupiter. *Icarus*, (submitted).

MEADOWS, V. AND CRISP, D. 1995 Impact Plume Composition from Near-Infrared Spec-
    troscopy. In *Proceedings of the European SL-9/Jupiter Workshop* (eds. R. West and
    H. Bohnhardt) p. 239. ESO.

MEADOWS. V., CRISP, D., ORTON, G., BROOKE, T., AND SPENCER, J. 1995 AAT IRIS Obser-
    vations of the SL-9 Impacts and Initial Fireball Evolution. In *Proceedings of the European
    SL-9/Jupiter Workshop* (eds. R. West and H. Bohnhardt) p. 233. ESO.

MORENO, F., MUNOZ, O., MOLINA, A., LOPEZ-MORENO, J. J., ORTIZ, J. L., RODRIGUEZ, J.,
    LOPEZ-JIMINEZ, A., GIRELA, F., LARSON, S. M. AND CAMPINS, H. 1995 Physical prop-
    erties of the aerosol debris generated by the impact of fragment H of comet P/Shoemaker-
    Levy 9 on Jupiter. *Geophys. Res. Lett.* **22**, 1609.

NICHOLSON, P. D., GIERASCH, P. J., HAYWARD, T. L., McGHEE, C. A., MOERSCH, J. E.,
    SQUYRES, S. W., VAN CLEVE, J., MATTHEWS, K., NEUGEBAUER, G., SHUPE, D., WEIN-
    BERGER, A., MILES, J. W., AND CONRATH, B. J. 1995a Palomar observations of the
    impact of fragment R of comet P/Shoemaker-Levy 9: I. Light curves. *Geophys. Res. Lett.*
    22, 1613.

NICHOLSON, P. D., GIERASCH, P. J., HAYWARD, T. L., McGHEE, C. A., MOERSCH, J. E.,
    SQUYRES, S. W., VAN CLEVE, J., MATTHEWS, K., NEUGEBAUER, G., SHUPE, D., WEIN-
    BERGER, A., MILES, J. W. AND CONRATH, B. J. 1995b Palomar observations of the impact
    of fragment R of comet P/Shoemaker-Levy 9: II. Spectra. *Geophys. Res. Lett.* **22**, 1617.

ORTIZ, J. L., MUNOZ, O., MORENO, F., MOLINA, A., HERBST, T. M., BIRKLE, K., BÖHNHARDT, H., AND HAMILTON, D. P. 1995 Models of the Shoemaker-Levy 9 Collision Generated Hazes. *Geophys. Res. Lett* **22**, 1605.

ORTON, G., A'HEARN, M., BAINES, K., DEMING, D., DOWLING, T., GOGUEN, J., GRIFFITH, C, HAMMEL, H., HOFFMANN, W., HUNTEN, D., JEWITT, D., KOSTIUK, T., *et al.* 1995 Collision of Comet Shoemaker-Levy 9 with Jupiter Observed by the NASA Infrared Telescope Facility. *Science* **267**, 1277.

SEKANINA, Z., CHODAS, P. W., AND YEOMANS, D. K. 1994 Tidal Disruption and the Appearance of Periodic Comet Shoemaker-Levy 9. *Astron. & Astrophysics* **289**, 607.

SKRUTSKIE, M. F., AND AAS, S. 1995 A 2.3 μm Light Curve of the "L" impact. *Icarus* (submitted).

SPRAGUE, A., BJORAKER, G., HUNTEN, D., WITTEBORN, F., KOZLOWSKI, R., AND WOODEN, D. 1995. Observations of H_2O following the R and W impacts of comet Shoemaker-Levy 9 into Jupiter's atmosphere. *Icarus* (in press).

TAKATA, T., O'KEEFE, J. D., AHRENS, T. J., AND ORTON, G. 1994 Comet Shoemaker-Levy 9: Impact on Jupiter and plume evolution. *Icarus* **109**, 3.

TAKEUCHI, S., HASEGAWA, H., WATANABE, J., YAMASHITA, T., ABE, M., HIROTA, Y., NISHIHARA, E., OKUMURA, S., AND MORI, A. 1995 Near-IR imaging observations of the cometary impact into Jupiter: Time variation of radiation from impacts of fragments C, D, and K. *Geophys. Res. Lett.* **22**, 1581.

WATANABE, J., YAMASHITA, T., HASEGAWA, H., TAKEUCHI, S., ABE, M., HIROTA, Y., NISHIHARA, E., OKUMURA, S., AND MORI, A. 1995 Near-IR Observation of Cometary Impacts to Jupiter: Brightness Variation of the Impact Plume of Fragment K. *Publ. Astron. Soc. Japan* **47**, L21.

WEAVER, H. A., A'HEARN, M. F., ARPIGNY, C., BOICE, D. C., FELDMAN, P. D., LARSON, S. M., LAMY, P., LEVY, D. H., MARSDEN, B. G., MEECH, K. J., NOLL, K. S., *et al.* 1995 The Hubble Space Telescope (HST) Observing Campaign on Comet Shoemaker-Levy 9. *Science* **267**, 1282.

WEST, R. A., KARKOSCHKA, E., FRIEDSON, A. J., SEYMOUR, M., BAINES, K. H., AND HAMMEL, H. B. 1995 Impact Debris Particles in Jupiter's Stratosphere. *Science* **267**, 1296.

WOODNEY, L. M., MEIER, R., A'HEARN, M. F., WELLNITZ, D., SMITH, T., VERVEER, A., AND MARTIN, R. 1995 Photometry of the Jovian Satellites During Comet D/Shoemaker-Levy 9's Impact with Jupiter. *DPS meeting abstract.*

ZAHNLE, K., AND MACLOW, M. M. 1994 The collision of Jupiter and Comet Shoemaker-Levy 9. *Icarus* **108**, 1.

ZAHNLE, K., AND MACLOW, M. M. 1995. A simple model for the light curve generated by a Shoemaker-Levy 9 impact. *J. Geophys. Res.* **100**, 16885.

HST imaging of Jupiter shortly after each impact: Plumes and fresh sites

By HEIDI B. HAMMEL

Department of Earth, Atmospheric, and Planetary Sciences, Massachusetts Institute of
Technology, Cambridge, MA 02139, USA

During the first few hours after each impact, numerous phenomena were observed with telescopes
on Earth, in orbit, and in space. The primary events in that time were: impacts themselves,
rise and fall of large plumes of ejected material, and atmospheric waves; also of interest were the
characteristic morphologies of fresh sites. Based on timing from Galileo instruments and ground-
based observations, the Hubble Space Telescope (HST) recorded actual impact phenomena for
fragments G and W, with the A and E impacts occurring just prior to the HST observation
window. For these four events, plumes were directly imaged; plume development and collapse
correlated with strong infrared emission near the jovian limb, supporting the interpretation that
the IR brightness was created by the fall-back of plume material from high altitude (see chapter
by Nicholson). For medium-to-large fresh impact sites imaged by HST within a few hours
of impact, expanding rings were detected, caused by horizontal propagation of atmospheric
waves (see chapters by Ingersoll and Zahnle). Initial site morphology at visible wavelengths was
similar for all medium-to-large impacts: a dark streak surrounded by dark material, dominated
by a large crescent-shaped ejecta to the southeast. Smaller impact sites typically only showed
a dark patch (no ejecta) which dissipated quickly. This chapter summarizes the most recent
measurements and interpretations of plumes and fresh impact sites as observed by HST.

1. Introduction

Because of its high spatial resolution, the Wide Field and Planetary Camera 2 (WFPC2)
on Hubble Space Telescope (HST) provided unique views of several phenomena occurring
in the initial hours after impact of the Shoemaker-Levy 9 fragments: in particular, the
dynamics of plume development were directly imaged, multiple rings were resolved surrounding the moderate and large impact sites, and detailed structure was discernible in
the fresh impact sites themselves (Hammel *et al.* 1995). In the months since the collision,
progress has been made both in quantifying these observations, and in interpreting them
in the context of the more complete record of ground-based and Galileo data. In this
chapter, I review the HST results, present new analyses of some data, and discuss the
current interpretation of the observations.

2. Data overview

The atmospheric imaging group of the HST Comet Science Team used a total of
forty-three orbits for WFPC2 observations of dynamical effects in Jupiter's atmosphere
generated by the impact of the comet fragments (Hammel *et al.* 1995). The team scheduled the majority of the orbits for the week of impacts (195 images), reserving six orbits
(75 images) for pre-impact characterization and nine orbits for observations of post-
impact evolution (177 images). The team defined a set of filters (see Table 1 of Hammel
et al. 1995), and cycled through them as often as possible.

The 0.045-arcsecond pixels of the Planetary Camera (PC) subtended about 171 km
at the sub-Earth point on Jupiter; pixels for Wide Field Camera detector 3 (WF3)
were 0.099 arcseconds or roughly 375 km (the pixel-to-km ratio varied with both time
and position on the planet). Both cameras were used, depending both on the expected

Impact [†]	Frame[‡]	Filter[¶]	Exp (s)	UT Description[∥]
A	0c01	FQCH4N	14.0	20:13:17	Emission in shadow
	0c02	FQCH4N	4.0	20:15:17	Nothing visible
	0c03	F953N	16.0	20:18:17	Plume in sunlight
	0c04	F547M	0.16	20:21:17	Plume spreading
	0c05	F410M	2.0	20:24:17	Plume falling
	0c06	F336W	3.5	20:27:17	Plume settled to disk
E	0g01	FQCH4P15	30.0	15:19:17	Plume in sunlight,thermal tail
	0g02	FQCH4P15	16.0	15:21:17	Larger plume
	0g03	F953N	35.0	15:24:17	Plume falling
	0g04	F555W	0.3	15:27:17	Plume settled to disk
G	0o02	FQCH4P15	30.0	07:33:17	Emission in shadow from meteor
	0o03	FQCH4P15	16.0	07:35:17	Emission in shadow
	0o04	F953N	16.0	07:38:17	Plume in sunlight, thermal tail
	0o05	F555W	0.3	07:41:17	Larger plume
	0o06	F410M	10.0	07:44:17	Even larger plume
	0o07	F336W	18.0	07:51:17	Plume settled to disk
W	1n05	F555W	0.3	08:06:17	Emission in shadow from meteor
	1n06	F410M	10.0	08:09:17	Plume in sunlight
	1n07	F336W	18.0	08:16:17	Larger plume, streak at base
	1n08	FQCH4P15	100.0	08:20:17	Plume settling to disk
	1n09	FQCH4P15	30.0	08:23:17	Plume settled to disk

[†] Images with initial plume phenomena are listed; see Table 3 of Hammel *et al.* (1995).
[‡] Frame prefix "u2fi" has been omitted.
[¶] Filters differ for A because the observations were made with WF3 rather than PC1.
[∥] Universal time on 1994 dates: A—16 July; E—17 July; G—18 July; W—22 July.

TABLE 1. Selected images from HST plume sequences

phenomena (*e.g.*, plumes or first looks at impact sites) and on circumstances of the HST orbit (sometimes event timing necessitated imaging during the South Atlantic Anomaly, forcing the choice of WF3).

3. Plumes

On all four targeted impact events (A, E, G, and W), HST images showed plumes of ejected material above the limb of Jupiter (Fig. 1); see Table 1, adapted from Hammel *et al.* (1995). The time scale from first detection to final settling into the upper atmosphere of Jupiter ranged from 15 to 20 minutes. Prior to impact, the optical properties of the plume phenomena were highly uncertain; therefore, filters were changed between each exposure in order to maximize the probability of detecting a plume. Plume images later in the impact week were also used to study impact sites created earlier, which required complete filter coverage.

3.1. *Detections of impacts*

In both the G and W plume sequences, bright pixels were seen above the limb of Jupiter in images taken at or very near to the time of impact (Fig. 1, Table 1). The initial G impact signal detected by the Galileo Photopolarimeter Radiometer (PPR) occurred at 7:33:32 UT at 945 nm (Martin *et al.* 1995). The Galileo Near-Infrared Mapping

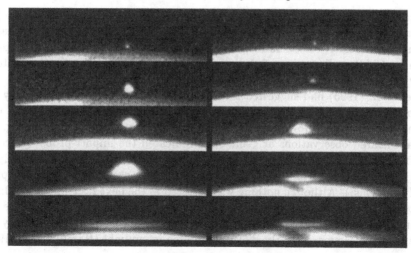

FIGURE 1. Selected plume images from the impacts of fragments G (left) and W (right). The full plume sequences are presented in Hammel *et al.* (1995); see Table 1 for filters and times. In both columns, the top image was taken at or very near the impact of impact; the subsequent temporal sampling differs in the columns. Both sequences were imaged with the PC. (*a*) The G impact created one of the largest impact sites; only K and L were bigger. In the third image shown here (taken 5 minutes after impact), bright pixels are seen both in emission in Jupiter's shadow, and in reflection above the shadow. The images are, from top to bottom, frames 0o02, 0o03, 0o04, 0o06, and 0o07; time from top to bottom is 15 minutes. (*b*) The W impact was the last and hence was closest to the jovian limb. Bright pixels appear to extend northward from the plume in the third image (10 minutes after impact). The images are, from top to bottom, frames 0n05, 0n06, 0n07, 0n08, and 0n09; time from top to bottom is 17 minutes.

Spectrometer also detected an event at 7:33:37 at several wavelengths between 0.7 and 5.3 μm (Carlson *et al.* 1995). An 889-nm image taken with HST from 7:33:17 to 7:33:47 UT (encompassing the Galileo first detection) clearly reveals bright pixels in the shadow of Jupiter. PPR data indicate that the initial G flash lasted for 50 seconds (Martin *et al.* 1995).

A similar phenomenon was detected in the HST sequence of W plume images. A 555-nm image of the W impact was taken by HST at 08:06:17. The image, which shows bright pixels in jovian shadow, was obtained within 0.5 second of a 559-nm image from the Galileo Solid-State Imaging (SSI) experiment at 08:06:16.67 (Chapman *et al.* 1995). The SSI image was second in a sequence of three images (each separated by 2 1/3 seconds) where emission first appears from the W impact event. The rapid rise of the SSI signal has been interpreted as radiation emitted by the initial meteor, suggesting a similar interpretation for the initial HST W signal.

It is unlikely that HST detected impact debris rising above the limb in these first G and W plume images. The G impact occurred beyond the horizon as seen from HST; thus material would have to have been 444 km above the 100-mbar limb to be visible (atmospheric refraction is not a significant effect, reducing the distance by only 45 km). Similarly, for the W impact, material would have to be 137 km above the 100-mbar surface to be visible from Earth (refraction reduces this distance by less than 25 km). In both cases, to be visible to HST so shortly after impact, ejected material would have to rise with a velocity higher than was inferred from later plume images (Hammel *et al.* 1995).

These images may have captured the incoming meteors for G and W. Alternatively, they may have detected light from the meteor scattering off infalling cometary dust or meteor light reflected off dust left in the atmosphere by the break-up of the fragment upon entry. It may even be a combination of these effects; further analysis has still not yet yielded conclusive answers for the source of these bright pixels. The chapter by Nicholson has further discussion of these initial events, placing them in the context of the many excellent ground-based lightcurves.

Hammel *et al.* (1995) had suggested that the image at 20:13:17 UT on 16 July 1994, which showed bright pixels in the vicinity of the A impact site, may have been a detection of the impact itself. However, subsequent verification of the timing from ground-based near-infrared observations at Calar Alto indicated the impact probably occurred at 20:11:59 ± 5 seconds (Herbst *et al.* 1995). Thus, the 20:13:17 image probably captured the subsequent rising plume in jovian shadow, as was seen for the G and W plumes (Hammel *et al.* 1995). Although nothing was detected in the next HST image at 20:15:17, that exposure time was much shorter than that of the previous image (at the same wavelength) and it is also probable that the plume was cooling (*i.e.*, fading) quickly. The third image in the sequence clearly shows the plume in sunlight (Hammel *et al.* 1995).

3.2. *Plume geometry*

Initial calculations of plume heights as a function of time for the A, E, G, and W plumes suggested that all observed plumes attained nearly identical terminal altitudes near 3000 km within 6–8 minutes of impact, and were falling and spreading within 10 minutes of impact (Hammel *et al.* 1995). Identical plume heights for explosions of different energies were not predicted (Ahrens *et al.* 1994a, Boslough *et al.* 1994a, Boslough *et al.* 1994b, Zahnle and Mac Low 1994).

The team has since refined the plume height measurements with more rigorous image navigation (determination of the planet's location) and added measurements of the widths of the plumes at their base as defined by the shadow of Jupiter (Fig. 2). Note that the plume widths must be used with care, since the altitude of the shadow of Jupiter with respect to the 100-mbar level varies both with differing geometry for different impacts and with time for any single impact.

The more rigorous measurements have confirmed the earlier findings, and provided more details for models. The G impact seems to have remained elevated longer and then collapsed faster than the other three observed plumes (Fig. 2). Sophisticated modeling may indicate whether this is an intrinsic property of a larger plume, or whether differing viewing geometry may be involved.

Hammel *et al.* (1995) pointed out that the seventh image of the W sequence showed an unexplained feature: material extended northward beyond the edge of the well-defined plume top (Fig. 1). They suggested that this may be debris "splashing" outward, as predicted in some pre-impact models (Mac Low and Zahnle 1994), and may be seen only in the W impact because of its position (it hit closest to the Jovian limb). However, because the W impact occurred on the site of the K impact (one of the largest impacts), there was also speculation that the material may have been residual elevated debris from the K impact. The team has subsequently checked other images where large sites (G and L) appear on the limb, searching for elevated debris. None has been seen, strengthening the suggestion that this material is related to the W impact itself.

4. Fresh Sites

4.1. *Timing based on site location*

Fifteen fragments left clear evidence of impact sites (Table 2); sites for smaller fragments (F, P2, T, U, and V, along with "missing" fragments J and M) were initially uniden-

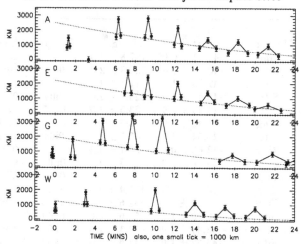

FIGURE 2. Plume height and width as a function of time. In each panel, the x axis indicates both time from impact (large ticks) and plume width (small ticks); the y axis indicates height above the 100-mbar level. For each image, the three points indicate the maximum height of the plume and the width of the plume measured at its apparent base, that is, where it rises above Jupiter's shadow into reflected sunlight (when the plume is in shadow, the width is the broadest part of the plume). The dashed lines indicate the calculated height of the jovian shadow above the 100-mbar level (P. Chodas, personal communication). Assumed impact times in this figure were A = 20:12:00, E = 15:12:00, G = 7:33:32, and W = 8:06:14 (all UT).

tified (Hammel *et al.* 1995). With more accurate timing from ground-based infrared lightcurves, it is probable that sites for several of these latter fragments will be recovered (*e.g.*, F, P2); this work is in progress. Table 2 gives latitudes and longitudes of the detected sites (the positions of the fresh impact sites and, if a ring was observed, the position of the ring's origin).

Impact times (Table 2) were inferred from the difference between predicted and observed longitudes. The HST times should be use with caution: subsequent comparisons with impact times inferred from ground-based and Galileo observations have shown a tendency for the HST times inferred from sites to be on average 2.5 minutes late, ranging from 10 seconds early for fragment D to more than 6 minutes late for fragment L (likely due to difficulty in defining a precise "site" for the large, complicated L impact scar).

4.2. *Site sizes*

Hammel *et al.* (1995) classified each impact site by its apparent size in the first available image after impact (Table 2); subsequent analysis has not significantly changed those results. Nicholson independently assessed the "size" of each impact based on its peak IR flux, finding a general correlation with the HST image classifications (see chapter by Nicholson). Similarly, comparisons of peak Galileo PPR signal values for several impacts agreed with the HST and ground-based IR assessments (Martin *et al.* 1995). Although these classes agree roughly with pre-impact fragment brightnesses (Weaver *et al.* 1995),there is some discrepancy: for example, see Figs. 1 and 2 of A'Hearn *et al.* (1995). The scatter in the correlation may suggest inhomogeneity in fragment strength or perhaps composition (though the latter is less likely).

Impact[†]	Class[‡]	Method[¶]	Lat.[¶]	Long.[¶]	Frame[¶]	Inferred Time (UT)[‖]
A = 21	2a	Ring	-43.41 ± 0.05	187.8 ± 0.3	—	$20{:}15{:}54 \pm 1$ min
		Site	-43.54 ± 1.0	186.3 ± 2.0	0i04	$20{:}13{:}24 \pm 3$ min
B = 20	3	Site	-42.79 ± 1.0	71.1 ± 2.0	0g05	$02{:}56{:}09 \pm 3$ min
C = 19	2a	Site	-43.41 ± 1.0	225.0 ± 2.0	0i04	$07{:}13{:}51 \pm 3$ min
D = 18	3	Site	-43.29 ± 1.0	33.5 ± 2.0	0g05	$11{:}52{:}50 \pm 3$ min
E = 17	2a	Ring	-43.48 ± 0.05	153.5 ± 0.2	—	$15{:}12{:}11 \pm 1$ min
		Site	-44.54 ± 1.0	153.5 ± 2.0	0i04	$15{:}12{:}11 \pm 3$ min
G = 15	1	Ring	-43.65 ± 0.04	25.7 ± 0.2	—	$07{:}33{:}17 \pm 1$ min
		Site	-43.66 ± 1.0	26.8 ± 2.0	0p03	$07{:}35{:}11 \pm 3$ min
		Plume	—	—	0o02	$07{:}33{:}16 \pm 0.5$ min
H = 14	2a	Site	-43.66 ± 1.0	101.4 ± 2.0	0v03	$19{:}33{:}21 \pm 3$ min
K = 12	1	Site	-43.29 ± 1.0	282.6 ± 2.0	1905	$10{:}30{:}58 \pm 3$ min
L = 11	1	Site	-42.79 ± 1.0	351.6 ± 2.0	6803	$22{:}21{:}44 \pm 3$ min
N = 9	3	Site	-43.41 ± 1.0	73.1 ± 2.0	1a03	$10{:}30{:}09 \pm 3$ min
Q2 = 7b	3	Site	-44.67 ± 1.0	47.5 ± 2.0	1a03	$19{:}46{:}31 \pm 3$ min
Q1 = 7a	2b	Ring	-44.37 ± 0.1	64.0 ± 0.5	—	$20{:}14{:}42 \pm 1$ min
		Site	-43.41 ± 1.0	66.3 ± 2.0	1a03	$20{:}18{:}24 \pm 3$ min
R = 6	2b	Ring	-44.17 ± 0.1	46.8 ± 0.5		$05{:}41{:}18 \pm 1$ min
		Site	-44.50 ± 1.0	43.6 ± 2.0	1a03	$05{:}36{:}06 \pm 3$ min
S = 5	2c	Site	-43.91 ± 1.0	34.0 ± 2.0	1l06	$15{:}17{:}46 \pm 3$ min
W = 1	2c	Site	-44.29 ± 1.0	284.8 ± 2.0	1o03	$08{:}08{:}46 \pm 3$ min
		Plume	—	—	1n05	$08{:}06{:}16 \pm 0.1$ min

[†] Impact sites for F=16, P2=8b, T=4, U=3, and V=2 were not detected. J=13, M=10, and P1=8a are omitted because they faded from view (the letters I and O were not used).

[‡] Impact site size based on first view after impact. Class 1 = dark region > 10000 km, large ejecta, probably multiple rings; Class 2a = 4000 < r < 8000 km, medium ejecta, possibly multiple rings; Class 2b = medium but slightly smaller ejecta, probably single ring; Class 2c (S, W) = < 6000 km, classification based on ground-based data, impacts occurred on earlier sites; Class 3 = < 3000 km, no ejecta, no ring; Class 4 = not detected.

[¶] Method of determining site location and time; multiple images were used for positions based on ring measurements. Latitudes are planetocentric; longitudes are System III. "Frame" is the image used to measure latitude and longitude of impact site (the prefix "u2fi" has been deleted).

[‖] To determine impact times for "sites" and "rings," the time was obtained by finding the difference in longitude between HST measurements and predictions, and assuming a rotation period of 9.92492 hrs (1.654153 mins/deg); for "plumes," the time was that of first image showing brightening.

TABLE 2. Relative sizes, locations, and times of impacts

4.3. *Fresh site morphology*

The medium-to-large fresh impact sites showed a consistent morphology (Hammel *et al.* 1995): a large crescent-shaped ejecta offset toward the southeast from a streak, with fainter traces of dark material extending all around the impact site (Fig. 3). The largest sites showed a sharp circular ring and sometimes a second faint inner ring (concentric with the main ring but visible mainly northwest of the ring center). The ring center presumably marks the point of maximum energy release. When a ring was seen, the streak extended from the ring center to the southeast toward the ejecta.

The crescent shape of the ejecta suggests that the range of elevation angles is small, although this is still under discussion (K. Jessup, personal communication; also see chapter by Crawford). The material is not highly collimated in azimuth: the crescent extends

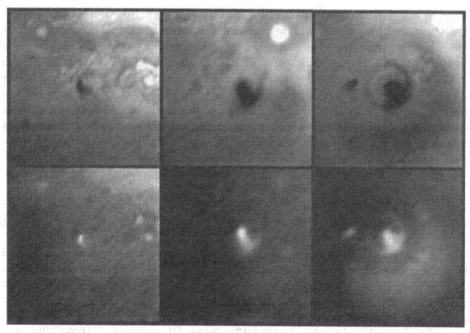

FIGURE 3. Fresh sites. These images show three fresh impact sites, ranging in size from small (D, left images, 2.03 hours after impact) to medium (E, middle images, 1.86 hours after impact) to large (G, right images, 1.66 hours after impact; the small site to the left of G is the D site about 10 hrs later). Upper D and E images are at 410 nm, the upper G image in at 555 nm, and all bottom row images are at 889 nm. The images of D (frames 0f01 and 0f05) and G (0p01 and 0p04) were made with the PC; the E images (0h01 and 0h05) were taken with WF3.

at least 180° around the impact site. As discussed in Hammel *et al.* (1995), the size of the G ejecta implies ejection velocities of more than 10 km/sec (see below), which is consistent with the observed plume heights.

Rays (linear features) were seen in the crescent of the G impact (no images were obtained soon enough after the K and L impacts to search for rays in their ejecta); the rays seemed to emanate from a point slightly (1000–2000 km) to the southeast of the ring center, near the end of the streak (Fig. 3). The apparent emanation point of the rays may refer to the center of collapse of the G plume, which is offset from the base of the plume. If the plume material was uniformly ejected, the rays may be indicative of inhomogeneities in the jovian atmosphere into which the plume collapsed. Alternatively, the linear features could instead be arcs of material that were thrown out by irregular events during the ejection process itself. The precise nature of these features is still unknown.

Smaller impact sites showed only the central streaks, *i.e.*, there was little obvious ejecta to the southeast (Fig. 3). However, the detection of IR "main events" for some of these smaller sites suggests that material was ejected (see chapter by Nicholson); presumably it was too optically thin to be detected in HST images.

4.4. *Site Color*

In the 889-nm methane absorption band, impact debris was brighter than the normal jovian clouds (Fig. 3), suggesting that it was at relatively high altitude, above most of the methane gas. At other wavelengths, including the 619-nm methane band, the impact debris appeared darker than the normal jovian clouds. The color of the ring material appeared to match that of the streak and ejecta blanket, suggesting similar composition, but that gave little insight into whether the material was cometary, jovian, or a mixture. One possibility may be that the material is primarily jovian gas with a small amount of cometary debris that has been heated to temperatures of 10,000 K or more (Martin *et al.* 1995), and has subsequently cooled, recombining to form complex hydrocarbons enhanced with residual non-organic cometary material. West *et al.* (1995) point out that the color of the material is consistent with organic material rich in sulfur and nitrogen. In addition, Noll *et al.* (1995) discuss chemistry of the features derived from HST spectra, which give more information than color alone. For more discussion about the nature of the dark material, see chapters by Lellouch, Moses, and West.

4.5. *Angle of Ejecta*

The azimuth of the ejecta's symmetry axis is determined by both the fragment's initial entry angle, and the subsequent flight of the ejected material through Jupiter's rotating atmosphere. Fragments entered the atmosphere at an elevation angle of 45°, with azimuth angle 16° counterclockwise from south in a reference frame rotating with Jupiter. Models of oblique impacts (Ahrens *et al.* 1994b, Boslough *et al.* 1994a) predicted initial ejection of material back along this same trajectory. Based on the G impact ejecta's horizontal extent of 13000 km and an assumed 45° exit angle for ballistic particles, the ejection velocity of that impact was estimated to be on the order of 17 km sec^{-1}. This would correspond to a total flight time of about 17 minutes (Hammel *et al.* 1995). Planetary rotation during that time would add an additional azimuth angle of about 7° (calculated as $\Omega t \sin\lambda$, where Ω is the angular velocity of the planet, t is the time, and λ is the planetocentric latitude of the impact site), giving a total azimuth angle from south of 23°. The observed azimuth (Fig. 3), 35° ± 5°, appears to indicate that the material was "in flight" for 45 minutes, although images show the plumes have almost fully collapsed within about half that time (Fig. 1; Table 1). One explanation is that ejected material slid or bounced along the top of the atmosphere following reentry; if friction with the underlying layer was low, the azimuthal rotation would have been the same as if the material followed a single ballistic trajectory for 45 minutes.

5. Waves

Images taken within a few hours of the larger impacts revealed expanding sharply defined "rings" that were almost certainly caused by atmospheric waves of some sort. The most dramatic (and hence most often shown) example was the multiple ring system created by the large G fragment (Fig. 3), although rings were also seen after the A, E, R, and Q1 impacts (Hammel *et al.* 1995).

These rings appear to expand with a speed independent of impact energy. Hammel *et al.* (1995) measured the radii of the circular features using several techniques. Figure 4 shows positions of the main rings seen after the A, E, G, Q1, and R impacts, as well as those of the inner rings from E and G.

For the main rings, a best-fit line yielded a slope (*i.e.*, propagation velocity) of 454 ± 20 m sec^{-1} and an initial radius of 586±125 km. A positive initial radius could arise from either nonlinear (faster) initial propagation or a finite source size. The inner rings' velocity was not well determined by the data, but was probably in the range 180–350 m sec^{-1}. Assuming inner and outer rings started at the same time and radius (*i.e.*, fixing the intercept at 586 km), Hammel *et al.* (1995) found a velocity of 189 ± 10 m sec^{-1} (formal error), roughly consistent with models suggesting a 3:1 ratio for the speeds of the two fastest modes of a linear tropospheric wave (Ingersoll and Kanamori 1995).

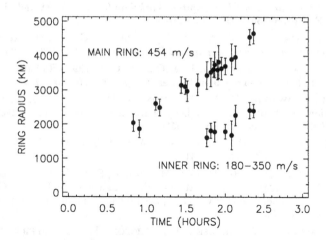

FIGURE 4. Wave propagation. Each datum is a measurement of the radius of the ring as a function of time from impact for images obtained after five impacts with different explosion energies (A, E, G, Q1, and R). The measurements for the main ring (upper cluster) fall on a straight line, indicating a speed independent of explosion energy. The slope of the inner ring seen in the E and G impacts (lower cluster) is less well constrained. See chapters by Ingersoll and Zahnle for discussions of the physical interpretation of the rings.

For a detailed discussion of the linear wave interpretation, see the chapter by Ingersoll. Alternatively, these rings may be a manifestation of nonlinear (breaking) waves, as discussed in the chapter by Zahnle.

6. Conclusion

The wealth of detail observed in HST images continues to yield insights into the phenomena resulting from the impact of the fragments of Shoemaker-Levy 9. The bulk of the burden of unraveling remaining puzzles now lies on modelers, who must not only create models that can reproduce the basic physics of a catastrophic explosion, but must also be able to do this successfully for a large number of events with subtly different initial conditions. Fortunately, the Hubble Space Telescope, along with many ground-based and space-based instruments, has provided a remarkable body of data to fuel these efforts.

HBH acknowledges the expertise and assistance of many people involved with the HST Comet Crash investigation, especially R. Beebe, A. Ingersoll, R. West, and G. Orton, along with J. Mills, A. Simon, T. Dowling, J. Harrington, R. Yelle, and others. Special acknowledgments go to HST Comet Science Team colleagues J. Clarke, H. Weaver, M. McGrath, and K. Noll for their help before, during, and after the crash. These observations were made with the NASA/ESA Hubble Space Telescope, with support provided through grant number GO-5624.08-93A from the Space Telescope Science Institute, which is operated by Association of Universities for Research in Astronomy, Inc., under NASA contract NAS5-26555.

REFERENCES

A'HEARN, M. F., MEIER, R., WELLNITZ, D., WOODNEY, L., MARTIN, R., SMITH, T. & VERVEER, A. 1995 Were any impact flashes seen in reflection from the satellites? In

Proceedings of the European SL-9/Jupiter Workshop (eds. R. West and H. Boehnhardt), pp. 113–118. European Southern Observatory.

AHRENS, T. J., TAKATA, T., O'KEEFE, J. D. & ORTON, G. S. 1994a Impact of Comet Shoemaker-Levy 9 on Jupiter. *Geophysical Research Letters* **21**, 1087–1090.

AHRENS, T. J., TAKATA, T., O'KEEFE, J. D. & ORTON, G. S. 1994b Radiative signatures from impact of comet Shoemaker-Levy 9 on Jupiter. *Geophysical Research Letters* **21**, 1551–1553.

BOSLOUGH, M. B., CRAWFORD, D. A., ROBINSON, A. C. & TRUCANO, T. G. 1994a Mass and penetration depth of Shoemaker-Levy 9 fragments from time-resolved photometry. *Geophysical Research Letters* **21**, 1555–1558.

BOSLOUGH, M. B., CRAWFORD, D. A., ROBINSON, A. C. & TRUCANO, T. G. 1994b Watching for fireballs on Jupiter. *EOS* **75**, 305–310.

CARLSON, R. W., WEISSMAN, P. R., SEGURA, M., HUI, J., SMYTHE, W. D., JOHNSON, T. V., BAINES, K. H., DROSSART, P., ENCRENAZ, T. & LEADER, F. E. 1995 Galileo infrared observations of the Shoemaker-Levy 9 G impact fireball: a preliminary report. *Geophysical Research Letters* **22**, 1557–1560.

CHAPMAN, C., MERLINE, W. J., KLAASEN, K., JOHNSON, T. V., HEFFERNAN, C., BELTON, M. J. S., INGERSOLL, A. P. & THE GALILEO IMAGING TEAM 1995 Preliminary results of Galileo direct imaging of S-L 9 impacts. *Geophysical Research Letters* **22**, 1561–1564.

HAMMEL, H. B., BEEBE, R. F., INGERSOLL, A. P., ORTON, G. S., MILLS, J. R., SIMON, A. A., CHODAS, P., CLARKE, J. T., DE JONG, E., DOWLING, T. E., HARRINGTON, J., HUBER, L. E., KARKOSCHKA, E., SANTORI, C. M., TOIGO, A., YEOMANS, D. & WEST, R. A. 1995 Hubble Space Telescope imaging of Jupiter: atmospheric phenomena created by the impact of comet Shoemaker-Levy 9. *Science* **267**, 1288–1296.

HERBST, T., HAMILTON, D. P., BOHNHARDT, H. & ORTIZ-MORENO, J. L. 1995 Near-infrared imaging and spectroscopy of the SL-9 impacts from Calar Alto. *Geophysical Review Letters*, submitted.

INGERSOLL, A. P. & KANAMORI, H. 1995 Waves from the collisions of comet Shoemaker-Levy 9 with Jupiter. *Nature* **374**, 706–708.

MAC LOW, M.-M. & ZAHNLE, K. 1994 The impact of Shoemaker-Levy 9 on Jupiter. *Bull. Amer. Astron. Soc.* **26**, 926.

MARTIN, T. Z., ORTON, G. S., TRAVIS, L. D., TAMPPARI, L. K. & CLAYPOOL, I. 1995 Observation of Shoemaker-Levy 9 impacts by the Galileo Photopolarimeter Radiometer. *Science* **268**, 1875–1879.

NOLL, K. S., McGRATH, M. A., TRAFTON, L. M., ATREYA, S. K., CALDWELL, J. J., WEAVER, H. A., YELLE, R. V., BARNET, C. & EDGINGTON, S. 1995 HST spectroscopic observations of Jupiter after the collision of comet Shoemaker-Levy 9. *Science* **267**, 1307–1313.

WEAVER, H. A., A'HEARN, M. F., ARPIGNY, C., BOICE, D. C., FELDMAN, P. D., LARSON, S. M., LAMY, P., LEVY, D. H., MARSDEN, B. G., MEECH, K. J., NOLL, K. S., SCOTTI, J. V., SEKANINA, Z., SHOEMAKER, C. S., SHOEMAKER, E. M., SMITH, T. E., STERN, S. A., STORRS, A. D., TRAUGER, J. T., YEOMANS, D. K. & ZELLNER, B. 1995 The Hubble Space Telescope (HST) observing campaign on comet Shoemaker-Levy 9. *Science* **267**, 1282–1288.

WEST, R. A., KARKOSCHKA, E., FRIEDSON, A. J., SEYMOUR, M., BAINES, K. H. & HAMMEL, H. B. 1995 Impact debris particles in Jupiter's stratosphere. *Science* **267**, 1296–1301.

ZAHNLE, K. & MAC LOW, M.-M. 1994 The collision of Jupiter and comet Shoemaker-Levy 9. *Icarus* **108**, 1–17.

Galileo observations of the impacts

By CLARK R. CHAPMAN[1]†

[1]Planetary Science Institute, 620 N. 6th Avenue, Tucson AZ 85705

Galileo observations in the UV, visible, and infrared uniquely characterize the luminous phenomena associated primarily with the early stages of the impacts of SL9 fragments—the bolide and fireball phases—because of the spacecraft's direct view of the impact sites. The single luminous events, typically 1 min in duration at near-IR wavelengths, are interpreted as initial bolide flashes in the stratosphere followed immediately by development of a fireball above the ammonia clouds, which subsequently rises, expands, and cools from $\sim 8000\,$K to $\sim 1000\,$K over the first minute. The brightnesses of the bolide phases were remarkably similar for disparate events, including L and N, which were among the biggest and smallest of the impacts as classified by Earth-based phenomena. Subsequent fireball brightnesses differ much more, suggesting that the similar-sized fragments were near the threshold for creating fireballs and large dark features on Jupiter's face. Both bolides and fireballs were much dimmer than had been predicted before the impacts, implying that impactor masses were small ($\sim$0.5 km diameter). *Galileo* data clarify the physical interpretation of the "first precursor," as observed from Earth: it probably represents a massive meteor storm accompanying the main fragment, peaking $\sim$10 s before the fragment penetrates to the tropopause; hints of behind-the-limb luminous phenomena, recorded from Earth immediately following the peak of the first precursor, may be due to reflection of the late bolide/early fireball stages from comet debris very high in Jupiter's atmosphere.

1. Introduction

The *Galileo* spacecraft was in a unique position to observe the crash of Comet Shoemaker-Levy 9 (SL9) into Jupiter. Well on its way past the asteroid belt en route to its successful injection into Jupiter orbit in December 1995, *Galileo* was only 1.6 AU from Jupiter and offset $\sim$40.5° to the line-of-sight from Earth. Thus instruments on *Galileo*'s scan platform had a *direct view* of the impact sites (at tropospheric level) at the time of the impacts.

It was remarkable, of course, that Earth-based observers saw prominent early phenomena, despite predictions that most of the events would occur behind Jupiter's limb. Observed phenomena included brightenings prior to bolide entry, bolide entry, early development of the fireball, the rise of the cooling fireball/plume into direct view from Earth and then into solar illumination, and the vastly brighter subsequent splashback (see review by Nicholson, this volume).

All of the early-stage phenomena seen from Earth, however, were observed in ways that are difficult to quantify, being plagued by foreshortened geometry (made all the more ambiguous by uncertain altitudes of phenomena), intervening jovian atmosphere, tangential solar illumination, etc. *Galileo*, in contrast, had an unimpeded, direct view of the initial impacts and early phases of fireball development against the dark side of Jupiter. Therefore, despite the modest apertures of *Galileo*'s instruments, which could not compete with the major groundbased and Earth-orbital observatories, *Galileo* provides a reliable baseline on the initial luminous phases of the comet fragment impacts. The later phases, such as the solar-illuminated plume and the splashback, were near or below the *Galileo* camera's detection threshold; eventually, analysis of *Galileo* infrared

† Now at: Southwest Research Institute, 1050 Walnut Street, Boulder CO 80302, USA

observations of the splashback may supplement the direct observations made from the Earth as Jupiter's rotation brought the impact regions into good viewing geometry.

The early-stage, luminous phenomena are crucial for characterizing and understanding the impacts and the development of later stages. Prior to striking Jupiter at 60 km s^{-1}, the comet fragments were—compared with the spatial scale of later-stage phenomena on Jupiter—very compact objects, or clusters of objects. They gave up most of their appreciable kinetic energies in seconds as they streaked down through Jupiter's atmosphere. Initially, only very small volumes of the atmosphere were heated and otherwise affected by the explosions. *Galileo* instruments helped to define the character of the conversion of the bolides into fireballs, and the early stages of fireball development, at times when there was particularly strong bias against Earth-based observations (because the fireballs were created deep in Jupiter's atmosphere, hundreds of kilometers below the projection of Jupiter's limb as seen from Earth).

The major *Galileo* contributions to SL9 observations came from three instruments— the Solid State Imaging system (SSI, the camera), the Photopolarimeter Radiometer (PPR), and the Near-Infrared Mapping Spectrometer (NIMS). In addition, the Ultraviolet Spectrometer (UVS) detected the initial seconds of at least the G impact. None of the instruments, except for the camera, had spatial resolution on Jupiter from this distance. So the PPR, NIMS, and UVS data on the impacts are essentially photometric observations of phenomena *added to* the signal from Jupiter itself. Despite this disadvantage, the signal-to-noise ratios were generally large for the major impacts, and the bolide and fireball phenomena were well characterized.

The SSI camera had significant spatial resolution on gibbous Jupiter, roughly equivalent to that of a modest groundbased telescope—59 pixels across Jupiter's diameter (2430 km pixel^{-1})—although direct images were acquired for only one impact, that of fragment W (Fig. 1). The other recorded events (K, N, and V) were observed in a drift-scanning mode that provided spatial resolution in one dimension and a time-history of the brightness in the other. (Tape-recorded data for V have not been returned to Earth, although there is a remote possibility that they may be read back before they are overwritten in spring 1996.)

Limitations on the spacecraft prohibited the other instruments on *Galileo*'s scan platform from recording data simultaneously with SSI. Thus they observed a complementary set of impact events. The PPR obtained good data on fragments G, H, L, and Q1; other events were too weak to be detected, were played back with infrequent sampling intervals, or (for those judged likely to be too faint to detect) were not played back at all. The relatively data-intensive NIMS obtained infrared spectra only for the G and R impacts (data for two other weaker events were not returned to Earth).

Because of uncertain predictions of what would be observed, most *Galileo* instruments used a variety of strategies, within operational constraints, to observe the different impacts—varying filters, sampling frequencies, etc. Only a tiny fraction of all recorded data could be returned due to limited antenna capability (some data still remain on the spacecraft tape recorder as of this writing, February 1996, but probably will soon be overwritten by data from the Jupiter system). Thanks to prompt reporting of observations by groundbased observers using the SL9 Exploder on the Internet during the week of the impacts, it was possible to identify the events and crucial intervals of time for which it would be most useful to return data during the available period, August 1994 through January 1995. The resulting data set is a gold mine of unique information on the impacts.

Only preliminary reductions of the *Galileo* data have been published so far. SSI data for K, N, and W are summarized by Chapman *et al.* (1995a), but more data have since been

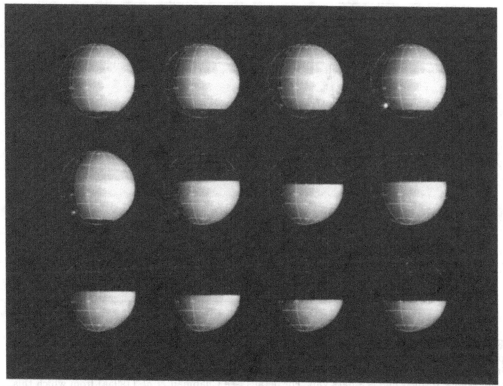

FIGURE 1. Galileo images of the W impact (0.56 μm). The first and last images were taken at 8:06:10 and 8:06:40 on 22 July 1994. The images are taken at intervals of 2.3 s except 7 s between images 5 and 6. The event is recorded on images 3 through 11. Missing portions of Jupiter are due to selective data return. A latitude/longitude grid is superimposed.

made available and refinements in the reductions are underway (Chapman and Merline, in preparation). The PPR observations are summarized by Martin *et al.* (1995), which includes refinements in interpretations developed at IAU Colloquium 156. A preliminary report on NIMS observations of the G event is by Carlson *et al.* (1995a); some of the NIMS R data are briefly discussed by Carlson *et al.* (1995b) as well as a light curve and spectrum for the beginning of the G splashback. Other preliminary reductions of NIMS data have been presented at various scientific meetings and in a recent book (Spencer, 1995). UVS observations of G are presented by Hord *et al.* (1995).

Intercomparison of the various *Galileo* data sets, among themselves and with ground-based data, has provided a clear picture of the early stages of the impacts, which I discuss in this chapter. The most important qualitative conclusions are that (1) the early luminous phenomena for the observed impacts had similar time histories, (2) they had unexpectedly similar luminosities despite wide variations in Earth-based observations of later phenomena, and (3) they were unexpectedly dim for the energies expected from the multi-km diameter fragment sizes that had been predicted prior to the events. Thus the *Galileo* observations support the conclusion of most researchers (see review by Mac Low, this volume) that the SL9 fragments were under 1 km in diameter. Furthermore, the NIMS observations of the G impact—a typical large event—characterize the history of the development of the fireball during the interval of time it was hidden from Earth-based observation. The G fireball started out very hot and small and was located in the upper

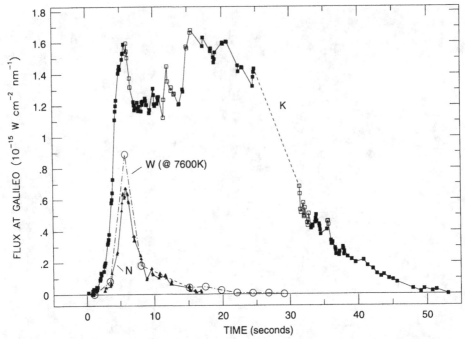

FIGURE 2. SSI light curves for K and N, observed at 0.89 μm. Data at 0.56 μm for W are plotted for comparison, assuming a 7600 K black body (likely to be approximately true near the initial spike, but not later in the event). Open symbols for K have larger uncertainties than solid points; dashed line is a data gap. For details, see Chapman *et al.* (1995a) from which this figure is adapted.

troposphere, just above the clouds; over the course of the next 70 s, it was observed to rise, expand, and cool, before fading from detectability by NIMS.

2. The *Galileo* observations

Galileo SSI data on impacts K, N, and W are shown in Fig. 2 (replotted from Chapman *et al.* 1995a, where more details may be found). Note that data for K and N were observed in the same mode (in the same 0.89 μm narrow-band methane filter, drift-scan mode); also note that the data for W (observed in a ∼2.3 s time-lapse direct imaging mode through the broader 0.56 μm green filter) have been rescaled to correct for the difference in wavelength, using the best available *Galileo* photometric calibrations (good to ±15%), assuming that a 7600 K black-body color temperature applies during the bolide phase.

The peak brightnesses are dim, compared with pre-impact predictions that they might rival the brightness of Jupiter itself. For instance, W reached only ∼1.5% of the brightness of Jupiter (at 0.56 μm). It is particularly noteworthy that the peak brightnesses of the events are similar, despite the very different characters of their later development as observed from Earth (K was one of the 3 biggest events, W intermediate, and N a minimal event). Impact N (class 3 in the Hammel *et al.* 1994, compilation) left only a tiny spot on Jupiter and was reported to be many hundreds of times fainter than K during the splashback phase. Nevertheless, N's peak brightness is down by only a factor of ∼2.5 from the brightness of K as observed by SSI.

Comparison of the light curves reveals that they represent two distinct physical phenomena. An abrupt initial rise of the light curve, mostly occurring in just 2 s, is seen in all three light curves. It then immediately decays, on a similar timescale, to less than 20% of peak brightness. This initial "spike" was interpreted by Chapman *et al.* (1995a) as the bolide flash in Jupiter's stratosphere. Despite the similarity of the bolide spikes for K, N, and W, the subsequent luminosity behavior was dramatically different. For K, there was a 45 s-long period of prominent luminosity, peaking about 10 s later than the bolide at a brightness slightly exceeding that of the bolide. This can be confidently interpreted (through analogy with NIMS data at the corresponding time during the similarly-shaped light curve for G—see below) as the hot, early stages of the rising fireball, which soon cooled to temperatures at which $0.89\,\mu$m radiation was minimal. (The possibility that a sudden onset of the second peak is enhanced by a second impact, occurring 10 s after the first, cannot be discounted.)

In sharp contrast, the N event shows little or no evidence of a fireball phase. Since a "main event" splashback was recorded from Earth with the characteristic 6 min delay after impact (Meadows *et al.* 1995; McGregor *et al.* 1996), we presume that there was a small fireball for N but that it was simply near or below the detection limit for SSI. The W light curve from SSI looks very similar to that of N, but recall that W was observed at a shorter wavelength. Presumably observations of W at $0.89\,\mu$m would have shown a fireball light curve intermediate between N and K. Luminosity from W was followed through nearly 30 s, a longer duration than for N due to the lower detection threshold of the direct imaging observing mode.

Galileo PPR data (at $0.945\,\mu$m) for impacts G, H, L, and Q1 are shown in Fig. 3 (from Martin *et al.* 1995). Note that data were sampled less frequently for G, which was being simultaneously observed by other instruments. Because PPR data represent excesses over the brightness of Jupiter, they are noisier than the SSI observations, especially for the weaker Q1 impact. Nevertheless, the well-defined curves for G, H, and L look very much like the SSI curve for K, showing a sharp initial onset, followed by a plateau, and then a decline into the noise. The durations ($\sim$30 s) are similar to what would have been observed for K above the same background.

Given that the PPR data were taken at a wavelength very similar to that of the SSI data for K, it is likely that the light curves have the same interpretation, despite the fact that none of the PPR curves clearly shows the dip seen between the two phases for K. The sharp onset evidently is the bolide phase, which merges into a more extended initial fireball phase. (The possibility that the light curves for G, H, K, and L represent only the fireball phase following an invisible "stealth" bolide, raised by Martin *et al.* [1995] and others, is discussed below.) Just as for the events observed by SSI, the brightnesses of the four PPR events are very similar, all within a factor of three of each other. From the Earth-based perspective, Q1 was an intermediate event like W while L was the strongest event of all.

Comparing Figs. 2 and 3 (and making corrections for the slightly different wavelengths), it appears that K was a little weaker than H while N and W were about half the brightness of Q1. The range in brightness between the brightest event (L) and the faintest (N) is less than a factor of 5.

The K event observed by SSI might be thought to be a little dim in comparison with the PPR events, based on expectations from pre-impact fragment magnitudes and from the relative strength of subsequent phenomena observed from Earth. Orton (1995) shows an alternative comparison of the SSI and PPR data sets that makes K as bright as L, but his scaling via an 18,000 K color temperature is based on a faulty understanding (for which I accept responsibility) of how the SSI data were calibrated. Apparently, there are *real* differences among the impactors resulting in scatter by a factor of two or more

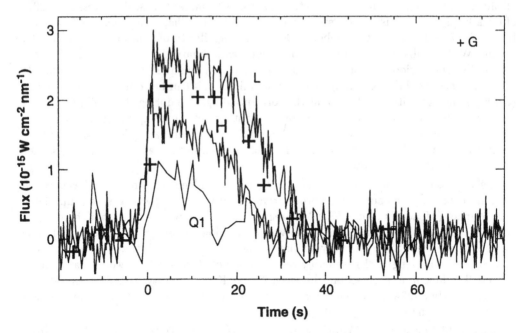

FIGURE 3. PPR light curves for L, H, Q1 (and G with lower sampling frequency), observed at 0.945 μm. (From Martin *et al.* 1995.)

away from any simple relationship between pre-impact inferred diameter and brightness of the bolide and early fireball phases.

PPR experimenters selected Q1 for observation in two wavelengths (in retrospect, we all wish that a brighter event had been observed this way). Data at the shorter wavelength (0.678 μm) show a sharp initial peak and more rapid decay than at the longer wavelength. A ratio of the two observations yields a color temperature of ∼18,000 K, but because of the noise for this weak event and use of preliminary instrumental calibrations, the uncertainty in the derived temperature is at least a factor of 2.

The best characterized impact is for the G fragment, observed by three instruments simultaneously. UVS (0.292 μm) detected G at a time simultaneous with the rapid initial rise of the PPR light curve. The UVS signal of $(4.3 \pm 0.9) \times 10^{-15}$ W cm^{-2} nm^{-1} compared with the PPR signal of $(1.1 \pm 0.2) \times 10^{-15}$ W cm^{-2} nm^{-1} allowed Hord *et al.* (1995) to calculate a color temperature of ∼7800 (±600) K for that moment (probably early bolide phase). The assumption of black-body radiation is supported, to a degree, by the roughly black-body spectrum obtained for G by NIMS 5 s later. Hord *et al.* interpret their data to apply to bolide passage through the atmosphere. During the sample of data from the three instruments taken 5 s later, G is roughly 2 times brighter as observed by PPR at 0.945 μm; but it had dropped below detectability to the UVS, implying a temperature <5500 K. The subsequent fireball phase was also not detected by UVS, implying low temperatures, consistent with those derived from NIMS data (see below).

The subsequent development of the G impact has been beautifully characterized by NIMS data, operating in its "fixed map" mode (Carlson *et al.* 1995a). Observations were taken at 17 wavelengths between 0.7 and 5.0 μm. Signal above background (reflected sunlight from Jupiter at short wavelengths, jovian thermal radiation at 5 μm) was detected at 10 wavelengths between 1.8 and 4.4 μm; several wavelengths fall within deep jovian

Time After Impact, s	Temp., K	Fireball Diam. km.	Altitude km.	Source
0	7800	6	?	UVS/PPR
5	~5500	7	20	NIMS/UVS
10	3000	25	30	NIMS
20	2200	40	40	NIMS
35	1500	80	80	NIMS

TABLE 1. G Fireball Development

stratospheric methane absorption bands, while about 6 of them approximately define the black-body continuum.

NIMS sampled the impact phenomena once every ~5.3 s, as did PPR and UVS. However, due to spatial scanning by NIMS, the impact site was not observed by NIMS at exactly the same times as the PPR observations. The first NIMS detection was ~ 5 s after the first detection by PPR and UVS; the previous NIMS sample, perhaps 1 s before the PPR/UVS detection, showed nothing above background. The NIMS data are noisy, due to sensitivity variations within the detector, but the light curve at 4.4 μm (in the continuum) peaks about 40 s after first detection by PPR/UVS, about the same time that the PPR light curve has faded into the background. The luminosity at 4.4 μm fades to background about 75 s after onset. (The NIMS light curve for R, returned with half the sampling frequency, has the same character as for G, but is down in intensity by about a factor of 4 during the fireball phase.)

Luminosity is again detected by NIMS 6 min after onset for both G and R, as the splashback—best characterized by groundbased observations—begins. The 6 min interval observed by NIMS strongly confirms the inference from analysis of groundbased data (see Nicholson's review, this volume) that the interval represents a real physical attribute of the plumes, common to all the larger impacts, rather than an aspect of geometric visibility from Earth. NIMS data-return for G and R cease ~9 min after event onset, or 3 min into the splashback phase while the intensity is still increasing. (The intensity of the splashback phase for R is down by about a factor of 2 from that for G. More R data may be played back from the spacecraft during spring 1996.)

Besides contributing photometric observations at different wavelengths from those of PPR and UVS, NIMS provides two additional types of constraints to understanding the development of the G fireball. First, its series of simultaneous measurements at multiple continuum wavelengths *defines* the black-body temperature of the source; the multiple wavelengths demonstrate a roughly black-body emission, rather than requiring an *assumption* of black-body emission. Secondly, from the strength of the methane absorption bands (and several lesser absorbers that have been approximately modelled), Carlson *et al.* (1995a) can specify the effective altitude in Jupiter's atmosphere at which the emission takes place. As expected, the methane absorptions fade as fireball development progresses, indicating that the effective radiating surface of the fireball is rising in Jupiter's stratosphere.

Various preliminary interpretations of NIMS data (Carlson *et al.* 1995a; Spencer 1995) provide a self-consistent portrayal of an expanding, rising, cooling fireball during the first half-minute of NIMS observations of G. Table 1 summarizes (with approximate numbers) the development, beginning with UVS/PPR data on the bolide phase (for which there is no altitude estimate) and continuing with NIMS characterization of the fireball. The altitude of the top of the fireball is referenced to the top of the ammonia clouds, about 200 mb pressure, about 20 km below the tropopause.

The NIMS cartoon of the event is of the explosion occurring in the troposphere, just above cloudtop level, with the center of the resulting spherical fireball rising 1 km s^{-1} and the upper surface rising $\sim 2 \text{ km s}^{-1}$, so that the bottom of the fireball remains near the cloud layer. Carlson *et al.* believe the NIMS data support an accelerating rise of the fireball, which would be the initial acceleration toward the $>10 \text{ km s}^{-1}$ velocities achieved by the plume. The cartoon is oversimplified, and it is not known whether the NIMS data pertain to the totality of the fireball or just to the part above the clouds (with much of the potentially luminous phenomena hidden beneath the opaque clouds). In all probability, however, it is this upper, most rapidly rising, part of the developing fireball that predominantly became the plume, whose expansion and later fallback was so well imaged by Hubble Space Telescope (Hammel *et al.* 1994) and whose thermal evolution was so spectacularly recorded by infrared telescopes on Earth.

Before the fireball phase faded below NIMS's detectability threshold more than a minute after impact, the temperature apparently fell to 1000 K; at about this time, the cooling plume became visible to Earth-based observers above the planet's limb and was detected (as the second precursor) at $2.3 \mu\text{m}$; an HST image of the G fireball, apparently in emission, before rising into sunlight, was taken shortly afterwards.

3. Discussion

The *Galileo* data have played a fundamental role in reconciling what at first appeared to be very confusing Earth-based observations of the early stages of the impacts. While there was a chance that a few impacts would fall in occasional gaps in *Galileo*'s observational records, it appears that the major luminous event (at visible and near-IR wavelengths) was recorded for every impact for which *Galileo* data were returned to Earth: G, H, K, L, N, Q1, R, and W. In contrast to some pre-impact predictions, the impact events (as distinct from much later splashbacks) were singular rather than being characterized by separate flashes from the bolide and fireball phases. The preferred interpretation of Chapman *et al.* (1995a) was that the phases were combined, as the upper parts of the trail heated by the passing bolide exploded to become the upper portion of the fireball (as predicted by Boslough *et al.* 1994; see also Boslough *et al.* 1995).

However, prior to IAU Colloquium 156, a view was advanced at the ESO Workshop in Garching (Orton, 1995, and Drossart *et al.* 1995; also see Martin *et al.* 1995) that the *Galileo* data, including the initial spikes observed by SSI, correspond only to the fireball phase and that the bolide phase was so weak as to be invisible. At IAU Colloquium 156, Chapman *et al.* (1995b) argued against the "stealth bolide" model. As asserted above, the similar initial spikes in the data for such disparate impacts as K and N have timescales (a few seconds) appropriate for bolide entry and are unlikely to pertain to rapid fireball development. Moreover, the SSI data are quite sensitive. To have escaped detection during the observing periods, a stealth bolide would have to be $<1\%$ the maximum brightness of K and $<0.5\%$ the maximum brightness of the fainter W event (the detection threshold is better for W). Although terrestrial meteors have spectra dominated by line emission, there is expectation of considerable continuum radiation from these large jovian bolides (Chevalier & Sarazin, 1994; see discussion by Mac Low, this volume). It seems inconceivable that the bolides for K and W would not have been detected by SSI.

The Garching discussions clearly had been confused by a variety of Earth-based "precursor" detections of events from seconds to minutes prior to the *Galileo* flashes. Discussions at IAU Colloquium 156, assisted by presentation of more (and corrected) ground-

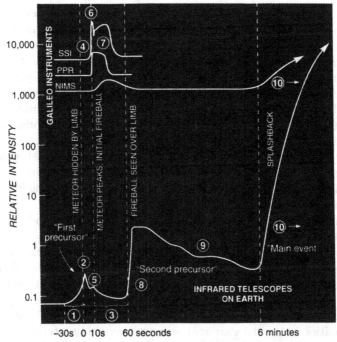

FIGURE 4. Schematic representation of a typical groundbased 2.3 μm SL9 light curve. Parts 1–3 are the first precursor. Shoulder (5) may be an indirect reflection of the sharp onset (4), initial bolide spike (6), and perhaps fireball phase (7) of the luminous phenomena observed directly by *Galileo* instruments. The second precursor (8, 9) and main splashback event (10) are also indicated; the latter was also detected by NIMS. Illustration developed during discussions at IAU Colloquium 156.

based data, clarified the situation, as shown in Fig. 4 (see Beatty and Levy, 1995). I now briefly summarize the chief elements of the synthesis. For more thorough discussion, especially for later stages of the light curves, see the review by Nicholson (this volume).

The initial rise of 2.3 μm groundbased light curves during the minute-or-so prior to the peak of the first precursor (#1 in Fig. 4, termed "leader emission" by Nicholson, this volume), seen for some events, varies considerably in different data sets, undoubtedly due to different sensitivities and time sampling intervals in addition to differences between the fragments and observing geometries. The most extreme reliable examples (*e.g.*, emission beginning ∼3.5 min in advance of the commencement of the SSI K event, seen by McGregor *et al.* 1995) cannot possibly be attributed to the comet fragment in the upper atmosphere of Jupiter since the fragment was still >10,000 km from Jupiter at the time. If a very modest fraction of a fragment's mass were distributed in coma dust, it would produce a meteor storm as the dust grains reached nanobar pressures in Jupiter's upper atmosphere, in direct view from Earth (especially for later impacts). The time history of the leader emission presumably defines (mostly) the much-elongated trail of dust associated with the fragment, including the changing cross-section and volume density of the dust cloud along its length. Coma dust lagging behind the fragment might also contribute to the declining portion of the so-called first precursor spike.

The peak of the first precursor (#2, Fig. 4) is identified as the time that the fragment passes behind the limb, as seen from Earth. The first precursor may simply be the peak of the meteor storm, as the densest part of the coma (nearest the fragment) strikes Jupiter's uppermost atmosphere. The meteor storm luminosity may be augmented by initial luminosity of the fragment itself, during the few seconds before it passes behind the limb (see calculation by Mac Low, this volume). The smooth rise of the leader to the peak of the first precursor, with no jump in the few seconds before the peak (shown best in high time-resolution data, such as observed for L by Hamilton *et al.* 1995) argues that the first precursor is due to the meteor storm alone, with no required contribution from the fragment itself. Note that the observed brightnesses of the first precursors indicate that the phenomena would be below *Galileo*'s detection limit, for reasonable color temperatures. For impact K, SSI detected the event just 5 s prior to the peak of the initial spike.

The 10 s duration between first precursor and the peak of the initial spike of luminous phenomena observed by *Galileo* corresponds to the time-of-flight of a fragment from passage behind the limb down into Jupiter's stratosphere (cf. Sekanina, 1995). This interpretation is compatible with all first precursor data of which I am aware, taking into account different sampling frequencies, timing uncertainties, and the shorter time-of-flights for later impacts in the series.

Light curves of several first precursors suggest non-uniform decrease in brightness following the peaks (#3, Fig. 4), as well as a shallower rate of decrease than the smooth initial rises. The best defined light curve for a first precursor is that for L by Hamilton *et al.* (1995). It shows a suspension in the drop of the first precursor, a plateau, lasting for at least 8 s. The 2.3 μm luminosity observed during this period, particularly if the rapid discontinuities in the data are real, is most reasonably interpreted as some kind of indirect detection of the rapidly changing luminous events taking place far below Jupiter's limb during the last phases of bolide entry and early fireball development (the "explosion"). The end of the plateau (#5, Fig. 4) may correspond to the brightest phase of the luminous phenomena in Jupiter's upper troposphere—the bolide and its explosive conversion into the initial fireball. Back-of-the-envelope calculations suggest that a plausible amount of comet dust deposited at high elevations during the meteor storm could conceivably reflect efficiently enough to account for the observations, but more analysis is needed.

As many researchers have commented, the second precursor observed from the Earth (#8, Fig. 4), occurring about a minute after impact (earlier for the later impacts), is readily interpreted as the fireball/plume rising into view from Earth above Jupiter's limb (and, for the shortest wavelength observations, by subsequent rise into sunlight). *Galileo* SSI detection of K at 0.89 μm fades out just as groundbased observers (*e.g.*, Watanabe *et al.* 1995) detected the onset of the second precursor. Preliminary studies of SSI imaging of K fail to show the sun-illuminated plume; calculations (Chapman and Merline, in preparation) suggest that it would be near the detection threshold. Six minutes after impact, re-impact of high velocity plume material generates the beginning of the main splashback thermal event (#10, Fig. 4).

4. Conclusions

The *Galileo* data define the early luminous phases of SL9 impacts, beginning with bolide penetration into Jupiter's atmosphere (but a few seconds after the groundbased first precursors, which are apparently due to a meteor storm at nanobar pressure levels) continuing through the period when a rising, expanding, cooling fireball was beginning to be directly visible from Earth. For the brightest events, the bolide and early (first 20 s) fireball luminosities were equal, indicating a continuity of the physical evolution of the hot bolide train into the upper part of the fireball. Perhaps a much brighter explosion occurred at depth in Jupiter's atmosphere, but its luminosity would have been hidden below the clouds. The portion of the superheated column of atmosphere above the clouds evidently developed into the upper, visible part of the fireball that evolved into the plume, later observed by HST and groundbased observers.

Smaller events (including those, like N, that evidently developed minor plumes/splashbacks and small dark spots) had bolides only slightly fainter than the largest impacts, but the fireball luminosities were much fainter. The "duds" or "fizzles" may have been clouds of dust without major fragments, or the fragments may simply have been below the threshold size required to produce a plume and splashback; from the Earth, fragment V (Nicholson, this volume) exhibits a classic first precursor light curve, but the *Galileo* SSI data—which might show a bolide spike, if a main fragment accompanied the dust—have not been returned to Earth and will soon be overwritten.

It is noteworthy that a wide variety of impacts, ranging from the mightiest (L) to one of the weakest non-fizzles (N) all have peak bolide-phase brightnesses within a factor of 5. Possibly this could imply that widely different size impactors give rise to similar bolide flashes (meteor physics for such large objects is not well understood). More likely, the sizes of most of the SL9 fragments ranged within roughly a factor of 2, despite larger differences inferred from pre-impact estimates and, especially, from extraordinary variations in the late-stage phenomena observed from Earth. It might, indeed, be regarded as fortuitous that the fragment sizes would thus seem to be near the threshold size capable of generating large dark splotches in Jupiter's upper atmosphere.

Equally noteworthy was the faintness of the luminosity from both the bolide and fireball phases. Mac Low (this volume) interprets both as evidence for small (~0.5 km diameter) fragment sizes, confirming the robust calculations from the physics of tidal break-up (cf. Asphaug and Benz, 1996). The apparent lack of SL9-like dark spots on Jupiter's face during the past century of nearly continuous observations by amateur and professional astronomers thus may be a constraint on the number of comets > 0.5 km diameter encountering Jupiter during the present epoch.

The *Galileo* results reported here are based on only a fraction of what will eventually be learned from complete reduction and interpretation of the data bases. With the *Galileo* orbiter now observing the Galilean satellites, the *Galileo* Science Team is necessarily deeply engaged in the intensive multiple-encounter activities. But eventually additional SL9 data will be analyzed, calibrations improved, and further insights about this remarkable astronomical event will emerge.

This research was supported by the *Galileo* Project and by grants from the National Science Foundation and NASA. I thank *Galileo* Project personnel for the extra work that made these observations possible. I thank Torrence Johnson and especially W. Merline for discussions.

REFERENCES

ASPHAUG, E. & BENZ, W. 1996 The tidal disruption of strengthless bodies: Lessons from comet Shoemaker-Levy 9. *Icarus*, in press.

BEATTY, J. K. & LEVY, D. H. 1995 Diary of a fireball. *Sky & Telescope* Oct. 1995, 22–23.

BOSLOUGH, M., CRAWFORD, D. A., ROBINSON, A. C., & TRUCANO, T. G. 1994 Watching for fireballs on Jupiter. *EOS* **75**, 305, 307 & 310.

BOSLOUGH, M. B., CRAWFORD, D. A., TRUCANO, T. G., & ROBINSON, A. C. 1995 Numerical modeling of Shoemaker-Levy 9 impacts as a framework for interpreting observations. *Geophys. Res. Lett.* **22**, 1821–1824.

CARLSON, R. W., WEISSMAN, P. R., SEGURA, M., HUI, J., SMYTHE, W. D., JOHNSON, T. V., BAINES, K. H., DROSSART, P., ENCRENAZ, TH. & LEADER, F. E. 1995a *Galileo* infrared observations of the Shoemaker-Levy 9 G impact fireball: A preliminary report. *Geophys. Res. Lett.* **22**, 1557–1560.

CARLSON, R. W., WEISSMAN, P. R., HUI, J., SEGURA, M., SMYTHE, W. D., BAINES, K. H., JOHNSON, T. V., DROSSART, P., ENCRENAZ, TH., LEADER, F. & MEHLMAN, R. 1995b Some timing and spectral aspects of the G and R collision events as observed by the *Galileo* Near Infrared Mapping Spectrometer. In *European SL-9/Jupiter Workshop Proceedings* (eds. R. West & H. Böhnhardt), pp. 69–73. European Southern Observatory.

CHAPMAN, C. R., MERLINE, W. J., KLAASEN, K., JOHNSON, T. V., HEFFERNAN, C., BELTON, M. J. S., INGERSOLL, A. P., & *Galileo* IMAGING TEAM 1995a Preliminary results of *Galileo* direct imaging of S-L 9 impacts. *Geophys. Res. Lett.* **22**, 1561–1564.

CHAPMAN, C. R., MERLINE, W. J., KLAASEN, K., JOHNSON, T. V., HEFFERNAN, C., BELTON, M. J. S., INGERSOLL, A. P., & *Galileo* IMAGING TEAM 1995b *Galileo* direct imaging of impacts K, N, and W. Abstracts for IAU Colloq. 156, 17.

CHEVALIER, R. A. & SARAZIN, C. L. 1994 Explosions of infalling comets in Jupiter's atmosphere. *Astrophys. J.* **429**, 863–875.

DROSSART, P., ENCRENAZ, TH., COLAS, F. & LAGAGE, P.-O. 1995 Interpretation of multi-wavelength infrared observations of selected impacts: What did we see? In *European SL-9/Jupiter Workshop Proceedings* (eds. R. West & H. Böhnhardt), pp. 417–421. European Southern Observatory.

HAMILTON, D. P., HERBST, T. M., RICHICHI, A., BÖHNHARDT & ORTIZ, J. L. 1995 Calar Alto observations of Shoemaker-Levy 9: Characteristics of the H and L impacts. *Geophys. Res. Lett.* **22**, 2417–2420.

HAMMEL, H. B. *et al.* 1994 HST imaging of atmospheric phenomena created by the impact of comet Shoemaker-Levy 9. *Science* **267**, 1288–1296.

HORD, C. W. *et al.* 1995 Direct observations of the comet Shoemaker-Levy 9 fragment G impact by *Galileo* UVS. *Geophys. Res. Lett.* **22**, 1565–1568.

MARTIN, T. Z., ORTON, G. S., TRAVIS, L. D., TAMPPARI, L. K. & CLAYPOOL, I. 1995 Observation of Shoemaker-Levy impacts by the *Galileo* Photopolarimeter Radiometer. *Science* **268**, 1875–1879.

MCGREGOR, P. J., NICHOLSON, P. D. & ALLEN, M. G. 1996 CASPIR observations of the collision of comet Shoemaker-Levy 9 with Jupiter. *Icarus*, in press.

MEADOWS, V., CRISP, D., ORTON, G., BROOKE, T. & SPENCER, J. 1995 AAT IRIS observations of the SL-9 impacts and initial fireball evolution. In *European SL-9/Jupiter Workshop Proceedings*, (eds. R. West & H. Böhnhardt), pp. 129–134. European Southern Observatory.

ORTON, G. S. 1995 Comparison of *Galileo* SL-9 impact observations. In *European SL-9/Jupiter Workshop Proceedings*, (eds. R. West & H. Böhnhardt), pp. 75–80. European Southern Observatory.

SEKANINA, Z. 1995 Collision of comet Shoemaker-Levy 9 with Jupiter: Impact study of two fragments from timing of precursor events. Abstracts for IAU Colloq. 156, 99.

SPENCER, J. R. 1995 A typical impact. In *The Great Comet Crash* (eds. J. R. Spencer & J. Mitton), pp. 74–86. Cambridge Univ. Press.

WATANABE, J. *et al.* 1995 Near-IR observation of cometary impacts to Jupiter: Brightness variation of the impact plume of fragment K. *Publ. Astron. Soc. Japan* **47**, L21–L24.

Models of fragment penetration and fireball evolution

By DAVID A. CRAWFORD

Experimental Impact Physics Department, Sandia National Laboratories, MS 0821,
Albuquerque, NM 87185, USA

A new analytical model that is calibrated against numerical simulations performed with the CTH shock physics code provides a useful description of the entry of Periodic Comet Shoemaker-Levy 9 into the Jovian atmosphere. Mass loss due to radiative heating of fragments larger than 100 m in diameter is insignificant because of energy conservation during the ablative process. Nevertheless, radiative ablation is a major contributor to atmospheric energy deposition at high altitude and plays an important role in early-time fireball evolution. The analytical model provides the initial conditions from which fireball and plume evolution can be calculated using CTH. The results from these simulations suggest that if the tops of the plumes originated from a specific level of the Jovian atmosphere then maximum plume heights are independent of fragment size provided the fragments penetrated at least 30 km below this level. If the tops of the plumes originated from the visible cloud tops, then fragment masses greater than 4×10^{12} g, corresponding to 200 m diameter fully dense water ice, are required to explain the observations. If the plumes originated from the NH$_4$SH layer then masses greater than 3×10^{13} g (400 m water ice) are required. The lateral extent and mass of the observable plume are functions of fragment size and contribute to the lateral extent and albedo of the debris patterns after re-impact with the atmosphere. The apparent gap between the central disturbance of the impact site and the inner front of the crescent-shaped ejecta may reflect the fragment's depth of penetration below the source layer of the visible ejecta.

1. Introduction

Models of Comet Shoemaker-Levy 9 fragments entering the Jovian atmosphere and subsequent fireball and plume evolution are strongly constrained by the wealth of consequences observed during the impact week. Some consequences were predicted. For example, three-dimensional, bilaterally-symmetric computational simulations performed prior to the impacts demonstrated that each event would produce a debris-laden fireball/plume that would expand explosively up the entry channel produced by the impacting fragment (Boslough et al. 1994a,b; Crawford et al. 1994, 1995b; Takata et al. 1994; Shoemaker et al. 1995). The simulations were consistent with the bilateral symmetry of the observed impact sites, the concentration of dark ejecta materials in the direction from which the fragments came and the plumes themselves, which exhibited a lateral offset during their evolution (Figure 1). Other consequences, such as the strength of the 'main' infrared event that correlated with the splashback of plume materials onto the atmosphere, were unanticipated, but consistent with pre-impact theoretical models (Figure 2). Still other consequences were entirely unexpected. For example, plumes observed by the Hubble Space Telescope all had approximately the same maximum altitude yet the dark ejecta they left behind varied considerably in albedo and lateral extent (Hammel et al. 1995). Several proposals have been made to explain this phenomenon (see the chapter by Zahnle and later in this chapter) but the matter is not yet settled.

Pre-impact and post-impact theoretical modeling can help provide a framework for interpreting the observations. Figure 3 depicts an idealized representation of the sequence of events inferred from Earth-based photometry data and is based, in part, on computational simulations. The figure is not intended to show the exact geometry, but

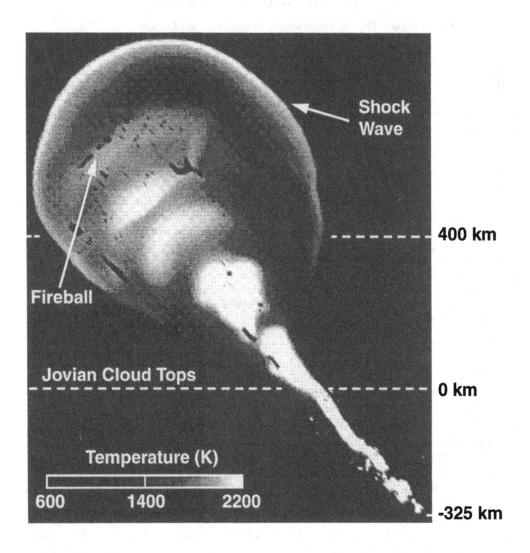

FIGURE 1. Pre-impact three-dimensional simulation of the fireball produced from a 3 km ice fragment entering the Jovian atmosphere at 60 km s^{-1}. This is a two-dimensional slice along the bilateral symmetry plane made 69 seconds after the passage of the fragment through the 100 km altitude referenced to the 1-bar pressure level of the atmosphere. Gray scale represents temperature. The simulation was performed using a parallel version of the CTH shock-physics code on the 1840-processor Intel Paragon at Sandia National Laboratories.

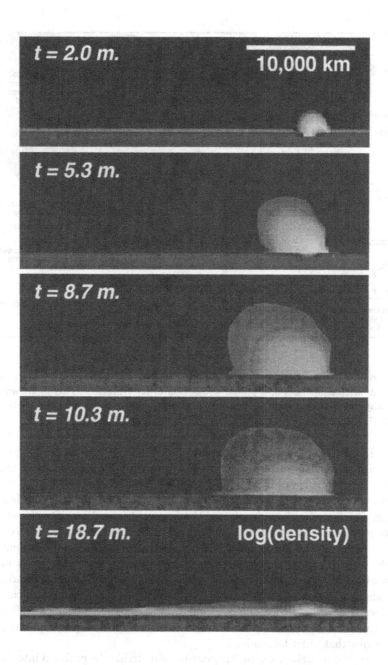

FIGURE 2. Post-impact computational simulation of 3-D fireball and plume evolution after the impact of a 3-km diameter fragment. Shading indicates log(density) with a visibility cutoff at 10^{-12} g cm^3. Here, the initial conditions and driving physics are the same as the pre-impact simulation but the spatial resolution has been reduced to allow the simulation to span a longer event time. The inelastic collision of the plume splashback (most dramatically illustrated at the end of the simulation) yields enough thermal heating over a broad area to account for the strength of the 'main' infrared event (from Boslough *et al.* 1995).

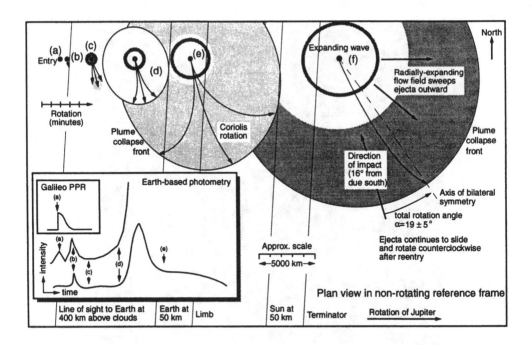

FIGURE 3. Plan view (map projection) of idealized impact site from a stationary (non-rotating vantage point, with snapshots of a planar projection of its evolution. This figure schematically represents features that were seen after several of the larger impacts and depicts an interpretation based, in part, on computational simulations (from Boslough *et al.* 1995).

is a composite of features observed from various events. It depicts an overhead view of an impact site from a non-rotating vantage point with map projections of the evolving impact sites at various times after impact.

Of course within the context of a qualitative theoretical framework, detailed questions still remain. For instance, how deeply did the fragments penetrate into the atmosphere? There is observational evidence that the largest fragments penetrated the NH_4SH layer but did not penetrate as far as the Jovian water table (Zahnle, this volume). This is a relatively narrow range of possible penetration depths but, what mass and/or size fragments will reach these depths? This is where our theoretical understanding is most vague. Aside from our poor understanding of the composition and mechanical properties of the comet fragments (after all, it is our hope that the impact of SL9 on Jupiter will help provide some of this information), the answer to this question depends on the approximations that must be made.

Most theoretical models of the SL9 impact usually divide the problem into at least two, perhaps more, phases. The first phase describes the first 10–30 seconds of the impact as a fragment penetrates the atmosphere. The second describes the early development of the fireball up to several minutes after impact. Sometimes, a third phase of the problem, the plume splashback, is analyzed. Each part of the problem presents unique challenges that require approximation in some form or another. Some modelers combine different parts of the problem or treat the same part in different ways to perform 'reality checks'.

The approximations used in modeling the penetration phase are discussed later in this chapter and in the chapter by MacLow. The plume splashback phase is discussed in the chapter by Zahnle. Perhaps of special note are the approximations required to model the early development of the fireball. Because the fragments penetrated the atmosphere at an angle of 45 degrees, the problem is intrinsically three dimensional. While a limited number of 3D calculations have been performed on supercomputers (*e.g.*, Figure 1 and the simulations of Takata *et al.* 1994), it is simply more practical to explore model dependences using two-dimensional approximations. There are two possible endmember 2D approximations that can be used. One is to approximate the fireball as resulting from a vertical impact and exploit cylindrical symmetry about the vertical axis (*e.g.*, Zahnle & MacLow, 1994). Another, which we will exploit later in this chapter, is to use cylindrical symmetry with respect to the entry trajectory (inclined at 45 degrees). Neither approach is completely correct. The real answer lies somewhere in between.

In the remainder of this chapter, an analytical model of fragment penetration into the Jovian atmosphere is derived and calibrated with the knowledge gained from earlier work (Crawford *et al.* 1995b). A series of detailed two-dimensional fireball calculations are then performed using, as input, the energy deposition curves derived from the analytical model. Plumes resulting from fragments with diameters of 125, 250, 375, 500, 750, 1000, 1500 and 2000 meters are presented. The plumes are ballistically extrapolated, their morphology at maximum altitude is presented, and the resulting ejecta patterns are shown.

2. Entry models

Understanding the mechanisms of energy loss during meteoroid traversal of planetary atmospheres is crucial for understanding the development of fireballs and plumes that were observed during the impact of Comet Shoemaker-Levy 9. Early-time evolution of the fireball is most dependent on the nature of energy deposition at relatively high altitudes (Boslough *et al.* 1995; Crawford *et al.* 1995a), yet final penetration depth is most dependent on energy deposition at relatively low altitudes (Crawford *et al.* 1994, 1995; MacLow & Zahnle, 1994; Takata *et al.* 1994; Zahnle & MacLow, 1994). Therefore, consideration of energy deposition at all altitudes is an important component for understanding fireball and plume development.

Analytical models of the deceleration, mass loss, hydrodynamic deformation and mechanical breakup of meteoroids during passage through planetary atmospheres have been proposed and refined by many researchers (Ivanov & Yu, 1988; Zahnle, 1992; Hills & Goda, 1993; Ceplecha *et al.* 1993; Sekanina, 1994). Ceplecha *et al.* (1993) describe the longest standing model, that of classical ablation. Ivanov & Yu (1988) describe some of the hydrodynamic deformations experienced by meteoroids in atmospheric flight and Zahnle (1992), Hills & Goda (1993) and Sekanina (1994) attempt to link the classical ablation model with hydrodynamic deformation and fragmentation models.

Here, we describe classical ablation from the point of view of large impactors such as Shoemaker-Levy 9. A modified ablation model, as suggested by Chevalier & Sarazin (1994) and Field & Ferrara (1995), satisfies conservation of energy during the ablative process and adds a further refinement to accomodate observations of terrestrial meteors. A new analytic hydrodynamic deformation model, based on an observation of O'Keefe *et al.* (1994), compares favorably with numerical simulations performed using the CTH shock-physics code (Crawford *et al.* 1994, 1995a,b). Coupling the new hydrodynamic model to the modified ablation model yields energy deposition curves resulting from fragment penetration through 1200 km of Jovian atmosphere.

2.1. *Classical ablation*

The classical ablation model can be expressed as a series of differential equations representing the deceleration and mass loss of a meteoroid entering a planetary atmosphere (after Zahnle, 1992):

$$m\frac{dv}{dt} = -\frac{C_d}{2}\pi r^2 \rho v^2 \tag{2.1}$$

$$Q\frac{dm}{dt} = -\frac{C_h}{2}\pi r^2 \rho v^3 \tag{2.2}$$

where m, v and r are the meteoroid's mass, velocity and radius, respectively, ρ is the atmospheric density and C_d and C_h are drag and heat transfer coefficients, respectively. Equation (2.1) is simply Newton's law of motion, whereas (2.2) represents mass loss due to radiative or frictional heating by the atmosphere. The ablated mass is assumed to rapidly decelerate to zero velocity relative to the atmosphere. The variable Q, usually equated to the heat of fusion or vaporization, can be combined with C_d and C_h to form an ablation coefficient $\sigma = \frac{1}{2}C_h/(C_d Q)$ (Ceplecha *et al.* 1993). Often, an additional equation is prescribed describing the deformation and/or fragmentation of the meteoroid through change of the radius, $r(t)$ (Zahnle, 1992; Chyba *et al.* 1993; Hills & Goda, 1993; Sekanina, 1993; Zahnle & MacLow 1994).

Strictly speaking, equations (2.1) and (2.2) do not conserve energy when accepted values of $C_d = 1$, $C_h = 0.01$–0.6 and $Q = 2.5 \times 10^{10}$ erg g^{-1} (Biberman *et al.* 1980; Zahnle & MacLow, 1994) or $\sigma = 0.01$–0.2 s^2 km^{-2} (Ceplecha *et al.* 1993; Sekanina, 1993) are used. This is evident when one considers that the power delivered to the system by the atmosphere flowing by the fragment at 60 km s^{-1}, $\frac{1}{2}C_d \pi r^2 \rho v^3 = 10^{20}\rho \pi r^2$ erg s^{-1}, is dwarfed by the power represented by loss of 'ablated mass', $\frac{1}{2}\frac{dm}{dt}v^2 = \frac{1}{2}C_h \pi r^2 \rho v^5/Q = 10^{23}\rho \pi r^2$ erg s^{-1}.

Because the ablated mass is not truly lost from the system until it has been decelerated to a small fraction of the impact velocity (Bronshten, 1983), the correct expression for Q is more appropriately represented by:

$$Q = Q_0 + \frac{1}{2}v^2 \tag{2.3}$$

where Q_0 is equated with the heat of fusion or vaporization in the usual sense and $\frac{1}{2}v^2$ is the energy required to accelerate a unit of ablated mass to the velocity v. The second term on the right side of (2.3) will dominate when $v^2 \gg 2Q_0$. For most materials of interest this will occur when $v > 5$ km s^{-1}.

During the impact of Comet Shoemaker-Levy 9 ($v = 60$ km s^{-1}), the second term on the right side of (2.3) was approximately 700 times greater than the first and (2.3) can be closely approximated by

$$Q \approx \frac{1}{2}v^2. \tag{2.4}$$

This leads to a modified ablation equation (2.2) of the form,

$$\frac{dm}{dt} = -C_h \pi r^2 \rho v, \tag{2.5}$$

to be used in conjunction with equation (2.1) and an equation for $r(t)$ to be described later. Field & Ferrara (1995), using a more sophisticated approach, derive a similar expression and note that C_h has a theoretical upper bound of 1.0.

Equation (2.5) predicts that a 100-m solid ice fragment entering Jupiter's atmosphere will lose less than 1% of its mass before reaching the 1-bar level. Because this redefinition of Q implies about 700 times less mass loss, it must be reconciled with the observational

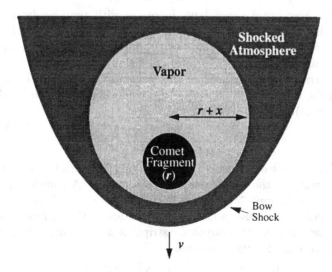

FIGURE 4. Schematic diagram of the analytical entry model that explicitly models the development and behavior of an ablative vapor layer. The increased interaction cross section ($S = \pi(r + x)^2$) accomodates energy conservation during the ablative process while retaining the large energy deposition rates of previous models. The thickness of the vapor layer, which changes dynamically as the impactor penetrates, is a function of impactor composition, the temperature at the bow shock and the opacity of the vapor and atmospheric gases.

data of small meteors entering Earth's atmosphere—for which (2.2) has been successfully applied many times—before it can be used to describe objects larger than 100 m entering Jupiter's atmosphere. This, apparently, will be determined by the only remaining free parameter, $r(t)$.

2.2. Vaporization

Ceplecha *et al.* (1993) find two independent parameters for equations 1(2.1) and (2.2): the ablation coeficient (σ) and the shape-density coeficient ($K = SC_d m^{-2/3}$) where S is described as the 'head cross-section'. In other words, the 'meteoroid cross-section', πr^2, as portrayed up to now, is actually an 'interaction cross-section' that may include fragments, ablated debris and vapor that is traveling in concert with the main meteoroid during its traverse of the atmosphere. The power deposited in the atmosphere by passage of the meteoroid can be written in terms of K and σ:

$$\frac{dE}{dt} = -\left(m\frac{dv}{dt} + \frac{1}{2}\frac{dm}{dt}v\right)v = K\rho m^{2/3}\left(v^3 + \frac{1}{2}\sigma v^5\right) \approx \frac{1}{2}\sigma K\rho m^{2/3}v^5. \qquad (2.6)$$

It is clear from (2.6) that a decrease in the accepted values of σ, as (2.3) and (2.5) imply, can be accomodated by an increase in K to yield the same dE/dt. In this case, a hundred-fold increase in S (*i.e.*, a ten-fold increase in r), for small meteoroids, produces the desired effect.

Consider the dynamics at high altitude, where small terrestrial meteors are observed. At these altitudes, spreading due to aerodynamic forces will be negligible because of the low atmospheric density (Zahnle *et al.* 1992; Hills & Goda, 1993). Spreading at these altitudes due to vapor pressure of the ablated meteoroid material can be substantial, however. The ablated vapor traveling in concert with the main fragment occupies a layer of thickness x surrounding the fragment and increases the interaction cross section (Figure 4).

The mass of the meteoroid vapor layer increases with time as described by equation (2.2) where the original definition of $Q = Q_0$, the heat of vaporization, is used. In the meteoroid's reference frame, the bulk of the vaporized mass does not possess the kinetic energy required to escape the immediate vicinity (as shown previously). It will accumulate and form a layer surrounding the meteoroid. Equating the vapor pressure of this layer with the kinematic pressure (ρv^2), provides an expression for the increasing layer thickness (dx/dt):

$$\frac{dx}{dt} = \frac{C_h R}{8 Q_0 A} v T_v \left(\frac{r}{r+x} \right)^2 \tag{2.7}$$

where R is the molar gas constant, r, the meteoroid radius, A, the meteoroid mean molecular weight and T_v, the average vapor temperature. Assuming $T_v = 10^4$ K for water ice impacting at 60 km s^{-1}, dx/dt is initially 1 km s^{-1} and the layer will thicken rapidly. At some point, the mass gained by vaporization at the surface of the meteoroid is balanced by mass lost from hydrodynamic stripping at the surface of the vapor layer. Adding a term to account for this produces:

$$\frac{dx}{dt} = \frac{C_h R}{8 Q_0 A} v T_v \left(\frac{r}{r+x} \right)^2 - \frac{1}{4} \frac{C_m R T_v}{A v} \tag{2.8}$$

where $C_m < 1$ is the efficiency of mass loss from the vapor layer. The layer will reach an equilibrium thickness, x_m, when $dx/dt = 0$:

$$x_m = \left(v \sqrt{\frac{C_h}{2 C_m Q_0}} - 1 \right) r \tag{2.9}$$

and $S = \pi(r + x_m)^2 = \pi r^2 C_h v^2 / (2 C_m Q_0) \approx 150 \pi r^2$ for $C_h = 0.1$, $C_m = 0.5$, $Q_0 = 2.5 \times 10^{10}$ erg g^{-1} and $v = 60$ km s^{-1}. For small meteoroids with vapor layers defined by (2.9), it is easy to show that the modified ablation equation (2.5) is equal to (2.2) provided $S = \pi(r+x_m)^2$; hence, explicitly modeling the vapor layer in this way reconciles the modified ablation equation (2.5) with the terrestrial observations.

For small meteoroids, x_m is important. For large meteoroids ($r > 10$–100 m), the layer reaches an equilibrium thickness (x_c) when opacity of the vapor is taken into account (Field & Ferrara, 1995):

$$x_c = \left[\frac{8(\gamma_v - 1)(\gamma + 1)^2}{3 \gamma_v (\gamma - 1)} \right]^{1/2} \left(\frac{\sigma T_s^4 l}{\rho v^3} \right)^{1/2} r^{1/2}. \tag{2.10}$$

In this expression, γ_v and γ are for vapor and atmosphere, respectively, $T_s \approx 4 \times 10^4$ K is the shock temperature and l is the photon mean-free-path, which is dependent on vapor temperature and density, generally. From Chevalier & Sarazin (1994), l is approximately $3 \times 10^{-8} \rho^{-2}$ cm for shocked Jovian atmosphere. Since the opacity of vaporized cometary material is considered to be larger than that of clean Jovian air (Chevalier & Sarazin, 1994; Zahnle & MacLow, 1994; Field & Ferrara, 1995), this prescription for l will give an upper bound for x_c.

2.3. Hydrodynamic spreading

Analytical models describing hydrodynamic deformation and mechanical breakup of meteoroids during passage through dense planetary atmospheres have been proposed by several researchers (Ivanov & Yu, 1988; Zahnle, 1992; Chyba *et al.* 1993; Hills & Goda, 1993; O'Keefe *et al.* 1994). Ivanov & Yu (1988) modeled the deformations experienced by a fluid spherical body passing through an atmosphere. Zahnle (1992) and Chyba *et al.* (1993) proposed the 'pancake model' in which the large differential hydrodynamic

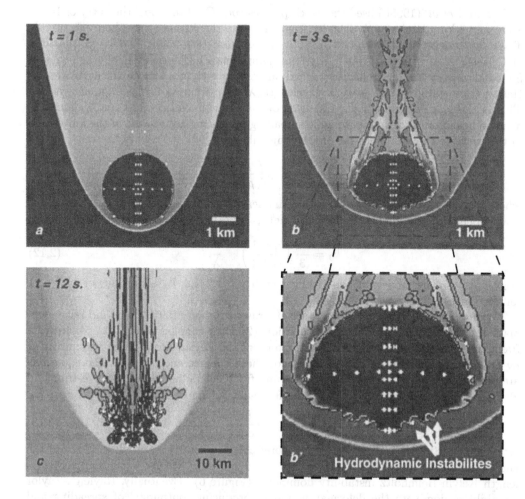

FIGURE 5. Sequential stages of projectile entry (a), deformation (b,b′) and breakup (c). During entry, the projectile forms a clean bow shock in the upper reaches of the atmosphere. Atmospheric temperatures at the leading edge of the projectile reach values as great as 40,000 K. During deformation, the projectile flattens at an average rate governed by the aerodynamic flow field (the growth rate of a long wavelength Kelvin-Helmholtz instability). Eventually, Rayleigh-Taylor hydrodynamic instabilities fragment the body (after Crawford *et al.* 1995b).

pressure experienced between the front and rear of an idealized cylindrical projectile is virtually unopposed by the ambient pressure along its sides. The meteoroid spreads and flattens, leading to a 'pancake' that rapidly decelerates, depositing the bulk of its energy across less than an atmospheric scale height. Hills & Goda (1993) proposed a model whereby a small asteroid traveling through an atmosphere suffers successive fragmentation wherever the aerodynamic pressure exceeds the material yield strength. The fragmentation continues until the aerodynamic pressure is lower than the yield strength. Like the pancake model, the spreading meteoroid deposits the bulk of its energy across less than an atmospheric scale height. O'Keefe *et al.* (1994) examined the role of Rayleigh-Taylor (R-T) and Kelvin-Helmholtz (K-H) instabilities on the hydrodynamic deformation and breakup of meteoroids. They proposed that K-H instabilities limit spreading and R-T instabilities initiate breakup.

Crawford *et al.* (1995b) used the shock-physics code CTH to model the entry of 1-km, 2-km and 3-km diameter spherical ice fragments into the Jovian atmosphere (Figure 5). These numerical simulations are best fit by an analytical model based on the idea of O'Keefe *et al.* (1994). Apparently, the hydrodynamic deformation of fragments of Shoemaker-Levy 9 is controlled by the exponential growth of a long wavelength Kelvin-Helmholtz instability with spatial wavelength $\lambda = 4r$ where r is the fragment radius. This is the longest wave that a fragment of diameter $2r$ can support and defines the average behavior of the hydrodynamic spreading. Equating the growth of the amplitude of this wave to the radial growth of the projectile yields:

$$\frac{dr}{dt} = nr = k_0 \left(\frac{\rho}{\rho_f}\right)^{1/2} v \qquad (2.11)$$

where n is the wave growth rate, ρ_f is the bulk density of the fragment and k_0 is a constant that depends on the dimensionless wavelength, λ/r (after Field & Ferrara, 1995):

$$k_0 = \frac{2\pi}{3} \left(\frac{\gamma - 1}{\gamma + 1}\right)^{1/2} \frac{r}{\lambda}. \qquad (2.12)$$

For an atmosphere modeled as an ideal gas with $\gamma = 1.2$–1.4, k_0 has a value of 0.16–0.21 for the longest supported wavelength.

To compare with the numerical results of Crawford *et al.* (1995b), who did not model mass loss from radiative ablation, Equation (2.11) is coupled with (2.1) to describe fragment deceleration during entry. A model of the Jovian atmosphere was provided by Orton (unpublished data) and scaled for the 45 degree entry angle. The equations are numerically integrated to find velocity and fragment radius as functions of time and altitude. The deposition of energy by the fragment is found from:

$$\frac{dE}{dt} = -m\frac{dv}{dt}v. \qquad (2.13)$$

A k_0 value of 0.2, consistent with the theoretical arguments above, adequately fits the energy deposition curves derived from the numerical results at least while the long wavelength Kelvin-Helmholtz instability dominates (Figure 6). Eventually, Rayleigh-Taylor instabilities dominate the deformation mode, producing 'outbursts' of spreading and fragmentation. The 'pancake' model (also shown in Figure 6) places the peak explosion altitude slightly higher. Neither model addresses the onset and growth of Rayleigh-Taylor instabilities and, consequently, does not well describe the fragmentation behavior near the terminus of the penetration. Nevertheless, the average energy deposition in the higher altitude region of the Jovian atmosphere appears to be better described by including the limiting effect of the long wavelength Kelvin-Helmholtz instability. It seems that much theoretical, computational and experimental effort will need to be expended in order to advance our understanding of this important process.

2.4. *Putting it all together*

Equations (2.1), (2.5), (2.8) and (2.11) provide a fairly complete theoretical description of the average energy deposition resulting from fragments of Shoemaker-Levy 9 penetrating the higher altitudes (greater than -100 km relative to the 1-bar reference level) of Jupiter's atmosphere. Putting the expressions in terms of entry angle (θ) and altitude in the atmosphere (z) and using the definition for the interaction cross-section, $S = \pi(r + x)^2$, yields:

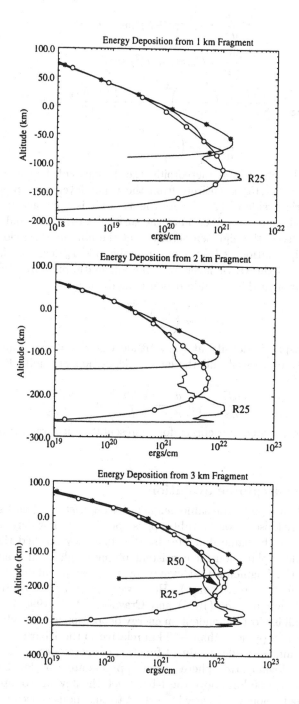

FIGURE 6. Energy deposition curves from 1-, 2- and 3-km fragments entering the Jovian atmosphere. The solid curves without symbols are from numerical simulations of Crawford *et al.* (1995). The curves are labeled with the resolution of the calculation (represented by the number of zones across the radius, R25 and R50). The results of our analytical model are super-imposed (open circles). The 'pancake' model of Zahnle (1992) places the peak energy deposition slightly higher (stars).

$$m\frac{dv}{dz} = \frac{C_d}{2}\pi(r+x)^2\rho v \sec\theta \tag{2.14}$$

$$\frac{dm}{dz} = C_m\pi(r+x)^2\rho\sec\theta \tag{2.15}$$

$$\frac{dx}{dz} = \frac{C_h R}{8Q_0 A}T_v\left(\frac{r}{r+x}\right)^2\sec\theta - \frac{1}{4}\frac{C_m R T_v}{Av^2}\sec\theta \tag{2.16}$$

$$\frac{dr}{dz} = k_0\left(\frac{\rho}{\rho_f}\right)^{1/2}\sec\theta \tag{2.17}$$

where it is understood that for large meteoroids, the thickness of the vapor (x) is limited by (2.10) and mass loss (2.15) is entirely from the vapor layer, as represented by the second term on the right side of (2.16). As the x_c limit is approached and the layer becomes optically thick, vapor production at the surface of the meteoroid slows because the surface no longer 'sees' the high temperatures at the bow shock. Below the altitude at which x_c is reached, equation (2.16) no longer applies and x_c decreases faster than the adiabatic compression of the vapor layer from the kinematic pressure, ρv^2. Hence, the compression of the vapor in this altitude region is modeled as:

$$x = x_c\left(\frac{\rho_c}{\rho}\right)^{1/3\gamma_v} \tag{2.18}$$

where ρ_c is the atmospheric density at the altitude where $x = x_c$. Equations (2.14)–(2.18) are numerically integrated and the energy deposition per unit altitude (dE/dz) is:

$$\frac{dE}{dz} = \left(m\frac{dv}{dz} + \frac{1}{2}\frac{dm}{dz}v\right)v \tag{2.19}$$

Figure 7. shows energy deposition curves for representative ice fragments entering the Jovian atmosphere.

3. Models of fireball/plume evolution

Why did all the plumes go to the same height? Boslough *et al.* (1995) suggested that the SL9 fragments were loosely-bound 'rubble piles', possibly with widely varying masses, that dispersed to about the same diameter by the time they reached the atmosphere. This leads to a simple explanation for consistent plume heights but places constraints on the properties of the fragments that cannot be easily accomodated by parent-body breakup models (Asphaug, pers. comm.). Here, we explore an alternative hypothesis. As emphasized in Boslough *et al.* (1995) and Crawford *et al.* (1995a), the early-time evolution of the fireball is most dependent on energy deposition at relativly high altitudes in the Jovian atmosphere (greater than −50 km relative to the 1-bar reference altitude). Perhaps maximum plume altitude is a direct function of energy deposition in a relatively narrow region of the Jovian atmosphere—from approximately 50 km below the 1-bar altitude to approximately 50 km above the 1-bar level, the Jovian tropopause.

The presence of the tropopause is very important to this model. Because the evolution of the fireball is strongly dependent on the density gradient in the atmosphere, and the gradient is greatest at the tropopause, the hot Jovian atmosphere and cometary debris left in this critical region of the atmosphere will accelerate faster than any other. This hypothesis can be tested by performing two-dimensional fireball/plume calculations based on the energy deposition curves derived previously. But first, it is useful to explain why

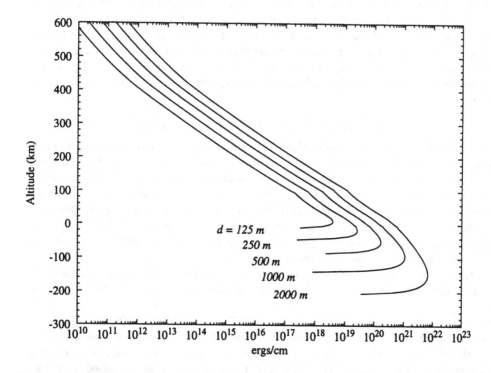

FIGURE 7. Energy deposition curves from our analytical model for 125–2000 m diameter ice fragments entering the Jovian atmosphere. The change of slope at an altitude of 100 km results from the opacity limit of Equation (2.10) and appears in about the same location as determined by Zahnle and MacLow (1994).

energy deposited from different size impactors in this 'favored' region of the atmosphere will always yield nearly the same maximum plume altitude.

3.1. *A simple model to explain why all of the observed plumes went to the same height*

Neglecting, for the moment, the interesting question of what happens if a fragment disintegrates and stops in the favored region of the atmosphere, consider the energy deposited by a fragment penetrating through the atmosphere before it has suffered significant deformation. This energy is initially deposited in a column with radius proportional to r. The energy deposited per unit length (dE/dz) is

$$\frac{dE}{dz} = \frac{C_d}{2} S \rho v^2 \sec \theta \qquad (3.20)$$

where S is the interaction cross-section defined previously. The total energy deposited across a scale height of atmosphere (H) is

$$E = H \frac{dE}{dz} = H \frac{C_d}{2} S \rho v^2 \sec \theta \qquad (3.21)$$

into a total atmospheric mass $(M = \rho S H)$. The specific energy (ϵ) of this column heated by the passage of the fragment is

$$\epsilon = \frac{E}{M} = \frac{C_d v^2}{2} \sec \theta \qquad (3.22)$$

and the characteristic velocity of the heated atmosphere, $u = \sqrt{2\epsilon}$ (Zel'dovich & Raizer, 1967), is independent of fragment size. Since maximum altitude (Z_{max}) will scale as u^2 g−1, plumes resulting from different size impactors should, to first order, go to the same height. Hence, the plumes reach the same altitude for much the same reason that a rifle bullet and artillery shell have approximately the same muzzle velocity. The masses of the projectiles are substantially different yet the specific energy of the driving gas is the same.

3.2. *Numerical models of fireball and plume evolution*

Fireball and plume calculations using the CTH shock-physics computational hydrocode were performed to test the hypothesis of Section 3.1 and to investigate the dynamics of small fragments exploding in the favored region of the atmosphere. CTH is a multi-material, multi-phase computational shock-physics code that solves mass, momentum and energy conservation and material constitutive relations on an Eulerian grid (McGlaun *et al.* 1990). It can realistically model equations-of-state of many materials simultaneously. An accurate equation-of-state (including dissociation and ionization) for a mix of 89% H_2, 11% He and free electrons (Kerley, unpublished data) was used to model the Jovian atmosphere. The atmosphere was constrained to be gravitationally stable with a thermal profile provided by Orton (unpublished data) and extended adiabatically at depth.

Computational simulations of 125, 250, 375, 500, 750, 1000, 1500 and 2000 meter diameter fragments entering the Jovian atmosphere were performed using a two dimensional cylindrical coordinate system. The resulting fireball retains axisymmetry with respect to the entry channel for much of the early-time plume evolution as described by Crawford *et al.* (1995b) and as shown in Figure 1. Hence, the atmosphere has been scaled to account for the 45 degree entry angle and the gravitational constant (g) has been reduced accordingly. The resolution of the computational grid varies from 0.6 km per zone to 2.5 km per zone for simulations of the smallest and largest impacting fragments, respectively. The energy deposited by each fragment's passage through the atmosphere is based on the analytical model derived in Section 2. This approach has the advantage of simulating the evolving wake and fireball at high altitude even as the fragment is penetrating to lower altitude. The energy of the decelerating and ablating fragment is added to the internal energy of the atmosphere in 2 km long by $10r$ radius cylindrical segments (where r is the fragment radius). Each segment receives the appropriate energy from equation (2.19) at the appropriate time. The simulation starts with the fragment (represented as a source of energy) at 640 km altitude (900 km on the computational grid).

Each simulation has 1000 Lagrangian tracer particles distributed in eight layers of 125 particles each. The topmost layer is located at the approximate location of the Jovian cloud tops (11 km altitude). The remaining layers, in descending order, are at 0, −11, −21, −32, −42, −71 and −106 km respectively. The particles are evenly distributed from 0 to 25 km radially for simulations of large fragments (750 m and larger) and from 0 to 5 km radially for simulations of small fragments (500 m and smaller). These particles trace the ejection of atmospheric materials by the fireball and cooling plume. The locations and velocities of the particles at the end of the simulations are ballistically extrapolated to determine maximum plume height and final ejecta morphology.

Figures 8 and 9 show the first three minutes of entry-wake and plume evolution for 250, 500, 1000 and 2000 meter fragments. In the temperature plots of Figure 8, the penetrating 'fragments' can be followed through the first 20 seconds. Near the end of the penetration phase, the 'entry fireball' is starting to evolve at the top of the entry channel.

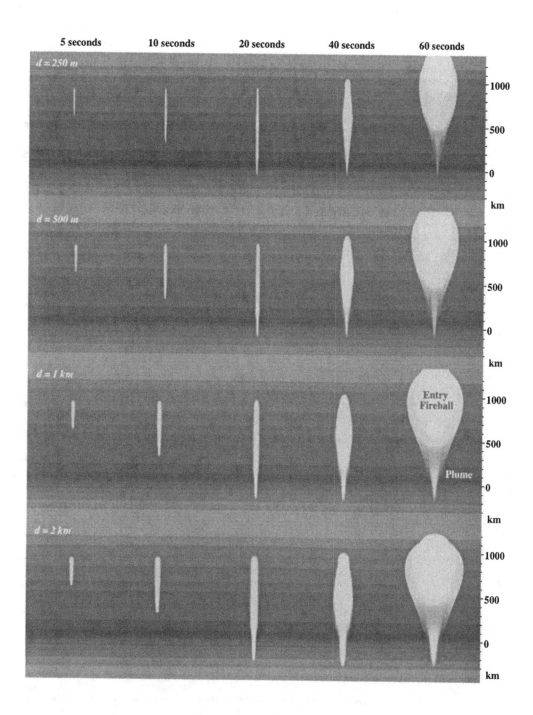

FIGURE 8. Simulations of the entry of 250, 500, 1000 and 2000 meter fragments entering the Jovian atmosphere: the first 60 seconds. Gray scale is proportional to log(T). The development of the 'entry fireball', beginning at 40 seconds, is clearly seen. The plume can be seen beginning to develop at 60 seconds near the Jovian tropopause. The plume contains opaque materials derived from the cloud layers, hence was probably the over-limb feature observed in the HST images.

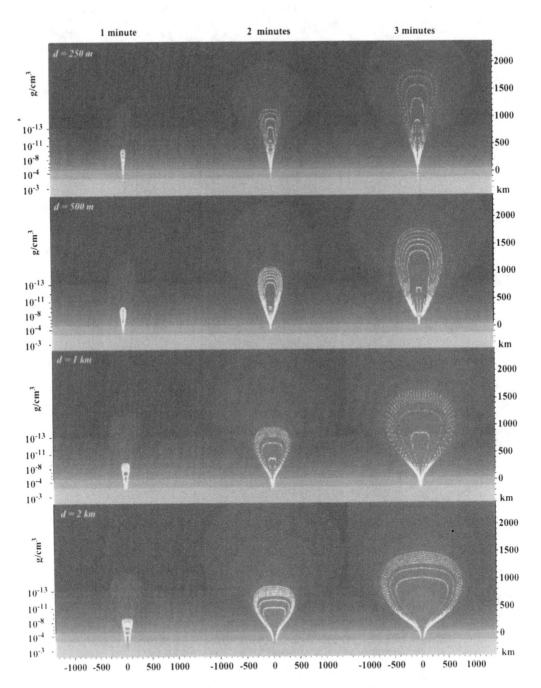

FIGURE 9. Simulations of fireballs from 250, 500, 1000 and 2000 meter fragments entering the Jovian atmosphere: the first 3 minutes. Gray scale is proportional to log(Density). One thousand tracer particles, representing the Jovian cloud layers, are superimposed (white dots). Material derived from the Jovian clouds reach nearly the same altitude at the end of the simulations whereas the isodensity contours do not.

This fireball has low mass, is probably ionized, and accelerates rapidly out of the Jovian thermosphere at velocities approaching 60 km s^{-1}. It could be a perturbing influence on the Jovian magnetosphere, perhaps leading to the increased auroral emissions observed at the magnetic conjugate point of the K impact site (Clarke *et al.* 1995). However, the entry fireball is extremely sensitive to the rate of ablation, hence the numbers quoted here should not be taken too literally.

At 40 seconds, the fireball containing the visible plume is seen starting at the tropopause and rising ballistically. No evidence is seen for an 'explosion' at the terminus of penetration although a weak shock can be seen propagating cylindrically. This shock degrades to a linear wave within a few minutes. This is consistent with the model of Chevalier & Sarazin (1994) and three-dimensional simulations of Crawford *et al.* (1994) whereby deeply exploding fragments do not generate strong shock blowouts at their terminus (Zahnle & MacLow, 1994) but generate a buoyant column of hot gas instead.

Figure 9 shows the density distribution of the fireballs and plumes 1, 2 and 3 minutes into the simulations. Overlain, are the locations of the eight layers of tracer particles. As is made apparent by the plots, the maximum altitude of a particular isodensity contour strongly increases as a function of fragment size, yet maximum altitude of a given layer of tracer particles (which originate from the same altitude in the atmosphere) does not. As an expression of the lower degree of radial confinement (represented by H/r), the maximum plume altitude slowly decreases with larger impactor size. Because the plume has cooled to $\approx$ 100 K by this time, material derived from the Jovian cloud layers will have condensed. This suggests that the source of opacity for the the plumes was simply the Jovian cloud layers, that the top of the plumes represent a vertical translation of altered cloud materials and that the plume outlined by a layer of tracer particles represents the 'visible plume'. The resolution of the numerical models, and their sensitivity to the rate of ablation do not allow an exact determination of which cloud layer corresponds to the top of the visible plume at this time. However, the morphology of the ejecta (as shown in the next sections) and spectroscopic data (Noll *et al.* 1995) suggest that the NH$_4$SH layer was probably involved.

The lateral extent and the tracer particle density (*i.e.*, opaque mass density) of the plumes are functions of fragment size. In fact, once the visible plumes are rotated to account for the 45 degree entry angle, the increased lateral extent of the plumes resulting from larger fragment sizes nearly compensates for the weakly decreasing height of the larger 2-D plumes. For fragments greater than approximately 500 m in diameter, the mass density of the opaque material (again, represented by the tracers) is proportional to the cross-sectional area of the impactor. This is consistent with a simple model whereby the impactor punches a hole with radius proportional to the fragment radius, r, in the Jovian clouds. The contents of the hole make up the visible plume. It is easy to show that for large impactors that penetrate completely through the opaque source layer, the volume of excavated opaque material is proportional r^2. For small impactors that do not penetrate completely through, but explode inside, the volume of excavated material is proportional to r^3.

3.3. *Ballistic plume extrapolation and ejecta emplacement*

By the end of the simulations, material located in the upper half of the plume is travelling with little force contribution to its motion except that due to gravity. At this point, the motion of the tracer particles is ballistically extrapolated in three dimensions. Figure 10 shows the configuration of the excavated atmospheric layers at the time of maximum visible plume height. All of the plumes rise to approximately the same maximum altitude, yet the layers within are bunched more tightly together for larger impactors. The former is consistent with HST observations (Hammel *et al.* 1995) and the latter has important implications for the ejecta emplacement process.

Even though the impact angle was 45 degrees from vertical, the line segment connecting the peak of the plume with the impact location is not necessarily 45 degrees from vertical and may actually be an indicator of fragment size. The simulated plume from the largest impacting fragment (2 km diameter) has an inclination angle of just 20 degrees from vertical. This requires a cautionary note, however. Because nature performed this impact experiment in three dimensions and here the simulations were performed in two dimensions, this observation may be qualitatively correct but not quantitatively accurate. For example, the inclination angle of a plume formed by a 2 km impactor will probably turn out to be somewhat less than 20 degrees. We can guess that this will be the case because the atmospheric density gradient will tend to direct the fireball in a more vertical direction during the early hydrodynamic expansion phase.

As shown in Figure 11, all of the simulated plume heights and times-to-maximum-altitude, except those from the smallest impactor (125 m in diameter), are consistent with measurements (3000 km and 400–600 s, respectively) made by Hammel *et al.* (1995). If the tops of the plumes originated from the tops of the NH_3 cloud layer, then this places a lower bound of 4×10^{12} g (200 m diameter solid water ice) on the mass of the fragments that caused the A, E, G and W plumes observed by HST. If the plumes originated from the NH_4SH layer, then masses greater than 4×10^{13} g (400 m diameter water ice) are required. Moreover, these mass estimates are consistent with those derived from the SL9 parent body breakup study of Asphaug & Benz (1994) and average fragment size estimates derived from crater chain measurements on Callisto and Ganymede (McKinnon & Schenk, 1995).

Figure 12 shows 'ejecta patterns' represented by the extrapolated impact locations of the tracer particles after their re-impact with the Jovian stratosphere at 100 km altitude. Each plot used the upper six layers of tracer particles (from -42 km to 11 km altitude). The location of each tracer particle was ballistically extrapolated assuming a flat planet approximation with constant g of 2500 cm s^2. In Figure 12, the relative albedo is proportional to the amount of opaque material produced in the wake of the entering fragment which is assumed to be proportional to the peak temperature (T_p) experienced by the tracer during the numerical simulation. This is represented by the empirical expression:

$$T_p = 40,000 \left(\frac{r}{r_t}\right)^2 \ K \qquad (3.23)$$

where r_t is the initial radial location of the tracer particle and r is the radius of the comet fragment. The albedo is an increasing function of the cross-sectional area of the impactor, hence in Figure 12, the relative albedo is scaled by r^{-2} in order to emphasize the morphology of the ejecta.

The simulated ejecta patterns have several features in common with the observed impact sites. They exhibit crescent-shaped ejecta patterns in the uprange direction with a dark central spot and a gap (seen for the larger fragments) in between. The gap appears to be determined by the degree of penetration below the opaque source layer. In the simulated ejecta patterns, the gap appears for fragments greater than 250 m in diameter, but depends on the choice for the opaque source region (the atmosphere above -42 km, in this case). A thicker, deeper source region fills in the gap whereas a thinner, shallower one broadens it out.

The simulated ejecta patterns fail to match the observations in one crucial aspect, however. The observed debris patterns were seen to extend for 360 degrees surrounding

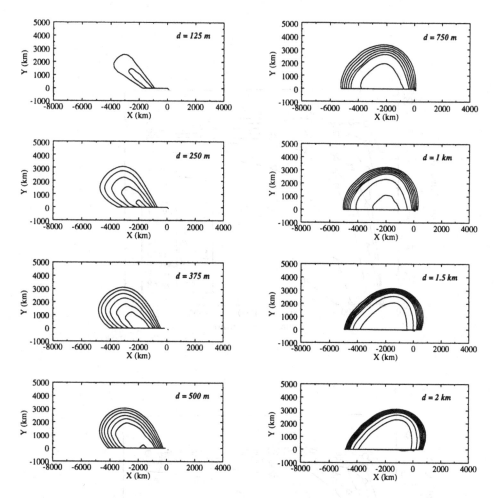

FIGURE 10. Morphology of the simulated plumes arising from the impact of 125–2000 meter diameter fragments entering the Jovian atmosphere. Each plot is made at the time of maximum height with each layer of tracer particles represented by curves in this cross-sectional representation. The topmost layer corresponds to material derived from 11 km altitude with the remaining layers at 0, −11, −21, −32, −42, −71 and −106 km. The origin corresponds to passage of the fragment through the 1-bar level of the atmosphere.

each impact site with a significant crescent-shaped enhancement seen (for the larger impacts) to the south and east. While the crescent-shaped enhancement is produced by the simulations, the surrounding ejecta pattern is not. The disagreement probably arises from the use of a 2D fireball simulation prior to ballistic extrapolation in 3D. As discussed in Section 1, there are only two possible endmember treatments of this problem using 2D simulations. One, is to treat the event as intrinsically vertical, in which case, the ejecta pattern will be constrained to fall in a 360 degree pattern surrounding the impact site but will not produce the crescent-shaped feature that was observed. The other approach, which is used here, is to treat the event as axially symmetric with respect to the impact direction (inclined at 45 degrees in this case), at least for the first few minutes. This approach can simulate the crescent-shaped feature but fails to produce the surrounding ejecta pattern. Obviously, the answer is in between.

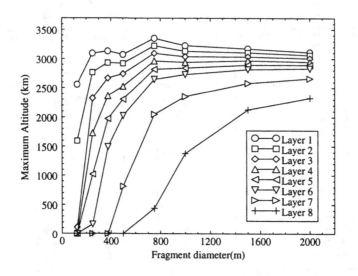

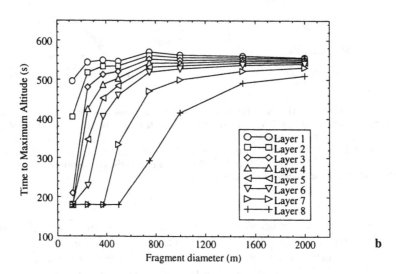

FIGURE 11. Maximum altitude (a) and time-to-maximum-altitude (b) vs. fragment diameter for atmospheric layers represented by tracer particles. Layers 1–8 correspond to initial altitudes of 11, 0, −11, −21, −32, −42, −71 and −106 km respectively.

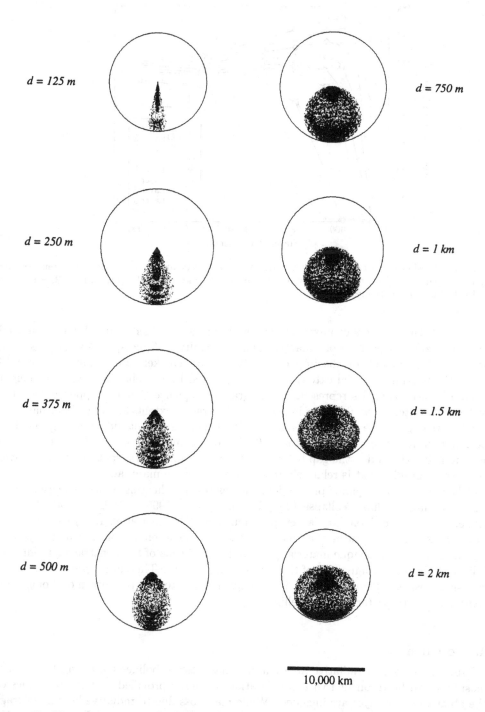

FIGURE 12. Ejecta patterns produced by ballistic extrapolation of the uppermost six layers of tracer particles. The albedo of each plot is proportional to peak temperature experienced by the tracers and has been scaled by d^{-2}. The circle surrounding each pattern represents the maximum radial extent of ejected material and is centered on the location of fragment passage through the 1-bar level of the atmosphere.

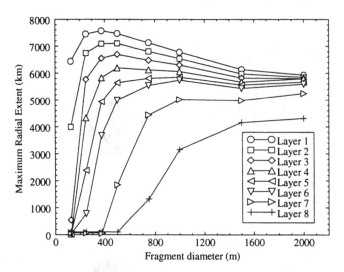

FIGURE 13. Maximum radial extent vs. fragment diameter for atmospheric layers represented by tracer particles. Layers 1–8 correspond to initial altitudes of 11, 0, −11, −21, −32, −42, −71 and −106 km respectively.

The fact that impacts can excavate material normally hidden at depth has been used to understand the history of lunar volcanism (Schultz & Spudis, 1979) and has been proposed as a model for the impacts of SL9 by Shoemaker *et al.* (1995). Figure 13 shows the maximum radial extent of material ejected by the plumes of SL9. The eight layers of tracer particles represent a stratigraphic sequence that is mapped out on the surface of Jupiter surrounding an impact site. The radial extents are only about half that required to explain the size of the G impact site, hence significant sliding, as suggested by Boslough *et al.* (1995), Hammel *et al.* (1995) and Shoemaker *et al.* (1995), has probably occurred. In this model, the gap between the crescent and central spot represents a layer in the stratigraphy that is relatively free of opaque source material.

Other processes of opaque production, such as re-entry heating of atmospheric/plume materials during plume collapse (*e.g.*, Boslough *et al.* 1995 or Zahnle *et al.* 1995) or models of gap formation from an evaporation wave (Hammel *et al.* 1995) can also lead to crescent-shaped features that may be less dependent on the nature of the source region. In any case, opaque material produced in the wake of the penetrating fragments is likely to be a key component of the dark ejecta patterns. This simple process by which wake-synthesized opaque material is subsequently ejected to produce morphologically consistent ejecta patterns is compelling.

4. Conclusions

Previous models of radiative ablation and mass loss of bolides traversing Earth's atmosphere can be reconciled with conservation of energy provided that the dynamics of the ablative vapor layer are included. While mass loss due to radiative heating of fragments larger than 100 m in the Jovian atmosphere is insignificant, ablation is a major contributor to energy deposition at high altitude and has an important role in early-time fireball evolution. Results from fireball and plume evolution models using the CTH shock-physics code demonstrate that constant maximum plume heights consistent with Hubble observations are observed as long as the tops of the plumes are derived from the same level of the atmosphere and that the fragments penetrated at least 30 km below this level. If the tops of the plumes originated from the visible cloud tops, then fragment masses greater than 4×10^{12} g, corresponding to 200 m diameter fully dense water ice, are required to explain the observations. If the plumes originated from the NH_4SH layer, then masses greater than 4×10^{13} g (400 m water ice) are required. The lateral extent of the plume at maximum height and the darkness and lateral extent of the ejecta pattern are determined by the cross-sectional area of the penetrating fragment. A simple model, whereby opaque material is synthesized in the wake of penetrating fragments and subsequently ejected by the fireball, produces ejecta patterns that are, in many ways, similar to the observed impact sites. A straightforward extension of this model suggests that the apparent gap between the central disturbance of the impact site and the inner front of the crescent-shaped ejecta may reflect the fragment's depth of penetration below the source layer of the visible ejecta.

This chapter benefitted greatly from the leadership and insight of Mark Boslough and from the critical analysis of the other members of Sandia's SL9 team, Tim Trucano and Allen Robinson, and from discussions with Kelly Beatty, Zdenek Ceplecha, Heidi Hammel, Mordecai MacLow, Glen Orton, Toshika Takata and Kevin Zahnle. Gerald Kerley provided the equation of state for Jupiter's atmosphere and Glen Orton provided the structure of the atmosphere based on Voyager data. This work was performed at Sandia National Laboratories under U.S. Department of Energy contract DE-AC04-94AL85000 and was funded by the Laboratory Directed Research and Development Program (LDRD) and the National Science Foundation under Agreement No. 9322118.

REFERENCES

ASPHAUG, E. & BENZ, W. 1994 Density of comet Shoemaker-Levy 9 deduced by modelling breakup of the parent 'rubble pile'. *Nature* **370**, 120–124.

BIBERMAN, L. M., BRONIN, S. YA. & BRYKIN, M. V. 1980 Moving of a blunt body through the dense atmosphere under conditions of severe aerodynamic heating and ablation. *Acta Astronautica* **7**, 53–65.

BOSLOUGH M. B., CRAWFORD, D. A., ROBINSON, A. C. & TRUCANO, T. G. 1994a Watching for Fireballs on Jupiter. *Eos* **75**, 305–310.

BOSLOUGH M. B., CRAWFORD, D. A., ROBINSON, A. C. & TRUCANO, T. G. 1994b Mass and penetration depth of Shoemaker-Levy 9 fragments from time-resolved photometry. *Geophysical Research Letters* **21**, 1555–1558.

BOSLOUGH M. B., CRAWFORD, D. A., TRUCANO, T. G., & ROBINSON, A. C. 1995 Numerical modeling of Shoemaker-Levy 9 impacts as a framework for interpreting observations. *Geophysical Research Letters* **22**, 1821–1824.

BRONSHTEN, V. A. 1983 *Physics of Meteoric Phenomena*. Reidel.

CEPLECHA, Z., SPURNY, P., BOROVICKA, J., & KECLIKOVA, J. 1993 Atmospheric fragmentation of meteoroids. *Astronomy and Astrophysics* **279**, 615–626.

CHEVALIER, R. A. & SARAZIN, C. L. 1994 Explosions of Infalling Comets in Jupiter's Atmosphere. *Astrophysical Journal* **429**, 863–875.

CHYBA, C. F., THOMAS, P. J. & ZAHNLE, K. J. 1993 The 1908 Tunguska explosion: atmospheric disruption of a stony asteroid. *Nature* **361**, 40–44.

CLARKE, J. T., PRANGE, R., BALLESTER, G. E., TRAUGER, J., EVANS, R., REGO, D., STAPELFELDT, K., IP, W., GERARD, J.-C., HAMMEL, H., BALLAV, M., JAFFEL, L. B., BERTAUX, J.-L., CRISP, D., EMERICH, C., HARRIS, W., HORANYI, M., MILLER, S., STORRS, A., & WEAVER, H. 1995 HST Far-Ultraviolet Imaging of Jupiter During the Impacts of Comet Shoemaker-Levy 9. *Science* **267**, 1302–1307.

CRAWFORD, D. A., BOSLOUGH, M. B., TRUCANO, T. G. & ROBINSON, A. C. 1994 The Impact of Comet Shoemaker-Levy 9 on Jupiter. *Shock Waves* **47**, 47–50.

CRAWFORD, D. A., BOSLOUGH, M. B., ROBINSON, A. C., & TRUCANO, T. G. 1995a Dependence of Shoemaker-Levy 9 Impact Fireball Evolution on Fragment Size and Mass. *Lunar and Planetary Science* **26**, 291–292.

CRAWFORD, D. A., BOSLOUGH, M. B., TRUCANO, T. G. & ROBINSON, A. C. 1995b The Impact of Periodic Comet Shoemaker-Levy 9 on Jupiter. *Int. J. Impact Engin.* **17**, 253–262.

FIELD, G. B. & FERRARA, A. 1995 The Behavior of Fragments of Comet-Shoemaker-Levy 9 in the Atmosphere of Jupiter. *Astrophysical Journal* **438**, 957–967.

HAMMEL, H. B., BEEBE, R. F., INGERSOLL, A. P., ORTON, G. S., MILLS, J. R., SIMON, A. A., CHODA, P., CLARKE, J. T., DE JONG, E., DOWLING, T. E., HARRINGTON, J., HUBER, L. F., KARKOSCHKA, E., SANTORI, C. M., TOIGO, A., YEOMANS, D., & WEST, R. A. 1995 HST Imaging of Atmospheric Phenomena Created by the Impact of Comet Shoemaker-Levy 9. *Science* **267**, 1288–1296.

HILLS, J. G., & GODA, M. P. 1993 The Fragmentation of Small Asteroids in the Atmosphere. *Astronomical Journal* **105**, 1114–1144.

IVANOV, B. A., & YU, O. 1988 Simple Hydrodynamic Model of Atmospheric Breakup of Hypervelocity Projectiles. *Lunar and Planetary Science* **19**, 535–536.

MACLOW, M. M., & ZAHNLE, K. 1994 Explosion Comet-Shoemaker-Levy 9 on Entry into the Jovian Atmosphere. *Astrophysical Journal* **434**, L33–L36.

MCGLAUN, J. M., THOMPSON, S. L., & ELRICK, M. G. 1990 CTH - A three-dimensional shock-wave physics code. *Int. J. Impact Engin.* **10**, 351.

MCKINNON, W. B., & SCHENK, P. M. 1995 Estimates of comet fragment masses from impact crater chains on Callisto and Ganymede. *Geophysical Research Letters* **22**, 1829–1832.

NOLL, K. S., MCGRATH, M. A., TRAFTON, L. M., ATREYA, S. K., CALDWELL, J. J., WEAVER, H. A., YELLE, R. V., BARNET, C., & EDGINGTON, S. 1995 HST Spectroscopic Observations of Jupiter After Collision of Comet Shoemaker-Levy 9. *Science* **267**, 1307–1313.

O'KEEFE, J. D., TAKATA, T., & AHRENS, T. J. 1994 Penetration of Large Bolides into Dense Planetary Atmospheres—Role of Hydrodynamic Instabilities. *Lunar and Planetary Science* **25**, 1023–1024.

SCHULTZ, P. H. & SPUDIS, P. H. 1979 Evidence for ancient mare volcanism. *Lunar and Planetary Science Conf. Proc.* **10**, 2899–2918.

SEKANINA, Z. 1993 Disintegration phenomena expected during the forthcoming collision of periodic comet Shoemaker-Levy 9 with Jupiter. *Science* **262**, 382.

SHOEMAKER, E. M., HASSIG, P. J., & RODDY, D. J. 1995 Numerical simulations of the Shoemaker-Levy 9 impact plumes and clouds: A progress report. *Geophysical Research Letters* **22**, 1825–1828.

TAKATA, T., O'KEEFE, J. D., AHRENS, T. J., & ORTON, G. S. 1994 Comet Shoemaker-Levy 9: Impact on Jupiter and Plume Evolution. *Icarus* **109**, 3–19.

ZAHNLE, K. J. 1992 Airburst Origin of Dark Shadows on Venus. *Journal Geophysical Research* **97**, 10,243–10,255.

ZAHNLE, K. & MACLOW, M. M. 1994 The Collision of Jupiter and Comet Shoemaker-Levy 9. *Icarus* **108**, 1–17.

ZAHNLE, K., MACLOW, M. M., LODDERS, K., & FEGLEY, B., JR. 1995 Sulfur chemistry in the wake of comet Shoemaker-Levy 9. *Geophysical Research Letters* **22**, 1593–1596.

ZEL'DOVICH, Y. B. & RAIZER, Y. P. 1967 *Physics of Shock Waves and High-Temperature Hydrodynamic Phenomena*. Academic Press.

Entry and fireball models vs. observations: What have we learned?

By MORDECAI-MARK MAC LOW[1,2]†

[1]Astronomy & Astrophysics Center, University of Chicago, 5640 South Ellis Avenue, Chicago, IL 60637, USA

[2]Also Department of Astronomy, University of Illinois at Urbana-Champaign

This review attempts to give a coherent explanation of the main observations of the entry Comet Shoemaker-Levy 9 and the aftermath of the resulting explosions by using models of the tidal breakup of the comet, the entry of individual fragments into the jovian atmosphere, and the resulting fireballs and plumes. A critical review shows that the models appear reasonably well understood. The biggest theoretical uncertainties currently concern how to best tie models of the entry to models of the resulting fireballs. The key unknown before the impact was the size and kinetic energy of the comet fragments. The evidence now available includes the behavior of the chain of fragments, the luminosity of the observed visible fireballs and later infrared emission, the chemistry of the spots, and the lack of seismic waves or perturbations at the water cloud pressure level. These observations point to the fragments having diameters under a kilometer, densities of order 0.5 $g\,cm^{-3}$, and kinetic energies of order 10^{27} erg.

1. Introduction

In this review and in the review by Zahnle (this volume; hereafter "the plume review"), we make the argument that the fragments of Comet Shoemaker-Levy 9 that hit Jupiter were quite small, with diameters of under a kilometer and densities of order 0.5 $g\,cm^{-3}$. The largest fragments probably had kinetic energies of order 10^{27} ergs.

The evidence for small impactors includes the length of the pre-impact chain of fragments, the dim entry flashes and fireballs observed at optical wavelengths from *Galileo* and *HST*, the lack of observed seismic waves, and the strength and duration of infrared emission from the reentry of the ejected plumes. The evidence for explosions above the water clouds includes the high carbon to oxygen ratio seen by the *Hubble Space Telescope* (HST) in the ejecta spots (see the review by Lellouch in this volume for discussion of the CO observations, which, although they yield a lower C/O ratio, apparently still do not indicate penetration of the water clouds), and the lack of perturbations in 3 cm radio observations penetrating through to the 5 bar level where the water clouds are expected to lie. Plume heights do not serve as good evidence, as I will show below that they are difficult to calibrate.

Chronologically, this review covers the period from the tidal breakup of the parent body on its final passage through perijove through a time about 30 minutes after impact when the last bounce in the IR light curve was detected, excluding the main peak in the IR light curve, which is discussed by the plume review.

2. Tidal breakup

Several different groups have modeled the initial tidal breakup of the comet. It determines the size and density of the impacting comet fragments, giving initial conditions for models of atmospheric entry. For this reason I consider these models in some detail.

† Present Address: Max-Planck-Institut für Astronomie, Königstuhl 17, D-69117 Heidelberg, Germany

The maximum stress on the comet as it passed through perijove was minuscule, so Scotti and Melosh (1993) explained both the low strength and the uniform visual magnitude of SL9's debris by suggesting that 21 gravitationally-bound, primordial "cometesimals" came apart at perijove in a "delta function" breakup, and that each object then followed a post-perijove trajectory independent of the rest. Tidal effects are reduced in their model to a single moment when interparticle forces (such as self-gravity and collisions) are switched off. They integrated cometesimal orbits from perijove to the time of the first observations to derive an initial comet diameter ≈ 1.6 km. They neglected self-gravity, so their model does not constrain density except to imply that the comet was deep inside the Roche limit. Their model has difficulty explaining the historical record of crater chains on Ganymede and Callisto, since longer chains contain larger craters (Schenk *et al.* 1995). This would imply that larger comets formed from larger cometesimals, in contrast to modern theories of primordial accretion (Weidenschilling 1994).

Sekanina *et al.* (1994) used a somewhat more sophisticated model, outlined in Sekanina's chapter in this volume. In this model, the comet is assumed to break up during perijove passage. The fragments then remain in a collisionally interacting pile or cloud that produces a distribution of particle sizes before drifting apart. However, they begin to calculate train properties only after an effective time of disruption, when the fragments begin to disperse. Their Figure 8 shows that the derived radius of the parent body in their model explicitly depends on the value adopted for the effective time of disruption.

In essence, this is also a "delta function" breakup model, with the time of breakup equal to their effective time of disruption. Similarly to Scotti & Melosh (1993), Sekanina *et al.* neglect self-gravity, although they do not assume that the parent comet consisted of 21 uniform cometesimals. They constrain the comet's rotation by comparison with the observed position angle of the fragment chain. If breakup occurs at perijove, they find the same small parent comet diameter as Scotti and Melosh (1993). However, they instead propose that the effective time of breakup was more than 2 hours after perijove, allowing them to find solutions more consistent with the 7 km diameter parent comet proposed by Weaver *et al.* (1994).

The assumption of a single moment of breakup or dispersion, however, oversimplifies the process. Sekanina *et al.* (1994) are correct in assuming that the comet will begin to be affected by the strain well before the final dispersion. However their assumption that the comet remains spherical appears incorrect. The tidal forces distort the body into a cylinder in a process that begins hours before perijove and continues for hours after. The distortion begins with the initial strain, followed by torquing of the distorted body by planetary tides. The resulting spin-up causes further distortion, as do tides acting across the increasingly elongated shape. Finally the fragments of the comet lose physical contact with each other, and finally disperse or clump due to self-gravity, depending on the ratio of self-gravity to tides. Asphaug and Benz (1994) used an N-body code incorporating self-gravity and collisions to demonstrate that the 21-grain "cometesimal" model of Scotti and Melosh (1993) cannot fit the observations. Self-gravity results in pairs or larger clusters of cometesimals sticking together for densities greater than 0.05 g cm^{-3}, so the resulting chain contains far fewer than 21 observable fragments.

On the other hand, Asphaug & Benz (1994) used Dobrovolskis' (1990) analysis of gravity and strength in tidal encounters to show that the comet was effectively strengthless when it encountered Jupiter. For tides to fragment an intact object into $\sim$21 pieces, the maximum body strength must have been lower than the peak stress at perijove by more than an order of magnitude, *i.e.*, less than 1 dyn cm^{-2} (for comparison, ice has a

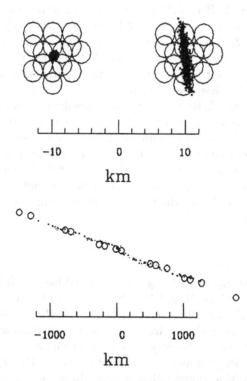

FIGURE 1. Comparison between evolution of a strengthless comet held together by self-gravity (Asphaug & Benz 1994) and evolution of a 21-grain, 9 km comet that breaks up instantaneously at a fixed time after perijove (following Sekanina *et al.* 1994). The first panel shows initial conditions. The second panel shows both models at the time of breakup. Note that, at this time, the *length* of the strengthless chain is equal to the *diameter* of the 21-grain comet. The final panel demonstrates that both models reach the correct position angle at late times.

strength of about 10^7 dyn cm^{-2}). This suggests the parent comet consisted of a number $N \gg 21$ of small particles interacting primarily through gravitational forces.

Asphaug & Benz (1994) then used their N-body code to show that the observations are best fit by a parent comet between 1.5 and 2.0 km in diameter, consisting of hundreds or thousands of grains. Although particles of fixed size were used for computational convenience, the results do not vary whether hundreds or thousands of particles are used in the computation, and so should not vary substantially if a realistic distribution of particle sizes is used. Figure 1 compares the development of such a comet to the development of a comet that breaks up at a fixed time after perijove. The bulk density of the parent body can be constrained to lie between 0.5 and 0.7 g cm^{-3}, because the number of clumps (fragments) observed in the chain is sensitive to density, due to the action of self-gravity.

Asphaug & Benz's numerical results agree with the analytical tidal breakup model of Sridhar & Tremaine (1992). These results were independently and almost simultaneously arrived at by Solem (1994), and were recently verified by Richardson, Asphaug, & Benner (1995) who use the exact orbit of the comet (including aspherical moments of the jovian field plus the influence of Saturn and the Sun) to follow the fragments until they arrive back at Jupiter two years later. The model appears robust. The discrepancy of comet diameter and position angle between the model of Sekanina *et al.* (1994) and the self-gravitating cluster models (Asphaug & Benz 1994, Solem 1994, Richardson *et al.* 1995) was resolved by Asphaug & Benz (1995). They demonstrate that, by the time Sekanina *et al.* allow breakup to begin (more than 2 hours after perijove), the 1.5 km diameter comet used by Asphaug & Benz has deformed into a 9 km *long* "cigar" (see Fig. 1) with a rotation period resulting from planetary torque that matches the rotational period that Sekanina *et al.* found for their 9 km diameter comet from observational constraints.

An analytic explanation has been advanced for the sensitivity of the number of the clumps to the density (Hahn & Rettig 1995). After breakup, comet material is distributed more or less uniformly along a cylinder. It then clumps due to Jeans instabilities driven by self-gravity. To get the observed number of fragments, assuming coagulation occurs when the cylinder crosses the Roche distance of the original parent, a bulk density of 0.6 g cm^{-3} is required, in good agreement with the numerical models.

A final point in favor of accepting the smaller comet is that it happens to agree with the median diameter derived from the record of crater chains on Ganymede and Callisto by Schenk *et al.* (1995).

3. Early stages of entry

Consider the first two peaks in the generic, ground-based, IR light curve, described by Nicholson (this volume) as P1 and P2. P1 appears to be caused by the meteor trail resulting from the initial entry of a comet nucleus, while P2 results from the appearance of the resulting fireball over the horizon as viewed from the Earth (see Fig. 2 of the plume review). A few nuclei showed a very faint precursor to the precursors that we could label P0, slowly rising for as much as a minute before P1. The form of P0, and its lack of detection by *Galileo*, suggest that it is produced by the massive meteor shower that occurs as the tidally stretched coma enters before the central nucleus.

The luminosity of the meteor trail can be estimated by considering the flight of the impactor through the upper jovian atmosphere. Graham *et al.* (1995) showed this to be a plausible explanation for P1 by assuming that drag dominated energy transfer to the atmosphere. At these very high altitudes, however, radiative ablation actually dominates the energy transfer. The mass loss due to radiative ablation of a spherical comet with radius r_c entering with velocity v_c is

$$\frac{dm_c}{dt} = -\frac{C_H}{2Q}\pi r_c^2 \rho(z) v_c^3, \tag{3.1}$$

where C_H is the heat transfer coefficient, $\rho(z)$ is the atmospheric density, and Q is the heat of ablation. The energy loss rate is then

$$\frac{dE}{dt} = \frac{1}{2}\dot{m}_c v_c^2 = -\frac{C_H}{4Q}\pi r_c^2 \rho(z) v_c^5 \tag{3.2}$$

$$= -(10^{31} \text{ erg s}^{-1})\rho(z)\left(\frac{r_c}{200 \text{ m}}\right)^2 \left(\frac{C_H}{0.1}\right)\left(\frac{v_c}{60 \text{ km s}^{-1}}\right)^5, \tag{3.3}$$

where $Q = 2.5 \times 10^{10}$ erg g^{-1} for an icy comet (Chyba *et al.* 1993). This is a factor of forty higher rate than Graham *et al.* (1995) used. However, the efficiency of conversion of this energy to visible radiation η remains unknown. Small meteors in the Earth's atmosphere have efficiencies $\eta \sim 10^{-4} - 10^{-2}$ (Bronshten 1983); larger objects may be somewhat more efficient as longer pathlengths increase the optical depth through hot gas. Graham *et al.* found that the peak flux of the first flash for the R impact was ~ 0.4 Jy at 2.3 μm, corresponding to 5×10^{19} erg s^{-1}. The detected trail would then require an atmospheric density $\rho = (5 \times 10^{-12} \text{ g cm}^{-3})\eta$. For an efficiency of $\eta = 0.01$ this corresponds to an altitude of 300 km above the 1 bar level, 1.4 s of flight time above the visible limb at

240 km. The trail could have been observed longer if the efficiency were higher, or if scattering from dust allowed its observation below the limb. Although the uncertainties are large, it appears plausible that P1 was indeed produced by the meteor trail through the upper atmosphere.

The most detailed published treatment of the meteor's passage through the deeper atmosphere is by Chevalier & Sarazin (1994), who attempt to describe the 60 $\mathrm{km\,s^{-1}}$ bow shock in some detail. They consider only the emission from shocked jovian air, neglecting any additional opacity added to the wake of the comet by ablated cometary material. At a pressure level of 100 μbar and an altitude of 200 km above the 1 bar pressure level, Chevalier & Sarazin (1994) find that hydrogen that has passed through a 60 $\mathrm{km\,s^{-1}}$ shock emits lines in the optical and UV. The opacity of atomic hydrogen at these temperatures is quite low, so they find that the lines remain optically thin until pressures increase to about 10 mbar, at an altitude of 100 km. At this level, the densest regions of hot air behind the bow shock become optically thick, and strong continuum radiation appears, with an optical luminosity $L_{opt} \sim (3 \times 10^{22}\ \mathrm{erg\ s^{-1}\ km^{-2}})A_c$ depending on the area of the comet nucleus A_c. Chevalier & Sarazin (1994) also find that shocked air in the wake will continue to emit in the optical for a few tens of km above the plunging comet, at a rate of $\sim 5 \times 10^{22}\ \mathrm{erg\ cm^{-2}\ km^{-1}}$.

The optical luminosity from the bow shock increases rather slowly with depth because more and more of the emission comes out in the UV as the shock temperature increases at greater densities (see Table 2 of Chevalier & Sarazin 1994). The UV pulse detected by the *Galileo UVS* probably occurred during this period. The optical luminosity reaches $L_{opt} \sim (9 \times 10^{22}\ \mathrm{erg\ s^{-1}\ km^{-2}})A_c$ at a depth of 100 km below 1 bar. The effective bolide diameter increases by a factor of five within a scale height when Rayleigh-Taylor instabilities become effective, as discussed in the next section (Mac Low & Zahnle 1994). The increase of bow shock area by a factor of 25 in a second probably produces the initial sharp rise in the *Galileo PPR* optical light curve. The initial peak has $L_{opt} \sim 5 \times 10^{23}\ \mathrm{erg\ s^{-1}}$ (Chapman, this volume). By assuming this factor of five increase in radius at the point of maximum energy release, I can estimate the initial bolide diameter based on the observed L_{opt}. Table 2 of Chevalier & Sarazin (1994) suggests that the luminosity at a pressure level of 2 bars is $L_{opt} \sim (8 \times 10^{22}\ \mathrm{erg\ s^{-1}\ km^{-2}})A_c$. Taking the increase of area into account, the observed value of L_{opt} suggests the initial diameter of the bolide was just over a half kilometer, supporting the argument for small impactors.

During the meteor phase, the bow shock surrounding the nucleus entering at a velocity v_c confines it and compresses it from its original density ρ_c to its compressible limit ρ_c', as shown in Figure 2. Field & Ferrara (1995) show that this occurs before the onset of the instabilities that ultimately tear the nucleus apart. They do this by directly computing the passage through the nucleus of the compression wave driven by the ram pressure of the bow shock. The front face of the nucleus decelerates to a velocity v_c', while the rest of the comet continues unaffected until the compression wave reaches it. The thickness of the compressed region h increases as more and more of the nucleus runs into it. The thickness can be found by equating the mass flux out of the uncompressed nucleus to the mass flux into the compressed layer to get

$$\frac{dh}{dt} = \frac{\rho_c}{\rho_c'}(v_c - v_c').$$

(3.4)

The velocity of the compressed layer can be found from the drag equation (eqn. 4.8) acting on it, so that the compressed layer grows as

$$h = \frac{H}{\rho_c'}[C_D \rho_c \rho(z)]^{1/2} \sec\theta.$$

(3.5)

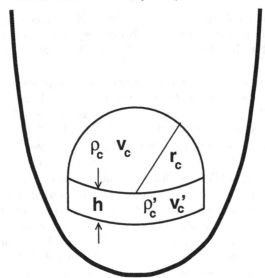

FIGURE 2. Initial compression of entering comet nucleus by bowshock. This will occur before other processes can affect the nucleus.

The final thickness of the layer, $h_f = (4\rho_c/3\rho'_c)r_c$ is reached at an altitude z_f, where the atmospheric density

$$\rho(z_f) = \frac{\rho'^2_c h^2_f}{C_D \rho_c H^2 \sec^2 \theta}. \tag{3.6}$$

For example, for $\rho_c = 0.5$ g cm^{-3}, $R = 300$ m, $\rho'_c = 1$ g cm^{-3}, and $\theta = 45°$, the altitude of complete compression is $z_f = +50$ km, more than a scale height above the level of final energy deposition, as the next section shows.

4. Final Deceleration and energy deposition

Where does the energy and mass of the incoming bolide get deposited in the atmosphere? The subsequent development of the fireball, partition of energy between the lower atmosphere and plume, and distribution of ejecta all depend on the initial conditions set up during this final entry phase. A number of different groups have used analytic and numerical models to try to answer these questions.

This section begins by describing the three main analytic approaches—ablation models, pancake models, and an instability model. I critique the simple ablation models, and then show that the pancake models succeed as well as they do because they are equivalent, to within factors of order unity, to the more physically detailed instability model. I then discuss the numerical models, why they disagree with each other, and attempt to show which ones are most reliable. Taking proper account of the continuing downward movement of the wake results in an altitude of peak energy deposition about half a scale height below the point of peak instantaneous energy deposition, the quantity predicted by the analytic models.

4.1. Analytic models

The first model published (Sekanina 1993) computed the energy deposition of a nucleus by following its radiative ablation. The mass loss due to radiative ablation is given by equation (3.1). Sekanina (1993) took his ablation coefficient (equivalent to C_H/Q

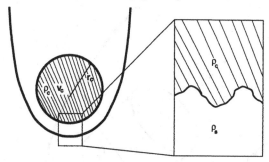

FIGURE 3. Growth of Rayleigh-Taylor instabilities along the front surface of an entering comet nucleus.

in eqn. [3.1]) from observations of terrestrial fireballs, getting an effective heat transfer coefficient $C_H = 0.5$, rather larger than the value usually assumed for meteors of $C_H = 0.1 - 0.2$ (Bronshten 1983). A constant heat transfer coefficient appears to explain observations of terrestrial meteors with diameters up to meters. Sekanina (1993) assumed that kilometer-sized objects would behave in the same way. For the smaller objects, the amount of ablation depends linearly on the flux of kinetic energy, $\rho v^3/2$, across the shock. However, radiative ablation of larger objects is limited by thermal emission from the hot, postshock gas, proportional to σT_s^4 (Biberman *et al.* 1980, Zahnle 1992). This does not increase as fast as ρ as the object moves deeper. Chevalier & Sarazin (1994) reach similar conclusions. Field & Ferrara (1995) go further by attempting to analytically model the advection of the vaporized surface layer. They conclude that even the rate given by Zahnle (1992) is an overestimate for these large objects because of the need to not only vaporize material off the surface, but also to decelerate it. For example, a nucleus with radius $r_c = 1.5$ km, and mass m_c loses only $3 \times 10^{-3} m_c$ as it falls through a scale height at the 1 bar pressure level.

The entering nucleus instead gets torn apart by the ram pressure from the bowshock. For large objects, this process determines the energy deposition profile. Energy gets transferred more efficiently to the atmosphere as the cross-section of the nucleus increases due to fragmentation. Fragmentation occurs because low-density, shocked gas decelerates the high density nucleus, causing the front of the nucleus to become Rayleigh-Taylor unstable as shown in Figure 3. (Note that, though Kelvin-Helmholtz instabilities occur, as argued in the chapter by Crawford, they grow about four times slower than the Rayleigh-Taylor instabilities, and so do not dominate the spreading of the impactor.) Svetsov, Nemtchinov & Teterev (1995) have analytically modeled the fragmentation due to Rayleigh-Taylor instabilities as follows.

The growth rate for these instabilities is (*e.g.*, Chandrasekhar 1961, p. 435)

$$\omega = \left(\frac{2\pi a}{\lambda}\right)^{1/2} \left(\frac{\rho_c - \rho_s}{\rho_c + \rho_s}\right), \tag{4.7}$$

where a is the acceleration of the interface, λ is the unstable wavelength under consideration, ρ_c is the density of the comet nucleus, and ρ_s is the density of the shocked gas. As long as $\rho_c \gg \rho_s$ the second factor on the right (the Atwood number) approaches unity. The acceleration can be derived from the drag equation,

$$a = \frac{C_D \pi r_c^2 \rho(z) v^2}{2 m_c}, \tag{4.8}$$

where the drag coefficient C_D is measured to be close to unity for a sphere, the atmospheric density at altitude z is $\rho(z)$, and the velocity is v_c. The relevant dynami-

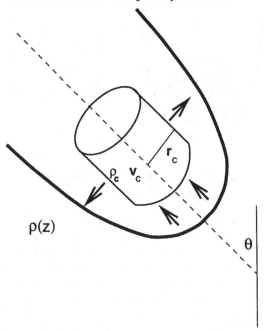

FIGURE 4. Pancake model: Quasistatic deformation of entering nucleus by the difference in ram pressure acting on the front and the side of the object.

cal timescale $t_{\rm dyn} \sim H \sec\theta/v_c$. While the smallest wavelengths grow the fastest, the wavelengths that fragment the bolide have wavelength $\lambda \sim r_c$. When the growth rate corresponding to that wavelength $\omega(\lambda,t)$ increases to the point that $\omega t_{\rm dyn} = 1$ the comet fragments, causing the energy deposition rate to increase dramatically. In fact, this is a simplification of the true physics, since shorter wavelengths that grow into the nonlinear regime begin to combine into longer wavelength perturbations (Read 1984, Youngs 1984), explaining why higher resolution is needed in gas dynamical computations to properly model the fragmentation. The atmospheric density at which fragmentation occurs is given by

$$\rho(z) = \frac{4}{3\pi C_D}\left(\frac{r_c}{H}\cos\theta\right)^2 \rho_c, \qquad (4.9)$$

where H is the scale height of the atmosphere.

A number of workers have modeled the entry and energy deposition profile using an even simpler model first proposed by Zahnle (1992) that has come to be called the pancake model (Zahnle & Mac Low 1994, Mac Low & Zahnle 1994, Chevalier & Sarazin 1994, Field & Ferrara 1995, Svetsov et al. 1995). The assumption used in this model is that once aerodynamic forces have overcome any material strength, the incoming object quasistatically deforms due to the difference in the ram pressure acting on the front and the sides, as shown in Figure 4. As the object flattens, its cross-section and its drag increase, bringing it to a rather abrupt halt accompanied by explosive energy release. The most elegant formulation of this model is given by Field & Ferrara (1995), who took a classical mechanical approach by deriving an equation of motion from Lagrange's equation. They draw the analogy to the response of a fluid in a dish when the sides of the dish are removed. Zahnle & Mac Low (1994) followed Chyba, Thomas & Zahnle (1993) in adding a term to account for radiative ablation in the upper atmosphere, and solving the resulting equation numerically. Here I present the simplest version of the pancake model in order to give analytic results.

The entering object is taken to be a right circular cylinder of radius r_c, density ρ_c, and height h, entering with initial velocity v_c, at an angle θ, as shown in Figure 4. The drag coefficient of a right circular cylinder is measured to be $C_D = 1.7$. The ram pressure on the front face $P_f = \frac{1}{2}C_D\rho(z)v^2 \sec^2\theta$, where v is the instantaneous velocity. The average internal pressure $P_i \sim \frac{1}{2}P_f$. It acts on the sides of the cylinder causing it to expand, so the force equation is

$$2\pi r_c h P_i = \pi r_c^2 h \rho_c \ddot{r}_c, \tag{4.10}$$

where the right-hand-side is the mass of the cylinder times its sideways acceleration. Converting from time derivatives to space derivatives, $d/dt = v \sec\theta(d/dz)$, and expanding, we find that the radius evolves as

$$r_c\frac{d^2 r_c}{dz^2} + \frac{r_c}{v}\frac{dv}{dz}\frac{dr_c}{dz} = \frac{C_D\rho(z)}{2\rho_c}\sec^2\theta. \tag{4.11}$$

The nonlinear term may be neglected to first order. In an exponential atmosphere, this equation may then be approximately solved,

$$r_c(z) \simeq H \sec\theta \left(\frac{2C_D\rho(z)}{\rho_c}\right)^{1/2}. \tag{4.12}$$

This is effectively equation (4.9), derived by Svetsov *et al.* (1995) by considering Rayleigh-Taylor instabilities. This agreement accounts for the surprisingly good performance of the pancake model when compared to numerical models (Mac Low & Zahnle 1994).

Substituting equation (4.12) into the drag equation (4.8) and integrating, the velocity of the impactor

$$v(z) = v_c \exp\left(-\frac{\pi C_D^2 \rho(z)^2 H^3 \sec^3\theta}{2m_c\rho_c}\right). \tag{4.13}$$

The energy release rate neglecting ablation is $dE/dz = mv\,dv/dz$, so the pressure level of maximum energy release is

$$P_0 = (9.0\text{ bars})C_D d_{km}^{3/2}\left(\frac{\rho_c}{1\text{ g cm}^{-3}}\right)\left(\frac{g}{2486\text{ cm s}^{-2}}\right)\left(\frac{H}{45\text{ km}}\right)^{-1/2}\left(\frac{\sec\theta}{\sqrt{2}}\right)^{-3/2}, \tag{4.14}$$

where g is the acceleration of gravity, and d_{km} is the diameter of the comet, measured in km. The jovian atmosphere actually does not have a constant scale height, so this equation should be solved iteratively, starting with a guess at the scale height in the region of maximum energy deposition, and then using on subsequent iterations the scale height H corresponding to the pressure level P_0 given by the previous iteration. Below, I will show that numerical models suggest that the final level of energy deposition is about half a scale height deeper than the instantaneous level given here, which should also be taken into account when using this equation. Figure 5 shows the height of maximum instantaneous energy deposition for impactors of different diameter and density. including the effects of high-altitude radiative ablation as described in Zahnle & Mac Low (1994), which shifts the curves slightly from the purely analytic solution.

4.2. Numerical models

In contrast to the analytic models, numerical models of the entry of a comet nucleus have come to a remarkably broad range of conclusions. A nucleus with diameter of 1 km, and density 1 g cm^{-3}, has been predicted to explode everywhere from the stratosphere to well below the water clouds. I explain the differences by examining the numerical

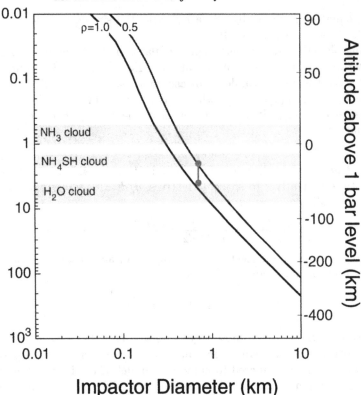

FIGURE 5. Altitude of maximum instantaneous energy deposition for impactors of density 0.5 and 1.0 $g\,cm^{-3}$ as a function of impactor diameter, using the model described in Zahnle & Mac Low (1994). The top of the dumbbell-shaped point shows the height of instantaneous energy deposition of an impactor of the size described by Asphaug & Benz (1994), while the bottom shows the height of final energy deposition after downward advection (see text).

methods, grid resolution, and diagnostic measurements used. It appears that downward advection of the wake lowers the height of maximum energy deposition about half a scale height below the instantaneous value given by the analytic models.

4.2.1. *Critical review*

I divide the models into four categories. First, are those groups that show maximum energy deposition at or above the predictions of the pancake model, including Mac Low & Zahnle (1994), Yabe *et al.* (1994), and Shoemaker, Hassig, & Roddy (1995). I also include in this category the computations of shocked clouds of interstellar gas by Klein, McKee & Colella (1994) that have been independently analyzed using a model equivalent to the pancake model. Second, the Sandia group (Boslough, Crawford, Trucano, and Robinson) has converged on a prediction of energy deposition about a scale height deeper than the analytic model in a series of important papers (Boslough *et al.* 1994, 1995; Crawford *et al.* 1994, 1995). Third, two groups predicted deep penetration, with a smooth energy deposition profile having no clear point of maximum energy release (Takata *et al.* 1994; Wingate, Hoffman, & Stellingwerf 1995). Finally, two groups presented preliminary results but did not pursue the problem far enough to come to solid conclusions, so I will not discuss them further (Vickery 1993; Moran & Tipton 1993).

First, I discuss the models that give results at or above the prediction of equation (4.14). I will discuss my own models more extensively below, but briefly describe here three other models. Hassig, Roddy & Shoemaker (1995) used a multi-material code that attempts to follow the fragmentation and vaporization of a solid body. Although, due to computational expense, they only followed the last 100 km or so of the entry path, starting at 31 km below 1 bar, they appear, with this rather different method, to get results in agreement with equation (4.14). Yabe *et al.* used a third-order, Eulerian code with a maximum resolution of 30 zones across the radius (I will hereafter denote the resolution by the notation Rn, where n is the number of resolution elements in the radius of the object, so the effective resolution here, for example, is R30). They tested several equations of state, and, in agreement with the Sandia group and Mac Low & Zahnle (1994) found little dependence. Their quantitative result—explosion at 10 bar for a 3 km diameter comet—is significantly higher than the 75 bar prediction of the pancake model (eq. 4.14) for their parameters. They did not use a realistic jovian atmosphere, instead using one with a constant scale height of 43 km and a comet hitting at normal incidence rather than an angle of 45°; however it is not clear why that should cause their simulations to show such high explosions. This result needs further examination.

An interesting point of comparison is to models of clouds of interstellar gas hit by supernova blast waves. This is a question of some interest for the state of the interstellar medium and has been attacked vigorously with analytic and computational methods. Klein *et al.* (1994) used an adaptive mesh refinement technique on an Eulerian grid to achieve effective resolutions as high as R240 on this problem. They performed a careful resolution study to ensure that they only quoted converged results. They independently derived an analytic model of the acceleration of the cloud by the blast wave that is equivalent to Zahnle (1992), as shown in Mac Low & Zahnle (1994), and showed that it fit their numerical results well.

Next let me discuss the computations performed by the Sandia group (Boslough *et al.* 1994, 1995; Crawford *et al.* 1994, 1995). They computed their entry models on an Eulerian grid large enough to capture the entire entry wake. To do this, they placed a moving region of fine zones at the position of the nucleus, and used a ratioed grid with slowly increasing zone sizes along the tail. By this method, they achieved resolutions of R25 to R50 at the nucleus, sufficient to capture its breakup, according to the resolution study done by Mac Low & Zahnle (1994). The resolution slowly degraded along the wake, with zone sizes reaching 5 km at a distance of 100 km (Crawford *et al.* 1995) and 25 km at the lowest resolution. The advantage of this grid is that they were able to directly measure energy deposition in the atmosphere at the end of their computation, taking account of the downward advection of energy in the moving wake swept up by the nucleus. In contrast, the pancake model computes the instantaneous loss of kinetic energy from the cometary material; that was also the quantity measured by Mac Low & Zahnle (1994) and Yabe *et al.* (1995). I'll show below that this explains part of the difference between their reported results and those of Mac Low & Zahnle (1994).

There was some confusion about how deep the Sandia group actually predicted maximum energy deposition for a 1 km object. This appears to have occurred because, in their first papers (Crawford *et al.* 1994, Boslough *et al.* 1994) they reported results for an object with a 100 bar yield strength (appropriate for a solid, stony asteroid) that deposited its energy at 180 km. However, later they quoted results for a strengthless, but incompressible, object probably more appropriate for the rubble piles suggested by Asphaug & Benz (1994), that deposited its energy at 120–130 km (Crawford *et al.* 1995).

Third, I discuss the models that found deep penetration and smooth energy deposition profiles. Takata *et al.* (1994) used a smooth particle hydrodynamics (SPH) code. SPH follows the fluid with particles, and computes intensive quantities such as density and pressure by averaging over the particles. Although it appears able to reach arbitrary resolution, in practice a minimum particle separation or smoothing length must be enforced to avoid extremely small timesteps and excessive computation time. Takata *et al.* used a smoothing length of $0.25 R_c$, giving an effective resolution of only R4. This resolution cannot resolve the Rayleigh-Taylor instabilities that bring the comet to an abrupt halt as discussed above. MacLow & Zahnle (1994) showed computations on an Eulerian grid at a resolution of R6 that behaved very similarly to the models shown in Takata *et al.* and showed that the behavior changes at higher resolution, where it converges on a different solution.

Wingate *et al.* (1995) also used an SPH code, but at much higher resolution, with effective resolutions reaching R50, easily sufficient to resolve the instabilities. Their computation apparently suffered from a more subtle problem, however. Because of the stochastic nature of SPH particles, defining the exact position of a shock front is difficult. As a result, shocks tend to be much broader than in grid-based codes of similar resolution. A shock front is, of course, stable against Rayleigh-Taylor instabilities, although the contact discontinuity between the shocked gas and the comet nucleus is unstable. In the computations shown by Wingate *et al.* (1995), the contact discontinuity was (at least intermittently) stable, probably because the shock was so broad that it sometimes overlapped the contact discontinuity and stabilized it. As a result, the nucleus penetrated much deeper than it would have otherwise. This may also explain the strong oscillations they observed, if the shock moved on and off the contact discontinuity.

4.2.2. *Computations*

For my own computations described here, I used ZEUS, a general purpose astrophysical MHD code, developed by M. L. Norman and his students at the Laboratory for Computational Astrophysics (LCA) of the National Center for Supercomputing Applications. A full suite of test problems is described by Stone & Norman (1992). The particular version of the code used here is called ZEUS-3D, and was developed by Norman and D. A. Clarke. It is publically available by registration with the LCA at lca@ncsa.uiuc.edu. The code uses second-order Van Leer (1977) advection on an Eulerian, moving grid in Cartesian, cylindrical or spherical geometry. The code is fully three-dimensional and includes magnetic fields, but for reasons of time we have only done two-dimensional, gas dynamical models to date. Shocks are resolved using a Von Neumann artificial viscosity. I have implemented both tracer fields and tracer particles for this problem.

For our models we use two different equations of state. One is just an adiabatic equation of state with adiabatic index $\gamma = 1.2$ at early times when dissociation and ionization is important (Chevalier & Sarazin 1994) or $\gamma = 1.4$ for computations extending to later time (*e.g.*, those discussed in the plume review). The other is a stiffened gas equation of state appropriate to ice, described by Mac Low & Zahnle (1994), which we use for the comet nucleus in our initial entry models. The entry models published in MacLow & Zahnle (1994) used a cylindrical grid 5 km high by 3 km in radius. The innermost 3 km by 1 km has the full resolution specified (up to a maximum of R100, corresponding to a zone size of 5 m); outside of that there is a layer of zones each a factor of 1.03 bigger than the one inside until zones ten times as large as the central zones are reached, and finally a layer of constant-size zones out to the edges of the grid, for a total of 0.2 megazones in our R100 computation.

The energy deposition of these models was determined by measuring the loss of kinetic energy by cometary material on the grid. Figure 6 shows that, measured this way, the numerical models agree rather well with the simple analytic model given by equation (4.14).

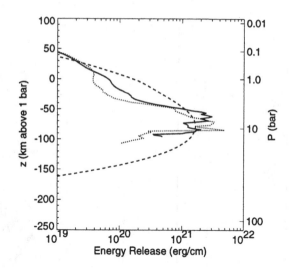

FIGURE 6. Comparison of energy release profiles for incompressible (*solid curve*) and compressible (*dotted curve*) equations of state run at a resolution of R50. The simple pancake model of equation (4.14) is also shown (*dashed curve*), with a drag coefficient $C_D = 1$, a pressure scale height $H = 45$ km, and a 1 km diameter nucleus. From Mac Low & Zahnle (1994).

However, this method of measurement has two problems. First, cometary material flows off the back of the grid while it still carries some kinetic energy, and second, kinetic energy transferred to the ambient atmosphere is not immediately thermalized, but instead is partitioned between kinetic and thermal energy in the atmosphere. The second effect appears to be far more important, as it causes significant downward advection of energy contained in the ambient atmosphere, as pointed out by the Sandia group.

To measure these two effects, I ran a new R50 computation, but instead of cutting off the wake after 5 km, I extended the region with coarse vertical grids (100 m zones) up for a full 100 km, and the region with coarse radial grids out to 10 km, for a total of 3.1 megazones. For comparison, at 100 km above the entering nucleus, the Sandia group has vertical zones 50 times as large. I forced the grid to follow the entering comet nucleus until it had fragmented to such an extent that no zone contained more than 90% comet material, and then allowed material to flow off the bottom of the grid. (In retrospect, this was not the best way to do this—following the downward flow for longer would have been better—but the computation took 40 Cray Y-MP hours so I have not yet run a better model.)

Figure 7 shows the entry wake just after the comet nucleus has passed through the altitude of maximum instantaneous energy deposition. The shape of the wake shows the explosive nature of the energy deposition when the nucleus begins to fragment. The piece of the nucleus seen penetrating deeper into the atmosphere carries about a quarter of the kinetic energy of the initial nucleus. This is an upper limit to the real value, as the deeply penetrating piece is badly underresolved. Comparison between the R50 and R100 models of Mac Low & Zahnle (1994) suggests that the central fragment gets torn apart more thoroughly than an R50 model shows, and so stops more quickly.

I tested for downward advection of energy as suggested by the Sandia group by integrating the actual thermal energy contained on the grid at several times after the nucleus reached the altitude of maximum energy deposition. The resulting profiles are compared to the instantaneous energy loss from the cometary material in Figure 8. The interpre-

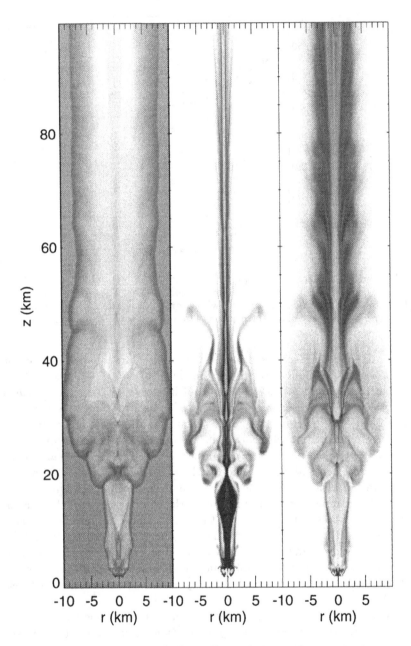

FIGURE 7. Greyscale images of 100 km of the wake of a 1 km comet nucleus, computed with a resolution at the nucleus of R50 (10 m) and in the tail of 100 m. The bottom of the grid lies 153 km below the 1 bar pressure level. The variables shown are the log of density, the specific energy (roughly equivalent to temperature), and the concentration of comet material. In each case, white is highest and black is lowest. This model is equivalent to the incompressible R50 model of Figure 6, except for the much larger grid.

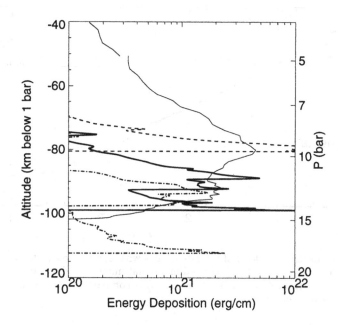

FIGURE 8. Profiles of thermal energy deposited in the atmosphere at times separated by one second (*dashed and solid lines*), compared to instantaneous energy loss by the comet nucleus (*dotted line*). The lowest peak in each of the thermal energy curves corresponds to the bow shock, while the higher one is the region of peak energy deposition, which can be seen moving downwards. Note that energy starts flowing off the sides of the grid at the altitude of peak energy deposition (see Fig. 7) after the second curve, so the peak energy is artificially low in the third curve.

tation of this figure is, unfortunately, complicated by the narrow radial extent of the grid, as the developing blast wave runs off the grid sideways, carrying a fair amount of energy with it. However, the downward advection of the energy peak can be seen. It appears that, in this high resolution model, the peak moves down about 30 km from the altitude predicted by the pancake model. This is within 20 km of the results obtained by the Sandia group. I believe that the remaining difference comes from the lack of resolution in the wake in the Sandia computation. The low resolution tends to suppress shear instabilities that will act to slow the downward moving wake. Nevertheless, the results are close enough that I conclude that we have reached agreement on the question of the altitude of energy deposition.

Another issue illustrated by this model is that the comet material ends up at the very highest temperatures, so it will not be confined in the deep atmosphere, as suggested by the Sandia group, Takata *et al.* (1994), and others, but will rise with the plume. At least half the mass of cometary material at the end of the computation has stopped moving downward or already begun to rise.

The filamentary distribution of high-temperature cometary material seen in Figure 7 suggests that its high temperatures would not be captured at lower numerical resolution. This becomes an important issue, as discussed in the next section. Every computational model published, aside from my own, attempts to directly transfer the numerical results of entry models to a larger, lower-resolution grid in order to compute the development of the fireball. The interpolation process involved makes it nearly impossible to maintain the high temperatures shown by the high resolution computation shown in Figure 7, and

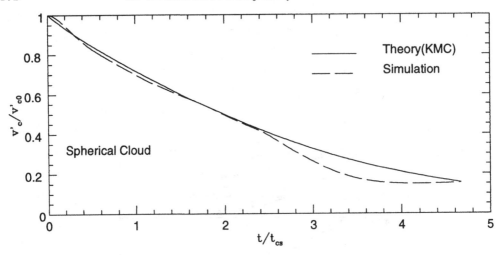

FIGURE 9. Acceleration of a cloud of interstellar gas with density contrast $\chi = 10$ hit by a shock wave with Mach number $M = 10$. A three-dimensional numerical simulation (*dashed line*) with resolution of R60 performed by Xu & Stone (1995) is compared to the analytic pancake model (*solid line*) of Klein *et al.* (1994).

so most models do not model the rise of the comet ejecta properly. (The next section discusses the different, but equally large, uncertainties in my own fireball models.)

4.2.3. *Three-dimensional models*

Adequately resolved, three-dimensional, models of the comet entry remain needed. The Sandia group showed an R10 computation (Crawford, this volume), and Takata *et al.* (1994) have done the R4 computation described above. On the basis of their R10 computation, Crawford speculated that three-dimensional models might show deeper penetration due to non-axisymmetric fluting seen on the edges of the object.

However, adequately resolved, three-dimensional models have been computed for the related problem described in § 4.2.1, of a supernova shock hitting an interstellar cloud. Xu & Stone (1995) have computed that problem at a resolution of R60, and find that the acceleration of the cloud has converged well at that resolution, as would be expected from the two-dimensional models. The fluting instability does indeed occur in the material torn off the sides of the cloud, as suggested by Crawford (this volume). The acceleration of the cloud can be compared to the analytic model of Klein *et al.* (1994), as shown in Figure 9. The behavior of the corresponding two-dimensional model at twice the linear resolution (R120) is shown in Figure 12a of Klein *et al.* (1994). Comparison of that figure to Figure 9 shows that the three-dimensional model agrees within 3% with the two-dimensional model; both models diverge slightly from the analytic model, in the same direction. This comparison suggests that three-dimensional effects can also be safely neglected in computing the energy deposition profile of an entering comet nucleus.

5. Fireball development

This section treats the fireball resulting from the energy deposition discussed in such detail in the previous section. I discuss the initial conditions, the brightness of the resulting fireball, and the difficulty of determining the energy of the explosion from the height of the observed plumes.

5.1. *Initial conditions*

The advantages of starting off a computation of the fireball by directly using the final result from an entry computation are obvious. Most modelers have adopted this approach. The difficulties are more subtle. I have taken a different approach that has a complementary set of advantages and problems, namely using the analytic pancake model to generate the initial conditions for the fireball.

Directly using an entry computation for the initial conditions of the fireball requires interpolation from the fine grid used in the entry computation to the coarser grid needed to compute the much larger scale explosion of the fireball. This inevitably requires averaging over much of the detail of the entry computation. Figure 7 shows that important variables such as temperature vary strongly on length scales of a few hundred meters. In my highest resolution, two-dimensional, fireball models, the smallest zones I have used are 0.5 km across; usually I use at least 1 km zones. Three-dimensional computations are even more demanding—Crawford *et al.* (1994) used 5 km zones for their model of a 3 km impactor, and only somewhat better (though unspecified) resolution for their model of a 1 km impactor. The result of this interpolation is to unphysically reduce the temperature of the cometary material and the most strongly shocked jovian atmosphere.

The cometary material in particular should also be hot because its composition gives it a high mean weight per particle μ compared to the jovian atmosphere; the temperature of gas shocked to a particular velocity is directly proportional to μ. Reduced temperatures in the plumes result in longer rise times, reduced transport of cometary material, and smaller plumes for the same energy deposition.

Instead, I treated the analytic energy deposition profile as a moving line charge (Zahnle & Mac Low 1995), which has a similarity solution. (My models of the fireball mostly neglect the correction to the analytic model due to the downward advection of the wake. Preliminary tests show that it does not qualitatively change my conclusions, but it should be included for quantitatively correct results.) I assume that the cometary material is mixed with its own weight of jovian atmosphere when computing densities, compositions, and temperatures.

The radius of the wake left by the moving line charge is a parabola in altitude z,

$$R(z) \approx \left(r_c^2 + \left(\frac{4}{\pi} \frac{(\gamma+1)^2(\gamma-1)}{3\gamma+1} \right)^{1/2} \frac{1}{\rho(z)} \frac{dE}{dz} \frac{z-z_0}{v_c} \right)^{1/2}, \quad (5.15)$$

where z_0 is the altitude of maximum energy deposition; r_c is the radius of the impactor and v_c is its velocity. The energy deposition rate dE/dz is computed directly from the pancake model, as described below equation (4.13). The wake is assumed to be well-mixed, so that energy and mass are uniformly distributed across it. In most of our models, I neglected the vertical velocity, since an ideal moving line charge has no momentum. (As discussed in § 4.2.2, this is somewhat inaccurate, but preliminary work shows only quantitative changes in our conclusions due to it.) The energy density in the wake is then

$$e(z) = \rho_{atm}(z) \frac{\eta_e}{R^2(z)} \frac{dE}{dz}. \quad (5.16)$$

The wake set up is only few zones wide, so clipping due to the finite size of the zones becomes quite important. I apply fudge factors η_e and η_d to bring the total energy and density deposited up to the correct values. The assumption of a moving line charge does not describe the deposition of mass in the wake by the comet. Instead, I assume that all the energy deposition comes from cometary material being deposited at rest, so that

$$\frac{dm}{dz} = 2v^2 \frac{dE}{dz}.$$ (5.17)

This is an approximation, since

$$\frac{dE}{dz} = \frac{1}{2}v^2 \frac{dm}{dz} + mv\frac{dv}{dz}.$$ (5.18)

The first term dominates at high altitude, while the second dominates during the final flare and explosion. Better wake models should take this into account. The mass density in the wake is then

$$\rho'(z) = \rho(z) + \frac{\eta_d}{R^2(z)}\frac{dm}{dz}.$$ (5.19)

Clearly, this description of the entry wake is even more simplified than that given by direct interpolation of entry models onto a larger grid. However, since it is analytic, it can be easily changed. Even though it gets the details wrong, it can capture the essential characteristics of the entry wake, such as the high temperature of the cometary material. Many different models can be run to understand which features are robust and which highly dependent on the details of the initial conditions, and indeed we have run many models in the course of our research. (See Zahnle & Mac Low 1994 for an example; since then we have more than tripled the number of models run.)

5.2. *Fireball luminosity*

It is difficult to compute the luminosity of the visible fireball—the optically thick gas emitting in visible light. The area of the emitting surface remains quite small during this period, so radiation carries away only a few percent of the total energy, and conservation laws cannot be called upon to help constrain the answer. Instead, a direct computation of the opacity of the hot gas (most likely due to H^-) must be done, which involves a careful model of the microphysics. Computing the luminosity of the infrared fireball has similar problems, compounded by our even poorer understanding of the formation of infrared opacity sources (most likely dust).

The connection to the observations appears qualitatively clear, even if it remains quantitatively opaque. The *Galileo PPR* observations begin when the entering object explosively expands at the altitude of peak energy deposition, as discussed in § 3, with a contribution from the hot, expanding wake above. As viewed from Earth, this occurs while the fireball is still behind the limb of the planet. The fireball is then observed by *Galileo* to rise, expand and adiabatically cool. By the time the observed fireballs rise over the limb of the planet into sight from the Earth, they have cooled enough that they emit strongly in the near-IR, but not in the visible, producing the peak designated P2 in the IR light curves (Nicholson, this volume).

Ahrens *et al.* (1994) attempted to compute the visible opacity of the fireball by using an arbitrary but plausible grey opacity for their models of the fireball. They showed that, for a 1 km object with energy of 10^{28} ergs, the *visible* light fireball should have been easily observable from the Earth as it rose from behind the limb. This conclusion appears to be fairly robust, since Zahnle & Mac Low (1995) arrived at a similar result by assuming that cometary material with solar abundances of all elements except H and He was initially mixed equally by mass with clean jovian atmosphere. They followed the advection of the cometary material using a tracer field, and computed the temperature and H^- opacity by solving a Saha equation including the easily ionized metals that can provide the free electrons necessary to produce the H^- ions. Zahnle & Mac Low (1995) present visible light curves of the fireball, as viewed from the side, including the effects

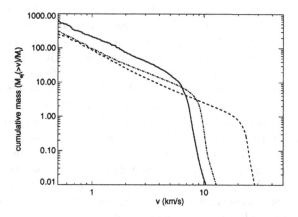

FIGURE 10. Cumulative mass above each velocity as a function of velocity for three different models with energy of 2×10^{27} ergs. One model uses the standard initial conditions described in § 5.1 (*dash-dotted line*), one model reduces the total density in the wake by a factor of four, increasing the temperature (*dashed line*), and one model gives the initial wake a downward velocity of 20 km s^{-1} (*solid line*).

of the horizon. The lack of a visible fireball observed from the ground therefore suggests that the impacting objects had energies significantly less than 10^{28} ergs.

I note that *HST* did observe an extremely faint visible fireball almost lost in the glare from the limb (Hammel, this volume). The *Galileo PPR* result that the visible fireball was only about 10% of the luminosity of Jupiter (Chapman, this volume) also supports the conclusion that the fireballs were far fainter than expected for 1 km objects with density of ice.

5.3. *Plume heights*

A number of groups have attempted to calibrate the size of the impactors by comparing the observed heights of the plumes to numerical models, including Takata, Ahrens, & Harris (1995), Crawford *et al.* (1995), and Shoemaker *et al.* (1995). These efforts have two problems, one minor, and one major. The minor problem is that the plumes are observed in sunlight reflecting off of dust, but no model exists for dust formation that can give a prediction for what density of gas will form enough dust to be observed. The major problem is that the velocity of upward expansion of the plume appears quite sensitive to the details of the initial conditions.

These problems can be demonstrated with fireball models using the initial conditions described above. The plume review also uses these models to calibrate the simple analytic models described there. These models use a ratioed grid in cylindrical coordinates, with the highest resolution region (1 km zones) extending from -150 km to 100 km above the one bar level, and for 75 km radially. The size of the zones then increases by 3% per zone until the zone size reaches 10 km, and then remains constant to the edges of the grid at an altitude of 3001 km and a radius of 5076 km, for a total grid size of just under 0.4 megazones. The adiabatic constant $\gamma = 1.4$. The model with energy of 2×10^{27} ergs that I will pay most attention to has an impactor with density $\rho_c = 1$ g cm^{-3}, diameter 600 m, and altitude of maximum energy deposition 25 km below 1 bar. Clipping requires that I set $\eta_e = 2.5$ and $\eta_d = 1.7$ to reach the correct input energy and mass. This particular model is also examined in detail in the plume review.

Let us examine the plume on its way up. Note that there is no sharp line between the end of the fireball phase and the beginning of the plume phase, though practically speaking, it occurs when the rising gas no longer emits detectable thermal emission, and only reflects sunlight. (At least until it falls back and gets reheated, as the plume review discusses.) A useful way of displaying the results of these models is by measuring the cumulative mass above each velocity, as shown in Figure 10. This figure shows the results from three different runs with energies of 2×10^{27} ergs, but with slight variations on the initial conditions. One run uses exactly the initial conditions described in § 5.1 and above. The next run uses those initial conditions, but with the density in the wake, $\rho'(z)$ in equation (5.19), reduced by a factor of four, increasing the temperature in the wake without changing the total energy. The third run uses the initial conditions of §5.1, but gives the wake a uniform downward motion of 20 $\mathrm{km\,s^{-1}}$, somewhat faster than the energy actually appears to be advected in Figure 8. This last run represents the maximum effect that the downward motion of the wake should have, both because it is moving downward so fast, and because the initial wake is quite underresolved, suppressing shear instabilities that would otherwise slow the wake.

Figure 10 shows that the main difference between the three different sets of initial conditions is the position of the knee where each of the cumulative velocity curves bends over. The shapes of the three curves remain quite constant, suggesting that this is a robust result. Zahnle & Mac Low (1995) discuss in detail why a power-law distribution of ejecta such as this is expected from a blowout. It is tempting to treat the knee as the natural edge of the plume, since relatively little gas rises faster than this velocity. The presence of the knee in all three models suggests that the problem of not knowing where the plume ends may indeed not be a major one, though dust formation models will be required to answer the question definitively.

The position of the knee, on the other hand, clearly depends sensitively on the details of the initial conditions. In order to reach the heights observed of about 3000 km above the 1 bar level (Hammel, this volume), a vertically rising plume must travel at 12 $\mathrm{km\,s^{-1}}$, while a plume rising at 45° must travel at 17 $\mathrm{km\,s^{-1}}$. The Sandia group has shown that, in fact, the plume does tend to follow the entry path initially. However, Jessup, Clarke & Hammel showed in a poster at this conference that the plumes did not travel sideways at 12 $\mathrm{km\,s^{-1}}$, so the plumes may begin to rise more vertically once they have left the atmosphere.

Clearly, gas in these models can reach the correct velocities, but calibration appears very difficult. I believe that it is quite likely that, in the short term, modelers may be forced to calibrate their models against the plume height observations, rather than being able to predict them, because of the difficulty of correctly modeling the details of things like the equation of state of jovian air mixed with varying amounts of cometary material. Therefore, I believe that no solid information about the size of the impactors can currently be derived from models of the plume height.

6. Plume bounces

I am now going to skip lightly over the period of plume fallback that generates most of the observed IR, the main event in Nicholson's terminology (this volume), and the chemistry of the plume and spots, all of which is discussed in the plume review. However, the last act of the plume dynamics deserves attention. The ballistic plume rises to its peak height, where all its kinetic energy has been converted into gravitational potential energy, and then falls, regaining kinetic energy until it hits the atmosphere at the same velocity it was ejected, converting its energy back to thermal energy in a shock wave. The strong IR radiation of the main event carries away much of this energy, but not all of it. The remaining energy gets converted back into kinetic energy by the expansion of the shocked, high-pressure, layer of plume material back up into space when infalling plume material no longer confines it. This material eventually comes crashing back down, releasing further IR radiation.

I noted these bounces in simulations well before the event. For example, Figure 11 shows a plot of material moving back up shown in my presentation at the Maryland Workshop in 1994 January. I discounted the appearance of bounces in the models, expecting the radiative cooling to be strong enough to suppress them, until I saw the first light curves from Nicholson (this volume) and Graham *et al.* (1995) that clearly show two bounces ten and twenty minutes after the main event. The alert reader may note what neither I nor anyone else realized at the time, namely that if the radiative cooling were strong enough to suppress the bounces, it would produce strong enough emission to be easily observable, as, indeed, it was. This, in my opinion, was the fundamental error that prevented a clear prediction of the main event from being made well before the impacts.

A one-dimensional model of the bounces, including radiative cooling, was presented by Deming *et al.* at this meeting. In this model, the infalling plume was modeled as a simple slab of gas falling from 3000 km onto the jovian atmosphere, and the radiative emission was modeled with a grey opacity of $0.4 \, \mathrm{cm}^{-2} \, \mathrm{g}^{-1}$. Figure 12 shows the emission from the series of bounces. The timing and relative brightness appear to match the observations convincingly.

7. Conclusions

In this article, I have presented a critical review of models of the tidal breakup, entry, and fireball, as these provide the basis for any deduction from the observations of the size and energy of the objects. I believe that the areas that are well understood are the initial entry and energy deposition, the behavior of the plume once it has been launched, and, although controversy continues, the tidal breakup models. The area that currently needs further work is the translation of the models for energy deposition into satisfactory initial conditions for models of the initial development of the fireball from the entry wake.

The evidence for the size and energy of the objects can be broken into three categories— evidence for shallow penetration of the jovian atmosphere, evidence for small impactor energies, and direct evidence for small impactor sizes.

Observations of sulfur and water appear to show that the explosions occurred above the water clouds. Zahnle *et al.* (1995) show that the observations of large amounts of S_2 and CS_2 and small amounts of SO_2 can be best explained by shock chemistry in dry jovian air, perhaps mixed with small amounts of cometary oxygen. Direct observations of H_2O and CH_4 in the G and K impact sites by Bjoraker *et al.* (1995) gave a ratio of around unity, which they interpreted as an observation of cometary water in quantities of order 10^{12} g, equivalent to spheres of ice with diameters of order 100 m, clearly a lower limit to the impactor sizes. (Note that Lellouch, in this volume, interprets CO observations to derive objects as large as 1 km at densities of $0.5 \, \mathrm{g} \, \mathrm{cm}^{-3}$, at the upper limit of our range.) Radio observations of thermal emission at 3 and 6 cm show no perturbations at the 5 bar level where the water clouds are expected to lie (Grossman *et al.* in a poster at this meeting), also suggesting energy deposition higher in the atmosphere, though beam dilution could hide small perturbations. Models of the entry show that only small impactors will deposit their energies at such high altitudes.

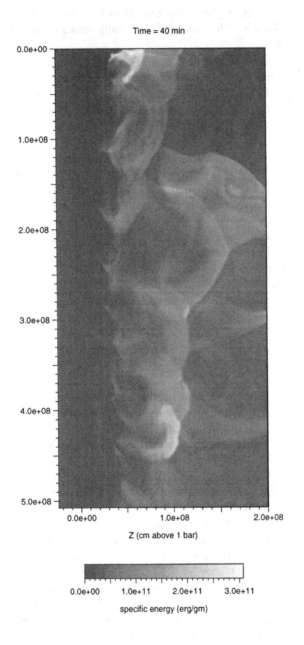

FIGURE 11. Specific energy (roughly equivalent to temperature) of gas at a time of 40 minutes after impact for the plume from a 1 km object in the absence of radiative cooling. Note that the center of the planet is to the left, and that the temperature minimum in the tropopause may just be visible at the left of the plot as a darker line. Regions of high specific energy trace hot shock waves as material bounces up.

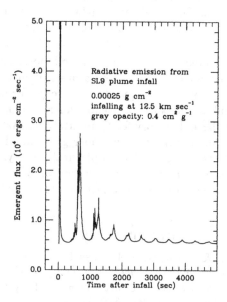

FIGURE 12. Model by Deming *et al.* of the emission from the bouncing plume, using a one-dimensional model with an infalling slab of gas and the parameters noted. Note that the emission from the first bounce goes far off scale on this linear plot. The second and third bounces were observed.

The luminosity of the visible fireballs and the infrared main events suggest low impactor energies, as does the lack of observed seismic waves. Kilometer-sized impactors with density close to 1 g cm^{-3} would have produced visible fireballs with luminosity comparable to Jupiter's as observed from *Galileo*, and easily visible from the Earth as they rose above the horizon (§ 5.2; Zahnle & Mac Low 1995). The luminosities and lifetimes of the infrared main events are best fit by semi-analytic plume models with low energies as well; for example, the R impact is best fit by an impact with energy 6×10^{26} erg (the plume review; Zahnle & Mac Low 1995).

Attempts were made to detect seismic waves generated by the impacts by Marley *et al.* (1994) and Logonné *et al.* (1994). Neither group was successful, giving upper limits on the energy of the impactor of under 10^{28} erg. These limits depend on the partitioning of energy into the seismic modes, which is not known well; however neither group assumed more than 30% efficiency. Higher efficiencies might be expected if most of the impact energy were buried in the atmosphere, which would lower the energy limits even further. These limits will probably also be reduced as the search images are processed more carefully.

Self-gravitating cluster models of the tidal breakup by Asphaug & Benz (1994) and Solem (1994) can explain the observed length and position angle of the comet train, and the number of fragments in it. These models conclude that the parent body had a diameter of 1.5 km and a density of 0.6 g cm^{-3}, suggesting that the fragments have radii in the half-kilometer range. Another piece of circumstantial evidence for a small parent body is the estimate of the total volume of dust in the stratosphere by West *et al.* (1995), who derived a mass equivalent to a sphere of diameter 1 km the day after the last impact.

From these various pieces of evidence, it appears that the most coherent picture emerges if the largest nuclei of Comet Shoemaker-Levy 9 had diameters under 1 km, densities of order 0.5 $g\,cm^{-3}$, and kinetic energies of order 10^{27} ergs.

All the original work presented here was done in collaboration with Kevin Zahnle. The section on tidal breakup models derives from a draft manuscript provided by Erik Asphang (Asphang and Beng 1995). Computations were performed at the Pittsburgh Supercomputing Center, and at the National Center for Supercomputing Applications, using software provided by the Laboratory for Computational Astrophysics. This work was supported by the NSF Aeronomy Program under grant number AST93-22509, and by the NASA programs in Exobiology and Astrophysical Theory. Too many people to list have been extremely generous in sharing results and preprints prior to publication, an openness that has been one of the joys of working on this problem.

REFERENCES

AHRENS, T. J., TAKATA, T., & O'KEEFE, J. D. 1994 Radiative signatures from impact of comet Shoemaker-Levy 9 on Jupiter. *Geophys. Res. Lett.*, **21**, 1551–1553.

ASPHAUG, E. & BENZ, W. 1994 Density of comet Shoemaker-Levy 9 deduced by modelling breakup of the parent 'rubble pile'. *Nature*, **370**, 120–124.

ASPHAUG, E. & BENZ, W. 1995 The tidal disruption of strengthless bodies: Lessons from comet Shoemaker-Levy 9. *Icarus*, submitted.

BIBERMAN, L. M., BRONIN, S. YA., & BRYKIN, M. V. 1980 Moving of a blunt body through a dense atmosphere under conditions of severe aerodynamic heating and ablation. *Acta Astronaut.*, **7**, 53–65.

BJORAKER, G. L., STOLOVY, S. R., HERTER, T. L., GULL, G. E., & PIRGER, B. E. 1995 Detection of water after the collision of fragments G and K of comet Shomaker-Levy 9 with Jupiter. *Icarus*, submitted.

BOSLOUGH, M. B., CRAWFORD, D. A., ROBINSON, A. C., & TRUCANO, T. G. 1994 Mass and penetration depth of Shoemaker-Levy 9 fragments from time-resolved photometry. *Geophys. Res. Lett.*, **21**, 1555–1558.

BOSLOUGH, M. B., CRAWFORD, D. A., TRUCANO, T. G., & ROBINSON, A. C. 1995 Numerical modeling of Shoemaker-Levy 9 impacts as a framework for interpreting observations. *Geophys. Res. Lett.*, **22**, 1821–1824.

BRONSHTEN, V. A. 1983 *Physics of Meteoric Phenomena* Reidel.

CHANDRASEKHAR, S. 1961 *Hydrodynamic and hydromagnetic stability.* Dover.

CHEVALIER, R. A. & SARAZIN, C. L. 1994 Explosions of infalling comets in Jupiter's atmosphere. *Astrophys. J.*, **429**, 863–875.

CHYBA, C. F., THOMAS, P. J., & ZAHNLE, K. J. 1993 The 1908 Tunguska explosion: Atmospheric disruption of a stony asteroid. *Nature*, **361**, 40–44.

CRAWFORD, D. A. BOSLOUGH, M. B., TRUCANO, T. G., AND ROBINSON, A. H. 1994 The impact of comet Shoemaker-Levy 9 on Jupiter. *Shock Waves*, 4, 47–50.

CRAWFORD, D. A. BOSLOUGH, M. B., TRUCANO, T. G., AND ROBINSON, A. H. 1995 The impact of periodic comet Shoemaker-Levy 9 on Jupiter. *Int. J. of Impact Engin.*, in press.

DOBROVOLSKIS, A. R. 1990 Tidal disruption of solid bodies. *Icarus*, **88**, 24–38.

FIELD, G. B. & FERRARA, A. 1995 The behavior of fragments of comet Shoemaker-Levy 9 in the atmosphere of Jupiter. *Astrophys. J.*, **438**, 957–967.

GRAHAM, J. R., DE PATER, I., JERNIGAN, J. G., LIU, M. C., & BROWN, M. E. 1995 The fragment R collision: W. M. Keck telescope observations of SL9. *Science*, **267**, 1320–1323.

HAHN, J. M. & RETTIG, T. W. 1995 Jeans instability in comet Shoemaker-Levy 9. *J. Geophys. Res.–Planets*, submitted.

KLEIN, R. I., McKEE, C. F., & COLELLA, P. 1994 On the hydrodynamic interaction of shock waves with interstellar clouds. I. Nonradiative shocks in small clouds. *Astrophys. J.*, **420**, 213–236.

LOGONNEÉ, P., BILLEBAUD, F., VAUGLIN, I., MERLIN, P., SYBILLE, F., MOSSER, B., LAGAGE, P. O., GAUTIER, D., & DROSSART, P. 1994 Seismic waves generated by the SL-9 impact. *Bull. Amer. Astron. Soc.*, **26**, 1580.

MAC LOW, M.-M. & ZAHNLE, K. 1994 Explosion of comet Shoemaker-Levy 9 on entry into the jovian atmosphere. *Astrophys. J.* **434**, L33–L36.

MARLEY, M. S., DAYAL, A., DEUTSCH, L. K., FAZIO, G. G., HOFFMANN, W. F., HORA, J. L., HUNTEN, D. M., SPRAGUE, A. L., SYKES, M. V., WALTER, C., & WELLS, K. W. 1994 A search for seismic waves launched by the impact of comet Shoemaker-Levy 9. *Bull. Amer. Astron. Soc.*, **26**, 1580.

MORAN, B. & TIPTON, R. 1993 Simulation of a comet impact on Jupiter. *Eos*, **74**, 43, 389.

READS, K. I. 1984 Experimental investigation of turbulent mixing by Rayleigh-Taylor instability. *Physica*, **12D**, 45–58.

RICHARDSON, D., ASPHAUG, E., & BENNER, L. 1995 Precise orbital integration of the self-gravitational clustering of SL9 debris. *Bull. Amer. Astron. Soc.*, in press.

SCHENK, P., ASPHAUG, E., McKINNON, W., MELOSH, H. J., & WEISSMAN, P. 1995 Cometary nuclei and tidal disruption: The geologic record of crater chains on Callisto and Ganymede. *Icarus*, submitted.

SCOTTI, J. V., & MELOSH, H. J. 1993 Tidal breakup and dispersion of P/Shoemaker-Levy 9: Estimate of progenitor size. *Nature*, **365**, 7333.

SEKANINA, Z. 1993 Disintegration phenomena expected during collision of comet Shoemaker-Levy 9 with Jupiter. *Science*, **262**, 382–387.

SEKANINA, Z. 1996 Preprint Astronomy & Astrophysics, in press.

SEKANINA, Z., CHODAS, P. W., & YEOMANS, D. K. 1994 Tidal disruption and the appearance of periodic comet Shoemaker-Levy 9. *Astron. Astrophys.*, **289**, 607–636.

SHOEMAKER, E. M., HASSIG, P. J., & RODDY, D. J. 1995 Numerical simulations of the Shoemaker-Levy 9 impact plumes and clouds: A progress report. *Geophys. Res. Lett.*, **22**, 1825–1828.

SOLEM, J. 1994 Density and size of comet Shoemaker-Levy 9 deduced from a tidal breakup model. *Nature*, **370**, 349–351.

SRIDHAR, S., & TREMAINE, S. 1992 Tidal disruption of viscous bodies. *Icarus*, **95**, 86–99.

SVETSOV, V. V., NEMTCHINOV, I. V., & TETEREV, A. V. 1995 Disintegration of large meteoroids in Earth's atmosphere: Theoretical models. *Icarus*, **116**, 131–153.

STONE, J. M. & NORMAN, M. L. 1992 Zeus-2D: A radiation magnetohydrodynamics code for astrophysical flows in two space dimensions. I. The hydrodynamic algorithms and tests. *Astrophys. J. Supp.*, **80**, 753–790.

TAKATA, T., O'KEEFE, J. D., AHRENS, T. J., & ORTON, G. S. 1994 Comet Shoemaker-Levy 9: Impact on Jupiter and plume evolution. *Icarus* **109**, 3–19.

TAKATA, T., AHRENS, T. J., HARRIS, A. W. 1995 Comet Shoemaker-Levy 9: Fragment and Progenitor Energy. *Geophys. Res. Lett.*, **22**, 2429–2432.

VAN LEER, B. 1977 Towards the ultimate conservative difference scheme. IV. A new approach to numerical convection. *J. Comp. Phys.*, **23**, 276–299.

VICKERY, A. M. 1993 Numerical simulation of a comet impact on Jupiter. *Eos*, **74**, 43, 391.

WEAVER, H. A., FELDMAN, P. D., A'HEARN, M. F., & ARPIGNY, C. 1994 Hubble Space Telescope observations of comet P/Shoemaker-Levy 9 (1993e). *Science*, **263**, 787–791.

WEIDENSCHILLING, S. J. 1994 Origin of cometary nuclei as 'rubble piles'. *Nature*, **368**, 721–723.

WEST, R. A., KARKOSCHKA, E., FRIEDSON, A. J., SEYMOUR, M., BAINES, K. H., & HAMMEL, H. B. 1995 Impact debris particles in Jupiter's stratosphere. *Science*, **267**, 1296–1301.

WINGATE, C. A., HOFFMAN, N. M., & STELLINGWERF, R. F. 1994 SPH calculations of comet Shoemaker-Levy 9/Jupiter impact. *Bull. Amer. Astron. Soc.*, **26**, 879–880.

XU, J. & STONE, J. M. 1995 The hydrodynamics of shock-cloud interactions in three dimensions. *Astrophys. J.*, in press.

YABE, T., XIAO, F., ZHANG, D., SASAKI, S., ABE, Y., KOBAYASHI, N., & TERASAWA, T. 1994 Effect of EOS on break-up of Shoemaker-Levy 9 entering jovian atmosphere. *J. Geomag. Geoelectr.*, **46**, 657–662.

YOUNGS, D. L. 1984 Numerical simulation of turbulent mixing by Rayleigh-Taylor instability. *Physica*, **12D**, 32–44.

ZAHNLE, K. J. 1992 Airburst origin of dark shadows on Venus. *J. Geophys. Res.*, **97**, 10243–10255.

ZAHNLE, K. & MAC LOW, M.-M. 1994 The collision of Jupiter and comet Shoemaker-Levy 9. *Icarus*, **108**, 1–17.

ZAHNLE, K. & MAC LOW, M.-M. 1995 A simple model for the light curve generated by a Shoemaker-Levy 9 impact. *J. Geophys. Res.–Planets*, in press.

ZAHNLE, K., MAC LOW, M.-M., LODDERS, K., & FEGLEY, B. 1995 Sulfur chemistry in the wake of comet Shoemaker-Levy 9. *Geophys. Res. Lett.*, **22**, 1593–1596.

Dynamics and chemistry of SL9 plumes

By KEVIN ZAHNLE[1]

NASA Ames Research Center, M.S. 245-3, Moffett Field, California 94035-1000, USA

The SL9 impacts are known by their plumes. Several of these were imaged by HST towering 3000 km above Jupiter's limb. The heat released when they fell produced the famous infrared main events. The reentry shocks must have been significantly hotter than the observed color temperature would imply, which indicates that the shocks were radiatively cooled, and that most of the energy released on reentry was radiated. This allows us to use the infrared luminosities of the main event to estimate the energy of the impacts; we find that the R impact released some $0.3 - 1 \times 10^{27}$ ergs. Shock chemistry generates a suite of molecules not usually seen on Jupiter. The chemistry reflects a wide range of different shock temperatures, pressures, and gas compositions. The primary product, apart from H_2, is CO, the yield of which depends only weakly on the comet's composition, and so can be used to weigh the comet. Abundant water and S_2 are consistent with a somewhat oxidized gas (presumably the comet itself), but the absence of SO_2 and CO_2 shows that conditions were neither too oxidizing nor the shocks too hot. Meanwhile, production of CS, CS_2, and HCN appears to require a source in dry jovian air; *i.e.*, the airbursts occurred above the jovian water table. Tidal disruption calculations and models of the infrared light curves agree on an average fragment diameter of about half a kilometer. Chemical products and atmospheric disruption models agree on placing the terminal explosions around the 1 to 4 bar levels. The plumes were spectacular because the explosions were shallow, not because the explosions were large.

1. Introduction

Nine months after P/Shoemaker-Levy 9 (SL9) struck Jupiter, the size of the fragments and the depth of their terminal airbursts remain controversial. In this chapter I will be concerned mostly with the rise and fall of the ejecta plumes. I will address six topics. The first is a brief overview of the observed light curves, with interpretation and foreshadowing as seems appropriate. This is followed by an illustrative summary of a detailed numerical simulation of a 2-D, axisymmetric ejecta plume. The numerical model shows several of the observed phenomena directly (had we heeded our model, our predictions would have been better), but as it lacks radiative cooling, it can't be used to calculate light curves. To explain the numerical simulations we developed a simple analytic approximation that can be used to calculate light curves (Zahnle & Mac Low, 1995)—how this is done is the third topic. The fourth topic is shock chemistry. Chemical products sample the composition of the most strongly shocked air, and thereby constrain the composition of the comet and the depth of the explosions. Fifth, we address the nature of the 450 m s^{-1} wave, which, if interpreted as a deep tropospheric gravity wave, could provide an observational argument that might favor deep explosions (cf., Ingersoll & Kanamori 1995). Finally, we offer an explanation of why the plumes were all the same height, and in the process make the point that big plumes do not require big impactors.

2. Light curves

As viewed at infrared wavelengths from Earth, a typical light curve had three peaks, the first two faint and brief and the third bright and long-lasting (Graham *et al.* 1995; Nicholson *et al.* 1995). Although first reported for the R impact (Graham *et al.* 1995),

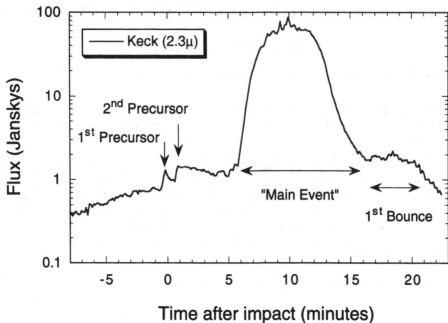

FIGURE 1. Annotated light curve of fragment R at 2.3μ, after Graham *et al.* (1995).

the three peaks were seen for all the well-observed events (Nicholson, this volume, provides what will probably remain the definitive summary of SL9 light curves). But because the first and second peaks were subtle, and so mostly overlooked at the time, they have since been called the first and second precursors. The spectacular third peak (or "main event") was seen by all, often in real time, thanks to the Internet. By contrast, the Galileo spacecraft, which was well placed to view the events directly, detected two peaks of comparable magnitude, the first corresponding roughly with the first precursor, and the second corresponding with the main event.

Figure 1 shows a typical SL9 light curve (shown is the Keck curve for fragment R at 2.3 microns, Graham *et al.* 1995). Figure 2 is a cartoon that illustrates the viewing geometries and evolution of a typical event. The first peak, or first precursor, signaled the entry of the fragment into Jupiter's atmosphere. A typical first precursor was detected on Earth some 10 seconds before the impact was first detected by the Galileo spacecraft. This apparent paradox was resolved at this conference, and is discussed in detail in the chapter by Chapman. In brief, it appears that the meteor trail was first seen from Earth while it was still high enough above Jupiter's cloudtops to be viewed directly by terrestrial observers. This early signal was too faint to be detected by Galileo. Only later, well after the meteor disappeared behind the limb as viewed from Earth, did it become bright enough to be detected by Galileo. From either vantage the characteristic time scale of brightening was short, consistent with the rise time of order $H/v \sim 1$ s expected for a meteor of velocity v penetrating an atmosphere of scale height H. That the first precursor was seen at infrared wavelengths implies a relatively low temperature for the emitting matter; this is most easily understood as the exploding meteor trail, in which gas cools by expansion on a timescale of tens of seconds.

Most of the meteor's initial kinetic energy is released in the last second of its existence. These matters are discussed in detail in the chapters by Mac Low and Crawford. What

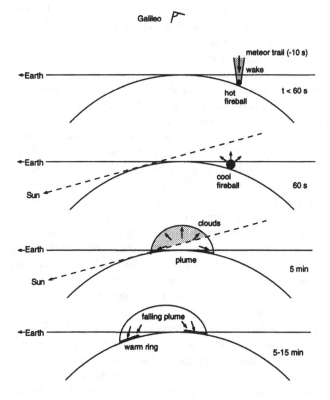

FIGURE 2. A cartoon that illustrates the viewing geometry of a typical SL9 impact. The first precursor corresponds to the fragment entering the atmosphere. The second precursor corresponds to the fireball rising into direct view from Earth. HST observed the plume in scattered sunlight, projected against the black sky of space. Finally, when the plume fell back to Jupiter, it produced a bright infrared event seen well from both Galileo and Earth.

is important here is that at the end of its flight the meteor releases enough energy in a small enough volume that the result is an explosion.

The second peak, or second precursor, occurs when the fireball from the terminal explosion rises above Jupiter's limb into direct view from Earth. That this would happen was predicted (Boslough *et al.* 1994; Ahrens *et al.* 1994). By the time ($\sim$60 s) the fireball cleared the limb it had cooled to $\sim$500–700 K (Carlson *et al.* 1995a). Accordingly, the second peak was more prominent at longer wavelengths, as is immediately apparent when the Palomar light curves (3.2 and 4.5 μm, Nicholson *et al.* 1995a) and the Keck light curve (2.3 μm, Graham *et al.* 1995) for the R impact are compared. In one of their numerical simulations, Boslough *et al.* (1994) get both the timing and the temperature of the debris front about right for a 1 km (diameter) impactor.

The onset of the second peak was abrupt (Graham *et al.* 1995), which requires the fireball to have a sharply defined photosphere when it rises above the limb. The observed rise time, $\sim$7 s, is what one would expect from a 100–200 km diameter fireball rising at 13 km s^{-1}. A well-defined surface is a general feature of numerical fireballs produced by terminal explosions (*e.g.*, Zahnle & Mac Low 1994; Boslough *et al.* 1994). The more evenly distributed line charge has an initial length scale of order H, which is mostly forgotten by the time the fireball inflates to > 100 km diameter.

As the fireball rises it continues to expand, cool, and fade. In the fireball are found a fair fraction of the former comet and a comparable mass of highly shocked Jovian air. The fireball begins as the hottest part of the ejecta plume. It rises the fastest, and eventually becomes the vanguard of the plume. A much larger mass of mostly jovian air follows behind. Because the fireball's radiating surface is relatively small, radiative losses are small, and the ejecta plume expands almost adiabatically. Most of the plume's initial thermal energy is converted to kinetic energy of expansion. As the fireball expands first silicates condense, later carbonaceous matter, then water (if present), ammonia, and so on. The condensates made the sunlit parts of the plume visible from Earth, there to be imaged by the Hubble Space Telescope (HST) (Hammel *et al.* 1995).

The third peak was caused by the ejecta plume falling back on the atmosphere (Zahnle & Mac Low 1995; Nicholson *et al.* 1995; Graham *et al.* 1995). The infrared events typically began about 5 or 6 minutes after the impact itself, and lasted another 10–15 minutes, somewhat less for the smallest events (*e.g.*, D), and somewhat more for the largest (*e.g.*, K). The timescales are those expected for a falling ballistic plume. It takes about $\sqrt{8z/g}$ for a plume to rise to a height z and fall back again. For the 3000 km plumes observed by the HST, this time is $\sim 10^3$ s, *i.e.*, 15 minutes. In the Palomar and Keck data the main light peak was followed by two smaller maxima about 10 and 20 minutes later (Nicholson *et al.* 1995), probably bounces of some sort (see also chapter by Mac Low). Presumably these too were universal features.

Because the plume falls over an enormous area, thermal radiation produced by the plume's reentry into the atmosphere accounts for a large fraction of the impact energy. In our numerical simulations the plume's kinetic energy is typically about 20–40% of the total impact energy; the higher fraction is seen in the models with the best energy conservation. That most of this was promptly radiated is implicit in the low effective radiating temperature, roughly 500 to 1000 K (Nicholson, this volume). These temperatures are much lower than shock temperatures obtained using the standard Rankine-Hugoniot jump conditions. Hot shocks were real enough: appropriately high temperatures (2000–5000 K) are seen in the 2.3 micron CO band (Knacke *et al.* 1994, Crisp & Meadows 1995, Kim *et al.* 1995). But most of the radiation is much cooler. Apparently radiative cooling of the shocked material was efficient. One plausible opacity source—dust—is clearly visible in the Hubble images. Another is line emission by very hot molecules. To some extent the latter surely occurred, as the aforementioned high CO temperatures indicate.

Although it is natural to think that the 5–6 minute delay between an impact and the main event was at least in part due to rotation of the impact site into better view, it now appears that viewing geometry was relatively unimportant. This is shown by the timing of the IR light curve recorded by Galileo for the R and G events, which is similar to that observed on Earth (Carlson *et al.* 1995b). This implies that something other than Jupiter's rotation delayed the third peak.

One possibility is that very little material was ejected with vertical velocities between 2 and 5 km s^{-1}. Although not a hypothesis I am fond of, it might be consistent with a rising fireball: we do see evidence in our 2-D numerical simulations for a modest lack of ejecta with velocities between 2 and 7 km s^{-1} (see below), but not the near-total absence required to account for the relatively abrupt onset of the main events. It is imaginable that a more realistic geometry might alter this in the desired direction.

But our favorite hypothesis is that emission required dust, and dust did not appear in the ejecta blanket until 5–6 minutes had passed. One possibility is that the dust is cometary debris, either organic (cf., West *et al.* 1995) or silicate (cf., Field *et al.* 1995). Cometary debris would be among the most highly shocked material and therefore it would be among the fastest moving material in the ejecta plume. Another is that the dust is shock-generated organic material (cf., West *et al.* 1995). Shocks faster than about 4–5 km s^{-1} will generate carbonaceous dust from jovian air (Zahnle *et al.* 1995; the relevant shock chemistry will be discussed below). Shock chemistry can occur in the explosion or on reentry into the atmosphere, but in either case the affected gas spends some time aloft. As it takes about $2v_z/g \approx 5$ minutes for 4 km s^{-1} ejecta to reenter the atmosphere, the shock-generation model predicts the timing of the third peak automatically.

A related possibility is that the radiating temperature was set directly by the condensation temperature of organic grains. The pressure and temperature of the reentry shock are roughly those of the graphite stability field in solar composition gas; *i.e.*, where CO and CH$_4$ are present in comparable abundance and the population of complex hydrocarbons peaks.

Any of these stories would be consistent with the morphology of the ejecta blanket: a dark outer crescent (or ring) encompassing a relatively dust-free interior. In a ballistic model the more distant material falls later, so that the inner edge of the outer crescent maps to the beginning of the main event.

3. Numerical plumes

Three dimensional numerical plume models have been presented by the Sandia group (Boslough *et al.* 1994; Crawford *et al.* 1994), Shoemaker *et al.* (1995), and by Takata *et al.* (1994). The former models use an Eulerian grid, the last uses smoothed particle hydrodynamics (SPH). Finite computational resources have restricted 3-D modelling to the first 2–3 minutes of an event (a 3-D plume model strains even the mammoth Paragon computer at Sandia). All three groups use initial conditions that place the explosions too deep, and as a consequence, the computed plumes are too small for a given impact. However, because these groups mostly modelled big impacts, many of the simulated plumes turned out to be about the right size; *e.g.*, Takata *et al.* (1994) even predicted that the SL9 plumes would be 3000 km high (which they were). Plume height will be addressed in § 7 below.

The 3-D models show a pronounced "wake effect": ejecta are launched preferentially along the bolide's entry path. This occurs because the wake is hot and rarefied, hence sound speeds (and shock speeds) are faster, and there is less inertia to slow expansion. That the ejecta would be preferentially channelled by the wake should be counted as a successful prediction. HST images clearly show that the ejecta blankets were azimuthally asymmetric, and thickest in the general direction from which the comets came (Hammel *et al.* 1995).

But the analogy to a cannon has been somewhat overdrawn. Even though mass was preferentially channelled through the wake, the velocity fields of the plumes were much more isotropic, and probably not greatly different from the velocity field of an axisymmetric plume. In particular, detailed examination of the HST images shows that the apex of the plume remained pretty much directly above the impact site throughout a plume's evolution (K. L. Jessup, pers. comm. 1995). This implies that the material shot straight up had the highest vertical velocity. By contrast, the apex of a cannonball plume moves downstream with time. Another argument that favors an isotropic velocity field is that the outer "crescent" appears to form a nearly complete, equidistant ring centered on the impact site, especially well seen in late observations of the fragment K event (McGregor *et al.* 1995).

Figures 3–6 show four stages in the evolution of a highly idealized 2-D axisymmetric numerical plume. The view is from the side. The gray scale indicates temperature, the contours indicate the fractional presence of cometary (fireball) material, and the arrows

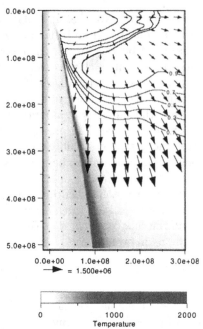

FIGURE 3. An axisymmetric numerical plume at 5.0, 6.7, 10.1, and 14.2 minutes (Fig. 3–6, respectively). The axes are labeled in cm and the velocities in cm s; the gray scale indicates approximate temperature in kelvins. The impactor's entry path is coincident with the left vertical axis; peak energy release (terminal explosion) is at $z = -25$ km (2.4 bars). Contours mark the location of fireball material ($\sim 10\%$ comet in this model). The shock wave from the explosion is prominent as a hot, wide cone, reaching 1000 km altitude 5000 km from the impact site. The former fireball is a discernable as a warm hemispheroid roughly 2000 km in radius. Underneath it is a cold plume of mostly jovian air.

indicate the velocity field. In this model the initial conditions presume a vertical line charge, coincident with the z-axis, extending from about $z = -50$ km to the top of the grid (the altitude z is measured from the 1 bar level), The resolution is 1 km near the site of the explosion, grading to 10 km in the most distant parts of the grid. The energy release profile is derived from an analytic approximation to large-body meteor flight (Zahnle & Mac Low 1994). Peak energy release is at -25 km (~ 2.4 bars) Most of the fragment's energy is released in the deepest scale height. The total energy release is 2×10^{27} ergs, equivalent to a 750 m diameter fragment of density 0.5 g cm^3. The size and density are consistent with the largest fragments produced by tidal disruption models of self-gravitating comets (Asphaug & Benz 1994; Solem 1994; Asphaug & Benz 1996). The detailed initial conditions and gridding for this model are described more fully in the chapter by Mac Low (this volume).

The contours mark the location of cometary material. However, the initial conditions were too coarse to retain this resolution, and as a result the comet is about ten-fold depleted in the fireball, so that a contour line of unity is only about 10% comet by mass. Although it seems reasonable to us that these highly shocked, turbulent gases would mix, we see little direct evidence for mixing in our numerical simulations (see Fig. 7 in chapter by Mac Low).

The hydrodynamic code computes mass density, energy density, and horizontal and vertical velocities: ρ, e, v_r, and v_z, respectively. These are the independent variables. Pressure p is required to evaluate force; it is obtained from an equation of state. For simplicity we have assumed an ideal gas with constant γ (ratio of specific heats), by

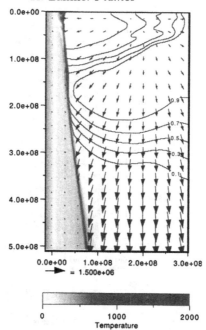

FIGURE 4. The axisymmetric numerical plume at 6.7 minutes. The reentry shock is prominent as an annulus $1000 < r < 2000$ km. It is warmest where the infalling material is richest in former comet.

which $p = e/(\gamma - 1)$. In this simulation, which focuses on the dynamics of the plume, we take $\gamma = 1.4$, which is appropriate to a cool diatomic gas. The dynamics do not appear to depend strongly on γ.

The hydrodynamics code computes neither temperature nor mean molecular weight; these require a more sophisticated equation of state, one that accounts for vaporization, dissociation, and ionization, for both jovian air and for cometary material. The latter presents myriad difficulties. The temperatures shown in Fig. 3–6 were computed assuming an ideal gas, $(\mu_c \gg \mu_J)$.

$$T \approx \frac{p}{\rho} \frac{m_H}{k} \left(\frac{1-f}{2.4} + \frac{f}{20} \right)^{-1}, \tag{3.1}$$

in which m_H is the mass of a proton; $\mu_J = 2.4$ and $\mu_c = 20$ are the mean molecular weights of jovian air and cometary gases respectively; and f is the fraction that is cometary (by mass).

Temperatures computed using Eq. 3.1 are reasonably accurate for clean, cool jovian air, but become highly suspect for hot, comet-rich gas. There are three problems: (1) the mean molecular weight drops due to dissociation and ionization (*e.g.*, μ_J for dissociated air is about 1.3); (2) the composition of the comet-rich fireball is unknown, both because the composition of the comet is unknown, and because the degree of mixing with ambient air is unknown; and (3) radiative cooling becomes important. In Eq. 3.1, the mean molecular weight of cometary material is unimportant, because $f \leq 0.1$ sets an upper limit of $\mu = 2.6$ on the mean molecular weight of the fireball. Shock temperatures reported by Kim *et al.* (1995) imply that the mean molecular weight was ~ 2.5. If mixing were much less efficient, then we would expect higher temperatures in the fireball material than shown on Figs. 3–6.

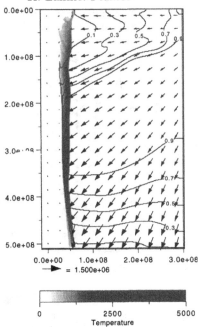

FIGURE 5. The axisymmetric numerical plume at 10.1 minutes. This is near peak light for the main event. The reentry shock is prominent as an annulus $2000 < r < 4000$ km; it is distinctly higher than the contact discontinuity between the plume material and the ambient jovian atmosphere. By this time the velocity field is almost precisely ballistic.

The most important omission from the numerical models is the exclusion of radiative cooling. As noted, radiative cooling was unimportant compared to adiabatic cooling during the explosion and during the expansion of the plume. The dynamics of the plume are reasonably well-described without it. But radiative cooling, either by molecules or by embedded or shock-generated particles, clearly dominated the energy budget of the reentry shock. We will return to radiative cooling in the next section, in the context of modelling the observed infrared light curves.

Figure 3 shows the plume 5 minutes after impact. The fireball material, found mostly in a thick hemispherical shell, by this time is no longer hot, but it has yet to fall back. Outside the shell are found hot but rarefied jovian gases that were shocked by the blast wave from the original explosion. Velocities in this gas are higher than predicted by homologous expansion; these high velocities recall the recent passage of the shock wave from the explosion. The shock itself is the prominent diagonal line of hot material (as a figure of revolution, it is really a wide cone). Inside the shell we find a fountain of extremely cold jovian air. This air had been heated by the explosion, but by this time has expanded adiabatically to a point where any condensibles within it have probably condensed.

Figure 4, at $t = 6.7$ minutes, shows the plume shortly after it has begun to fall back on the atmosphere. When the plume falls back onto the atmosphere, it liberates the vertical component of its kinetic energy in a reentry shock. There are actually two shocks, one driven into the Jovian atmosphere and the other into the falling plume. Warm shock temperatures are seen, especially between 1000 and 2000 km from the impact site. The highest temperatures occur in cometary gases, which have higher molecular weight. Warm temperatures are also seen at the epicenter, where jovian air reentering

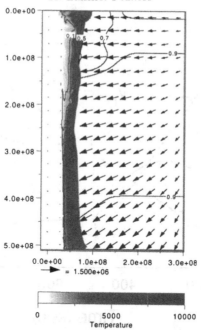

FIGURE 6. The axisymmetric numerical plume at 14.2 minutes. By this time the plume is fading. Because reentry velocities rise monotonically with time, the reentry shock is hotter than it has been, but so little material is falling that the shock is very high in the atmosphere, and no longer especially luminous. The contact discontinuity, marked by the lower contours, indicates the initial height of the reentry shock. The annulus as a whole is spreading rapidly. Not directly apparent is that it is also spreading homologously.

the atmosphere has been compressed and heated. By this time the velocity field in the plume is practically homologous and ballistic.

$$v_r = \frac{r}{t}; \quad v_z = \frac{z}{t} - \frac{1}{2}gt \tag{3.2}$$

Note that the reentry shock has risen well above the the contact discontinuity (marked by the contours) between the shocked ejecta plume and the ambient jovian atmosphere.

By $t = 10.1$ minutes (Fig. 5), the reentry shock has produced high temperatures in an annulus $1500 < r < 4000$ km from the epicenter. Comet-rich material rains down on this annulus. Because reentry velocities get progressively higher as time passes, $v_z \approx \frac{1}{2}gt$, temperatures are higher.

Figure 6, at $t = 14.2$ minutes, shows further evolution. Particularly clear in Fig. 6 is that (1) the shock temperatures are very high but that (2) the reentry shock has risen some 500 km above its original altitude, which is recalled by the position of the contact discontinuity between falling matter and ambient air. Relatively little material is actually falling by 14.2 minutes, which is why the shock is so high; indeed, previously shocked air is springing back at considerable velocity. These "bouncing shocks" may be responsible for the bouncing light curves seen so well in the Palomar data for the R impact (Nicholson *et al.* 1995; Nicholson, this volume; Deming *et al.* 1995; Mac Low, this volume).

Horizontal velocity is preserved across the shock. Therefore the shocked plume continues to expand radially for a considerable time after it reenters the atmosphere. This can be seen in Fig. 6; it is explicit in Fig. 7, which shows the position and velocity history of a "typical" parcel. One might expect the plume to expand horizontally until it has

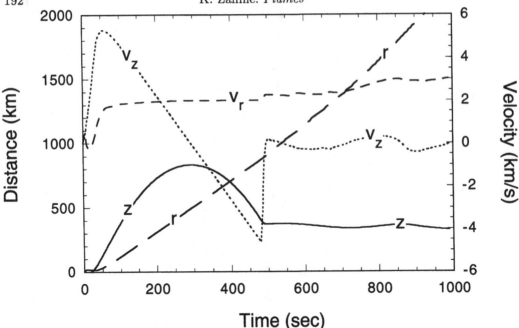

FIGURE 7. Location and velocity history of a representative strongly shocked parcel of jovian air. The parcel originated at the 2 bar level, near the site of the airburst. It was shocked to $T > 9000$ K, expanded, cooled, and ultimately ejected at about 6 km/sec; it reentered the atmosphere ~ 500 s later, where it was shocked. Note that the parcel retains its horizontal velocity across the shock.

swept up its own mass in jovian air. The characteristic crescent-shaped footprint of the plume was observed to have rotated through a larger angle than can be accounted for by Coriolis force acting only while the plume was in orbit (Hammel *et al.* 1995). The additional rotation of the footprint was caused by Coriolis force continuing to act on the radially expanding plume for another 20–30 minutes after reentry. An alternative perspective on the same process is that Jupiter rotated beneath the spreading plume until the plume coupled with the planet.

4. A plume model

As a first model for the infrared light curve, we (Zahnle & Mac Low 1995) devised a 2-D axisymmetric plume model to describe the decline and fall of the ejecta plume. The "toy plume" model assumes that the velocity field of the ejecta plume is ballistic and that the mass-velocity distribution in the ejecta plume follows a power law that we have calibrated to our detailed numerical models. The ballistic approximation is valid to a factor of order $c_s^2/v^2 \ll 1$, where c_s is the sound speed and v a typical velocity in the ejecta plume, and is in excellent agreement with the results of detailed numerical models. For simplicity we also assumed that the plume is axisymmetric with an opening angle θ'.

The mass-velocity distribution of ejecta from hypervelocity impacts generally obeys a power law of form

$$M(>v) \propto v^{-\alpha}, \tag{4.3}$$

in which the notation $M(>v)$ refers to the cumulative mass ejected at velocities greater than v. Our numerical experiments of impacts on Jupiter give $\alpha \approx 1.6$ as an average over the range 0.01–10 km s^{-1}, varying somewhat depending on explosion parameters. The distribution truncates at a maximum velocity, v_{max}, that can be regarded as determining the height of the plume; typical SL9 plumes reached 3000 km, so that $v_{max} \approx 12$ km s^{-1}. The normalized form of Eq. 4.3 is

$$M(>v) \approx 340\,\eta\,m_i \left(\frac{1 \text{ km s}^{-1}}{v}\right)^{1.55}, \tag{4.4}$$

where η is the fraction of impact energy invested in the ejecta plume. Numerical simulations indicate that $\eta \approx 0.4$. Equation 4.4 has been evaluated for $\alpha = 1.55$ and $v_{max} = 12$ km s^{-1}. These parameters describe the mass-velocity distribution we used in Zahnle & Mac Low (1995), and in the subsequent discussion which is based on the paper.

However, application of a single power law to both strong $(v > c_s)$ and weak $(v < c_s)$ shocks is suspect, and when we examine our numerical models in detail we find that the plume is better described by two powers, for which $\alpha \approx 1.2$ for $v > c_s$ and $\alpha \approx 1.8$ for $v < c_s$. This distribution is not too far removed from momentum scaling for high velocities, and energy scaling for low velocities. As strong shocks and high velocity ejecta are our concern here, the lower value of α would have been a better choice. Such a distribution is relatively deficient in material launched with velocities $c_s < v < v_{max}/2$. which may have contrtibuted somewhat to the observed shapes of the main event light curves.

We subdivide the plume three dimensionally (r, θ, ϕ) into a vast number of mass elements. Each element is launched on its own unique ballistic trajectory. We tile Jupiter's "surface" by distance and azimuth to produce a kind of dartboard centered on the impact site. We then count up where and when the mass elements reenter the atmosphere (*i.e.*, where and when they hit the dartboard). The effective radiating temperature of the reentry shock is determined by balancing the energy supplied by infalling ejecta against thermal radiation by opacity sources (dust, soot, molecules, darts, etc.) embedded in or generated by the reentry shock. The approach is closely analogous to that used by Melosh *et al.* (1990) and Zahnle (1990) to model thermal radiation following the K/T impact.

4.1. *Pressure*

The pressure level of the reentry shock, if strong, is

$$p' = \frac{\gamma + 1}{2} \rho v_z^2. \tag{4.5}$$

This can be evaluated directly to give p' as a function of position and time, but this is not as useful or as accurate a measure of "the" shock pressure as one might hope. To first approximation, the greatest mass flux of material to arrive at any point is the first material to get there. Thereafter the mass flux declines monotonically and the shock rises to progressively lower pressures. Meanwhile the heated atmosphere both cools radiatively and also rises to seek a new scale height. The accumulated mass fallen at a given place (multiplied by g) provides a measure of the pressure level of the contact discontinuity between the plume and the Jovian atmosphere, which suggests that this is probably a good measure of the reentry shock.

An illustrative analytic approximation to the pressure at the contact discontinuity can be obtained by replacing v^2 in Eq. 4.4 by rg, which gives

$$p(r) = \frac{g}{2\pi r}\frac{dM}{dr} \tag{4.6}$$

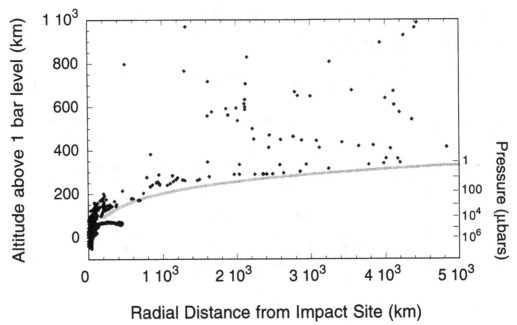

FIGURE 8. The altitude and pressure of the reentry shock 15 min after impact, for a 2×10^{27} erg SL9 impact (the model shown in Figs. 3–6). The solid curve shows the location of the reentry shock as approximated by Eq. 4.7. The diamonds show tracer particles that were swept up by the plume. The deepest diamonds trace the boundary between Jupiter and the fallen plume.

When evaluated for $\alpha = 1.55$ and $v_{\mathrm{max}} = 12$ km s^{-1}, this becomes

$$p(r) \approx 90\,\eta \left(\frac{m_i}{10^{14}\mathrm{g}} \right) \left(\frac{1000\ \mathrm{km}}{r} \right)^{2.8} \mu\mathrm{bars}. \qquad (4.7)$$

Shock pressures are directly proportional to the mass of the impactor. Pressures at large distances are quite low. At $r = 2000$ km, a typical shock pressure produced by a 10^{14} g impact with $\eta = 0.4$ would be 5 μbars.

Figure 8 compares the pressure level given by Eq. 4.7 (faded solid curve) to results obtained directly from the numerical model shown in Figs. 3–6 (diamonds). As this is the same numerical model by which we calibrated Eq. 4.4, good agreement might be expected. The diamonds indicate where tracer particles swept up in the plume have gone to after 15 min. The lowermost diamonds trace the location of the boundary between Jupiter and the ejecta.

4.2. *Temperature*

The ordinary Rankine-Hugoniot relations for a strong shock require that the infalling material reach a temperature of

$$T' \geq \frac{\gamma - 1}{2} \frac{m_{\mathrm{H}}}{k} v_z^2 = 1500\ \mathrm{K} \left(\frac{\mu}{2.5} \right) \left(\frac{v_z}{5\ \mathrm{km/s}} \right)^2, \qquad (4.8)$$

where m is the mean molecular mass. The temperature in Eq. 4.8 is the temperature of the gas immediately after it passes through the shock. It can be high. Temperatures would be especially high in material that originated from the comet because the mean molecular weight of the vaporized comet is relatively large, probably of the order of $10 < \mu < 20$. The detection of hot (> 2000 K) carbon monoxide emission at 2.3 μm (Knacke *et al.* 1994; Meadows *et al.* 1994; Crisp & Meadows 1995, Kim *et al.* 1995) offers direct evidence that former cometary material became very hot on reentering the atmosphere. Figure 9 is a schematic picture of the shocks while the plume is still falling.

The effective radiating temperature, between 500 and 1000 K (Nicholson *et al.* 1995b), was much lower than implied by Eq. 4.8. Evidently radiative cooling by the shocked material was efficient. HST images clearly revealed that the ejecta blankets were dusty (Hammel *et al.* 1995). The dust is an obvious opacity source for radiative cooling. Another plausible source is cooling by molecular lines.

The dust could be cometary (ice, organic, or silicate), or it could be generated when the plume strikes the atmosphere. The latter requires a sufficiently strong shock. (Zahnle *et al.* 1995) show that reentry velocities must exceed 4–5 km s^{-1} for shock heating to be strong enough to burn jovian air. This could explain the relative transparency of the inner parts of the ejecta blanket as observed with the HST (Hammel *et al.* 1995), and the late onset of infrared radiation as observed by the Galileo spacecraft (Carlson *et al.* 1995a,b).

A simple expression for the radiating temperature of a dusty layer heated by falling ejecta is to assume an instantaneous energy balance between the kinetic energy of falling matter ($\dot{e}$) and radiative cooling, buffered by the heat capacity of the falling matter (Zahnle 1990):

$$\dot{e} = \frac{1}{\gamma - 1} \frac{kT}{m} \dot{m} + 2\sigma T^4 \left(1 - e^{-2\tau}\right). \tag{4.9}$$

The factor $2\sigma T^4$ assumes a grey radiator that emits both up and down (Fig. 9). The factor $1 - e^{-2\tau}$ allows for the optical depth τ of the radiating layer; the factor of 2 gives the right limit as $\tau \to 0$. We solve Eq. 4.9 for the temperature T by Newton's method.

We consider two models for the optical depth τ. Both assume a gray absorber that is uniformly embedded in the shocked air. In one model, τ is proportional to the total integrated mass of material falling at radius r, so that all the dust in the column contributes to radiative cooling. In the second model we limit τ to the most recently arriving material (specifically, only the most recent 90 s contribute). We do this because later arriving material shocks at progressively higher altitudes, for which the cumulative optical depth of deeper material may not be relevant. The temperatures that result are much more nearly constant, both in radius and time (Fig. 10).

The flux at Earth is calculated using $\tau(r, t)$ and $T(r, t)$, accounting for the rotation of Jupiter and the projected surface area of the plume's footprint. Figure 11 shows calculated light curves at 2.3, 3.2, and 4.5 microns, as seen from Earth. The Keck light curve for fragment R (Graham *et al.* 1995) at 2.3 microns is shown for comparison. The match at 3.2 microns and 4.5 microns is comparably good (Nicholson *et al.* 1995). The important free parameters in the model are impact energy and opacity. The radiating temperature depends most strongly on opacity; impact energy scales the absolute flux. These models use a 450 m diameter impactor of density 0.5 g cm^3, $v_{max} = 11$ km s-1, and opening angle $\theta' = 45°$. The total impact energy is 5×10^{26} ergs, 40% of which is invested in the ejecta plume. The G, K, and L impacts were some 5–10 times brighter (Nicholson, this volume).

5. Chemistry

The impacts produced strong shocks, both promptly at the impact site and again, later, and over thousands of kilometers, when the ejecta plume reentered the atmosphere. The chemistry in these shocks was distinctive. The most surprising report was of a huge

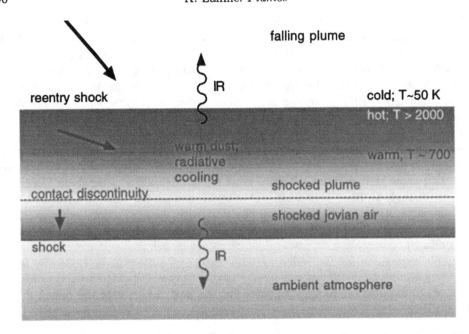

FIGURE 9. Radiative cooling in the reentry shocks. Velocities are shown schematically. There are two shocks, one driven into the ambient atmosphere and the other in the falling plume. The highest temperatures are in freshly shocked gas. Thereafter the gas cools radiatively. The contact discontinuity between air and plume is also a velocity discontinuity, as the ejecta continues to flow away from the impact site. Shear instabilities will cause mixing between atmosphere and ejecta, and eventually help brake the flow.

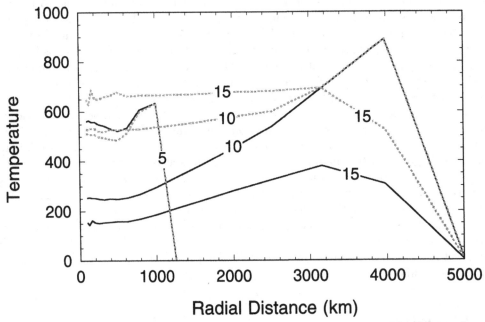

FIGURE 10. Effective radiating temperatures at 5, 10, and 15 min produced by the toy plume with constant opacity $\kappa = 100$ cm^2 g-1. The solid curves use the total column optical depth. The dotted curves assume that only recently fallen material contributes to the effective optical depth.

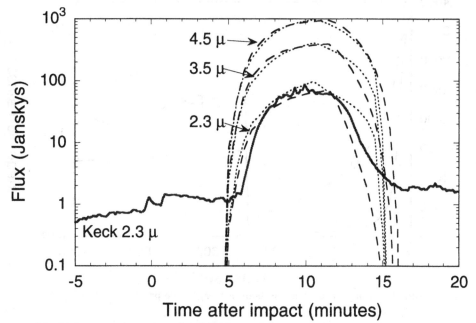

FIGURE 11. Light curves at selected infrared wavelengths generated by the toy plume model, as seen from Earth (Zahnle & Mac Low 1995). The dashed curves assume that the total optical depth is the effective optical depth; the dotted curves assume that only recently fallen material contributes to the effective optical depth. The Keck light curve (Graham *et al.* 1995) at 2.3 μm is shown for comparison (solid curve).

amount of diatomic sulfur (S_2) at the site of the G impact (Noll *et al.* 1995). Other reported products unusual to Jupiter include CS, CS_2, OCS, H_2S, SO_2, HCN, CO, and H_2O (Lellouch *et al.* 1995; Lellouch, this volume); although H_2S and H_2O are doubtless abundant below the visible clouds.

A general rule of shock chemistry is that CO forms until either C or O is exhausted. If O>C, the other products are oxidized, and excess O goes to H_2O. If C>O, the other products are reduced, and excess C goes to HCN. C_2H_2, and a wide variety of more complicated organics. Ultimately, given time, the carbon would react all the way to graphite, but in practice the reactions are incomplete. The dark ejecta debris is probably composed in part of carbonaceous particles generated by the shocks. In a sense, the SL9 impacts performed the famous Miller-Urey experiment on a grand scale, with one result being the production of a lot of complex brown organic solids (called "tholins").

We use a straightforward chemical kinetics model for the H,N,C,O,S system (Zahnle *et al.* 1995) to follow the nonequilibrium chemistry behind the shocks. The model traces the evolving chemical composition of a parcel of gas by directly integrating the web of chemical reactions. Pressure and temperature histories of the parcels are patterned after those calculated by numerical hydrodynamic simulations of the ejecta plume. A given plume parcel is generally shocked twice; *i.e.*, a parcel shocked near the impact site is ejected at high velocity and is shocked again when it reenters the atmosphere. The final state of the gas depends mostly on the second shock, provided that the latter is hot enough. Figures 7 and 12 show an illustrative parcel history taken directly from the numerical model illustrated in Fig. 3–6.

We also used a thermochemical equilibrium model (Fegley & Lodders 1994; Zahnle *et al.* 1995). The key difference between a kinetics model and a thermochemical equilib-

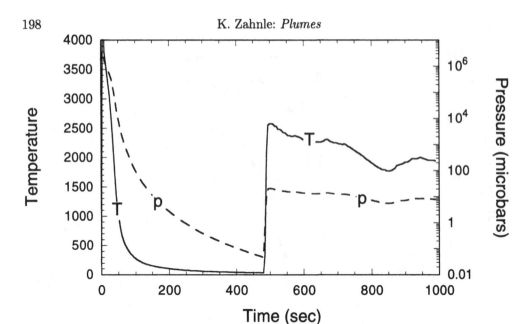

FIGURE 12. Temperature and pressure history of the parcel shown in Fig. 7.

rium model is that in a kinetics model every chemical reaction must be specified. If no favorable reaction exists, a thermochemically favored species need not form, and the system can settle into a metastable disequilibrium. There are two sorts of kinetic inhibition. One occurs when there really is no favorable reaction. This often happens in a rapidly cooled parcel, because many chemical reactions become very slow at low temperatures. Such a system would be described as quenched. The other, unique to models, occurs when important favorable reactions are unknown, unguessed, or omitted. As relatively few reactions have been accurately measured and tabulated, the modeller needs to invent a great many plausible reactions that, if neglected, would leave an incomplete system producing skewed results. As tables of measured reaction rates are at best incomplete, a great many reactions are guesswork, and the detailed results must be regarded with caution. Thermochemical equilibria calculations are relatively immune from these problems, but their relevance to disequilibrium systems is questionable.

A good, albeit sensitive, example of the limitations of a kinetics model is my previous conclusion that S_2 would be the major sulfur species in strongly shocked dry jovian air, rather than CS, as predicted by thermochemical equilibrium (Zahnle *et al.* 1995). This result now appears to be more wrong than right. The missing pathway to CS was $CH_3 + S \longrightarrow H_2CS$, an invented, but crucial, reaction. We have since added HCS and H_2CS to our list of species calculated; these additions are important as they provide the main channel for CS and CS_2 formation. We now find that CS and CS_2 are usually the major products of sulfur shock chemistry in dry jovian air, as predicted by thermochemical equilibrium.

Thus, again, we must address the presence of S_2 as a major product. Our earlier finding that S_2 can be a disequilibrium byproduct of hydrogen recombination in the low pressure reentry shock (Zahnle *et al.* 1995) remains valid for a range of chemical compositions. Fig. 13 follows the chemistry of a strongly shocked parcel of plume gas (the parcel's temperature and pressure follow Fig. 12). The parcel's composition is assumed to be 50% cometary, 50% jovian air by mass. "Comet" is here defined as a mixture of C_2H_2, H_2O, NH_3, and H_2S, with C, O, N, and S present in solar proportions. Silicates are

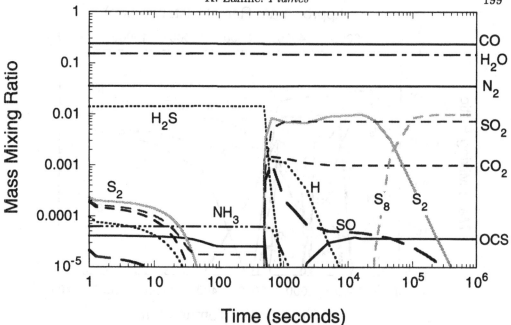

FIGURE 13. Time-dependent chemistry in a representative parcel of shocked plume gas. The gas is assumed 50% comet, 50% jovian air, by mass. Elemental sulfur is a major product. H_2S survives the fireball, but is consumed on reentry. As a solar composition gas is relatively oxidizing, other major products include SO_2 and CO_2.

ignored—it is assumed that they will reform to stochiometrically equivalent silicates. In this parcel S_2 is a major sulfur compound after reentry, and remains so while the gas remains warm. Eventually the gas cools to the point where S_2 polymerizes to S_8, which probably condenses.

Figure 13 is concerned with the fate of a single parcel shocked to $T' = 2500$ K and $p' = 20$ microbars. In this particular parcel, the final sulfur products are S_8 and SO_2; some OCS forms, as well. As SO_2 was not observed to be a major product, this parcel is probably too oxidizing to fit SL9. Other major products of this O-rich gas are H_2, CO, H_2O, N_2, and CO_2, the latter also unreported.

Figures 14 and 15 generalize results from parcels like that shown in Fig. 13 to a range of different peak shock temperatures and pressures. They show the final products in the parcel after the parcel has cooled. These figures are prepared for dry and wet jovian air (*i.e.*, comet-free). The peak shock temperature in the fireball is assumed to be twice T' (*i.e.*, $v_z^2 \approx v^2/2$). For specificity we use Eq. 4.7 with $\eta = 0.5$ to relate T' to p'. The different T' correspond to a range of different reentry velocities, which in turn correspond to a range of distances ($r \sim 2v_z^2/g$) from the impact site; approximate distances are also indicated.

The products of shocked dry jovian air (Fig. 14) are of bewildering complexity. For shocks in the range $2000 < T' < 2500$ K, especially complex products are formed from pieces of CH_4 and NH_3. The products shown here, C_2H_n, C_4H_2, and HCN are just the simplest; a wide-range of more complicated hydrocarbons, nitriles, and amines are to be expected in nature. At higher shock temperatures the products are simpler, with N_2 favored among nitrogen compounds and C_2H_2 favored among the hydrocarbons. In reality it is probable that much of the carbon we assign to C_2H_2, C_2H_4, and C_4H_2 actually ends up in more complicated compounds or particulates that would qualify as

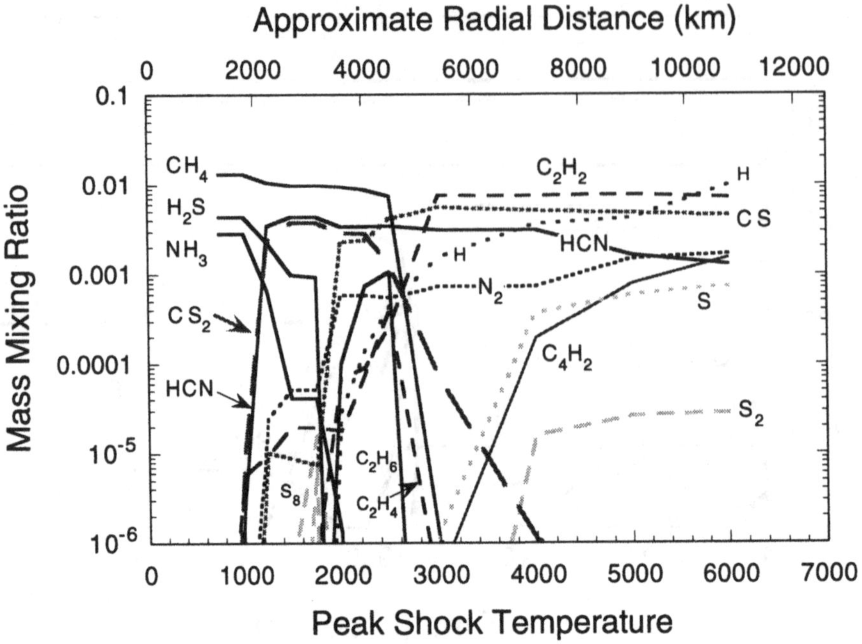

FIGURE 14. Reentry products in dry jovian air as a function of peak shock temperature. Also shown is the corresponding distance from the impact site. This C-rich gas produces a bewildering array of chemical products, but in this case neither S_2 nor S_8 are prominent among them. Products for $T' < 2000$ K are mostly determined by quenching of the fireball rather than the reentry shock.

a kind of tholin; HCN may have the same fate. Whether CS or S_n forms depends on the availability of reactive carbon. If most C goes into refractory grains, S_n would be more favored than shown here. The same applies to HCN and N_2; N_2 is more favored if C gets tied up in grains. CS is also known to polymerize readily, even explosively (Moltzen *et al.* 1988), so it too may have been quickly incorporated into grains. Note that carbonaceous particulates would not be expected to form for peak shock temperatures less than about 2000 K, which requires reentry velocities greater than about 4 km s^{-1}.

Figure 15 shows reentry products in wet jovian air as a function of peak shock temperature. Here we do find elemental sulfur as a major product, especially for $2000 < T' < 3000$ K, while at higher temperatures SO_2 supplants it. At very high T', CO_2 and even O_2 become major products. (Recall that T' is the peak shock temperature, which is not the same as the temperature when these products form.) Under other circumstances (not shown here), shocked wet jovian air can produce a range of reduced products, principally CS, CS_2, and HCN—these form if the pressures are high enough and the temperatures are in the narrow range (~ 2000 K) where CH_4 and H_2S react but H_2O does not. Thus shock production of tholins from jovian air is not necessarily precluded by the presence of abundant water.

Figure 16 gives the final sulfur products of a shocked 10^{14} g comet mixed with an equal mass of jovian air, shown as functions of the C/O ratio in the reacting gas. The total chemical product is computed by summing the products of all shock temperatures, weighted by the mass of gas shocked to that temperature. When taken at face value, the report that the K impact produced 2.5×10^{14} g of CO (Lellouch, this volume) would imply a K fragment mass as high as 5×10^{14} g; this mass is about two or three times larger than that obtained from the infrared light curve.

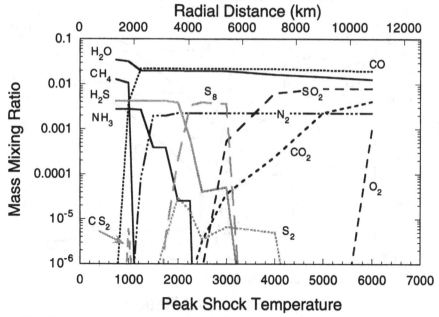

FIGURE 15. Reentry products in wet jovian air as a function of peak shock temperature. Stairstepping by NH_3 and H_2S is an artifact. Products for $T' < 2000$ K are mostly determined by quenching of the fireball rather than the reentry shock. Note that elemental sulfur is favored by middling shock temperatures, while the highly oxidized products are favored by high shock temperatures.

Sulfur compounds provide a measure of the oxidation state of the shocked gas. Where O>C, the primary products are SO, SO_2, and S_n. Where C>O, the products are CS, CS_2, H_2S, and S_n. Interestingly, in these calculations OCS appears to be an indicator of an oxidized gas, but this result could well be model-dependent. Elemental sulfur is exceptional in that it forms at all O/C ratios. Results are qualitatively similar for reentry shock chemistry of wet jovian air, as can be inferred by comparing Figs. 14 (dry jovian air) and Fig. 15 (wet jovian air) to Fig. 16. Addition of jovian H_2 favors S_8 at high O/C and disfavors S_8 at low O/C; and of course favors H_2S generally.

At this point it no longer appears that chemistry by itself can distinguish between shallow and deep explosions. The redox state of the reactants is ambiguous, and in any event most of the CO, H_2O, and S-species probably derive from the comet. The early idea that much of the observed sulfur was jovian was driven by a perceived need to produce 10^{15} g of S_2. This need has vanished as the estimates for the mass of S_2 at the impact sites have been revised downward.

Despite the presence of abundant H_2O, there is no evidence that the reacting gases were O-rich. On the contrary, neither SO_2 nor CO_2 appear to have been unambiguously detected, while CS, CS_2, and HCN were prominent products. There is also the possibility that the dust is carbonaceous. On the other hand, the inferred high abundance of S_2 is consistent with a mildly oxidized composition, and the presence of abundant H_2O would appear to require O>C.

Hot water and CO are almost certainly cometary. Excavation of modest amounts of jovian water requires that the water be lifted from within the clouds, where the water mixing ratio is limited by the saturation vapor pressure, and would therefore have a very small scale height, on the order of 2 or 3 km. The amount of water excavated would

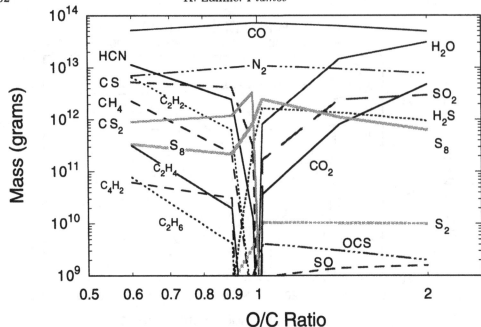

FIGURE 16. Total product of a plume 50% air, 50% comet by mass, produced by a 10^{14} g comet, as a function of the comet's C/O ratio. It is assumed that silicates form grains that do not participate in the chemistry. The product is averaged over reentry shocks from 1500 K to 5000 K. The chemical products are not indicative of any particular pressure or temperature. But is also plain that no single O/C ratio can explain the full range of products seen in the aftermath of SL9.

necessarily vary drastically from impact to impact; its roughly comparable presence following the large K impact and the smaller R impact (Sprague *et al.* 1996; Bjoraker *et al.* 1995) cannot be understood as jovian. An analogous argument should apply to nitrogen and sulfur species if the explosions took place within the putative NH_4SH clouds; I am unaware of evidence that argues either way. Nonetheless, as Fig. 14 shows, shocked dry jovian air offers a good potential source of HCN, CS_2, and CS that might complement the more oxidized products of the more highly shocked, comet-fouled plume.

An intriguing possibility is that water condensed in the plume and that the ice grains passed through the rarefied gases of the reentry shock intact, only to flame out as meteors a few seconds later at somewhat higher pressures, in a gas that was more purely jovian. Here the O/C ratio may have been locally high, or temperatures too low for H_2O to react. Such a scenario could physically separate reacting sulfur from oxygen, and might also leave the reacting gas in the shock oxygen-depleted, thereby producing two redox regimes from the same O-rich cometary ejecta. And if both C-rich grains and O-rich ice failed to fully vaporize in the shock, the odds in favor of S_2 are raised dramatically.

6. Rings

One of the more intriguing features observed by HST were rings, encircling several of the impact sites, that expanded radially at constant velocities of order 450–500 m s^{-1} (Hammel *et al.* 1995). The rings moved too slowly to be shock waves or acoustic waves. Ingersoll & Kanamori (1995) suggest that the rings are stratospheric manifestations of gravity waves propagating in a wet troposphere. In their model the speed of the wave is set by the static stability of the troposphere, which in turn depends on its water content. Conventional Jupiters are constructed with 2–3 times solar water, for which Ingersoll & Kanamori (IK) calculate leisurely gravity wave speeds of order 150 m s^{-1}. To bring the velocity up to the observed 450 m s^{-1} requires raising the water content to 10 times solar. The model predicts that all impacts would produce waves of the same speed (as observed), and because the waves are linear they do not slow down (as observed).

A weakness of the IK model is that it doesn't actually account for the ring. There are two problems. The first is that the ring itself appears to be high in the stratosphere and, judging by its color, its composition is not obviously different from that of other impact debris. In the IK model the ring would be a condensation cloud, but what is the condensible? The obvious candidates, water and ammonia, are white, while a suite of complex organics would not in general be expected to condense and subsequently evaporate with anything like a well-defined phase change. Perhaps the most viable candidate is molecular sulfur (S_8), but here again its color works against it.

A second argument against a condensation cloud is more telling: where are the sound waves? Strong acoustic waves moving at velocities of order 1 km s^{-1} are inevitable. Strong stratospheric gravity waves with velocities of order 900 m s^{-1} are also generated (IK suggest that the explosions were below 10 bars to minimize stratospheric waves). These waves quickly develop into shock waves with strong temperature and pressure amplitudes. We see these waves in abundance in our numerical simulations. Condensibles would respond to these 1 km s^{-1} waves at least as strongly as they would to the distant pulse of a tropospheric gravity wave. That 1 km s^{-1} acoustic waves were not seen seems clear, and hence it is difficult to accept the hypothesis that the 450 m s^{-1} ring marks the passage of a wave through a condensible material.

An alternative hypothesis is that the rings are nonlinear stratospheric gravity waves. Breaking waves can sweep material along with them. We have suggested that the rings are actual rings of impact debris flowing outward from the impact site (Young *et al.* 1995). This hypothesis is suggested to us by the numerical models, in which we see just such a ring of airburst debris propagating outwards at 500 m s^{-1} at the base of the stratosphere. (Fig. 17 and 18).

The numerical ring is an attractive candidate for several reasons. It has the right velocity. It has the right location, centered on the stratospheric waveguide at 10–50 mbar. Its velocity happens to be the same as the critical velocity $u = NH$ (490 m s^{-1} in our numerical model), where N is the Brunt-Väisäla frequency; *i.e.*, the numerical ring appears to propagate at a Froude number of 1. It is far from clear that this is more than simple coincidence, but it is an intriguing coincidence, nonetheless.

But as IK point out, the nonlinear wave has a severe disadvantage: it slows down. Our numerical wave illustrates precisely this defect. After maintaining a constant velocity of 490 m s^{-1} for 700 s, it slows. The real rings lasted at least 10^4 s. Numerical experiments indicate that the failure to propagate is not a numerical effect. The energy of the numerical wave is only about 10^{25} ergs; *i.e.*, only about 1% of the impact energy. Numerical experiments indicate that doubling the energy in the wave doubles the area it covers. Hence we seek an additional energy source.

A possible source of energy is the latent heat of condensation of water vapor lifted by the impact. Stoker (1986) shows that, because it is dense, moist jovian air tends to be stable against convection, but if lifted by more than about 7 km the cumulative effects of condensation (heating, water depletion) are sufficient to cause the parcel to rise rapidly and indefinitely.

The explosion leaves a transient crater in the atmosphere. Air rises from below to fill the hole. This can be seen in the lowest lines of tracer particles in Fig. 17. The

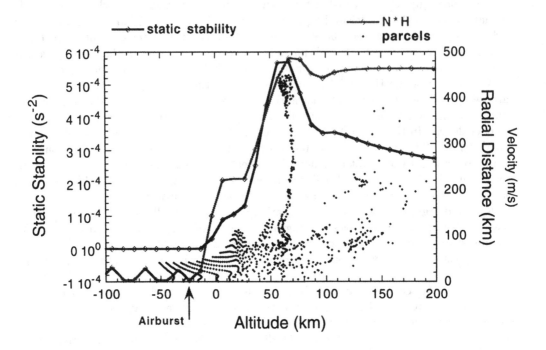

FIGURE 17. This figure shows the locations of several hundred tracer particles 950 s after impact in the general vicinity of the impact site, as calculated by ZEUS-3D. The particles were initially packed in a cylinder 50 km radius and extending in z from -65 km to $+100$ km, where $z = 0$ at 1 bar. The explosion released 2×10^{27} ergs at $z = -25$ km. Many of the most strongly shocked particles are thousands of kilometers distant, well beyond the frame of this picture. The feature at $z = 65$ km and $r = 450$ km is a ring of moderately strongly shocked jovian air driven from deep inside the original cylinder. We also show the static stability of the jovian atmosphere that was used in this numerical simulation. It is interesting to note that the numerical ring travels within the waveguide defined by the static stability. The velocity of the ring is shown in Fig. 18.

lowest line of parcels began at $z = -65$ km. In the numerical experiment they are lifted 25 km along the z-axis, and about half as much at $r = 50$ km. These parcels were probably water-rich when they began to rise, but on being lifted > 10 km effectively all the water they carry condenses, with concomitant release of latent heat of condensation. The amount of energy available in latent heat can rival that released by the impact itself.

As a specific example, assume that jovian water is enhanced 5-fold over cosmic. The corresponding lifting condensation level would be at 8.5 bars and 308 K ($z \approx -75$) km. The available latent heat of condensation in a cylinder of this air 50 km radius and 30 km thick is $\sim 2 \times 10^{26}$ ergs. These dimensions are in all likelihood smaller than the actual volume of lifted moist air, and hence this estimate is, in all likelihood, smaller than the actual amount of energy that would be available to the convective plume. This energy is mostly delivered to the base of the stratosphere, where the plume spreads as a large but otherwise conventional anvil (Note that the anvil, because it is cold, would contain very little water. The water that fueled the plume falls out deep in the troposphere.) It is this additional energy that we speculate powers the ring, and keeps it going well beyond the predictions of early numerical experiments.

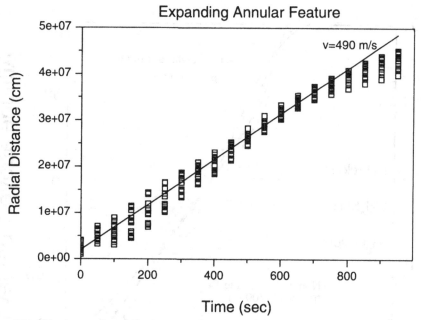

FIGURE 18. The expansion velocity as a function of time of the annular feature shown in Fig. 17, as defined by 13 representative tracer particles. The ring maintains a steady velocity of about 490 m s^{-1} for some 700 sec, although individual particles cycle through slower and faster velocities. This is a breaking wave. After 700 s it slows down, and the wave ceases to break.

A third interesting possibility is a hybrid model, in which the master wave is a linear disturbance in the wet troposphere, but which manifests itself as a nonlinear, breaking wave in the stratosphere, where the amplitude of the wave might well be larger than in the troposphere. In the hybrid model the tropospheric wave serves both as energy source and pacemaker, while the visible wave remains a material ring of impact debris and/or charred jovian air travelling along the stratospheric waveguide.

7. Plume heights

Several of the SL9 ejecta plumes were observed by the HST to reach approximately the same height, about 3000 km above the jovian cloudtops (Hammel *et al.* 1995). The duration of the infrared events, produced by the plume falling back on the atmosphere, measures time aloft and hence provides a second, more sensitive measure of plume height. The light curves (Nicholson, this volume) indicate that the largest impacts produced modestly higher plumes (compare K to D), but the difference was not large. Evidently all the plumes were launched with about the same vertical velocity, roughly 10–13 km s^{-1}. As the impactors themselves were not all the same, nor the impacts equally luminous, nor the plumes equally opaque, the similar plume heights has been seen as a puzzle needing explanation.

Figure 19, which summarizes our explanation, compares loci of constant plume height to calculated airburst altitudes for impacts of different energy. It is immediately apparent that the curves are approximately parallel. But where do these curves come from?

The characteristic radius R_s of a strong point explosion of energy E_i in a homogeneous gas with ambient pressure p_a is roughly

$$R_s \approx 0.5 \left(E_i / p_a \right)^{1/3} . \tag{7.10}$$

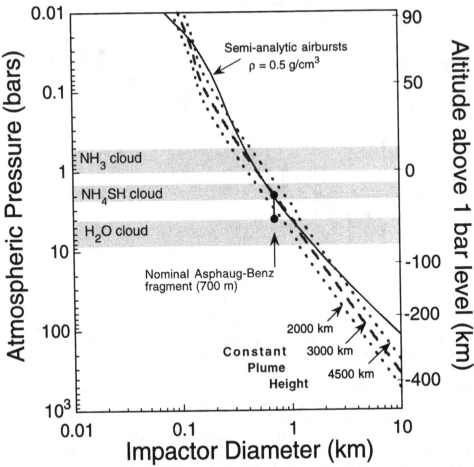

FIGURE 19. Airburst altitude, defined as the altitude of peak kinetic energy loss, for comets of density 0.5 g cm^3, is shown by the solid curve. The intermittent curves show loci of constant plume height. The labels, in kilometers, give relative plume heights. We assume that the airburst curve is correct, and use the observed 3000 km plume height to calibrate the curves of constant plume height. The diameter of the largest fragment, consistent with the tidal breakup analysis of Asphaug & Benz (1994), is marked by a dumbbell. The top of the dumbbell is placed on the airburst curve. As material can continue to move downward after it is shed by the comet, the effective explosion altitude is probably deeper. The bottom of the dumbbell, based on numerical results of Mac Low and Crawford *et al.*, allows for this.

This relation can be obtained by equating the ambient energy within a sphere of radius R_s to the energy of the explosion. The characteristic length scale of the atmosphere is its scale height H. Where $R_s \ll H$ the explosion is spherically symmetric and smothered by the atmosphere. Where R_s and H are more nearly equal the explosion is distorted by the gradient in the background atmosphere, and if $R_s > H$ the explosion is expected to blow out into space. Accordingly, one expects the dimensionless ratio R_s/H, or the functionally equivalent dimensionless ratio $E_i/p_a H^3$, to determine the shape of the transient crater and the shape of the plume, and in particular its opening angle.

We showed elsewhere (Zahnle & Mac Low 1994) that ejecta velocities should scale as $\sqrt{E_i/\rho_a H^3}$. This is agreeable with R_s/H being the key dimensionless parameter, because $\sqrt{E_i/\rho H^3} \propto c_s (R_s/H)^{3/2}$, where the sound speed c_s is the natural velocity in the problem. In its most reduced form, the argument compares the energy released by the explosion to the kinetic energy of the entrained air. As a simple example, assume that the explosion is isotropic, with energy $E_i/4\pi$ per steradian. The mass of air in the cone above the explosion, per unit solid angle, is $\int_0^\infty \rho_a e^{-z/H} r^2 dr = \rho_a H^3$. The asymptotic ejection velocity is therefore

$$v = \sqrt{\frac{E_i}{4\pi\rho_a H^3}} = \sqrt{\frac{E_i g}{4\pi p_a H^2}} \tag{7.11}$$

Although Eq. 7.11 gives the right dimensional form, the velocity it predicts is much too slow. For what we have taken to be typical SL9 parameters ($E_i = 2 \times 10^{27}$ ergs, $p_a = 2$ bars), Eq. 7.11 gives $v \approx 2$ km s^{-1}, which is much less than the ~ 10 km s^{-1} produced by numerical simulation (*e.g.*, Fig. 3–6) of the same event. In our 1994 paper, we suggested that the analytic argument fails to get the velocity right because it does not allow for the low density and high temperature of the wake (the wake provides a highway to space); other possibilities are that the assumption of an isotropic explosion is grossly in error, or that the assumption of radial trajectories is badly violated.

Given that ejecta velocities scale as $\sqrt{E_i/p_a H^2}$, the plume height z_{max} goes as

$$z_{max} \approx \frac{v^2_{max}}{2g} \propto \frac{E_i}{p_a H^2}, \tag{7.12}$$

where we have identified z_{max} with v_{max}, the maximum velocity in the toy plume. The curve labeled "constant plume height" in Fig. 19 traces $p \propto E_i/H^2$.

Airburst altitudes are calculated using the simple semi-analytic model for the deceleration and destruction of large meteoroids presented by Chyba *et al.* (1993) and Zahnle & Mac Low (1994). In brief, we numerically integrate standard equations for drag and ablation, supplemented by our nonstandard equation for the effective cross-section. The equations are cast in terms of altitude z and entry angle θ, measured from the zenith. The drag force is

$$m\frac{dv}{dz} = \frac{1}{2}C_D S \rho_a v \sec\theta, \tag{7.13}$$

where S is the effective cross-section, ρ_a is the atmospheric density, and C_D is the drag coefficient ($C_D \approx 0.9$ for a sphere). Mass loss by ablation is given by

$$Q\frac{dm}{dz} = \frac{1}{2}C_H S \rho_a v^2 \sec\theta, \tag{7.14}$$

where Q is a characteristic latent heat of ablation, which we identify with the heat of vaporization ($Q \approx 2.5 \times 10^{10}$ ergs g^{-1} for ice), and C_H (≤ 1) is the heat transfer coefficient. Radiative ablation, in which the meteor is evaporated by thermal radiation from the hot bow shock, is important for small bodies and low ρ_a, but becomes unimportant for large bodies at high ρ_a (Tauber & Kirk 1976; Biberman *et al.* 1980; Zahnle 1992; Zahnle & Mac Low 1994; Chevalier & Sarazin 1994; Field & Ferrara 1995). We use

$$C_H = \min\left(0.1, \frac{2\sigma T^4}{\rho_a v^3}\right), \tag{7.15}$$

where $C_H = 0.1$ is typical for terrestrial meteors at high altitudes. The temperature attained by the shocked gas is strongly regulated by thermal ionization to a value of the order of $30,000$ K (Biberman *et al.* 1980; Zahnle & Mac Low 1994).

The third, nonstandard equation describes how the shattered impactor spreads in response to aerodynamic forces. Chyba *et al.* (1993) approximate the impactor's radius by

$$\frac{d^2r}{dt^2} = \frac{1}{2}\frac{C_D \rho_a v^2}{\rho_i}. \tag{7.16}$$

Written in terms of altitude z, this becomes

$$r\frac{d^2r}{dz^2} + \frac{r}{v}\frac{dv}{dz}\frac{dr}{dz} = \frac{C_D \rho_a}{2\rho_i}\sec^2\theta. \tag{7.17}$$

where $\rho_i = 0.5$ is the density of the comet. Equation 7.17 is solved numerically in concert with Eq. 7.13 and Eq. 7.14 using $S \equiv \pi r^2$. The curves labeled by impactor density in Fig. 19 are the loci of maximum energy deposition. As material can continue to move downward after it is shed by the comet, the effective explosion altitude is probably deeper (see chapter by Mac Low), perhaps by as much as a scale height.

As noted at the beginning of this section, what Fig. 19 shows is that, when plotted against impact energy, loci of constant plume height are nearly parallel to the loci of calculated airburst altitudes. What this means is that the impact of SL9 fragments with diameters ranging from 100–1000 m generated plumes of similar size and shape. Different plumes differ mostly by their optical depth, by plumefall luminosity, and by the pressure level of atmospheric reentry, all of which are proportional to impactor mass. The essential reason that plumes tend to be the same size and shape is that smaller fragments exploded at higher altitudes, and so had less jovian air to lift, while the larger events, penetrating more deeply, spread their energy over a larger mass of jovian air. But the near constancy of $z_{\max}$ also owes something to coincidence: the greater role of radiative ablation in the flight of smaller objects raises explosion altitudes. If the SL9 fragments had been much larger, the three curves would not have been so nearly parallel, and the differences between the largest and smallest event would have been more pronounced.

8. Conclusions

Tidal disruption models show that the parent comet was about 1.5 km diameter and had a density of $\sim 0.6 \pm 0.1$ g cm^3 (Asphaug & Benz 1994, 1996; Solem 1994). These models are relatively strongly constrained by observations, and they have relatively few free parameters (diameter, density, and rotation). [However, see chapter by Sekanina for an alternative view. Eds.] The diameter of the parent body is directly proportional to the distance between the leading and trailing fragments, *i.e.*, the distance between A and W. The density of the parent body is determined by the number of fragments, *i.e.*, the number of letters between A and W. If the largest fragment is assigned 1/8 of the total mass, the largest fragments were some 700–800 m across, had mass of $\sim 1 \times 10^{14}$ g, and released energies of $\sim 2 \times 10^{27}$ ergs.

A first analysis of the infrared light curves agrees with this size. We estimate that the energy of the R impact plume was about 4×10^{26} ergs, about half of which is invested in the ejecta plume. Nicholson (this volume) suggests that the G, K, and L events were about five or ten times more luminous than R, which would place them at $\sim 2 - 5 \times 10^{27}$ ergs. When cast as the diameter of a 0.5 g cm^3 comet, a spherical G would have been 750–1000 m across.

The chemical evidence is ambiguous, but most indications are that C>O in the shocked, reacting gas. Telltale signatures of abundant oxygen—SO_2, SO, CO_2, O_2—were not seen, while signatures of abundant carbon—CS, CS_2, and HCN—were. On the other hand, abundant H_2O would appear to require O>C, and two other observed sulfur species, S_2 and OCS, appear to form more easily in a somewhat oxidized gas. Since on general principles one expects the comet to have had a more-or-less cosmic composition, *i.e.*,

O>C, the production of CS, CS_2, and HCN probably requires C>O in the shocked jovian air. This in turn implies that even the largest fragments released the bulk of their energy above the jovian water table, in all likelihood above 5 bars. There is no evidence in favor of the proposition that a significant amount of wet jovian air was shocked strongly enough to coax water to react; *i.e.*, wet jovian air saw only temperatures significantly below 2000 K.

A more direct measure of the impactor masses comes from the mass of CO. The reported mass of CO, some 2.5×10^{14} g for K (Lellouch, this volume), would imply impact energies as large as 10^{28} ergs, *i.e.*, two or three times higher than the energies estimated by other means. The CO measurements are probably more model dependent, and hence more uncertain, than are the estimates made from tidal disruption and infrared radiation, and probably less model dependent than explosion depths estimated from the apparent absence of jovian water.

The simple pancake model places maximum energy deposition by a 700 m diameter, 0.5 g cm^3 comet at 2 bars (Fig. 19). The melange—hot, swept up jovian air, plus disintegrating comet—continues downward some distance before it explodes, perhaps getting as deep as 4 bars before it turns around (see chapters by Crawford and Mac Low for more on these matters). Although some of the comet's mass may plunge deeper, little of its energy does. Concord with the observations is as good as could be hoped.

The outstanding argument in favor of a deep explosion is the identification of expanding rings as a condensation wave excited by a deeply seated tropospheric wave. This model requires deep energy deposition solely in order not to excite analogous stratospheric waves that would also be visible as condesation waves, but which are not observed. We offer as a speculative alternative that the expanding rings were a kind of nonlinear debris wave, in which the dark feature is analogous to the flotsam and jetsam carried by water waves. If they are not condensation waves, the perceived need for deep explosions is removed.

In sum, the SL9 impacts appear to have been small and shallow. Tidal disruption calculations and models of the infrared light curves agree on an average diameter of about half a kilometer, with the largest fragments approaching 1 km; the reported mass of CO fattens the largest fragment to 1.2 km. The chemical evidence implies that the explosions occured in the presence of sulfur but above the jovian water table. The inferred size, density, and the chemical limit on explosion depth are consistent with pre-impact predictions made using simple semi-analytic models. The spectacular plumes were spectacular mostly because the explosions were shallow, not because the explosions were large.

I thank, among others, E. Asphaug, B. Bézard, D. Deming, B. Fegley, J. Graham, J. Harrington, A. Ingersoll, T. Johnson, K. Lodders, J. I. Moses, P. Nicholson, K. Noll, M. Norman, I. de Pater, R. Yelle, R. Young, and of course M.-M. Mac Low, the co-author with all the numbers. K. Z. is supported by the NASA Exobiology Program.

REFERENCES

AHRENS, T., TAKATA, T., O'KEEFE, J. D., & ORTON, G. 1991 Impact of Comet Shoemaker-Levy 9 on Jupiter. *Geophys. Res. Lett.* **21**, 1087–1090.

ANDERS, E. & GREVESSE, N. 1989. Abundances of the elements: meteoritic and solar. *Geochim. Cosmochim. Acta* **53**, 197–214.

ASPHAUG, E. & BENZ, W. 1994. Density of comet Shoemaker-Levy 9 deduced by modelling breakup of the parent 'rubble pile'. *Nature* **370**, 120–124.

ASPHAUG, E. & BENZ, W. 1996. Size, density, and structure of Comet Shoemaker-Levy 9 inferred by the physics of tidal breakup. *Icarus* (in press).

BOSLOUGH, M., CRAWFORD, D., ROBINSON, A., & TRUCANO, T. 1994. Mass and penetration depth of Shoemaker-Levy 9 fragments from time-resolved photometry. *Geophys. Res. Lett.* **21**, 1555–1558.

BOSLOUGH, M., CRAWFORD, D., TRUCANO, T., & ROBINSON, A. 1995. Numerical modelling of Shoemaker-Levy 9 impacts as a framework for interpreting observations. *Geophys. Res. Lett.* **22**, 1821–1824.

CARLSON, R., WEISSMAN, P., HUI, J., SMYTHE, W., BAINES, K., JOHNSON, T. V., DROSSART, P., ENCRENAZ, T., LEADER, F., & MEHLMAN, R. 1994. Galileo NIMS observations of the impact of comet Shoemaker-Levy 9 on Jupiter. *EOS Trans. AGU* **75**, 401.

CARLSON, R., WEISSMAN, P., SEGURA, M., HUI, J., SMYTHE, W., JOHNSON, T. V., BAINES, K., DROSSART, P., ENCRENAZ, T., & LEADER, F. 1995. Galileo infrared observations of the Shoemaker-Levy 9 G impact fireball: A preliminary report. *Geophys. Res. Lett.* **22**, 1557–1560.

CARLSON, R., WEISSMAN, P., HUI, J., SMYTHE, W., BAINES, K., JOHNSON, T. V., DROSSART, P., ENCRENAZ, T., LEADER, F., & MEHLMAN, R. 1995. Some timing and spectral aspects of the G and R collision events as observed by the Galileo near-infrared mapping spectrometer, In *Proceedings: European Shoemaker-Levy 9 Conference* (ed. R. West). pp. 69–73.

CHAPMAN, C., MERLINE, W., KLAASEN, K., JOHNSON, T., HEFFERNAN, C., BELTON, M., INGERSOLL, A., & THE GALILEO IMAGING TEAM. 1995. Preliminary results of Galileo direct imaging of SL9 impacts. *Geophys. Res. Lett.* **22**, 1561–1564.

CHEVALIER, R., & SARAZIN, C. 1994. Explosions of infalling comets in Jupiter's atmosphere. *Astrophys. J.* **429**, 863–875.

CHYBA, C., THOMAS, P., & ZAHNLE, K. 1993. *Nature* **361**, 40–44.

CRAWFORD, D., BOSLOUGH, M., TRUCANO, T., & ROBINSON, A. 1994. Numerical simulations of fireball growth and ejecta distribution during Shoemaker-Levy 9 impact on Jupiter. *EOS Trans. AGU* **75**, 404.

CRAWFORD, D., BOSLOUGH, M., TRUCANO, T., & ROBINSON, A. 1994. The impact of comet Shoemaker-Levy 9 on Jupiter. *Shock Waves* **4**, 47–50.

CRISP, D., & MEADOWS, V. 1995. Near-infrared imaging spectroscopy of the impacts of SL9 fragments C, D, G, K, N, R, V, and W with Jupiter. *IAU Colloquium 156: The Collision of Comet P/Shoemaker-Levy 9 and Jupiter held at the Space Telescope Science Institute*, p. 25.

DE PATER, I. 1991. The Significance of Radio Observations for Planets, *Physics Reports* **200**, 1–37.

DROSSART, P., ENCRENAZ, T., LECACHEUX, J., COLAS, F., & LAGAGE, P. 1995. The time sequence of SL9/impacts H and L from infrared observations. *Geophys. Res. Lett.* **22**, 1769–1772.

FEGLEY, B. & LODDERS, K. 1994. Chemical models of the deep atmospheres of Jupiter and Saturn. *Icarus* **110**, 117–154.

FIELD, G. & FERRARA, A. 1995. The behavior of fragments of Comet Shoemaker-Levy 9 in the atmosphere of Jupiter. *Astrophys. J.* **438**, 957–967.

FIELD, G., TOZZI, G., & STANGA, R. 1995. Dust as the cause of spots on Jupiter. *Astron. Astrophys.* **294**, L53–L55.

FRIEDSON, A. J., HOFFMAN, W., GOGUEN, J., DEUTSCH, L., ORTON, G., HORA, J., DAYAL, A., SPITALE, J., WELLS, W. K., & FAZIO, G. 1995. Thermal infrared lightcurves of the impact of Comet Shoemaker-Levy 9 fragment R. *Geophys. Res. Lett.* **22**, 1569–1572.

GRAHAM, J., DE PATER, I., JERNIGAN, J., LIU, M., & BROWN, M. 1995. W. M. Keck telescope observations of the comet P/Shoemaker-Levy 9 fragment R Jupiter collision. *Science* **267**, 1320–1323.

HAMMEL, H., BEEBE, R., INGERSOLL, A., ORTON, G., MILLS, J., SIMON, A., CHODAS, P., CLARKE, J., DE JONG, E., DOWLING, T., HARRINGTON, J., HUBER, L., KARKOSCHKA, E., SANTORI, C., TOIGO, A., YEOMANS, D., & WEST, R. 1995. Hubble Space Telescope imaging of Jupiter: atmospheric phenomena created by the impact of comet Shoemaker-Levy 9. *Science* **267**, 1288–1296.

INGERSOLL, A., AND KANAMORI, H. 1995. Waves from the impacts of Shoemaker-Levy 9 with Jupiter *Nature* **374**, 706—708.

INGERSOLL, A., KANAMORI, H., & DOWLING, T. 1994. Atmospheric gravity waves from the impact of Shoemaker-Levy 9 with Jupiter *Geophys. Res. Lett.* **21**, 1083–1086.

KIM, S., RUIZ, M., RIEKE, G., RIEKE, M., MAC LOW, M.-M., & ZAHNLE, K. 1995. The re-entry shock of the R fragment of Comet Shoemaker-Levy 9. *Nature* (submitted).

KNACKE, R. F., GEBALLE, T. R., NOLL, K. S., & BROOKE, T. Y. 1994. Infrared spectra of the R post-impact events of comet Shoemaker-Levy 9. *Bull. Amer. Astron. Soc. Special Sessions on SL9* **26**, 3.25.

KNACKE, R. F., FAJARDO-ACOSTA, S. B., GEBALLE, T. R., & NOLL, K. S. 1995. *IAU Colloquium 156*, p. 59.

LELLOUCH, E., PAUBERT, G., MORENO, R., FESTOU, M., BÉZARD, B., BOCKELÉE-MORVAN, D., COLOM, P., CROVISIER, J., ENCRENAZ, T., GAUTIER, D., MARTEN, A., DESPOIS, D., STROBEL, D., & SIEVERS, A. 1995. Chemical and thermal response of Jupiter to the impact of comet Shoemaker-Levy 9. *Nature* **373**, 592–595.

MAC LOW, M.-M. & ZAHNLE, K. 1994. Explosion of Comet Shoemaker-Levy 9 on entry into the Jovian atmosphere, *Astrophys. J. Lett.* **434**, L33–L36.

MCGREGOR, P., NICHOLSON, P., & ALLEN, M. 1995. CASPIR observations of the collision of Comet Sheomaker-Levy 9 with Jupiter. *Icarus* (submitted).

MCKINNON, WM. B., & SCHENK, P. 1995. Estimates of comet fragment masses from impact crater chains on Callisto and Ganymede. *Geophys. Res. Lett.* **22**, 1829–1832.

MEADOWS, V., CRISP, D., ORTON, G., BROOKE, T., & SPENCER, J. 1994. AAT observations of Shoemaker-Levy 9 collisions with Jupiter. poster presented at the 26th Ann. Mtg., Div. Planet. Sci., Bethesda MD, Oct. 31–Nov. 4, 1994.

MELOSH, H. J., SCHNEIDER, N., ZAHNLE, K., & LATHAM, D. 1990. Ignition of global wildfires at the Cretaceous/Tertiary boundary. *Nature* **343**, 251–254.

MOSES, J. I., ALLEN M., & GLADSTONE, R. 1995. Post-SL9 sulfur photochemistry on Jupiter, *Geophys. Res. Lett.* **22**, 1597–1600.

NICHOLSON, P., GIERASCH, P., HAYWARD, T., MCGHEE, C., MOERSCH, J., SQUYRES, S., VAN CLEVE, J., MATTHEWS, K., NEUGEBAUER, G., SHUPE, D., WEINBERGER, A., MILES, J., & CONRATH, B. 1995a. Palomar observations of the impact of the R fragment of comet P/Shomeaker-Levy 9: Light curves. *Geophys. Res. Lett.* **22**, 1613–1616.

NICHOLSON, P., GIERASCH, P., HAYWARD, T., MCGHEE, C., MOERSCH, J., SQUYRES, S., VAN CLEVE, J., MATTHEWS, K., NEUGEBAUER, G., SHUPE, D., WEINBERGER, A., MILES, J., & CONRATH, B. 1995b. Palomar observations of the impact of the R fragment of comet P/Shomeaker-Levy 9: Spectra. *Geophys. Res. Lett.* **22**, 1617–1620.

NOLL, K., MCGRATH, M., TRAFTON, L., ATREYA, S., CALDWELL, J., WEAVER, H., YELLE, R., BARNET, C., & EDGINGTON, S. 1995. HST spectroscopic observations of Jupiter after the collision of Comet Shoemaker-Levy 9. *Science* **267**, 1307–1313.

PRINN, R., & FEGLEY, B. 1987. Bolide Impacts, acid rain, and biospheric traumas at the Cretaceous-Tertiary boundary. *Earth Planet. Sci. Lett.* **83** 1–15.

SHOEMAKER, E., HASSIG, P., & RODDY, D. 1995. Numerical modelling of Shoemaker-Levy 9 impacts as a framework for interpreting observations. *Geophys. Res. Lett.* **22**, 1825–1828.

SOLEM, J. C. 1994. Density and size of comet Shoemaker-Levy 9 deduced from a tidal breakup model. *Nature* **370**, 349–351.

SPRAGUE, A., BJORAKER, G., HUNTEN, D., WITTEBORN, F., KOZLOWSKI, R., & WOODEN, D. 1996. Water brought into Jupiter's atmosphere by fragments R and W of Comet SL-9. *Icarus* (in press).

STOKER, C. 1986. Moist convection: A mechanism for producing the vertical structure of the Jovian equatorial plumes. *Icarus* **67**, 106–125.

TAKATA, T., O'KEEFE, J. D., AHRENS, T. J., & ORTON, G. 1994. Comet Shoemaker-Levy 9: Impact on Jupiter and plume evolution. *Icarus* **109**, 3–19.

TAKATA, T., AHRENS, T. J., & HARRIS, A. 1995. Comet Shoemaker-Levy 9: Fragment and progenitor impact energy. *Geophys. Res. Lett.* **22**, 2433-2436.

WATANABE, J., YAMASHITA, T., HASEGAWA, T., TAKEUCHI, S., ABE, M., HIROTA, Y., NISHI-HARA, E., OKUMURA, S., & MORI, A. 1995. Near-IR observation of cometary impacts to Jupiter: Brightness variation of the impact plume of fragment K. *Publ. Astron. Soc. Jpn.* **47**, L21–L24.

WEST, R., KARKOSCHKA, E., FRIEDSON, A., SEYMOUR, M., BAINES, K., & HAMMEL, H. 1995. Impact debris particles in Jupiter's stratosphere. *Science* **267**, 1296–1301.

YABE, T., XIAO, F., ZHANG, D., SASAKI, S., ABE, Y., KOBAYASHI, N., & TERASAWA, T. 1994. Effect of EOS on break-up of Shoemaker-Levy 9 entering jovian atmosphere. *J. Geomag. Geolectr.* **46**, 657–662.

YOUNG, R., ZAHNLE, K., & MAC LOW M.-M., 1995. Nonlinear propagating features in the stratosphere of Jupiter generated by the impact of SL-9. *Bull. Amer. Astrom. Soc.*, (in press).

ZAHNLE, K. 1990. Atmospheric chemistry by large impacts. In *Global Catastrophes in Earth History* (eds. V. Sharpton & P. Ward). *Geol. Soc. Am. Spec. Pap. 247* pp. 271–P288.

ZAHNLE, K., & MAC LOW, M.-M. 1994. The collision of Jupiter and Comet Shoemaker-Levy 9. *Icarus* **108**, 1–17.

ZAHNLE, K., & MAC LOW, M.-M. 1995. A simple model for the light curve generated by a Shoemaker-Levy 9 impact. *J. Geophys. Res.* **100**, 16885–16894.

ZAHNLE, K., MAC LOW, M.-M., LODDERS, K., & FEGLEY, B. 1995. Sulfur chemistry in the wake of Shoemaker-Levy 9. *Geophys. Res. Lett.* **22**, 1593–1596.

ZEL'DOVICH, IA. B., & RAIZER, YU. P. 1967. *Physics of Shock Waves and High Temperature Hydrodynamic Phenomena* Academic.

Chemistry induced by the impacts: Observations

By EMMANUEL LELLOUCH

Observatoire de Paris, 92195 Meudon, France

This paper reviews spectroscopic measurements relevant to the chemical modifications of Jupiter's atmosphere induced by the Shoemaker-Levy 9 impacts. Such observations have been successful at all wavelength ranges from the UV to the centimeter. At the date this paper is written, newly detected or enhanced molecular species resulting from the impacts include H_2O, CO, S_2, CS_2, CS, OCS, NH_3, HCN and C_2H_4. There is also a tentative detection of enhanced PH_3 and a controversial detection of H_2S. All new and enhanced species were detected in Jupiter's stratosphere. With the exception of NH_3 (and perhaps H_2S and PH_3), apparently present down to the 10–50 mbar level, the minor species are seen at pressures lower than 1 mbar or less, consistent with a formation during the plume splashback at 1–100 microbar. NH_3 may result from upwelling associated with vertical mixing generated by the impacts. The main oxygen species is apparently CO, with a total mass of a few 10^{14} g for the largest impacts, consistent with that available in 400–700 m radius fragments. The observed O/S ratio is reasonably consistent with cometary abundances, but the O/N ratio (inferred from CO/HCN) is much larger, suggesting that another N species was formed but remained undetected, presumably N_2. The time evolution of NH_3, S_2, CS_2 shows evidence for photochemical activity taking place during and after the impact week.

1. Introduction and brief overview of measurements

The comet Shoemaker Levy-9 impacts on Jupiter on July 16–22, 1994 have produced a variety of impressive phenomena observable at different wavelengths that provided insight into the various phases of the event (meteor, fireball, plume evolution and re-entry) and the subsequent dynamical evolution of Jupiter's atmosphere. These observations and their interpretation are reviewed in several chapters of this volume. The present chapter is concerned with a particular aspect of the phenomenon, namely the observed chemical modifications of Jupiter's atmosphere induced by the impacts. The ultimate goal is to provide a list of the molecular species that were detected (either detected for the first time in Jupiter, or observed to be enhanced) subsequent to the impacts, with preliminary, but as reliable as possible, baseline estimates of their mixing ratios and total abundances.

The chapter is organized as follows. In the rest of this introductory section, a brief overview of the different types of measurements is given, including some remarks on their respective capabilities and the possible complications associated with their analysis. Then, the bulk of the paper (Sections 2 to 5) is dedicated to spectroscopic measurements performed during the impact week and within the next few weeks. For each minor species, the relevant measurements are described and compared, and differences are discussed. In some instances, a few suggestions are proposed which may help reconcile the different measurements. The long-term evolution is addressed in Section 6. In Section 7, a synthesis of the observations of minor species is attempted, and some elementary conclusions are drawn. The entire chapter is focussed on molecular gaseous species. In particular, observations of atomic species at visible and UV wavelengths are not addressed, neither is the compositional characterization of aerosols. These two topics are discussed in the chapters by Crovisier and West, respectively. In addition, although some description on the modification of Jupiter's thermal structure (as inferred from CH_4

213

and C_2H_2 infrared measurements) is occasionally needed when discussing abundance retrievals, this is not done here in a systematic way; we refer the reader to the chapter by Conrath for a complete description of the subject. Similarly, H_2 and H_3^+ observations, which are relevant to magnetospheric rather than chemical phenomena, are not covered at all here (see chapter by Ip).

Measurements of minor species after the impacts have been successful at all wavelengths. In the UV, the most useful measurements come from the HST/FOS and GHRS spectra, which cover altogether the 1250–3300 Å range. A preliminary analysis of these data was given by Noll *et al.* (1995) and more detailed modelling of the 1800–2300 Å range was performed by Atreya *et al.* (1995) and Yelle and McGrath (1996). The HST spectra show absorption features, from which S_2, CS_2 and NH_3 have been unambiguously identified, and emission features, due to CS and to a number of atomic lines. IUE observations (Harris *et al.* 1995) also indicate the presence of ammonia absorption lines, but these measurements have not been analyzed yet. Modelling the absorption lines is primarily complicated by (i) the influence of scattering by the aerosols produced by the impacts; in particular the aerosol unit optical depth level in the UV is uncertain by about one order of magnitude in pressure (ii) line saturation effects: laboratory data do not resolve individual lines in the electronic bands, and curve-of-growth effects may be difficult to assess; in some cases (*e.g.*, S_2), available laboratory data are still preliminary. On the other hand, the determination of molecular abundances from UV absorption lines (*i.e.*, in the solar reflected component) is not directly sensitive to the temperature, although the absorption cross sections may be significantly temperature-dependent (*e.g.*, S_2). For the emission lines, the chief problem at the present time is to ascertain the excitation mechanism. Noll *et al.* (1995) have assumed that they are produced by resonant fluorescence, but other mechanisms (notably non-resonant fluorescence (Carpenter *et al.* 1995) and thermal emission) may be at work and need to be further explored.

Infrared spectroscopic observations can be conveniently classified into two types of measurements: (i) the "plume/splashback phase" observations, *i.e.*, the measurements obtained within $\sim 1/2$ hour from impact time, and (ii) the "impact site" observations, that are performed hours or days after the impacts, when the site material is dragged by Jovian rotation. All detections of minor species at infrared wavelengths were obtained in thermal, as opposed to solar reflected, radiation. Plume/splashback observations were achieved notably in the 2–3 μm range by Galileo/NIMS and many Earth-based telescopes (*e.g.*, Anglo-Australian Telescope (AAT), Steward Observatory, Calar Alto) and at 7–20 microns from the KAO. These measurements are particularly difficult to analyze in terms of minor constituent abundance because: (i) they are extremely sensitive to temperature, as they correspond to the tail of the Planck function (especially at 2–3 μm), (ii) their very special geometry (a relatively warm plume at the limb expanding in a colder environment) requires elaboration of unusual and complex models, and (iii) they are affected by complications associated with departure from local thermodynamical conditions (LTE); for example, in the case of the widely observed CO 2.3 μm bands, non-LTE effects become important at pressures as deep as ~ 1 μbar (Hooker and Millikan 1963). Site observations do not pose these problems so acutely and their analysis is similar to that of "classical" Jupiter observations, although knowledge of the temperature structure is an important aspect that must be coped with. Infrared site observations were successfully obtained at 4.7 μm (CFHT/FTS, UKIRT/CGS-4, IRTF/CSHELL) and in the 8–12 μm window. The latter observations have the advantage that the temperature profile can be measured "simultaneously" from the emission of hydrocarbons such as CH_4 and C_2H_2, whose abundance is predicted to remain roughly unchanged during the fireball/splashback shock chemistry and subsequent photochemistry, and which therefore

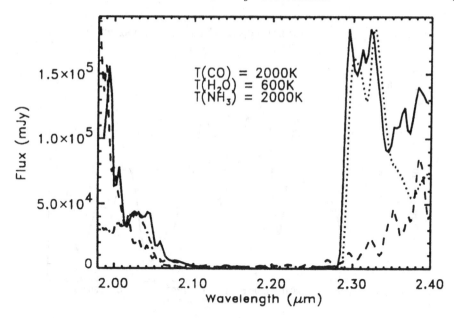

FIGURE 1. AAT/IRIS spectrum of fragment K splashback (continuum subtracted) along with a model including H_2O, CO, and NH_3. From Meadows and Crisp (1995a).

can be used as thermal probes of the stratosphere. Many successful results have been obtained at 8–12 μm from various types of experiments. Among these, one can mention (i) the heterodyne observations (IRTF/IRHS, Mount Wilson), which resolve the line profiles, thereby constraining the vertical profile of the minor compounds, although some complications arise because rotational broadening due to velocity smearing within the sites contributes to the lineshape and (ii) the spectro–imagers (in particular IRSHELL at the IRTF), which combine high spectral (~ 10000) and high spatial (1–1.5″) resolution and provide unique information on the horizontal distribution of the minor species.

Finally, millimeter and centimeter heterodyne observations, performed at IRAM (30-m), JCMT (15-m), SEST (15-m) and Medicina (32-m) have also allowed the detection of several new molecular species in Jupiter's stratosphere. In spite of the well-known weak (linear) dependence of the flux with temperature at these wavelengths, the analysis of these data is not insensitive to the thermal profile, because the lines (emission or absorption) are observed against a continuum whose brightness temperature is unfortunately very close (*e.g.*, 170 K at 1.3 mm) to the temperature of Jupiter's stratosphere. These observations are also hampered by the lack of spatial resolution (10″ at best, more typically 20–30″). As for the 10 μm measurements, the lineshapes are resolved, but again rotational effects within the sites may be important; in addition, the spectra often mix the contribution from different sites.

2. Oxygen species

2.1. *Water*

One of the last molecules whose detection was announced, water has in fact been reported independently by five different teams. Four of these detections were obtained in the plume/splashback phase. Galileo/NIMS observed emission from the 2.7 μm water band on impacts G and R (Carlson *et al.* 1995a). Observations with AAT/IRIS de-

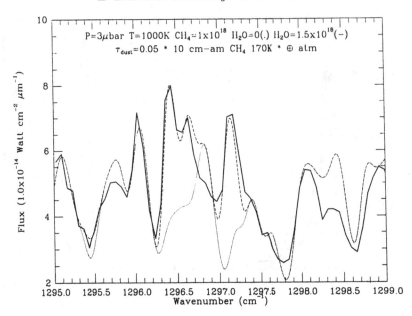

FIGURE 2. KAO/KEGS spectrum of impact G splashback, 14 minutes after impact (solid line). The H_2O lines at 1296.4, 1296.7, and 1297.2 cm^{-1} are visible. Dotted line: model fit (see text) with $H_2O = 0$. Dashed line: model fit, with a H_2O column density $= 1.5 \times 10^{18}$ cm^{-2}. From Bjoraker *et al.* (1995a).

tected H_2O at 1.99–2.02 and 2.35–2.40 micron on several (C, D, G, K, R, W) impacts (Meadows and Crisp 1995a; Fig. 1). In all cases, the observations correspond to 10 to 20 minutes after impact time (T_i, as defined, for example, by the moment of first signal on the Galileo instruments), and the water signal appeared to be delayed with respect to hot methane emission at 2.2 and 3.3 μm. Analysis of the NIMS data on impact G, using a model described below (Bjoraker *et al.* 1995a), indicated a water mixing ratio of 100 ppm at p$\leq$3 μbar and T = 2000 K, corresponding to a total mass of 10^{12} g; a factor-of-ten lower mass was inferred for impact R (Carlson *et al.* 1995b). Finally, the NIMS data on impact G also suggested a tentative detection of H_2O at 2.7 μm in the late fireball phase, $\sim$ 30–40 sec after impact (Carlson *et al.* 1995c).

At longer wavelengths, H_2O was unambiguously detected by the two experiments flown onboard the KAO. The high resolution (R = 9000) instrument KEGS detected several H_2O lines at 22.6 and 23.9 μm and three others near 7.7 μm (1296.4, 1296.7, and 1297.2 cm^{-1}) on impacts K and G (Bjoraker *et al.* 1995a, 1995b; Fig. 2.). All of these lines have high lower energy levels (500–700 cm^{-1} at 22–24 μm and 1500–2000 cm^{-1} at 8 μm). The lower resolution (R = 300) instrument HIFOGS observed part of the water ν_2 band near 6.6 μm on impacts R and W (Sprague *et al.* 1996). The 22–24 μm lines were visible in emission for about an hour after impact G. At 6.6 μm and 7.7 μm, the emissions appeared $\sim$5 minutes after impact, peaked at $\sim T_i +$ 12–14 minutes, and disappeared after another five minutes or so (Fig. 3). The duration of the emission correlates with the excitation level of the transition (Bjoraker *et al.* 1995b), strongly suggesting that the disappearance is primarily caused by the cooling of the stratosphere following the splashback-induced heating to several thousands of degrees. The 7.7 μm range also showed hot lines of CH_4. Modelling of the 6.6 and 7.7 micron spectra was performed in limb geometry, by assuming that the methane and water emissions are produced in a

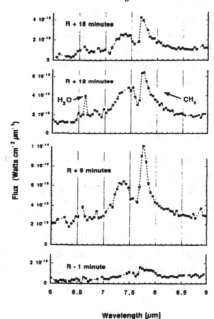

FIGURE 3. A sequence of KAO/HIFOGS spectrum of impact R splashback showing the 6.6 μm H_2O band and the enhanced 7–8 μm CH_4 emission. From Sprague *et al.* (1996).

portion of the stratosphere located above a pressure level p_0 and heated to a temperature T_0, and accounting for re-absorption of the radiation by cold CH_4 (Bjoraker *et al.* 1995a, Sprague *et al.* 1996). Best fits were obtained for $p_0 \sim 3 \mu$bar (Fig. 2). From the relative intensities of the three individual water lines near 7.7 μm, T_0 was inferred to be about 1000 K. The hot methane and water column densities were found to be similar and of order 10^{18} cm^{-2}. The corresponding water masses are of order 2×10^{12} g for the large impactors G and K, 3–14 $\times 10^{10}$ g for impact R, and 2–9 $\times 10^{10}$ g for impact W. These determinations may be regarded as lower limits to the total amount of water, as the data are sensitive only to the highest ($\sim$ 1000 K) temperature, although the fact that smaller masses are found for the smallest impactors suggest that a non-trivial fraction of the total water may have been actually detected.

Besides these various splashback phase measurements, Montebugnoli *et al.* (1995) reported the detection of water on impact site E, 52 hours after impact, from centimetric observations at 22 GHz performed on July 19, 1994 with the 32-m antenna in Medicina, Italy. They also mentioned the tentative observation of the same line on sites A and C on the same day, and on site E again on September 9, 1994. The site E, July 19, line is surprisingly narrow and strong. Because of the effects of rotational smearing, the linewidth ($\sim$ 40 kHz, *i.e.*, 0.5 km s^{-1} FWHM) implies an emitting region less than 1″ (3500 km) in diameter. With such a diameter, absolute calibration of the antenna temperature scale implies a brightness temperature line contrast $\Delta T_B \geq 4000$ K, which rules out LTE thermal emission as the mechanism. Cosmovici *et al.* (1995) suggest a maser-type emission (with radiative pumping by solar or jovian IR photons) taking place at pressures less than 1 nbar. Simple modelling suggests an emitting region 1500-km wide, 130–300 km thick, a masing optical depth of −4 and an H_2O column density of 10^{18} cm^{-2}, *i.e.*, a water mass of 5×10^{11} g. A few checks can be made on the plausibility of the detection. Maser emission of the 22 GHz water line in jovian conditions, not to be collisionally quenched, must, in fact, take place at $\sim 10^{-11}$ bar (J. Crovisier, priv.

comm.), *i.e.*, above the homopause. The diffusion time at this pressure level is ~ 10 minutes, implying that the observed water is in diffusion equilibrium. Since this pressure level also lies within the ionosphere, it must be wondered whether water can survive ionization. Examination of ion/neutral reactions involving water suggests that the three dominant reactions are (P. Colom, priv. comm.):

$$H_2O + H^+ \longrightarrow H_2O^+ + H$$
$$H_2O^+ + H_2 \longrightarrow H_3O^+ + H$$
$$H_3O^+ + e \longrightarrow H_2O + H$$

Equilibrium between these reactions, using typical ionospheric densities at 10^{-11} bar (Atreya 1986) and rate constants (Anicich and Huntress 1986), leads to concentration ratios of $H_2O/H_3O^+ \sim 40$ and $H_2O/H_2O^+ \sim 4000$, *i.e.*, neutral water remains the dominant species. Finally, water appears to be also stable against photolysis. The photolytic lifetime of an isolated water molecule at Jupiter's distance is ~ 1 month, and this time may be increased by factors of several by shielding effects, making the observation of water days or even two months after impact plausible. In summary, we do not see any obvious physical reason that would contradict this observation, although a complete model of the pumping mechanism remains to be elaborated.

2.2. *Carbon monoxide*

2.2.1. *Infrared observations*

Carbon monoxide has been searched for in many observations and successfully observed in both the infrared (2.3 μm and 4.7 μm) and the millimeter ranges.

Although far from being fully understood at the present time, the 2.3 μm observations of CO in the plume/splashback phase are among the most spectacular results of the SL9 campaign. Spectra at 2.3 μm were obtained from a variety of telescopes and instrumentation (Steward/FSPEC (Ruiz *et al.* 1994), AAT/IRIS (Fig. 1), UKIRT/CGS-4 (Knacke *et al.* 1995), Calar Alto/MAGIC (Herbst *et al.* 1995a), altogether covering most medium and large impacts (C, D, G, H, K, L, R, W). These spectra show CO emission features (generally unresolved, except in the Steward observations) belonging to the 2–0, 3–1 and 4–2 bands. These emissions, essentially concommitant with those seen for H_2O, appear ~ 10 minutes after impact time (*i.e.*, approximately at or slightly after the IR peak flux, which might indicate that the (O-bearing) cometary material transported in the plume hits the atmosphere after the jovian material), evolve rapidly in shape and intensity, and last about 10 minutes. The CO hot bands are diagnostic of high temperatures (1500 to 5000 K, increasing with time [Kim *et al.* 1995]), consistent with strong heating during the splashback phase. The AAT/IRIS observations (Crisp and Meadows 1995) also indicate that the relative intensities of the CO and H_2O emissions vary from impact to impact. (A preliminary analysis of these data (Meadows and Crisp 1995a) suggested that CO was seen in absorption prior to appearing in emission, but this conclusion was apparently erroneous and caused by confusion with methane bands). Preliminary fits to the spectra suggest total CO masses of order 3×10^{12} g for the large impactors (Knacke *et al.* 1995, Crisp and Meadows 1995), but this determination is very uncertain because (i) model fitting assumes no absorption of radiation by colder background atmosphere and (ii) non-LTE effects are not considered. A potentially even more serious problem may arise from the fact that the CO cooling rate from ~ 2000 K is so high that any CO heated up to this temperature will cool off and become invisible to the observations in some seconds. Therefore, the observations might in fact sample a "production rate" rather than a total mass (D. Crisp, priv. communication). The same problem may conceivably affect the H_2O measurements in the mid-infrared (see above), and further

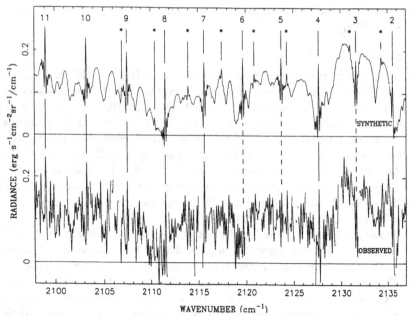

FIGURE 4. CFHT/FTS 4.7 μm spectrum of site L 4.5 hrs after impact. Resolution is 0.08 cm^{-1}. Lines from the P-branch of the (1-0) band of ^{12}CO are marked with their rotational number. A synthetic spectrum, assuming an isothermal CO layer at 274 K (see text), is shown for comparison. The asterisks indicate the position of the ^{13}CO lines. From Maillard *et al.* (1995).

progress will require consistent modeling of the splashback spectra at different wavelengths.

No plume/splashback spectroscopic measurements were reported at 4.7 μm, and the CO observations at these wavelengths pertain to evolved impact sites. The most convincing detections were obtained from quasi-simultaneous observations of site L, $\sim$ 4.5 hours after impact, from CFHT/FTS (Maillard *et al.* 1995) and UKIRT/CGS4 (Brooke *et al.* 1995). The high-resolution (R = 25000) CFHT/FTS observations allowed the detection of 12 emission lines of the (1–0) band of ^{12}CO (superimposed on the normal tropospheric CO absorption features), and a marginal detection of ^{13}CO (Fig. 4). These observations in fact primarily constrain the stratospheric temperature. The ^{12}CO lines are saturated and indicate T = 274$\pm$10 K at the level where the CO column density is $\sim 10^{16}$ cm^{-2}. The ^{13}CO lines can be fit either (i) with a constant stratospheric temperature of 274 K and a CO column density of $(1.5\pm0.8)\times10^{17}$ cm^{-2} (which thus appears as a lower limit to the CO) or (ii) with a downward temperature decrease and a larger CO column density (Maillard *et al.* 1996) In particular, the data are fully consistent with the CO abundances derived from the millimeter spectra (see below). Similar results were obtained from the UKIRT/CGS4 observations, which cover only 6 lines (and at moderate spectral resolution), but for which the 1-spatial dimension allows an estimate of the size of the emission and an inference of the total CO mass (Brooke *et al.* 1995). These CO emissions were found to have disappeared one day after impact, as a result of the temperature decrease (Maillard *et al.* 1995).

Another 4.7 μm observation, performed from IRTF/CSHELL on August 1–2, 1994, was reported by Orton *et al.* (1995). A preliminary analysis suggested that some 4.7 μm CO emission was still present $\sim$10 days after the impacts, but this result was apparently caused by an instrumental artifact (K. Noll, *et al.* 1996). CSHELL data recorded on

July 19 confirm that the CO absorption lines are back to normal 20 hours after impact (K. Noll, *et al.* 1996).

2.2.2. *Millimeter observations*

CO has been observed in many impact sites (C, E, G, H, K, L, Q1) in millimeter-wave spectroscopy. Most detections were obtained from the IRAM 30-m telescope through the CO rotational (2–1) line at 230 GHz (Lellouch *et al.* 1995). Observations of this line extend from minutes to months after the impacts. The CO (1–0) line was also observed on a few occasions. The (2–1) line was also detected at the SEST-15m on July 23, on a beam position principally encompassing sites G, Q1 and R (Bockelée-Morvan *et al.* 1995). Given their poor spatial resolution, all millimeter-wave measurements performed on July 21, 1994 and later include the contribution of several impact sites, which can be identified on the overall spectra from their individual velocity (see example in Bockelée-Morvan *et al.* 1995).

The CO rotational lines were originally detected as narrow emission lines (FWHM $\sim$ 2.5 MHz $\sim$ 3 km s^{-1}). This linewidth implies that CO is located in the stratosphere, at pressures of 1 mbar or less. After July 28, they were observed in absorption (see Sec. 6.2). Preliminary modelling of the CO (2–1) line on site G, $\sim$10 hours after impact, indicated CO column densities of $\sim 1.5 \times 10^{18}$ cm^{-2}, for a total CO mass of $\sim 10^{14}$ g (Lellouch *et al.* 1995). However, this modelling had important limitations: (i) the velocity dispersion within the site, which must contribute to the observed linewidth to some degree, was not considered (ii) a crude treatment of the geometry was adopted, with an average airmass of 2 (iii) an arbitrary thermal profile was used. This profile assumed uniform warming of the upper statosphere (at p$\leq$ 2 mbar) by 30 K above a pre–impact Jupiter reference profile. It turns out that this assumed thermal profile is inconsistent with other data. Specifically, IRTF/IRSHELL observations of strong and weak 8 μm CH$_4$ lines indicate that no warming larger than a few degrees occurred at pressures deeper than a few tenths of a millibar (Bézard *et al.* 1995a).

Improved modelling of the CO(2–1) line has been recently performed (Lellouch *et al.* in preparation). The line observed on site K, July 20, $\sim$ 33 hours after impact was selected, as independent information on the thermal profile and spatial extent of the emission was available from the IRTF/IRSHELL observations of the same site at $T_i + 23$ hours. (The spatial extent of the CO was assumed to follow that of the thermal perturbation). The modelling now includes the line smearing due to velocity dispersion within the site. In addition, taking advantage of the fact that the observations were recorded during a relatively long integration time (2.5 hours), they were split into three parts depending on the site position on Jupiter's disk ($\phi \leq 40°$; $40° \leq \phi \leq 60°$; $60° \leq \phi \leq 80°$, where ϕ is the site longitude measured from the central meridian), and the three parts were separately analyzed with an improved treatment of the geometry, including variation of the secant within a site. The lines corresponding to the three parts have very different aspects and contrasts (Fig. 5), primarily a result of (i) the variation of the continuum level with the airmass and (ii) the less important contribution of the velocity smearing near the limb. Yet, all three lines of Fig. 5 can be fit with a single model, in which CO is confined to pressures less than 10^{-4} bars and its mixing ratio somewhat increases with altitude, from 2×10^{-4} at 10^{-4} bars to 2×10^{-3} at 10^{-6} bars. The column density is about 4×10^{18} cm^{-2} and the total mass is 2.5×10^{14} g. Although this number is not final, as variations due to uncertainties in the thermal profile have not been considered yet (which means in particular that the conclusion that CO increases with altitude must not be taken too strongly at present time), it should be regarded as more reliable than that given in Lellouch *et al.* (1995). The combination of the lines observed in the three different

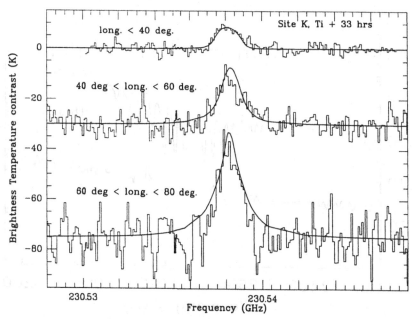

FIGURE 5. IRAM 30-m CO (2-1) spectra (histograms) of site K (July 20, UT=17:00–19:30) as a function of apparent longitude of site center with respect to central meridian. Solid lines: model fits (see text for parameters).

airmass ranges also allows us to relatively well constrain the pressure region where the CO is. Essentially, a significant fraction of the CO must be present at $p \geq 10^{-5}$ bar, otherwise the synthetic line near the limb is too narrow if the line near central meridian is well fitted. Conversely, CO must also decrease very quickly at pressures $\geq 10^{-4}$ bar, otherwise the synthetic line is too broad. From this, it can be inferred that the bulk of the CO lies between 10^{-5} and 10^{-4} bar.

Millimeter-wave observations are not very well suited to the study of the plume/splash-back phase because the amount of signal in the beam is severely penalized by the geometrical effect. Nevertheless, IRAM observations of impact H (impact time: July 18, 19:32 UT), taken at a time resolution of about 1.5 min., seem to reveal an intriguing result. For the first 12 minutes after impact, no CO emission is seen. At 19:45 UT, a strong emission is apparently detected, which fades within 3 minutes. The timing of this emission coincides with that of the plateau in the Calar Alto lightcurve of the same impact (Herbst *et al.* 1995b). The striking feature is that the line is not at the limb velocity (-8.8 km s^{-1}), but at an approaching velocity (~ -4 km s^{-1}) lower by ~ 5 km s^{-1}. Modelling of the ballistic trajectory of a plume for the initial velocity, azimuth, and elevation conditions given in Hammel *et al.* 1995 indicates that the velocity towards the observer remains at ~ -11 km s^{-1} during the entire plume flight. The observation of a ~ -4 km s^{-1} velocity is thus not understood, although it can be speculated from the timing that it might be linked to the bouncing of the plume after its impact with Jupiter's stratosphere.

2.2.3. *UV upper limit and reconciliation of measurements*

CO was not detected in the HST/GHRS spectra of site G and stringent upper limits on the CO column density were inferred from the absence of its fluorescence bands at 1419, 1447, 1478 and 1510 Å. Noll *et al.* (1995) find an upper limit of 6×10^{14} cm^{-2}, ~ 5000 times less than the millimeter value. Trafton *et al.* (1995) find that with the

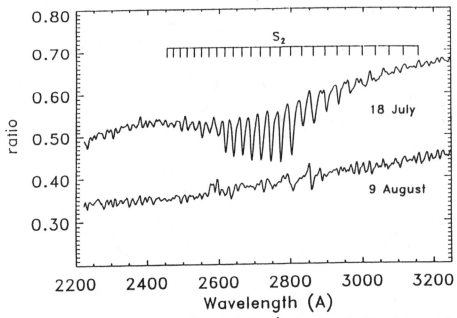

FIGURE 6. HST/FOS spectra of site G at 2200–3200 Å taken on July 18 and August 9, 1994 (ratioed to a pre-impact spectrum taken on July 14) showing ∼ 16 bands of the S_2 B–X system. From Noll *et al.* (1995).

column density given by Lellouch *et al.* (1995), the CO emission in the UV bands should be ∼ 8000 Rayleigh, whereas the HST spectra indicate an upper limit of ∼ 100 R. However, both these calculations assumed optical thinness, whereas it is clear that for the relevant column densities, the individual lines of the CO UV bands are heavily saturated. Independent calculations by J. Crovisier and D. F. Strobel indicate that when saturation effects are considered, the emission is in fact photon-limited to ∼ 30 R, regardless of the CO abundance. These estimates are confirmed by the brightness of the Venus CO airglow (Feldman *et al.* 1995). There is thus no contradiction between the CO column density inferred from the millimeter and its non-detection in the UV.

As mentioned earlier, the 4.7 μm observations are, in fact, more sensitive to the temperature than to the CO abundance, and are anyway consistent with the CO determinations from the millimeter. The 2.3 μm spectra indicate CO masses ∼ 100 times lower, but their analysis is fraught with complexity, and so far, it is not certain if these measurements sample the entire CO produced during the splashback. We therefore believe that the most reliable estimate of the CO resulting from the impact is given by the millimeter data and that a total mass of 2.5 × 10^{14} g for the large impactors (G, K) can be provisionally adopted as a baseline value. The CO mixing ratio at p≤ 10^{-4} bar, several times 10^{-4}, is typically 10^5 times larger than the normal Jupiter CO abundance (*e.g.*, Noll *et al.* (1988), implying in itself that the observed CO does not result from jovian tropospheric CO transported by the plume or convectively upwelling to the stratosphere.

3. Sulfur species

3.1. *Molecular sulfur* (S_2)

S_2 was unambiguously detected in HST/FOS spectra of site G on July 18 and 21, 1994 (T_i + 3 hours and T_i + 3.5 days respectively; Figs. 6 and 8), through its numerous

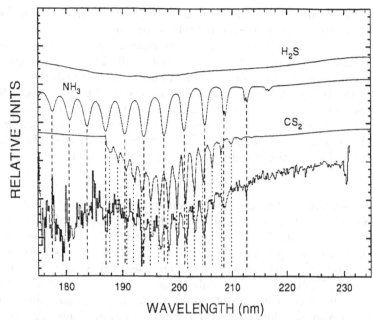

FIGURE 7. Bottom curve: HST/FOS spectrum of site G at 1750–2300 Å taken on July 18, 1994 (ratioed to a pre-impact spectrum). Top three curves. Absorption cross sections of H_2S, NH_3, and CS_2 (arbitrary scale), allowing the spectral identification of NH_3, CS_2, and perhaps H_2S. From Yelle and McGrath (1996).

bands belonging to the B–X system between 2500 and 3000 Å (Noll *et al.* 1995). The bands had disappeared when HST reobserved site G, on August 9, 1994. Preliminary modelling of five of these bands (5–0 through 9–0) was performed, assuming an isothermal homogenous atmosphere at 300 K. Matching of the equivalent widths in the July 18 spectrum indicated a column density of 3×10^{18} cm^{-2}, corresponding to a total mass of 2.5×10^{13} g in the FOS aperture. This value was considered as a lower limit to the total S_2 mass because (i) at this column density, individual S_2 lines are strongly saturated, and (ii) the areal extent of S_2 is probably larger the the FOS aperture. Noll *et al.* (1995) proposed a baseline value of about 10^{14} g. However, further modelling of all of the S_2 bands indicated that the observed spectrum cannot be fit at 300 K, and that much higher temperatures ($\sim$ 1250 K) are required, implying that S_2 is located in Jupiter's thermosphere at p$\leq 10^{-7}$ bar (Yelle and McGrath 1995). The resulting S_2 column density is $\sim 5 \times 10^{15}$ cm^{-2}, giving a mass of only 4×10^{10} g in the FOS aperture.

3.2. *Carbon disulfide*

CS_2 was detected in HST/FOS spectra of site G, on July 18, July 21, August 9, and August 23, 1994 (ratioed to a pre-impact spectrum taken on July 14) from its numerous bands between 1850 and 2100 Å, belonging to the $^1\Sigma_g^+ - {}^1B_2(^1\Sigma_u^+)$ system (Fig. 7). (The same spectra also show bands due to NH_3 (see Sec. 4.1.1), which become progressively more prominent relative to CS_2 after July 18). The emphasis of the analysis so far has been put on the July 18 spectrum. The preliminary analysis by Noll *et al.* (1995) indicated a CS_2 column density of 0.5–2×10^{15} cm^{-2}. Further determinations relied on an improved modelling of the scattering properties of aerosols at UV wavelengths. Following West *et al.* (1995), Atreya *et al.* (1995) assume that the aerosol unit optical depth level is located at 0.3 to 10 mbar, and find a column density of 0.43–1.1×10^{15} cm^{-2} at this level. Yelle and McGrath (1996) infer essentially the same number (0.8×10^{15} cm^{-2}) from the same spectrum, and find a decrease of the CS_2 abundance with time, by factors of 2–3 on August 9 and $\sim$5 on August 23, compared to July 18. There is, therefore, a consensus on the CS_2 abundance. In addition, Yelle and McGrath (1996) strongly emphasize that a good fit of the HST/FOS spectra can be obtained only if CS_2 lies above NH_3 and the bulk of the aerosols (*i.e.*, at pressures $\leq$1 mbar for July 18 and $\leq$0.1 mbar for August 9), a condition that Atreya *et al.* do not require. It can be finally noted that CS_2 has a very strong band at 6.5 μm (ν_3; $S = 2300$ cm^{-2} atm^{-1}), which may affect the KAO/HIFOGS spectra and provide additional constraints on the CS_2 distribution.

3.3. *Hydrogen sulfide*

The tentative detection of hydrogen sulfide was announced during the impact week and in the first HST spectroscopy paper (Noll *et al.* 1995), on the basis of the broad absorption centered at 2000 Å in the HST/FOS spectra of site G. A preliminary estimate of the H_2S column density was given as 2–5 $\times 10^{16}$ cm^{-2}. Further investigations by the HST team lead to divergent conclusions about the presence of H_2S. Yelle and McGrath (1996) note that the HST/FOS spectrum of July 18 has a positive slope at 2100–2300 Å, while the reflectivity in the August 9 and 23 spectra is flat in this region. Since the H_2S cross sections exhibit a similar slope longward of 2100 Å, they attribute this behaviour to the presence of H_2S on July 18 and its absence on subsequent dates (Fig. 7). They also moderately favor the scenario in which H_2S is mixed with NH_3 and the aerosols but below the CS_2 layer. In this case, for July 18, (and assuming an aerosol imaginary index of refraction (k) constant over the FOS range), they infer a mixing ratio of $\sim 5 \times 10^{-8}$ at $p \geq 5$ mbar. The mixing ratio is not well determined because of the uncertainties in the aerosol properties and could be 4 times smaller. Although H_2S in this model is uniformly mixed at 5–100 mbar, Yelle and McGrath (1996) recognize that the UV observations do not probe deeper than $\sim$10–20 mbar and stress that the mixing ratio is far better constrained than the total column density (R. Yelle, priv. comm.). A mixing ratio of only 10^{-8} is inferred if H_2S is assumed uniform at 0–100 mbar. On the other hand, Atreya *et al.* (1995) claim that the entire HST/FOS July 18 spectrum can be fit without invoking the presence of H_2S and obtain an upper limit on the column density of 1.2×10^{16} cm^{-2} at the $\tau = 1$ aerosol level. This corresponds to a mixing ratio $\leq 10^{-8}$ (resp. 3.5×10^{-7}) if this level is at 10 mbar (resp. 0.3 mbar). Finally, Caldwell *et al.* (1995), adopting a semi-log extrapolation of the West *et al.* (1995) values for the aerosol imaginary index of refraction, find that the slope at 2100–2300 Å in the FOS spectrum can be matched by a uniform H_2S mixing ratio of 5 ppb, in general agreement with the Yelle and McGrath (1996) homogeneous case. However, they prefer to view this number as an upper limit, as slightly modified aerosol properties would allow a fit to the entire spectrum with no H_2S at all. Strong opinions have been expressed about the presence or absence of H_2S, ranging from "the arguments for a significant abundance (of H_2S) in the atmosphere are strong" (Yelle and McGrath 1996) to "H_2S is not detected" (Atreya *et al.* 1995), possibly confusing the occasional reader, and an intermediate position such as Caldwell *et al.* seems the most reasonable. Yelle and McGrath argue that the variation of the spectrum slope between July 18 and August 23 suggests the the absorber on July 18 was indeed H_2S. However, since the NH_3 and CS_2 cross sections also have a slight positive slope at 210–230 nm, this variation could be plausibly due to the established decrease of CS_2 and NH_3 (see Sec. 4.1.1) with time, possibly coupled with variations of aerosol properties.

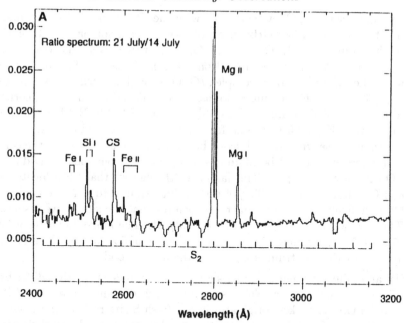

FIGURE 8. HST/FOS spectrum at 2400–3200 Å taken on July 21, 1994 (ratioed to the pre-impact spectrum). The FOS aperture encompassed site G and 45-min old site S. Absorption bands of S_2 and emissions lines of CS and several atomic species are identified. From Noll *et al.* (1995).

H_2S has also been unsuccessfully searched for at millimeter (IRAM, JCMT) and infrared (IRTF/IRSHELL) wavelengths. Unfortunately, this does not seem to provide a test of the claimed UV detection. Specifically, synthetic spectra of the H_2S 216 GHz line based on the H_2S distribution suggested by Yelle and McGrath (1996) are fully consistent with the non-detection of this line from IRAM observations of sites G and H on July 19, as H_2S in this model is essentially too deep to be detectable in the millimeter range. The millimeter upper limit only constrains the H_2S at lower pressures (*e.g.*, $H_2S \leq 8 \times 10^{-7}$ if assumed uniform at p$\leq$ 5 mbar). The upper limit from the infrared has still to be derived, but is expected to be even less stringent than from the millimeter.

3.4. *Carbon monosulfide*

CS has been observed at UV and millimeter wavelengths. In the UV, the detection comes from a limb spectrum (emission angle 79°) taken by HST/FOS on July 21 and showing the CS(0–0) transitions near 2580 Å and a number of metallic lines (Mg I, Mg II, Si I, Fe I, Fe II, and perhaps Ni I and Ti II) (Noll *et al.* 1995; Carpenter *et al.* 1995; Fig. 8). The aperture of the instrument for this observation encompassed site G and 45-min old site S. Noll *et al.* (1995) assumed that the excitation mechanism for all these features was solar resonant fluorescence and estimated the g-factors (fluorescence efficiency factor) using a temperature of 1000 K. In the case of CS, a column abundance of 2×10^{14} cm^{-2} was derived, corresponding to a total mass of 9×10^9 g in the FOS aperture. This estimate assumes optical thinness, and therefore should be viewed as a lower limit, although reasonably accurate, according to Noll *et al.* It is noteworthy that CS was not detected on the spectrum taken on July 18 ~3.5 hours after the G impact. On the other hand, weak emission features seem to appear near the position of the CS(0–0) transition on spectra through 23 August.

In the millimeter range, CS was detected with the IRAM–30m telescope, first through its $J = 5$–4 transition at 244.9 GHz, originally seen in emission on July 21 on site K and complexes including sites (L, G) and (G, Q, R, S) respectively (Lellouch *et al.* 1995). This line has since then been monitored on a regular basis (see Sec. 6.2). Preliminary modelling of the line observed on complex (G, Q, R, S) indicated that CS is confined to pressures ≤ 0.7 mbar with a mixing ratio (assumed uniform) of 4.5×10^{-8}, corresponding to a total mass, rescaled to site G, of 3×10^{11} g. The CS 244.9 GHz line appears slightly broader than the CO 230.5 GHz line for simultaneous observations. In particular, in the most frequent case when the telescope beam contained several sites, the contribution from the different sites could be more easily separated (from their individual velocity) on the CO that on the CS lines. This larger width suggests that CS is located at deeper levels than CO. Improved modelling of the CS line, similar to that presented for CO in Sec. 2.2.2, is difficult, because the method requires that the site be well isolated in the beam. The only available case is site K, on July 21. The model indicates a CS mixing ratio of 5×10^{-8} at $p \leq 1$ mbar, *i.e.*, a column density of 5×10^{15} cm^{-2} and a total mass of 5×10^{11} g, but the spectrum is noisy and the fit not good.

The UV and millimeter measurements give CS column densities that differ by a factor of ~ 20. There are several possible reasons for this discrepancy. First, the CS emission seen by HST/FOS most likely originates in the fresh S site rather than from the 3-day old G site. (Support for this comes from the metallic emissions observed simultaneously, reminiscent of the atomic emissions at visible wavelengths that lasted for ~ 1 hour after impact [*e.g.*, Roos-Serote *et al.* 1995, Fitzsimmons *et al.* 1995]). On the other hand, the millimeter measurements sample sites produced hours or days earlier. Since CS is not only produced in shock chemistry (Zahnle *et al.* 1995, Zahnle, this volume) but also builds up photochemically from CS_2 or S_2 on a ~ 1 day time scale (Moses *et al.* 1995a), it is conceivable that the largest abundance seen at millimeter wavelengths results from a difference in observing time. The second point is that the UV estimate relies on the assumption that the UV CS(0–0) emission is due to solar fluoresence. There is, in fact, still considerable ignorance on the mechanism responsible for the visible and UV emissions, some of which cannot be explained by resonant fluorescence (see Crovisier, this volume). Third, are optical depths effects properly estimated in the case of CS (cf. the case of CO above)? Finally, the CS abundance measurement from the millimeter-wave observations is not yet satisfactory; an improved determination will require detailed modelling of the spectra on the (G, Q, R, S) complex, involving separate modelling of the contribution of the different sites.

3.5. Carbonyl sulfide

The detection of OCS has been reported from IRAM observations at 219 GHz of complex (W, K) performed on July 22 (Lellouch *et al.* 1995). The detection is significant at the 6-σ level for the line area. Preliminary modelling indicates a OCS mixing ratio of 2×10^{-7} at $p \leq 1$ mbar. The total OCS mass, rescaled to the G site, is $\sim 3 \times 10^{12}$ g, *i.e.*, approximately 10 times that of CS. The inferred mixing ratio is consistent with an upper limit obtained from the HST/FOS spectrum of site G, July 18, namely OCS$\leq 10^{-6}$ at p ≤ 1 mbar (Yelle and McGrath 1996). Nonetheless, this single detection was not confirmed by observations at other wavelengths; an attempt from IRAM on August 19, 1994, when the CO and CS were observed in absorption, was also unsuccessful.

4. Nitrogen species

4.1. *Ammonia*

Unlike most other molecular species detected after the SL9 impacts (and like CO, and PH$_3$), ammonia is normally observable in Jupiter's atmosphere in the ultraviolet, visible and infrared (5 and 10 μm) ranges. These observations have established that the vertical distribution of ammonia is governed by condensation at 300–600 mbar and photolysis at upper levels, leading to unobservable amounts above the tropopause (see *e.g.*, Atreya 1986). UV and 10 μm observations of the impact sites have revealed a profound modification of this vertical distribution, with the injection of considerable amounts of ammonia in the stratosphere.

4.1.1. *Ultraviolet observations*

Disk center ultraviolet observations of Jupiter normally detect ammonia bands at 1900–2200 Å which originate from the upper troposphere ($\sim$100–300 mbar) (*e.g.*, Wagener *et al.* 1985). However, in spectra taken close to Jupiter's limb, which probe higher levels, these bands may become invisible, as exemplified by a pre-impact spectrum taken by HST/FOS on July 14 at an emission angle of 73°, and which only shows spectral signatures of acetylene (Yelle and McGrath 1996). HST/FOS observations of impact site G at 1650–2300 Å were taken on July 18, July 21, August 9 and August 23, at emission angles of 54°, 73°, 79° and 67° (Noll *et al.* 1995). Ratioing these spectra to the pre-impact spectrum of July 14 makes the ammonia bands at 1900–2200 Å clearly appear and indicates the presence of NH$_3$ in the stratosphere (Fig. 7). As for CS$_2$, and using the same approach, most of the analyses have been dedicated to the July 18 spectrum. Noll *et al.* (1995) infer a vertical column density of $\sim 10^{16}$ cm^{-2}, within a factor of 2. Atreya *et al.* (1995) obtain a similar number (0.25–1.3 $\times 10^{16}$ cm^{-2}) at the $\tau = 1$ aerosol level, assumed to lie at 0.3–10 mbar. If this level is located at 1 mbar, this corresponds to a mixing ratio of 0.2–1 $\times 10^{-7}$. Yelle and McGrath (1996) find that NH$_3$ is confined to pressures $\geq$5 mbar and determine a mixing ratio of 10^{-7} rather than a column density. As for H$_2$S, they mention that the NH$_3$ mixing ratio could be 4 times smaller. Yelle and McGrath further analyze the August 9 and 23 spectra for which they find ammonia mixing ratios of 10^{-7} and 3×10^{-8}, respectively. This decrease is consistent with a photochemical destruction on a month timescale, although the HST/FOS observations only sample the ammonia in the dense core region, so that part of the effect may be due to horizontal spreading.

4.1.2. *Ten–micron observations*

Most of the infrared observations of ammonia in the impact sites was obtained at 10 microns. In normal (pre-impact) conditions, ammonia 10 μm lines are formed in Jupiter's troposphere and thus appear as broad absorption features (*e.g.*, Kunde *et al.* 1982). Observations of impact sites detected emission cores in the center of these absorptions, demonstrating the presence of ammonia in Jupiter's stratosphere. The most detailed results were obtained from the IRTF/IRSHELL spectro-imaging observations of site K 23 hours, 5 days and 10 days after impact (Griffith *et al.* 1996). These measurements have a spectral resolution of 15000 and a spatial resolution of 1.4″ ($\sim$ 5000 km), so that they provide vertical, spatial and temporal information on the ammonia distribution (Fig. 9). In addition, thermal profile information is retrieved from quasi-simultaneous observations of methane lines near 1234 cm^{-1}. The site K observations at 908 and 948 cm^{-1}, 23 hours after impact, indicate that NH$_3$ is mostly located at p$\leq$10 mbar. For the pixel of maximum emission, the column density is 2–5 $\times 10^{17}$ cm^{-2} above 40 mbar.

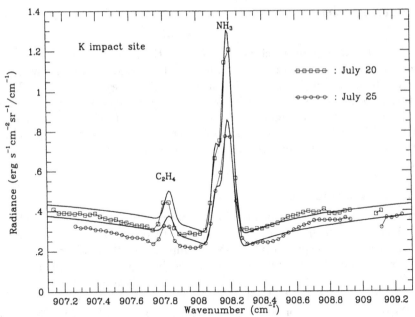

FIGURE 9. IRTF/IRSHELL spectra near 908 cm^{-1} of site K, 23 hours and 10 days after impact, showing stratospheric emission from NH$_3$ and C$_2$H$_4$. Note the decrease in time of the NH$_3$ emission. Adapted from Griffith *et al.* (1996) (courtesy C. Griffith, B. Bézard).

The mass of ammonia integrated over the K site is $\sim 10^{13}$ g. Observations 5 and 10 days after impact indicate that NH$_3$ spreads horizontally with time. Its vertical distribution is depleted with time as a function of altitude. Specifically, the column density at the central pixel decreases by 60% over 10 days, but given the lateral spreading, the total mass over the site decreases by only 25%, in excellent agreement with predictions from photochemical modelling (Moses *et al.* 1995b). Similar observations were obtained with the Palomar/SpectroCam 10 spectro-imager (Conrath *et al.* 1995). The calibration of these data is still preliminary, but first results suggest a stratospheric mixing ratio of 2×10^{-7}, roughly consistent with the IRTF/IRSHELL results.

Ammonia 10μm emission was also detected in heterodyne observations performed at Mt. Wilson/ISI telescope (Betz *et al.* 1995) and at the IRTF (IRHS spectrometer; Fast *et al.* 1995a). Observations at Mt. Wilson were recorded between July 16 and August 8, 1994. Narrow line emission was observed on impact sites E, G, H, K, L and Q1, and upper limits were obtained on sites A and C and on regions adjacent to the large sites. The strongest emission was seen on July 21 from site G, $\sim$3 days after impact, where the aQ(2,2), aQ(6,6) and aQ(9,8) lines of the ν_2 band were detected (Fig. 10). The first two lines appear to be saturated and constrain the stratospheric temperature to be $\sim$200 K, while the third one is weaker and provides a preliminary estimate of the column density of 1.5–2.0 $\times 10^{17}$ cm^{-2}. The linewidths ($\sim$1.7 km s^{-1} FWHM at the beginning of the observing period and 2–3 km s^{-1} towards the end) are mostly due to velocity dispersion within the sites, and from the absence of Lorentzian linewings, Betz *et al.* (1995) conclude that ammonia is primarily seen at p $\leq$ 2–3 mbar. At deeper levels, NH$_3$ is either absent or at a temperature below the detection threshold of the experiment ($\sim$ 165 K).

Similar data were obtained from IRTF/IRHS measurements of sites K, Q1 and R on July 23, 26 and 29. From a lineshape analysis, including deconvolution of the rotational

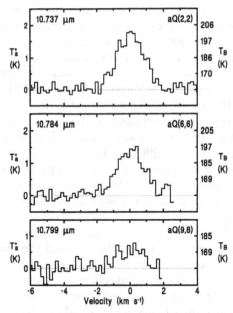

FIGURE 10. Mt. Wilson/ISI heterodyne observations of 3 NH_3 emission lines on site G, July 21, 1994. From Betz *et al.* (1995).

broadening, Fast *et al.* (1995a) find that Lorentz broadening is visible in their data and indicates that NH_3 is mostly present at 1–10 mbar. For sites Q1 and R, 8 days after impact, assuming uniform mixing in the stratosphere, they determine mixing ratios of 8×10^{-9} and 4×10^{-9} corresponding to column densities of 3×10^{16} and 1.5×10^{16} cm^{-2}, respectively. Kostiuk *et al.* (1995) find that the best fit of the Q1 data is obtained for $NH_3 = 2 \times 10^{-8}$ at p$\leq$ 10 mbar, giving a column density of $\sim 2 \times 10^{16}$ cm^{-2}.

Finally, broad-band IRTF/MIRAC2 images show enhanced NH_3 emission over nearly every impact site (Orton *et al.* 1995). This emission appears to decay from July 17 to August 6, consistent with photochemical destruction.

4.1.3. *Ammonia in the plume/splashback*

The AAT/IRIS spectra of the plume/splashback phase of many impacts, in which H_2O and CO were unambiguously detected (Meadows and Crisp 1995a), also contain a relatively clear signature of ammonia at 2.04 μm (Fig. 1). As for the other species, the associated abundance remains to be determined. If real, this observation is very important because it shows that some NH_3 is dragged or formed in the plume, whereas all previous observations see ammonia in the mid-stratosphere (1–50 mbar), which presumably results from slower vertical mixing following the impacts. In addition, the ammonia emission at 2.04 μm appears earlier than the CO and H_2O emissions seen in the same spectra (Sec. 2), suggesting that NH_3 is ejected at lower energies and to lower altitudes than CO and H_2O (Meadows and Crisp 1995b). If ammonia is detected at 2.04 μm, one might expect to see it also in the stronger 10 μm band observed during the splashback phase of impacts R and W by KAO/HIFOGS. This issue, however, has not been investigated yet (D. Hunten, priv. comm.).

4.1.4. *Reconciliation of the site measurements*

Intercomparison of the different ammonia site measurements shows no big contradiction, although some differences can be noted. The IRTF/IRSHELL column density of 2–5

$\times 10^{17}$ cm^{-2} above 40 mbar, measured one day after the K impact (Griffith *et al.* 1996), is largely consistent (1–3 times larger) with that determined by Betz *et al.* (1995) on site G at $T_i + 3$ days, but $\sim$10 times larger than the value obtained by Fast *et al.* (1995a). Comparison with the HST/UV determinations is difficult because the level probed by these measurements is unknown. If ammonia is assumed to extend down to 40 mbar, the average mixing ratio determined by Yelle and McGrath (1995) (10^{-7}) corresponds to a column density of 5×10^{17} cm^{-2}, consistent with IRSHELL. Clearing out the residual differences between the various measurements will require detailed comparisons, including fitting of the different datasets with a single model. One point that can be made is that different measurements may see different abundances, in part because ammonia varies with time. This argument is true but should not be pushed too much because both the IRSHELL and the UV measurements show that the ammonia column at the site center decreases by a factor of $\sim$2 only over a week or so. Specifically, the NH$_3$ column on site K, 10 days after impact inferred by Griffith *et al.*, is still $\sim$6 times larger than the value obtained by Fast *et al.* on site Q1, 8 days after impact. The second aspect is that the 10 micron measurements depend on the assumed thermal profile. In this respect, the IRSHELL determination (Griffith *et al.* 1996) seems to be the most reliable, because it makes use of a thermal profile simultaneously determined from methane or acetylene lines. For example, the thermal profile used by Fast *et al.* (1995a), 8 days after impact, is several degrees warmer than indicated by IRSHELL measurements at $T_i + 10$ days (B. Bézard, priv. comm.).

A key parameter for understanding the origin of ammonia is its vertical distribution. In this respect, the situation is not completely clear. Griffith *et al.* find that most of the ammonia is present at p $\leq$ 10 mbar, but cannot exclude that it be entirely located at much lower pressures. The two heterodyne measurements give diverging results. Unlike Betz *et al.*, Fast *et al.* find evidence for pressure broadening in their data, implying that ammonia is relatively deep (1–10 mbar). The reason for this discrepancy is obscure; it might in part be due to a vertical diffusion of NH$_3$ with time, or to the fact that the highest sensitivity of the IRTF/IRHS measurements of Fast *et al.* allows them to probe colder (*i.e.*, deeper) levels. The presence of NH$_3$ at pressures $\geq$ 5 mbar is inferred independently by Yelle and McGrath (1996), who, from the HST/FOS observations, essentially find that ammonia must be mixed with aerosols. Atreya *et al.* (1995), on the other hand, do not believe that height distribution can be extracted from these spectra (S. Atreya, priv. comm.). They mention that since the observations do not probe below levels where the aerosol optical depth exceeds the value of several, all that can be determined is the column density down to the $\tau = 1$ level. Apparently, the key questions regarding the analysis of the UV spectra are: (i) at what pressure level does the $\tau = 1$ aerosol level lie? and (ii) are the observations more sensitive to a mixing ratio or to an integrated column density? In favor of the conclusion that a large fraction of the ammonia is indeed mixed with the aerosols, is the fact that the Griffith *et al.* column density of NH$_3$ is $\sim$50 times larger than that inferred by Atreya *et al.* (1995) at the $\tau = 1$ aerosol level. In summary, there are several indications for the presence of ammonia at deeper levels than other compounds (*e.g.*, CO), but it seems prudent to wait until the discrepancies mentioned above are solved to accept this result as definitive.

4.2. *Hydrogen cyanide*

Prior to the SL9 impacts, the detection of hydrogen cyanide on Jupiter had been announced by Tokunaga *et al.* (1981) from 13-μm spectroscopy. A mixing ratio of $\sim 2 \times 10^{-9}$ had been inferred. Further searches for HCN in the same and other spectral

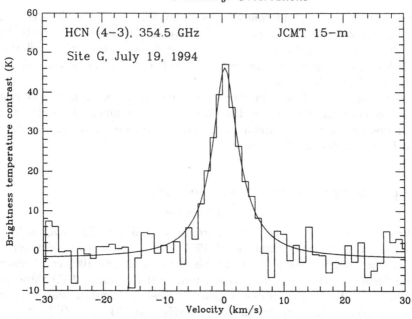

FIGURE 11. HCN (4-3) rotational line, observed on site G, July 19 from JCMT heterodyne observations (histograms). The model fit corresponds to a uniform HCN mixing ratio of 5×10^{-8} at p$\leq$ 0.5 mbar. The inset shows the associated thermal profile. This solution is non-unique, however (see text). From Marten *et al.* (1995).

ranges did not confirm this detection, however, and from a reassessment of the Tokunaga *et al.* observations, Bézard *et al.* (1995a) concluded that this detection was incorrect.

Following the cometary impacts, HCN has been observed in the millimeter/submillimeter range and at 13 μm. Observations at the JCMT have detected the HCN J = 4-3 line at 354.5 GHz on sites A, C, F, G, H and R, hours and days after impact, and the J = 3-2 line at 266 9 GHz on site R (Marten *et al.* 1995; Fig. 11). As for CO and CS, these lines were originally detected as narrow emission features, but had turned to absorption when they were reobserved in September 1994 and later. Preliminary modelling of this line was accomplished using a model similar to that of Lellouch *et al.* (1995) for CO, *i.e.*, using *a priori* assumptions on the thermal profile and a simplified geometrical treatment, and neglecting any velocity broadening within the sites. From such modelling, applied to J = 4-3 line observed on site G, July 19, a HCN mixing ratio of 5×10^{-8} at p$\leq$0.5 mbar was found, *i.e.*, a column density of 3×10^{15} cm^{-2} (Marten *et al.* 1995), which corresponds to a $\sim 3 \times 10^{11}$ g mass of HCN if a 3″ (11500 km) site diameter is assumed. However, this solution is non-unique, as the thermal profile is unconstrained. For example, a virtually identical fit of this line can also been obtained with another thermal profile, and a gas distribution in which HCN is restricted to p $\leq$ 0.1 mbar with a 10^{-6} mixing ratio, giving a 10^{16} cm^{-2} column density. Examination of the thermal profiles in both cases show that they are in fact inconsistent with other data (they are too warm in the 0.1–1 mbar region). Nevertheless, these two solutions, which differ in the total HCN mass by a factor of $\sim$3, give an idea of the precision that can be expected from the analysis of millimeter-wave data alone.

HCN was also successfully detected at 744.5 cm^{-1} in IRSHELL/IRTF observations of several impact sites (A, E, H on July 20, G, L, K+W on July 30 and G+Q+R on July 31). As for NH$_3$, the HCN emission was imaged with a 1.5″ resolution (Bézard *et al.* 1996). Thermal profile information was obtained from CH_4 or C_2H_2 emission. For site E on July 20, 2.6 days after impact, a HCN column density of $0.9 \pm 0.2 \times 10^{16}$ cm^{-2} is inferred. The HCN mixing ratio is $\sim 7 \times 10^{-7}$ at $p \leq 0.1$ mbar (assuming uniform mixing above this level), and the total mass over site E is 6×10^{11} g. For site K, 10 days after impact, the HCN mass is 7×10^{11} g.

Infrared and millimeter/submillimeter observations thus appear to give consistent results on the location and total mass of HCN. It must be emphasized that the analysis of the JCMT observations is still preliminary, and that further modelling, similar to that presented for CO in Sec. 2.2.2, is required.

5. Phosphine, hydrocarbons and upper limits

5.1. *Phosphine*

A tentative detection of PH_3 has been reported from Palomar/SpectroCam 10 observations of site L at 10.05 μm (Conrath *et al.* 1995). As compared to a spectrum taken on an adjacent impact-free region, the spectrum of site L at 987–1002 cm^{-1} shows three distinctive features: (i) an increased continuum level (by about 2 K) throughout the spectral range, probably due to dust (silicate) emission (Conrath, this volume; West, this volume), (ii) a strong emission line due to NH_3 near 993 cm^{-1}, (iii) extra emission, near 991.5 cm^{-1}, tentatively attributed to the Q branch of the ν_2 band of PH_3. A preliminary fit of this emission suggests a phosphine mixing ratio of 2×10^{-8}, assuming uniform mixing in the stratosphere. This abundance should conservatively be regarded as an upper limit (B. Conrath, priv. comm.). A more stringent upper limit can, in fact, be derived from the HST/FOS spectrum of site G. These data indicate a maximum PH_3 column density of 3.3×10^{14} cm^{-2} at the aerosol unit optical depth level (Atreya *et al.* 1995). Assuming this level is at 1 mbar, the PH_3 mixing ratio is $\leq 3 \times 10^{-9}$.

5.2. *Hydrocarbons*

5.2.1. *Ethylene*

Emission from C_2H_4 has been detected at 907.5 cm^{-1} in IRTF/IRSHELL spectra of site K (Griffith *et al.* 1996; see Fig. 9). Like for NH_3, this emission was observed 23 hours, 5 days and 10 days after impact. Orton *et al.* (1995) gave a rough estimate of 3 ppb for the C_2H_4 mixing ratio, assumed to be uniform in the stratosphere. Further analysis of the July spectrum ($T_i + 23$ hours), making use of line strengths measured for that purpose, indicated that the ethylene emission could be fit with a distribution similar to HCN (*i.e.*, restricted to $p \leq 0.1$ mbar) and a total mass of $\sim 3 \times 10^{12}$ g (Griffith *et al.* 1996). Although ethylene had been detected previously in Jupiter's stratosphere (in auroral regions (Kim *et al.* 1985) and subsequently at equatorial latitudes [Kostiuk *et al.* 1993]), its "normal" mixing ratio is of order 10^{-9} only. The observation of a ~ 1000–5000 times enhanced C_2H_4 abundance implies formation from shock chemistry, as is the case for CO (see above).

5.2.2. *Other hydrocarbons*

Observational evidence for a variation in the distribution and abundance of other hydrocarbons is not clear. As mentioned several times in this review, considerable enhancements of the emissions of CH_4 at 8 μm and C_2H_2 at 13.7 μm have been observed, but they can be attributed to temperature effects. Entry and fallback models (Zahnle, this volume) predict that the amount of methane, acetylene and ethane produced from shock chemistry is a small fraction (percent or less) of the material preexisting in Jupiter's

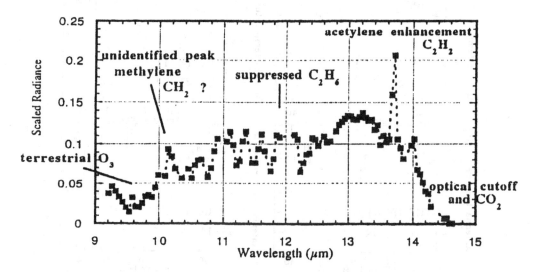

FIGURE 12. KAO/HIFOGS spectrum of fragment R splashback at 9–14 microns. A spectrum from the Northern hemisphere has been subtracted and the telluric absorptions have not been fully corrected. Enhanced acetylene and depressed ethane emission are seen, along with at least one unidentified peak. Courtesy A. Sprague.

stratosphere. If true, this fully justifies the use of the IR emissions as thermometers of the jovian stratosphere, but it makes searches for variations of C_2H_2 and C_2H_6 at other wavelengths hopeless. In the UV, in contrast to the pre-impact spectrum of July 14, the HST/FOS July 18, site G, spectrum does not show the acetylene bands that are masked by the ammonia bands. From these data, upper limits on the C_2H_2 mixing ratio of 3×10^{-7} and 10^{-6} have been inferred by Yelle and McGrath (1996) and Caldwell et al. (1995) respectively, several times larger than the normal jovian value (see review of the measurements in Gladstone et al. [1996]).

Variations of the C_2H_6 ν_9 band contrast at 820 cm^{-1} have also been observed, and appear to be un- or even anti-correlated with that of the C_2H_2 13.7 μm emission. For example, a KAO/HIFOGS spectrum of the R impact region shows strongly enhanced acetylene region, yet a suppressed C_2H_2 band (Sprague et al. 1996; Fig. 12). This does not necessarily imply a true decrease of the C_2H_6 abundance, however. The ethane emission probes deeper levels than the acetylene emission does (typically 1 mbar vs. 20 μbar; Drossart et al. 1993, Livengood et al. 1993), therefore it may be unaffected by a temperature increase at p$\leq$0.1 mbar; and the apparent disappearance of the ethane band may plausibly be due to an increase of the continuum due to enhanced dust opacity rather than to a decrease of the C_2H_6 mixing ratio. The same KAO/HIFOGS spectrum show a number of new peaks between 9 and 15 μm, that are not identified as yet, although one of them (at 10.2 μm) might be due to methylene (CH_2).

5.3. Upper limits

Searches for many other molecular species have been unsuccessfully conducted in various spectral ranges, notably in the UV and in the millimeter. In addition to that already mentioned on CO, H_2S, PH_3, and C_2H_2, upper limits are available on CH_3OH, H_2CO, SO_2, SiO, SO, HC_3N, CH_3CN, C_6H_6, and CO^+ (see Noll et al. 1995, Lellouch et al. 1995,

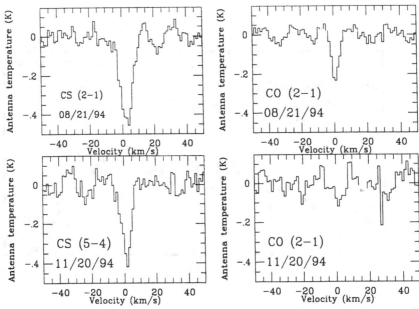

FIGURE 13. Long-term evolution of the CS (5–4) and CO (2–1) rotational lines, as observed from IRAM-30 m. Note the strong decrease of the CO (2–1) line. Courtesy R. Moreno.

Atreya *et al.* 1995, Yelle and McGrath 1996). Rather than give the associated values here, it seems more useful to mention that the derivation of these upper limits is generally based on *a priori* assumptions on the vertical distribution of the species. For example, the upper limits derived from the millimeter wave observations of Lellouch *et al.* (1995) implicitly assumed that the corresponding species are colocated with CO at $p \leq 0.3$ mbar. This assumption is probably valid for species that are directly produced by shock chemistry, but certainly not for molecules that result from subsequent photochemistry or that at updrafted by vertical mixing associated with the impacts. The case of H_2S is instructive, as the "raw" upper limit placed by Lellouch *et al.* (3×10^{-6} g cm^{-2}) is several times smaller than the actual column integrated mass of H_2S claimed by Yelle and McGrath (1995) (2.5×10^{-5} g cm^{-2}). However, as mentioned in Sec. 3.3, the Yelle and McGrath distribution for H_2S is, in fact, perfectly consistent with its non-detection at millimeter wavelengths. Also, in the case of photochemical species, simple assumptions on their vertical distributions (*e.g.*, uniform mixing) may not be very meaningful in the context of photochemical models of these species. Therefore, underlying assumptions must be carefully examined before drawing too firm conclusions from the upper limits. With these caveats in mind, the most significant negative result is perhaps the non-detection of SO_2 at UV wavelengths. Under the plausible hypothesis that SO_2 has the same altitude distribution as CS_2, Yelle and McGrath (1996) infer an upper limit of $\sim 10^{15}$ cm^{-2}, implying that less SO_2 is produced than CS_2.

6. Long-term evolution

Observations of Jupiter at millimeter, infrared (10 micron) and UV wavelengths, performed during the weeks and months following the SL9 impacts, have provided insight on the chemical evolution of Jupiter's stratosphere on a long timescale. Analysis of these data is often incomplete at the date this review is written (November 1995), but a number of qualitative conclusions can still be drawn.

6.1. *UV spectrum*

HST/FOS observations of Jupiter at the impact latitude and near the central meridian were performed on March 3, April 7, and again in September 1995 (McGrath *et al.* 1995). These spectra show a number of interesting features: (i) the continuum reflectivity is back to its pre-impact value, (ii) NH_3 bands dominate the spectrum at 1800–2200 Å, (iii) these bands do not show any significant variability over a 4° latitude range, and (iv) CS_2 bands are still present, although much weaker than during the impact week. Since tropospheric ammonia bands are normally visible in low emission angle spectra of Jupiter (see Sec. 4.1.1.), a detailed modelling of these observations will be needed to deduce whether some ammonia was still present in Jupiter's stratosphere 8 and 14 months after the impacts.

6.2. *10-micron spectrum*

IRTF/IRSHELL 10-micron observations of Jupiter's southern hemisphere were performed on May 15–18, 1995. These observations show: (i) no sign of NH_3 emission, implying an ammonia column density $\leq 10^{16}$ cm^{-2} above 40 mbar (*i.e.*, 30 times less than in July 1994) (Bézard *et al.* 1995b), and (ii) strong persistent HCN emission, with a latitudinal distribution extending roughly from 20° S to 60° S (comparable to that of aerosol particles at the same time), and a heterogeneous longitudinal distribution at a 1.5″ spatial resolution, with maximum HCN column densities of $\sim 5 \times 10^{15}$ cm^{-2} (Griffith *et al.* 1995). The total HCN mass seen in these data is ≥ 10 times larger than that created by a large, K-sized, impact, suggesting a possible photochemical production of HCN after the impacts week.

Finally, Fast *et al.* (1995b) report that ammonia stratospheric emission is at most marginally seen in 10 μm heterodyne observations performed in April 1995, suggesting an ammonia stratospheric mixing ratio of 10^{-10} or less.

6.3. *Millimeter-wave spectrum*

The millimeter-wave lines have dramatically changed since the impact period. CO, CS and HCN were originally observed as strong emission lines of comparable intensity, typically 40 K in brightness temperature. (OCS was detected only on one occasion and was weaker; the non-detection of OCS on August 19 is consistent with an expected photochemical decay by a factor ~ 3 over 1 month; Moses *et al.* 1995c). The contrast of the CO(2–1) and CS(5–4) lines decreased over the week following the impacts until around July 28, when the lines turned into absorption (Lellouch *et al.* 1995). Monitoring of these absorption lines at IRAM-30m on a week/month basis showed that, on the large impact sites, the CS absorption increased to reach a steady depth of ~ 0.6 K (in antenna temperature), while the CO absorption reached a maximum of ~ 0.3 K mid-August 1994, then decreased to a current depth of ~ 0.15 K (Moreno *et al.* 1995; Fig. 13). CS is now observed also in its (3–2) line at 147 GHz. JCMT observations performed since September 1994 show that the HCN (4–3) line has a similar behavior as CS (Matthews *et al.* 1995). The turnover to the absorption regime must primarily be a temperature effect, with Jupiter's thermal profile in the 0.1–0.01 mbar region probably becoming colder than in pre-impact conditions (Lellouch *et al.* 1995, Conrath, this volume). The increase of absorption line contrast is certainly due to an increase of the emitting area in the telescope beam (the observations themselves show evidence for horizontal spreading of the material). More relevant to this review is the question of the peculiar evolution of the

CO line. The comparison of CO and HCN is particularly puzzling, as these two species were probably formed at the same level and their vertical distribution must have followed the same evolution. As the observations have not been modelled yet, only qualitative suggestions can be proposed for the time being. Three explanations may be considered:

• Instrumental effect. If the CO line was less optically thick than the other two just after impact, it may now have become optically thin while the CS and HCN lines may still be optically thick. In the model presented in Sec. 2.2.2, the CO line has an optical depth of ∼9. A complete modelling of the CS and HCN emission lines is necessary to test this possibility.

• Chemical effect with the increase of HCN and CS. If the CO column density is assumed to remain constant with time, the decrease of the CO line must be due to a combination of temperature and CO vertical profile evolution. As the same must happen for HCN and CS, the fact that the HCN and CS lines are still very strong implies that fresh amounts of HCN and CS must have been created well after the impact period, perhaps from NH_3 and N_2 (via cosmic ray impact) and from CS_2 and S_2, respectively. This hypothesis may seem viable as the initial mass of NH_3 (and N_2?) is larger than that of HCN (see Table 1); similarly, enough sulfur might be originally available in S_2 to appreciably modify the CS amount with time. For HCN, recent photochemical models (Moses *et al.* 1995c) suggest that the HCN mass may increase by a factor about 2 over the initial abundance after 1 year.

• Chemical effect with a decrease of CO. If the HCN and CS are assumed approximately constant with time, the decrease of the CO line could be due to a true decrease of the CO mass. However, CO is notoriously stable and a mechanism must be found. OH radicals, liberated from the photolysis of H_2O on a month timescale may react with and consume CO molecules (as is the case in Mars and Venus), but again, photochemical models suggest this effect is negligible (J. Moses, priv. comm.)

Recent analyses of the millimeter lines (Moreno *et al.* 1995, Matthews *et al.* 1995) suggest the following column densities 6–12 months after impact: CS: 5×10^{14} cm^{-2}; HCN: 1×10^{15} cm^{-2}; CO: 2.5×10^{16} cm^{-2}, *i.e.*, 10, 10, and 160 times less than measured in July 1994 for the large impacts. The total extent of the chemical perturbation (from 10° S to 80° S) is ∼100 times larger than the total area of the fresh sites. Thus the total CS and HCN masses have remained approximately constant (or have moderately increased) since July 1994, while the total CO mass may have decreased by a factor ∼10. This would tentatively favor the third explanation.

7. Synthesis and conclusions

The observation of many molecular species in the plume/splashback phases and at the impact sites, most of which had never been observed in Jupiter before, provide a fairly detailed picture of the chemistry induced by the SL9 impacts. From the synthesis of the observations, one can draw several conclusions:

(*a*) All new or enhanced species were detected in the stratosphere or the thermosphere of Jupiter. There is no evidence for any variation of the chemical composition of the troposphere.

(*b*) Although the vertical distribution of some molecules is still uncertain (NH_3, C_2H_4), the observations suggest a distinction between two types of species (Yelle and McGrath 1996):

• Those present at pressures less than (0.1–1) mbar. These include CO, H_2O, S_2, HCN, CS, OCS, CS_2, C_2H_4, and a fraction of NH_3). Since fallback models (Zahnle

Species	Total mass (g)
CO	2.5×10^{14}
H_2O	$\geq 2 \times 10^{12}$
S_2	7×10^{11}
CS_2	1.5×10^{11}
CS	5×10^{11}
OCS	3×10^{12}
HCN	6×10^{11}
C_2H_4	3×10^{12}
NH_3	1×10^{13}
PH_3	??
H_2S	??

TABLE 1. "Baseline" masses for detected molecular species

et al. 1995, Zahnle, this volume) predict that plume material re-enters the atmosphere at pressure levels of 10–100 μbar, the simplest interpretation is that these species are either formed locally during plume re-entry or formed during the primary explosion, transported by the plume, and deposited in the upper stratosphere. Shock chemistry is in fact expected to take place both at explosion and during plume splashback. One aspect which remains to be explained is the apparent segregation of S_2 (present at $p \leq 0.1$ μbar only) with respect to the other species.

• Those present at $p \geq (1-10)$ mbar, must result from upwelling caused by the heating of jovian air generated by the impacts. Ammonia would be the prime example. This could be also the case for PH_3 and H_2S, if their detections are confirmed.

Note that ammonia apparently falls into the two categories. Carbon monosulfide is also an exception as a significant fraction of it may result from photochemistry rather than from hot/shock chemistry.

(c) Abundance determinations are available for most observations, especially for the "site" measurements. Except in the case of a number of plume/splashback observations, the comparison of retrieved abundances indicates an overall reasonable consistency. Plausible explanations can be found to account for the remaining discrepancies (*e.g.*, CS, NH_3) and future intercomparisons will allow to refine estimations. In almost no case (except for water) have the variations of molecular abundances from fragment to fragment been investigated. Future studies may allow to correlate the mass of the chemicals produced by the individual impact with the brightness/mass of the fragments. For the time being, "baseline" masses for the different molecular species, valid for a "large" impactor (*i.e.*, G, K), are gathered in Table 1. Some points must be made regarding Table 1: (i) the value for H_2O can be taken as a reasonably close lower limit (Sec. 1.1), (ii) the S_2 and CS_2 masses were obtained by using the column densities derived from HST/FOS and assuming an areal extent identical to that of CO. Whether S_2 is indeed more abundant than CS is an important question to test the possibility that CS is photochemically produced from S_2, (iii) the question mark for H_2S and PH_3 indicates that the detection is tentative. No mass estimate can be given since the depth to which they extend is unknown. For reasons discussed in Sec. 5.3, upper limits are not included in this Table.

(d) The molecular masses indicated in Table 1 imply $\sim 1.5 \times 10^{14}$ g of atomic O and $\sim 2.7 \times 10^{12}$ g of atomic S (excluding H_2S). The mass of atomic N is $\sim 3 \times 10^{11}$ g if only HCN is included and $\sim 1 \times 10^{13}$ g if NH_3 is considered. Assuming relative atomic abundances for the cometary fragments similar to those measured in comet P/Halley

(Jessberger *et al.* 1986), a 1.5×10^{14} g mass of O is available in a $\sim 3 \times 10^{14}$ g fragment. Tidal disruption models show that the parent comet was about 1.5 km diameter and had a density of $\sim 0.6 \pm 0.1$ g cm^{-3} (Asphaug and Benz 1995, Solem 1994), *i.e.*, a mass of $\sim 10^{15}$ g. The largest fragments may be assigned 1/8 of the total mass, or 1.3×10^{14} g. This is 2.5 times less than required from the observed oxygen mass, but the agreement may be considered as reasonable given the various uncertainties. In addition, the O/S ratio implied by the measurements is nominally 56, in plausible consistency with the value for comet P/Halley (14) (especially given the uncertainty in the mass of S_2). For O/N, the observations give a value of 500 if only HCN is considered, and 15 if ammonia is included, while the P/Halley value is 25. Therefore, the amounts of O, N, and S present in a large fragment can probably account for all the observed molecular abundances, so on this basis only, one might conclude that the shock chemistry involves jovian H and C (from H_2 and CH_4) and purely cometary O, N, and S. This view is probably incorrect, however, notably because most entry models (see Zahnle, Mac Low, Crawford, this volume) predict that km-size fragments penetrate down to or below the NH_4SH cloud at 2 bars, so that jovian sulfur must be involved in the chemistry. Another remark that can be made is that if, as suggested by some observations, most of the ammonia is updrafted from Jupiter's troposphere, then NH_3 must not be included when calculating the observed O/N ratio. In this case, the comparison with the P/Halley value indicates that another important N-bearing species has been produced but not observed, presumably N_2. Finally, if one accepts Yelle and McGrath's (1996) conclusion that H_2S is detected in the HST/FOS spectra, then their observed N/S ratio is approximately 2, in reasonable agreement with but somewhat smaller than the solar value (about 6; Grevesse and Anders 1991), which might suggest that sulfur is more enriched than nitrogen below the jovian NH_4SH clouds.

(*e*) The observed presence of water and the absence of SO_2 is a problem for models (see Zahnle, this volume), because shock chemistry in an oxygen-rich environment (O/C$\geq$1) leads to abundant production of H_2O and SO_2, while neither H_2O nor SO_2 is produced in oxygen-poor conditions (O/C$\leq$1). A possibility is that the observed H_2O results from cometary water that survived the explosion. The case for jovian H_2O excavated from below the water cloud does not seem very likely as this region has O/C $\sim$ 2-10, while the production of CS, CS_2 and HCN requires O/C $\leq$1 in the shocked jovian air (Zahnle, this volume).

(*f*) The ammonia stratospheric mixing ratio is of order 0.5–1$\times 10^{-7}$ (at a $\sim$ 5000 km spatial resolution), which is the nominal (pre-impact) value at $\sim$0.2 bar, and well below the abundance below the clouds($\sim 3 \times 10^{-4}$). If one admits that the large fragments exploded below the clouds, this suggests some very large horizontal dilution of the convected tropospheric jovian air at stratospheric levels.

(*g*) There is observational evidence for photochemistry occuring after the impacts. This is attested in particular by the decay of S_2, CS_2, NH_3 and the possible increase of CS on a week or month timescale.

Illuminating discussions with many colleagues are acknowledged, particularly S. Atreya, A. L. Betz, B. Bézard, J. Caldwell, B. Conrath, D. Crisp, C. Griffith, D. Hunten, T. Kostiuk, V. Meadows, J. Moses, R. Moreno, K. Noll, A. Sprague, R. Yelle, and K. Zahnle. This paper is dedicated to the memory of Jan Rosenqvist, deceased on July 20, 1995.

REFERENCES

ANICICH, V. G., & HUNTRESS, W. T. 1986 A survey of bimolecular ion-molecule reactions for use in modeling the chemistry of planetary atmospheres, cometary comae, and interstellar clouds. *Astrophys. J. Suppl. Ser.* **62**, 553–672.

ASPHAUG, E., & BENZ, W. 1995 The tidal disruption of strengthless bodies: lessons from comet Shoemaker-Levy 9. *Icarus* (submitted).

ATREYA, S. K. 1986 Atmospheres and ionospheres of the outer planets and their satellites. In *Physics and Chemistry in Space, Vol. 15*, (eds. L. J. Lanzerotti and D. Stöffler). Springer-Verlag.

ATREYA, S. K., EDGINGTON, S., TRAFTON, L. M., CALDWELL, J. J., NOLL, K. S. & WEAVER, H. A. 1995 Abundances of ammonia and carbon disulfide in the Jovian stratosphere following the impact of comet Shoemaker-Levy 9. *Geophys. Res. Lett.* **22**, 1625–1628.

BETZ, A. L., BOREIKO, R. T., BESTER, M., DANCHI, W. C., & HALE D. D. 1995 Stratospheric ammonia in Jupiter after the impact of comet Shoemaker-Levy 9., *Bull. Amer. Astron. Soc.*, **26**, 1590–1591.

BÉZARD, B., GRIFFITH, C. A., LACY, J., & OWEN, T. 1995a Non-detection of hydrogen cyanide on Jupiter. *Icarus*, **118**, 384–391.

BÉZARD, B., GRIFFITH, C. A., GREATHOUSE, T., KELLY, D., LACY, J., & ORTON, G. 1995b Jupiter ten months after the collision of comet SL9: stratospheric temperatures and ammonia distribution *Bull. Amer. Astron. Soc.* **27**, 72.

BÉZARD, B., GRIFFITH, C. A., GREATHOUSE, T., KELLY, D., LACY, J., & ORTON, G. 1996 Infrared spectral images of comet SL9 impact sites: temperature and HCN retrievals. *Icarus*, submitted.

BJORAKER, G. L., STOLOVY, S. R., HERTER, T. L., GULL, G. & PIRGER, B. E. 1995a Detection of water after the collision of fragments G and K of comet Shoemaker-Levy 9 with Jupiter 1995. *Science*, submitted.

BJORAKER, G. L., HERTER, T., GULL, G., STOLOVY, S. & PIRGER, B. 1995b Detection of water in the "splash" of fragments G and K of comet Shoemaker-Levy 9. In *Abstracts for IAU colloquium 156: The collision of comet P/Shoemaker-Levy 9 and Jupiter*, p. 8.

BOCKELÉE-MORVAN, D., LELLOUCH, E., COLOM, O., PAUBERT, G., MORENO, R., FESTOU, M., DESPOIS, D., SIEVERS, A., CROVISIER, J., GAUTIER, D. & MARTEN, A. 1995 Millimetre observations of the Shoemaker-Levy 9 impacts on Jupiter in July 1994 at the IRAM and SEST telescopes: CO, CS and OCS. In *Proceedings of European SL-9/Jupiter Workshop* (eds. R. West and H. Bohnhardt), pp. 251–255.

BROOKE, T. Y., ORTON, G. S., CRISP, D., FRIEDSON, A. J. & BJORAKER, G. 1995 Near-infrared spectroscopy of the Shoemaker-Levy 9 impact sites with UKIRT: CO emission from the L site. In *Abstracts for IAU colloquium 156: The collision of comet P/Shoemaker-Levy 9 and Jupiter*, p. 12.

CALDWELL, J. & 9 CO-AUTHORS. 1995 Upper limits on SiO, H_2S, C_2H_2 and H_2O on Jupiter from SL-9. *Bull. Amer. Astron. Soc.* **27**, 64.

CARLSON, R. W., WEISSMAN, P. R., HUI, J., SEGURA, M., SMYTHE, W. D., BAINES, K. H., JOHNSON, T. V., DROSSART, P., ENCRENAZ, T., LEADER, F. & MEHLMAN, R. 1995a Some timing and spectral aspects of the G and R collision events as observed by the Galileo near-infrared mapping spectrometer. In *Proceedings of European SL-9/Jupiter Workshop* (ed. R. West and H. Bohnhardt), pp. 69–74.

CARLSON, R. W., WEISSMAN, P. R., HUI, J., SEGURA, M., SMYTHE, W. D., BAINES, K. H., JOHNSON, T. V., DROSSART, P., ENCRENAZ, T., LEADER, F. & MEHLMAN, R. 1995b Galileo infrared observations of the Shoemaker-Levy 9 G and R fireballs and flash. In *Abstracts for IAU colloquium 156: The collision of comet P/Shoemaker-Levy 9 and Jupiter*, p. 15.

CARLSON, R. W., WEISSMAN, P. R., HUI, J., SEGURA, M., T. V., DROSSART, P. & EN-CRENAZ, T. 1995c Galileo infrared observations of the Shoemaker-Levy 9 G and R splash phases. *Bull. Amer. Astron. Soc.* **27**, 66.

CARPENTER, K. G., MCGRATH, M. A., YELLE, R. V., NOLL, K. S. & WEAVER H. A. 1995 Formation of atomic emission lines in the atmosphere of Jupiter after the comet Shoemaker-Levy 9 S impact. *Bull. Amer. Astron. Soc.* **27**, 64.

CONRATH, B. J., GIERASCH, P. J., HAYWARD, T., MCGHEE, C. NICHOLSON, P. D. & VAN CLEVE, J. 1995 Palomar mid-infrared spectroscopic observations of comet Shoemaker-Levy 9 impact sites. In *Abstracts for IAU colloquium 156: The collision of comet P/Shoemaker-Levy 9 and Jupiter*, p. 24.

COSMOVICI, C. B., MONTEBUGNOLI S., POGREBENKO, S. & COLOM, P. 1995 Water MASER detection at 22 GHz after the SL-9/Jupiter collision. *Bull. Amer. Astron. Soc.* **27**, 79.

CRISP, D. & MEADOWS, V. 1995 Near-infrared imaging spectroscopy of the impacts of SL9 fragments C, D, G, K, N, R, V, and W with Jupiter. In *Abstracts for IAU colloquium 156: The collision of comet P/Shoemaker-Levy 9 and Jupiter*, p. 25.

DROSSART, P., BÉZARD, B., ATREYA, S. K., BISHOP, J., WAITE, JR., J. H. & BOICE, D. 1993 Thermal profiles in the auroral regions of Jupiter. *J. Geophys. Res.* **98**, 18803–18811.

FAST, K. E., KOSTIUK, T., ESPENAK, F., ZIPOY, D., BUHL, D., LIVENGOOD, T. A., BJORAKER, G., ROMANI, P. & GOLDSTEIN, J. J. 1995a Infrared heterodyne observations of NH_3 and C_2H_6 after the collision of comet P/Shoemaker-Levy 9 with Jupiter. In *IAU colloquium 156: The collision of comet P/Shoemaker-Levy 9 and Jupiter*, p. 25.

FAST, K. E., LIVENGOOD, T. A., KOSTIUK, T., BUHL, D., ESPENAK, F., BJORAKER, G. L., ROMANI, P. N., JENNINGS, D. E., SADA, P., ZIPOY, D., GOLDSTEIN, J. J. & HEWEGAMA, T. 1995b NH_3 in Jupiter's stratosphere within the year following the SL9 impacts *Bull. Amer. Astron. Soc.* **27**, 72.

FELDMAN, P. D., DURRANCE, S. T. & DAVIDSEN, A. F. 1995 Far-ultraviolet spectroscopy of Venus and Mars at 4 Å resolution with the Hopkins Ultraviolet Telescope on Astro-2 *Bull. Amer. Astron. Soc.* **27**, 1079.

FITZSIMMONS, A., LITTLE, J. E., WLATON, N., CATCHPOLE, ANDREWS, P. J., WILLIAMS, I. P., HARLAFTIS, E. & RUDD, P. 1995 Optical imaging and spectroscopy of the impact plumes on Jupiter. In *Proceedings of European SL-9/Jupiter Workshop* (ed. R. West and H. Bohnhardt) pp. 197–201.

GLADSTONE, G. R., ALLEN, M. & YUNG, Y. L. 1996 Hydrocarbon photochemistry in the upper atmosphere of Jupiter. *Icarus*, **119**, 1–52.

GREVESSE, N, & ANDERS, E. 1991 Solar element abundances. In *Solar interior and atmosphere* (eds. A. N. Cox, W. C. Livingston and M. S. Matthews), pp. 1227–1234. The University of Arizona Press.

GRIFFITH, C. A., BÉZARD, B., GREATHOUSE, T., KELLY, D., LACY, J., & ORTON, G. 1995 Jupiter ten months after the collision of comet SL9: spectral maps of HCN and NH_3 *Bull. Amer. Astron. Soc.* **27**, 72.

GRIFFITH, C. A., BÉZARD, B., GREATHOUSE, T., KELLY, D., LACY, J., & NOLL, K. 1996 Infrared spectral images of comet SL9 impact sites: spatial and vertical distributions of ammonia, ethylene, and the enhanced 10 μm continuum emission, 21 hours, 6 and 12 days following collision. *Icarus*, submitted.

HAMMEL, H. B. & 16 CO-AUTHORS. 1995 HST imaging of atmospheric phenomena created by the impact of comet Shoemaker-Levy 9. *Science* **267**, 1288–1296.

HARRIS, W., M. & 15 CO-AUTHORS. 1995 A comprehensive IUE study of UV phenomena related to the collision of comet Shoemaker-Levy 9 (1993e) with Jupiter. *Science*, submitted.

HERBST, T. M., HAMILTON, D. P., BOEHNHARDT, H. & ORTIZ-MORENO, J.-L. 1995a SL-9 impact imaging, spectroscopy and long-term monitoring from the Calar Alto Observatory. In *Proceedings of European SL-9/Jupiter Workshop* (eds. R. West and H. Bohnhardt) pp. 119–122.

HERBST, T. M., HAMILTON, D. P., BOEHNHARDT, H. & ORTIZ-MORENO, J.-L. 1995b Near-infrared spectroscopy and long-term monitoring of the SL-9 impacts. In *IAU colloquium 156: The collision of comet P/Shoemaker-Levy 9 and Jupiter*, p. 50.

HOOKER, W. J. & MILLIKAN, R. C. 1963 Shock-tube study of vibration relaxation in carbon monoxide for the fundamental and the first overtone. *J. Chem. Phys.* **38**, 214–220.

JESSBERGER, E. K. & KISSEL, J. 1986 Chemical properties of cometary dust and a note on carbon isotopes. In *Comets in the Post Halley Era*, Vol. 2, (eds. R. L. Newburn, Jr., M. Neugebauer and J. Rahe), pp. 1075–1092. Kluwer Academic.

KIM, S.-J., CALDWELL, J., RIVOLO, A. R., WAGENER, R. & ORTON, G. S. 1985 Infrared polar brightening on Jupiter, III, Spectrometry from the Voyager 1 IRIS experiment. *Icarus* **64**, 233–248.

KIM, S. J., RUIZ, M., RIEKE, G. H. & RIEKE, J. 1995 Thermal history of the R impact flare of comet Shoemaker-Levy 9. *Bull. Amer. Astron. Soc.* **27**, 66.

KNACKE, R. F., FAJARDO-ACOSTA, S. B., GEBALLE, T. R., & NOLL, K. S. 1995 Near-infrared spectroscopy of the R site of comet Shoemaker-Levy 9. In *IAU colloquium 156: The collision of comet P/Shoemaker-Levy 9 and Jupiter*, p. 59.

KOSTIUK, T., ROMANI, P., ESPENAK, F., LIVENGOOD, T. & GOLDSTEIN, J. J. 1993 Temperatures and abundances in the Jovian auroral stratosphere, 2, Ethylene as a probe of the microbar region. *J. Geophys. Res.* **98**, 18823–18830.

KOSTIUK, T., ESPENAK, F., BUHL, D., ROMANI, P. N., FAST, K. E., LIVENGOOD, T. A., ZIPOY, D., HEWEGAMA, T. & GOLDSTEIN, J. J. 1995 Altitude distribution of ammonia deposited in Jupiter's stratosphere by the SL9 impacts. *Bull. Amer. Astron. Soc.* **27**, 72.

LELLOUCH, E. & 13 CO-AUTHORS 1995 Chemical and thermal response of Jupiter's atmosphere following the impact of comet Shoemaker-Levy 9. *Nature* **373**, 592–595.

LIVENGOOD, T. A., KOSTIUK, T., ESPENAK, F. & GOLDSTEIN, J. J. 1993 Temperatures and abundances in the Jovian auroral stratosphere, 1, Ethane as a probe of the millibar region. *J. Geophys. Res.* **98**, 18813–18822.

MAILLARD, J.-P., DROSSART, P., BÉZARD, B., DE BERGH, C., LELLOUCH, E., MARTEN, A., CALDWELL, J., HILICO, J.-C. & ATREYA, S. K. 1995 Methane and carbon monoxide einfrared emissions observed at the Canada-France-Hawaii telescope during the collision of comet SL-9 with Jupiter. *Geophys. Res. Lett.* **22**, 1573–1576.

MARTEN, A. & 16 CO-AUTHORS. 1995 The collision of the comet Shoemaker-Levy 9 with Jupiter: detection and evolution of HCN in the stratosphere of the planet. *Geophys. Res. Lett.* **22**, 1589–1592.

MATTHEWS, H. E., MARTEN, A., GRIFFIN, M. J., OWEN, T., & GAUTIER, D. 1995 JCMT observations of long-lived molecules on Jupiter in the aftermath of the comet Shoemaker-Levy 9 collision *Bull. Amer. Astron. Soc.* **27**, 67.

MCGRATH, M. A., YELLE, R. V., NOLL, K. S., WEAVER, H. A. & SMITH, T. E. 1995 Hubble Space Telescope spectroscopic observations of the jovian atmosphere following the SL9 impacts. *Bull. Amer. Astron. Soc.* **27**, 64.

MEADOWS, V. & CRISP, D. 1995a Impact plume composition from near-infrared spectroscopy. In *Proceedings of European SL-9/Jupiter Workshop* (eds. R. West and H. Bohnhardt) pp. 239–244.

MEADOWS, V. & CRISP, D. 1995b Near-infrared imaging spectroscopy of the impacts of SL9 fragments C, D, G, K, N, R, V, and W with Jupiter. *Bull. Amer. Astron. Soc.* **27**, 73.

MONTEBUGNOLI, S. & 16 CO-AUTHORS 1995 Detection of the 22-GHz line of water during and after the SL-9/Jupiter event. In *Proceedings of European SL-9/Jupiter Workshop* (eds. R. West and H. Bohnhardt) pp. 261–266.

MORENO, R., MARTEN, A., LELLOUCH E., PAUBERT, G. & WILD, W. 1995 Long-term evolution of CO and CS in the Jupiter stratosphere after the comet Shoemaker-Levy 9 collision: millimeter observations with the IRAM 30-m telescope. *Bull. Amer. Astron. Soc.* **27**, 75.

MOSES, J. I., ALLEN, M. & GLADSTONE, G. R. 1995a Post SL9 sulfur photochemistry on Jupiter. *Geophys. Res. Lett.* **22**, 1597–1600.

MOSES, J. I., ALLEN, M. & GLADSTONE, G. R. 1995b Nitrogen and oxygen photochemistry following SL9. *Geophys. Res. Lett.* **22**, 1601–1604.

MOSES, J. I., ALLEN, M., FEGLEY, B., JR. & GLADSTONE, G. R. 1995c Photochemical evolution of the post-SL9 jovian stratosphere. *Bull. Amer. Astron. Soc.* **27**, 65.

NOLL, K. S., KNACKE, R. F., GEBALLE, T. R. & TOKUNAGA, A. T. 1995 The origin and vertical distribution of carbon monoxide on Jupiter. *Astrophys. J.* **324**, 1210–1218.

NOLL, K. S., MCGRATH, M. A., TRAFTON, L. M., ATREYA, S. K., CALDWELL, J. J., WEAVER, H. A., YELLE, R. V., BARNET, C. & EDGINGTON, S. 1995 HST spectroscopic observations of Jupiter after the collision of comet Shoemaker-Levy 9. *Science* **267**, 1307–1313.

ORTON, G. & 57 CO-AUTHORS. 1995 Collision of comet Shoemaker-Levy 9 with Jupiter observed by the NASA Infrared Telescope facility. *Science* **267**, 1277–1282.

ROOS-SEROTE, M., BARUCCI, A., CROVISIER, J., DROSSART, P., FULCHIGNONI, M., LECACHEUX, J. & ROQUES, F. 1995 Metallic emission lines during the impacts L and Q_1 of comet P/Shoemaker-Levy 9 in Jupiter. *Geophys. Res. Lett.* **22**, 1621–1624.

RUIZ, M., RIEKE, G. H., RIEKE, M. J., MEANS, D. & FRAWLEY, P. 1994 Near infrared spectroscopy of SL-9 impacts. *Earth Moon Plan.* **66**, 91–98.

SOLEM, J. C. 1994. Density and size of comet Shoemaker-Levy 9 deduced from a tidal breakup model. *Nature* **370**, 349–351.

SPRAGUE, A. L., BJORAKER, G. L., HUNTEN, D. M., WITTEBORN, F. C., KOZLOWSKI, R. W. H. & WOODEN, D. H. 1996. Water brought into Jupiter's atmosphere by fragments R and W of comet SL-9. *Icarus*, in press.

TOKUNAGA, A. T., BECK, S. C., GEBALLE, T. R., LACY, J. H., & SERABYN, E. M. 1995 The detection of HCN on Jupiter. *Icarus* **48**, 283–289.

TRAFTON, L. M., ATREYA, S. K., MCGRATH, M. A., GLADSTONE, G. R., CALDWELL, J. J., NOLL, K. S., WEAVER, H. A., YELLE, R. V., BARNET, C. & EDGINGTON, S. 1995 FUV spectra of SL9 impact site G with HST/GHRS. In *IAU colloquium 156: The collision of comet P/Shoemaker-Levy 9 and Jupiter*, p. 112.

WAGENER, R., CALDWELL, J., OWEN, T., KIM, S.-J., ENCRENAZ, T. & COMBES, M. 1985 The jovian stratosphere in the ultraviolet. *Icarus* **63**, 222–236.

WEST, R. A., KARKOSCHKA, E., FRIEDSON, A. J., SEYMOUR, M., BAINES, K. H. & HAMMEL, H. B. 1995 Impact debris particles in Jupiter's stratosphere. *Science* **267** 1296–1301.

YELLE, R. V. & MCGRATH, M. A. 1995 Results from HST spectroscopy of the SL9 impact sites. *Bull. Amer. Astron. Soc.* **27**, 64.

YELLE, R. V. & MCGRATH, M. A. 1996 Ultraviolet spectroscopy of the SL9 impact sites, I: The 175–230 nm region. *Icarus*, in press.

ZAHNLE, K., MAC LOW, M.-M., LODDERS, K., & FEGLEY, JR., B. F. 1995 Sulfur chemistry in the wave of comet Shoemaker-Levy 9. *Geophys. Res. Lett.* **22**, 1593–1596.

SL9 impact chemistry: Long-term photochemical evolution

By JULIANNE I. MOSES

Lunar and Planetary Institute, 3600 Bay Area Blvd., Houston, TX 77058-1113, USA

One-dimensional photochemical models are used to provide an assessment of the chemical composition of the Shoemaker-Levy 9 impact sites soon after the impacts, and over time, as the impact-derived molecular species evolve due to photochemical processes. Photochemical model predictions are compared with the observed temporal variation of the impact-derived molecules in order to place constraints on the initial composition at the impact sites and on the amount of aerosol debris deposited in the stratosphere. The time variation of NH$_3$, HCN, OCS, and H$_2$S in the photochemical models roughly parallels that of the observations. S$_2$ persists too long in the photochemical models, suggesting that some of the estimated chemical rates constants and/or initial conditions (e.g., the assumed altitude distribution or abundance of S$_2$) are incorrect. Models predict that CS and CO persist for months or years in the jovian stratosphere. Observations indicate that the model results with regard to CS are qualitatively correct (although the measured CS abundance demonstrates the need for a larger assumed initial abundance of CS in the models), but that CO appears to be more stable in the models than is indicated by observations. The reason for this discrepancy is unknown. We use model-data comparisons to learn more about the unique photochemical processes occurring after the impacts.

1. Introduction

The Shoemaker-Levy 9 (SL9) impacts generated strong shocks in the jovian atmosphere in two distinct altitude regions: in the deep stratosphere or troposphere where strong deceleration of the comet fragments and maximum energy release occurred, and in the upper stratosphere where the impact plumes splashed back down into the atmosphere. Shocked, thermochemically processed cometary material and tropospheric jovian air were deposited in Jupiter's stratosphere after the impacts. The new impact-derived sulfur-, nitrogen-, oxygen-, and carbon-bearing molecules evolve with time due to solar ultraviolet photolysis, chemical reactions, vertical and horizontal transport, and condensation/aerosol formation. Because continuous observational coverage of the impact sites was not possible and because many of the potential impact-derived species are difficult to observe, photochemical models play an essential role in assessing the chemical state of the jovian atmosphere in the hours, days, and months following the impacts. We compare photochemical model predictions with observations in order to better define chemical abundances immediately after the impacts and to evaluate the long-term evolution of the impact-derived species.

The molecular species either detected for the first time in Jupiter's stratosphere or found to be enhanced after the SL9 impacts include S$_2$, CS$_2$, CS, OCS, NH$_3$, HCN, C$_2$H$_4$, H$_2$O, CO, and possibly H$_2$S (see Lellouch 1996). Of these observed species, S$_2$, OCS, H$_2$S, and NH$_3$ are found to be transient—observations taken months (and in some cases, days) after the impacts demonstrate that these molecules have disappeared or have been substantially reduced after the impacts (Noll et al. 1995, Yelle & McGrath 1996, Lellouch et al. 1995, Bézard et al. 1995, Griffith et al. 1995b, Fast et al. 1995). The signatures of two other molecules, CS$_2$ and CO, are observed to weaken over time, but spectra taken ~1 year after the impacts show evidence for the continued existence of both CS$_2$ and CO in the jovian stratosphere (McGrath et al. 1995, Matthews et al. 1995,

R. Moreno *et al.* 1995). In contrast, spectral signatures for HCN and CS remain strong throughout the year following the impacts, indicating that the CS and HCN abundances have remained constant or even increased with time (R. Moreno *et al.* 1995, Matthews *et al.* 1995).

In the following sections, we describe how photochemical models can be used to simulate the evolution of these and other vapor-phase species at the impact sites; in particular, we recount the details of the photochemical models of Moses *et al.* (1995a,b,c). Because the models rely on accurate descriptions of the chemical state of the atmosphere just after the impacts, we also delve briefly into the details of some of the observations and theoretical predictions regarding the impact chemistry. More thorough discussions are presented in other chapters in this volume: observational results are reviewed by Lellouch (1996), and thermochemical models are described by Zahnle (1996). We then discuss the important photochemical reactions that control the abundances of the major observed species and compare model predictions with observational data.

2. Photochemical model: Initial conditions

Without accurate initial conditions, photochemical models cannot present a reasonable description of the time variation of the chemistry at the impact sites. Fortunately, observational coverage of the impact events was extensive. Unfortunately, complete spectral, temporal, and spatial coverage of each impact site was impossible. Different observations refer to different impact sites at different times and are sensitive to different altitudes; these issues must be considered carefully before initial chemical profiles can be developed. In addition, the impact and plume re-entry mechanics were complicated—different regions of the impact sites map back to diverse temperature and compositional regimes in the impact plume/fireball and hence probably contain different final compositions. Theoretical thermochemical models can help shed light on these issues, but they, too, are hampered by a lack of knowledge of physical and chemical conditions in the fireball and plume stages.

Moses *et al.* (1995a,b,c) have attempted to consider some of these complexities by splitting the photochemical modeling into two parts. First, Moses *et al.* (1995a,b) have created a model designed to reproduce the photochemical evolution in the dark central core region of an impact site (*e.g.*, a region that may contain fewer molecules derived from the comet). Then, (Moses *et al.* 1995c) have created a model designed to be more representative of extended regions of the impact sites, or the impact sites as a whole. The most obvious difference between the two models is that the dark-core model assumes a very low abundance of oxygen species while the whole-region model assumes a much higher abundance of oxygen species. As discussed below, the results of this splitting were not entirely satisfactory with regard to the actual situation on Jupiter, and better descriptions of the initial chemical state of the atmosphere at the impact sites should be possible now that more observational analyses are being published.

Table 1 shows the initial conditions assumed in the photochemical models. Column abundances at three different pressure levels are included to aid in comparisons with observations. For the dark-core low-oxygen case (hereafter called Model A), Moses *et al.* (1995a,b) assume that S_2 dominates the sulfur compounds, $N_2/NH_3/HCN \approx 10/10/1.6$, and oxygen compounds are very minor. For the whole-region high-oxygen case (Model C), Moses *et al.* (1995c) assume that most of the oxygen is tied up in CO, with $[H_2O] \approx 0.3\,[CO]$, $[OCS] \approx 6 \times 10^{-3}[CO]$, and minor additional amounts of SO_2 and CO_2 present. The sulfur and nitrogen abundances are taken from observations, except that N_2, S_2, and H_2S are assumed to be present in quantities (perhaps) greater than

	Model A Moses *et al.* (1995a,b)			Model C Moses *et al.* (1995c)		
Species	above 12 mbar	above 0.11 mbar	above 0.009 mbar	above 12 mbar	above 0.11 mbar	above 0.009 mbar
S_2	1.5×10^{18}	1.5×10^{17}	1.2×10^{16}	9.4×10^{16}	3.0×10^{16}	2.4×10^{15}
H_2S	7.7×10^{16}	7.4×10^{15}	6.1×10^{14}	2.0×10^{17}	7.4×10^{15}	6.1×10^{14}
CS	5.4×10^{11}	3.0×10^{10}	2.4×10^{9}	2.2×10^{14}	6.0×10^{13}	4.8×10^{12}
CS_2	2.0×10^{15}	1.8×10^{14}	1.5×10^{13}	7.9×10^{14}	2.2×10^{14}	1.8×10^{13}
N_2	1.5×10^{17}	1.5×10^{16}	1.2×10^{15}	7.1×10^{16}	2.2×10^{16}	1.8×10^{15}
NH_3	1.5×10^{17}	1.5×10^{16}	1.2×10^{15}	3.5×10^{17}	1.2×10^{16}	9.7×10^{14}
HCN	2.4×10^{16}	2.2×10^{15}	1.8×10^{14}	5.0×10^{15}	1.5×10^{15}	1.2×10^{14}
H_2O	1.8×10^{14}	1.5×10^{13}	1.2×10^{12}	1.4×10^{18}	4.5×10^{17}	3.6×10^{16}
CO	1.8×10^{14}	1.5×10^{13}	1.2×10^{12}	4.0×10^{18}	1.3×10^{18}	1.1×10^{17}
CO_2	2.3×10^{12}	1.5×10^{11}	1.2×10^{10}	3.8×10^{12}	6.0×10^{11}	4.8×10^{10}
SO_2	1.0×10^{13}	7.4×10^{11}	6.1×10^{10}	3.1×10^{13}	7.4×10^{12}	6.0×10^{11}
OCS	2.3×10^{12}	1.5×10^{11}	1.2×10^{10}	2.4×10^{16}	7.4×10^{15}	6.1×10^{14}

TABLE 1. Initial column abundances (molecules cm^{-2})

observed, and their abundances are estimated by assuming a comet-like O/S/N ratio of roughly 25/2/1 (see Lellouch 1996) in the upper stratosphere ($P < 0.3$ mbar). Additional amounts of NH_3 and H_2S are assumed to be present in the lower stratosphere ($P > 0.3$ mbar) based on the suggestions of Yelle & McGrath (1996). Initial conditions for both of these models are out-of-date and need to be updated based on new observational and theoretical analyses.

2.1. *Constraints based on observations*

Moses *et al.* (1995a,b,c) selected initial abundances for their photochemical models in such a way as to remain as consistent as possible with the available observational reports. A thorough and detailed discussion of the observations is presented in Lellouch (1996). We will not attempt to repeat that information here but will simply provide a summary of some of the sources of the observational constraints. Table 2 contains a list of several early papers that have been particularly useful in constraining initial abundances for the photochemical models. Along with the observed molecules listed in this table, useful upper limits to many additional species are presented by Noll *et al.* (1995), Lellouch *et al.* (1995), Atreya *et al.* (1995), Yelle & McGrath (1996), Caldwell *et al.* (1995), and Sprague *et al.* (1996).

The observations of the chemical species taken at different wavelengths with different instruments are occasionally inconsistent; even within one dataset (*e.g.*, the Hubble Space Telescope ultraviolet spectra), different investigators derive dissimilar results. The inconsistencies probably derive from the different assumptions that have gone into the observational analyses. The major stumbling blocks to deriving chemical abundances are uncertainties about the stratospheric temperatures (for the infrared and even millimeter and sub-millimeter observations) and uncertainties about the absorbing properties of the aerosols (for the ultraviolet observations). However, most of the recent reports appear to exhibit a convergence with regard to chemical abundances (see Lellouch 1996), and it is hoped that these results can be used to place robust constraints on the initial conditions for the photochemical models.

Species	References used to constrain initial conditions
S_2	Noll *et al.* (1995), Yelle & McGrath (1996)
H_2S	Atreya *et al.* (1995), Caldwell *et al.* (1995), Noll *et al.* (1995), Yelle & McGrath (1996)
CS	Lellouch *et al.* (1995), R. Moreno *et al.* (1995), Noll *et al.* (1995)
CS_2	Atreya *et al.* (1995), Caldwell *et al.* (1995), McGrath *et al.* (1995), Noll *et al.* (1995), Yelle & McGrath (1996)
NH_3	Atreya *et al.* (1995), Betz *et al.* (1995), Bézard *et al.* (1995), Caldwell *et al.* (1995), Conrath *et al.* (1995), Fast *et al.* (1995), Griffith *et al.* (1995a,b), Kostiuk *et al.* (1996), McGrath *et al.* (1995), Noll *et al.* (1995), Orton *et al.* (1995), Yelle & McGrath (1996)
HCN	Griffith *et al.* (1995a), Marten *et al.* (1995), Matthews *et al.* (1995)
H_2O	Bjoraker *et al.* (1995), Carlson *et al.* (1995), Cosmovici *et al.* (1995), Meadows & Crisp (1995), Sprague *et al.* (1996)
CO	Brooke *et al.* (1995), Knacke *et al.* (1995), Lellouch *et al.* (1995), Maillard *et al.* (1995), Matthews *et al.* (1995), R. Moreno *et al.* (1995)
OCS	Lellouch *et al.* (1995)

TABLE 2. Sources of observational constraints

2.2. *Constraints based on thermochemical modeling*

When observations are unavailable, uncertain, or contradictory, theoretical models of the thermochemical processing that takes place in the fireball, plume, and plume re-entry shock can be used to help determine the initial chemical composition of the impact sites. Thermochemical models such as those presented by Zahnle (1996), Zahnle *et al.* (1995), Lyons & Kansal (1996), and Borunov *et al.* (1995) that predict how the different elements are partitioned among their constituent molecules are particularly useful in estimating the abundances of species that are difficult or impossible to observe on Jupiter. For example, molecular nitrogen is not observable in the jovian stratosphere, but theoretical models (*e.g.*, Zahnle 1996, Lyons & Kansal 1996) predict that it should be a major product under most shock pressure/temperature regimes relevant to the impacts. Similarly, H_2O is difficult to observe once it has cooled to ambient stratospheric temperatures; therefore, the observed H_2O abundance (*e.g.*, Bjoraker *et al.* 1995, Sprague *et al.* 1996) may represent a lower limit. Because H_2O photolysis drives the oxygen photochemistry, estimating its initial abundance is important for photochemical models. Thermochemical models are useful not only for supplying the photochemical models with initial conditions, but also for helping to determine the elemental compositions of the comet and/or jovian troposphere and for constraining the physical conditions of the impacts.

The temperatures and pressures in the fireball and re-entry shock play a critical role in determining the chemical makeup at the impact sites (cf. Zahnle 1996, Lyons & Kansal 1996). Although infrared observations are used to pin down the shock temperatures during plume re-entry (*e.g.*, Kim *et al.* 1995), the conditions in the initial shock are less certain. Numerical models of the impacts can help with this task. Zahnle (1996) uses hydrodynamic plume models to define the physical conditions for his thermochemical modeling; however, he finds that different gas parcels experience very different thermal histories depending on where they were positioned during the initial shock. Currently, no one has "added up" the different parcels in such a way as to map out compositional changes at the impact sites (due to different shock conditions) so that a three-dimensional description of the impact-site composition can be developed.

Even more important to the thermochemical models may be the relative elemental abundances in the plume/fireball; in particular, the C/O ratio controls much of the thermochemistry. For moderate or high shock temperatures, Zahnle (1996) and Lyons & Kansal (1996) show that plumes with C/O > 1 allow most of the oxygen to be tied up in CO, with H_2O, SO_2, SiO, NO, and SO playing a minor role; the excess carbon can then be used to form species such as carbon sulfide and hydrocarbons. Plumes with C/O < 1 have H_2O being more important than in the C/O > 1 case, C–S species being minor, and SO and SO_2 becoming increasingly important as shock temperatures become higher. What was the C/O ratio in the plumes during the SL9 impacts? That question may not be a reasonable one to ask because different portions of the plume may have contained different elemental compositions due to a variable amount of mixing between the vaporized comet and jovian air. Observations are somewhat contradictory regarding this point. Both H_2O and OCS were observed in non-trivial quantities, suggesting that C/O < 1. However, CS and CS_2 were also observed in moderate quantities while SO_2 is not reported, suggesting that C/O > 1. The problem is clearly complex. We seem to be seeing more than one type of chemistry in different regions of the impact sites. Until model parameter space has been explored more fully and until the observational constraints become increasingly secure, the initial conditions at the impact sites will remain uncertain.

3. Photochemical model: Other details

Moses *et al.* begin their modeling approximately one half hour after the impacts, after the plumes have splashed back down and spread into the upper atmosphere and after the surrounding air has cooled to a significant degree. Thermochemical processes cease to dominate the chemistry at this point, and solar radiation takes over. The models are one-dimensional in the vertical direction. Molecular and eddy diffusion are considered, but horizontal spreading is neglected. Because of this and other oversimplifications, the models are designed to illustrate general trends in the evolutionary behavior of the molecules introduced by the impacts; the results should be regarded as illustrative rather than quantitative.

The photochemical models simulate the variation of 145 different vapor-phase sulfur, nitrogen, oxygen, and hydrocarbon compounds at the impact sites. Over 900 chemical reactions are included. Most of the reaction rates are taken from the NIST Chemical Kinetics Database (version 6.0, Mallard *et al.* 1994) and literature published within the last couple of years; however, several potentially important reactions used in the model have never been studied in the laboratory. Moses *et al.* estimated the pathways, rates, and products for these reactions. The sulfur reactions, in particular, are poorly studied. A complete reaction list is available from J. Moses upon request.

The background atmosphere, temperature profile, and diffusion coefficients of Gladstone *et al.* (1996) are used in the photochemical models. Solar flux values from 1983 data are used in Model C; values from 1980 data, which are probably higher than those relevant to the 1994 impact date, are used in Model A. The exact choice of solar flux does not significantly affect the results because much of the photochemistry is initiated by longer wavelength ultraviolet radiation that does not vary noticeably with the solar cycle. The fluxes are diurnally averaged for 44° S latitude using the Jupiter-Sun geometry at the time of the impacts. A steady-state model that just considers hydrocarbon photochemistry is first created to simulate the pre-impact atmosphere. The resulting hydrocarbon-species profiles are then used as initial conditions for the post-impact model.

For boundary conditions, zero flux is assumed for all species except atomic H at the top of the atmosphere (10^{-6} mbar) and a zero concentration gradient at the bottom (6 bar) so that the species flow into the troposphere at a maximum possible rate. Atomic hydrogen, which is produced in the thermosphere of Jupiter, is given a downward flux of 4×10^9 cm^{-2} s^{-1} at the top (see Gladstone *et al.* 1996). Because the impact-derived species are not being continually replenished in the stratosphere, they will eventually diffuse or rain out into the troposphere. The eddy diffusion time scale at 1 mbar is 5 years for the unperturbed stratosphere but may be shorter at the impact sites if the events stirred up the atmosphere to any degree. Condensation is not included explicitly in the Moses *et al.* models, but potential condensates are noted.

4. Photochemical model: Results

The photochemical models of Moses *et al.* (1995a,b,c) demonstrate that the post-SL9 jovian stratosphere is a very dynamic place, with chemical abundances changing on time scales ranging from minutes to years. In the following sections, we will discuss the major photochemistry results in terms of the fate of the sulfur-, nitrogen-, and oxygen-bearing molecules. Although the initial conditions adopted in these early models are unlikely to be correct from a quantitative standpoint, the qualitative results can be used to illuminate the behavior of several of the observed species and to make predictions concerning the long-term evolution of the major impact-derived molecules. We will also use the models to identify molecular species that are photochemically stable and can be used to trace temperature changes or atmospheric dynamics, to identify molecules that might condense and affect atmospheric haze properties, to identify important as-yet-unobserved molecular species, and to use the predicted temporal variation in conjunction with observations to learn more about the unique photochemical processes occurring in the post-SL9 jovian stratosphere.

4.1. *Sulfur*

The photochemistry of sulfur in a reducing atmosphere is poorly understood. Very few rate constants are available in the literature, and Moses *et al.* were forced to estimate the rates for several important reactions. The best-studied reactions are those of sulfur compounds with sulfur, oxygen, and some hydrocarbon species. Significantly absent are reactions of sulfur species with organic radicals and nitrogen species, and possible formation mechanisms for sulfanes (H_2S_x). Despite uncertainties in reaction rates and initial conditions, one robust conclusion can be surmised from the photochemical models: the ultimate sink for the sulfur compounds is condensed S_8. Figures 1 and 2 illustrate the time variation in the partitioning of the sulfur compounds in the dark-core (Model A) and whole-disk (Model C) models.

Although many sulfur molecules were detected after the impacts, their relative abundances are not well known due to difficulties with the observational analyses. One very important molecule, both because of its high initial abundance and its low photochemical stability, is diatomic sulfur (S_2). Because of its short lifetime under laboratory conditions, S_2 has not been well studied. In the photochemical models, S_2 is lost due to rapid photolysis in the upper atmosphere and to the formation of larger sulfur molecules in the lower stratosphere. The S_2 (B–X) band system lies at 240–360 nm and is clearly observed by the Hubble Space Telescope (HST) Faint-Object Spectrograph (FOS) (Noll *et al.* 1995, Yelle & McGrath 1995). At wavelengths shorter than $\sim$278 nm, the S_2 dissociates into two ground-state (^{3}P) sulfur atoms. The diurnally averaged lifetime against photolysis for the S_2 molecule at 10^{-3} mbar and 44° S latitude on Jupiter is $\sim$6 hours.

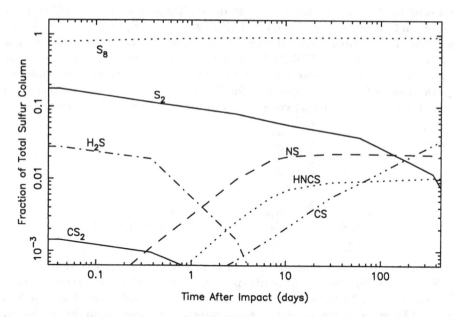

FIGURE 1. The time variation of the partitioning of sulfur compounds in Model A (Moses *et al.* 1995a). The ordinate represents the fraction of the total sulfur column (in terms of S atoms) that is contributed by the various molecules.

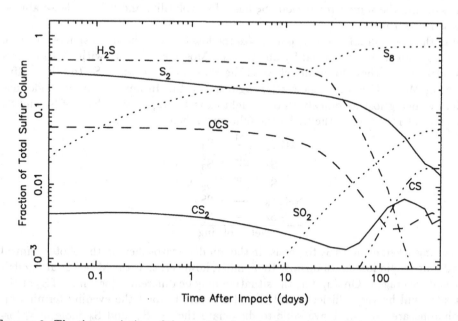

FIGURE 2. The time variation of the partitioning of sulfur compounds in Model C (Moses *et al.* 1995c). The ordinate represents the fraction of the total sulfur column (in terms of S atoms) that is contributed by the various molecules. Not shown, but also important sulfur reservoirs at various times, are S, S_3, S_4, H_2S_4, SO, $(SO)_2$, S_2O, NS, and HNCS.

Ground-state $S(^3P)$ atoms do not insert into H_2 or alkanes (*e.g.*, CH_4 and C_2H_6), but they do insert into alkenes (*e.g.*, C_2H_4) and alkynes (*e.g.*, C_2H_2) to form ring species that are stable in the former case and transient in the latter case. However, if S_2 is as abundant as Yelle & McGrath (1995) indicate (*e.g.*, 5×10^{15} cm^{-2}), the most probable fate of the sulfur atoms is to recycle S_2. Chemical loss processes other than photolysis dominate the loss of S_2. The major sink for S_2 in the photochemical models is the production of molecular sulfur (S_8) via a variety of three-body and other reactions:

$$S_x \xrightarrow{h\nu} S_{x-1} + S \qquad \text{for x = 2, 3, 4}$$
$$2S \xrightarrow{M} S_2$$
$$S + S_2 \xrightarrow{M} S_3$$
$$2S_3 \xrightarrow{M} S_6$$
$$S_2 + S_x \xrightarrow{M} S_{x+2} \qquad \text{for x = 2, 4, 6}$$
$$S + S_x \longrightarrow S_2 + S_{x-1} \qquad \text{for x = 3, 4, 5, 6}$$
$$2S_4 \xrightarrow{M} S_8$$

where M represents any third molecule or atom and $h\nu$ represents an ultraviolet photon. The $S + S$ and $S_2 + S_2$ three-body reactions have been found to be extremely rapid in the laboratory (*e.g.*, Nicholas *et al.* 1979); the other three-body rates have not been measured, and Moses *et al.* conservatively adopt rates that are lower than that for S_2+S_2. Even so, the formation of S_8 proceeds extremely rapidly in the lower stratosphere. Over 70% of the initial column budget of S_2 molecules in Model A has been converted to S_8 in the first hour of the calculations, and another $\sim$ 10% is converted to other sulfur molecules during the same time period. S_8 has a low volatility and will condense almost as soon as it is formed.

Because the reaction $2S_2 \xrightarrow{M} S_4$ dominates the loss of S_2 in the stratosphere, the rate of loss is dependent on the initial S_2 abundance. Moses *et al.* start with 16 times less S_2 in Model C than they do in Model A, so the rate of conversion of S_2 to other sulfur molecules in Model C is much slower (cf. figures 1 & 2). However, neither model has S_2 molecules being lost as quickly as one might expect. Rapid photolysis of S_3 and S_4 ensures efficient recycling of the S_2 by the following scheme:

$$2(2S_2 \xrightarrow{M} S_4)$$
$$S_4 \xrightarrow{h\nu} S_3 + S$$
$$S_3 \xrightarrow{h\nu} S_2 + S$$
$$S + S_3 \longrightarrow 2S_2$$
$$\underline{S + S_4 \longrightarrow S_2 + S_3}$$
$$\text{Net: nothing}$$

This recycling scheme allows S_2 to persist in the middle stratosphere in the photochemical models despite rapid loss mechanisms. Remember, however, that the Moses *et al.* models are diurnally averaged. On Jupiter, the situation may be different. Recycling of S_2 at the impact sites will be very efficient until the sites rotate beyond the evening terminator. Once photons are no longer available to dissociate the S_2, S_3, and S_4 molecules, formation of S_8 will proceed uninhibitedly. Therefore, the diurnally averaged models may considerably overestimate the S_2 abundance after a full Jupiter rotation. Future models should allow diurnal variation. In addition, the S_2 may have been deposited higher in the atmosphere and be at a higher temperature (Yelle & McGrath 1995) than has been assumed in the photochemical models. In that case, the formation of S_8 and other large sulfur molecules will be inhibited.

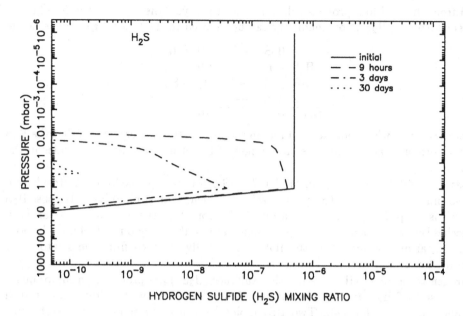

FIGURE 3. The photochemical evolution of H_2S in Model A (from Moses *et al.* 1995a) in terms of the temporal variation of the H_2S mixing ratio profile.

Hydrogen sulfide (H_2S) has been inferred from the July 18 HST FOS spectra of the G impact site (Noll *et al.* 1995, Yelle & McGrath 1996) because of the presence of a broad absorption feature in the 180–210 nm region. However, this spectral signature is not unique to H_2S; Atreya *et al.* (1995) suggest that the absorption slope in this wavelength region is due to aerosols rather than to H_2S (although they, too, seem to include H_2S in their modeling). There are many reasons why one might expect hydrogen sulfide to be present after the impacts. Thermochemical models (*e.g.*, Zahnle 1996, Lyons & Kansal 1996) show that H_2S tends to form under similar shock temperature and pressure conditions as NH_3, and NH_3 was observed to be enhanced in the jovian stratosphere after the SL9 impacts. If the comet fragments penetrated to at least the putative NH_4SH clouds and/or if sulfur was present in the comet in a solar-type abundance, then Zahnle (1996) and Lyons & Kansal (1996) predict that H_2S will be present after the plume splashdown in regions in which plume re-entry temperature did not exceed 2000 K (*e.g.*, near the central region of the plume re-entry scars). Similarly, if the bulk of the NH_3 comes from a later "upwelling" of tropospheric material, as Yelle and McGrath (1996) and others have suggested, then one might also expect H_2S to be present along with the NH_3. Because of these theoretical predictions, Moses *et al.* (1995a,b,c) have included H_2S in their photochemical models, and have even added "extra" H_2S in the lower stratosphere in Model C based on Yelle & McGrath's suggestion that the bulk of both the ammonia and hydrogen sulfide may be located at pressures greater than 5 mbar. Because the presence of H_2S is not clear-cut, both models may greatly overestimate the initial abundance of hydrogen sulfide.

The photochemical models demonstrate that H_2S has a very short lifetime in the jovian stratosphere (see figure 3). The H_2S lifetime against photolysis is ~2 days at 10^{-3} mbar. H_2S is also lost from the upper atmosphere by reaction with atomic hydrogen; both processes act to remove H_2S from the upper atmosphere on time scales of hours. The SH formed from the H_2S loss processes does not act to recycle the H_2S. In the middle and upper stratosphere, H_2S photochemistry can be reduced to the following simple scheme:

$$
\begin{array}{rcl}
H_2S & \xrightarrow{h\nu} & SH + H \\
H_2S + H & \longrightarrow & SH + H_2 \\
SH + H & \longrightarrow & H_2 + S \\
SH + S & \longrightarrow & S_2 + H \\
\hline
\text{Net: } 2H_2S & \longrightarrow & 2H_2 + S_2
\end{array}
$$

The above scheme, which operates rapidly in the jovian stratosphere, shows that the H_2S photolysis products act to increase the budget of the elemental sulfur molecules in the atmosphere.

Moses *et al.* (1995c) find that in Model C, H_2S can persist in the lower stratosphere if UV-absorbing dust is present to act as an effective shield against photolysis. Because H_2S photolysis is the primary source of H atom production at altitudes below 0.1 mbar, both loss mechanisms described in the above scheme are inhibited by dust shielding. In both photochemical models, however, the H_2S is eventually removed from the stratosphere (cf. figures 1 & 2).

Hydrogen tetrasulfide (H_2S_4) and other sulfanes (H_2S_x) are produced in minor quantities as result of H_2S and S_2 photolysis. When SH reacts with sulfur radicals such as S_3 and S_4, HS_2 is formed. Two HS_2 molecules can react along with a stabilizing third molecule or atom to form H_2S_4. However, this process is not the dominant loss mechanism for HS_2 in the photochemical models, and H_2S_4 and H_2S_x never become very abundant. H_2S_4 and H_2S_x are relatively involatile and can condense in the jovian stratosphere.

Carbon disulfide (CS_2) is observed unambiguously in the HST FOS spectra (Noll *et al.* 1995, Atreya *et al.* 1995, Yelle & McGrath 1996). Both Yelle & McGrath (1996) and Atreya *et al.* (1995) derive CS_2 column abundances of $\sim 10^{15}$ cm^{-2}. CS_2 has strong (resolved) absorption bands in the 185–230 nm region. At these wavelengths, the dissociation pathway is CS + S, with $\sim 80\%$ of the sulfur atoms being formed in the ^1D excited state and the rest in the ^3P ground state (Yang *et al.* 1980). The $S(^1D)$ will then react with H_2 to produce SH + H and the SH reacts to reform S and S_2. CS_2, like S_2, is recycled in the middle or lower stratosphere and is not lost as fast as its photolysis rate ($\sim$ 9 hours at 10^{-3} mbar) would indicate. Instead, the CS produced from CS_2 photolysis can react with S_2 or SH to recycle CS_2. On the other hand, photolysis is effective at removing CS_2 from the upper stratosphere (*e.g.*, $P < 10^{-3}$ mbar in Model A or 10^{-2} mbar in Model C). The late increase in the CS_2 abundance shown in figure 2 is due to reactions of sulfur and hydrocarbon radicals (see the CS discussion below).

Carbon monosulfide (CS) is a primary thermochemical product in the plume splashdown if the C/O ratio in the plume is grater than unity (Zahnle 1996, Lyons and Kansal 1996). CS is also produced directly by CS_2 photolysis and indirectly by several different reaction schemes whose ultimate source revolves around S_2. Laboratory studies of CS indicate that it is a short-lived radical that seems to self-polymerize rapidly (Moltzen *et al.* 1988). However, this polymerization reaction, which is not well understood, seems to require the presence of a container wall or other solid surface. Moses *et al.* (1995a,b,c) therefore concluded that CS might not polymerize in Jupiter's atmosphere (unless the reactions could somehow take place on aerosol surfaces) and so did not include polymerization schemes for CS. The apparent long lifetime of CS in the jovian stratosphere (R. Moreno *et al.* 1995) seems to have justified this conclusion.

Although Moses *et al.* (1995a,b,c) did not begin with much CS in their models, they found that the modeled CS abundance increases dramatically with time. The CS photochemistry is complex and difficult to follow, but it appears that reactions of S_2 and S with hydrocarbon radicals are responsible for the delayed increase in abundance shown in figures 1 & 2. The greater the initial S_2 abundance, the more CS that will be produced. Several reaction schemes such as the following contribute to CS production in the months following the impacts:

$$
\begin{array}{rcl}
S_2 + C_2H_3 & \longrightarrow & HCS + H_2CS \\
HCS + H & \longrightarrow & CS + H_2 \\
H_2CS & \xrightarrow{h\nu} & CS + H_2 \\
\hline
\text{Net: } S_2 + C_2H_3 + H & \longrightarrow & 2CS + 2H_2
\end{array}
$$

This scheme is the dominant source of CS production in both Model A & C. A similar scheme involving $S + CH_3 \longrightarrow H_2CS + H$ is also important in Model A. The low abundance of C_2H_3 and the relatively slow estimated rate for $S_2 + C_2H_3$ cause a delay in the production of CS by this and other similar mechanisms, especially in Model C, which has a smaller initial S_2 abundance. Note that thioformaldehyde (H_2CS) is an important intermediate in most of the schemes that convert S and S_2 into CS and CS_2 in the late stages of the sulfur photochemistry. The H_2CS abundance increases noticeably within a week of so after the impacts in the photochemical models, but its predicted peak column abundance (10^{14}–10^{15} cm^{-2}) may not be sufficient to allow H_2CS to be observable. Other organo-sulfur molecules such as methyl mercaptan (CH_3SH) and ethylene episulfide (C_2H_4S) are predicted to be even less abundant than H_2CS.

Because the limited available laboratory data indicate that CS reactions with other radical species are relatively slow, Moses *et al.* (1995a,b,c) find that very few reactions seem to be effective at removing CS from Jupiter's stratosphere. The only reactions that permanently remove CS from the modeled atmosphere are the hypothetical reactions $NH_2 + CS \longrightarrow HNCS + H$ and $CS + SO \longrightarrow S + OCS$; all others tend to recycle the CS. Moses *et al.* estimate relatively small rate constants for both these unmeasured reactions; however, both reactions eventually become important.

Once NH_3 photolysis begins, nitrogen-sulfur species become important reservoirs for both sulfur and nitrogen in the photochemical models. This result is highly speculative because very few kinetic studies of reactions between sulfur and nitrogen radicals are available in the literature. The two main nitrogen-sulfur molecules in the photochemical models, nitrogen sulfide (or sulfur nitride, NS) and isothiocyanic acid (HNCS) have both been detected in interstellar space (Turner 1989). Under laboratory conditions on Earth, NS (like CS) is a radical species that tends to polymerize rapidly (Heal 1972). In Jupiter's stratosphere, NS polymerization may not be as important, and photolysis or reactions such as $NH_2 + NS \longrightarrow N_2 + H_2S$ may be responsible for NS loss. In the photochemical models, NS is produced by $NH_2 + S \longrightarrow HNS + S$ followed by HNS photolysis or $HNS + S \longrightarrow NS + SH$. Photolysis of NS supplies a source of N atoms to the jovian stratosphere.

As already mentioned, Moses *et al.* find that HNCS is produced by the reaction of NH_2 with CS. Photolysis is the dominant loss mechanism—HNCS has a strong absorption band in the 210–270 nm region. However, if the dissociation pathway is predominantly H + NCS, the HNCS is likely to be recycled. Figure 1 shows that the abundances of NS and HNCS increase dramatically with time in Model A; these species also increase with time in Model C, but the NS and HNCS curves were deleted from figure 2 to keep the figure from becoming too confusing.

Carbonyl sulfide (OCS), which was detected at millimeter wavelengths by Lellouch *et al.* (1995), has a photolysis lifetime of $\sim$ 24 days at 10^{-3} mbar. The predominant photodissociation pathway is $CO + S(^1D)$. Recycling of OCS does not occur in the photochemical models; however, some OCS production takes place by the reaction CS

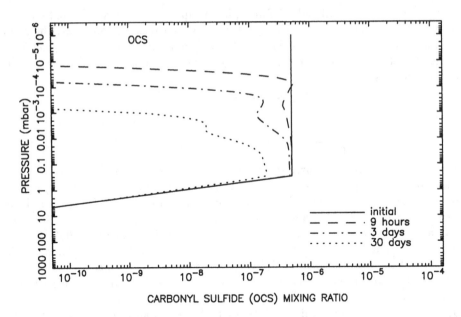

FIGURE 4. The photochemical evolution of OCS in Model C (from Moses *et al.* 1995c) in terms of the temporal variation of the OCS mixing ration profile.

$+ SO \longrightarrow OCS + S$, thus increasing the apparent OCS lifetime. Figure 4 shows the variation of the OCS mixing ratio with time in Model C. Note that OCS is more stable against photolysis in the middle and upper atmosphere than H_2S, CS_2, or S_2.

Sulfur dioxide (SO_2) and sulfur monoxide (SO) are also important reservoirs for the sulfur in models in which the initial water abundance is high (see figures 2 & 5). In fact, SO_2 is responsible for 5.8% and SO for 2.5% of the total column of sulfur after 1 year in Model C. SO and SO_2 formation depend on H_2O photolysis. Water photodissociates primarily into OH + H, and the OH radicals react with atomic sulfur to form SO, which can then react with itself to form SO_2 + S. Although the chemical rate constants and other input parameters to the photochemical models are uncertain, any late build up in the SO_2 abundance observed in the jovian stratosphere long after the impacts would most certainly be due to H_2O photolysis; therefore, a detection or even upper limit on the SO_2 abundance many months after the impacts would greatly aid in constraining the water abundance (if the measured H_2O abundance is indeed a lower limit).

Most of the sulfur molecules produced after the impacts are volatile enough or not abundant enough to condense in the models. Exceptions are S_8, H_2S_4, and other sulfanes (H_2S_x). Elemental sulfur, in particular, should be condensing proliferously about any pre-existing aerosols if S_2 was at all an important component at the impact sites.

4.2. *Nitrogen*

The kinetics of nitrogen compounds in a reducing atmosphere are much better studied than the corresponding case for sulfur, and Moses *et al.* (1995a,b,c) were able use laboratory measurements of reaction rates for most of the nitrogen reactions in their model. The nitrogen compounds formed during the SL9 impacts tend to be more stable than the sulfur compounds. In fact, of the three major nitrogen species introduced by the impacts—ammonia (NH_3), hydrogen cyanide (HCN), and molecular nitrogen (N_2)— only ammonia is photochemically active. Thus, NH_3 drives the nitrogen photochemistry

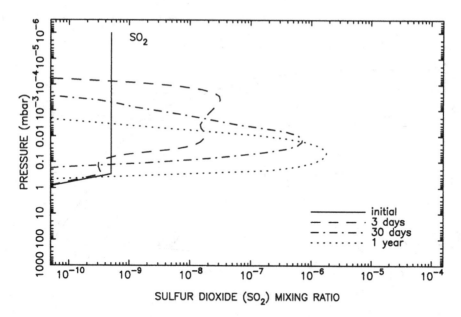

FIGURE 5. The photochemical evolution of SO_2 in Model C (from Moses *et al.* 1995c) in terms of the temporal variation of the SO_2 mixing ratio profile.

in the post-SL9 jovian stratosphere. Figures 6 & 7 illustrate the temporal variation of the partitioning of nitrogen among the different nitrogen compounds. Note that in both models, N_2 is the major reservoir for the nitrogen after several months.

The $(\tilde{A}–\tilde{X})$ band system of NH_3 lies in the 170–220 nm region. At these wavelengths, the primary photolysis pathway is $NH_2 + H$. The photolysis lifetime of NH_3 is ~ 5.7 days at 10^{-3} mbar—a time scale considerably longer than that for most of the active sulfur species. In the photochemical models, the NH_2 radicals formed by ammonia photolysis react with sulfur radicals such as S, SH, CS, and NS to either recycle the ammonia, to form N_2 and HCN, or to form nitrogen-sulfur species such as HNS, NS, and HNCS (see Section 4.1). In Model C, NS radicals are allowed to cluster together to form $(NS)_2$ and $(NS)_4$, which also become important reservoirs for the nitrogen and sulfur (see figure 7). The reactions of NH_2 with sulfur radicals cause the ammonia to be recycled less efficiently than otherwise might be the case and cause hydrazine (N_2H_4) formation to be suppressed.

As mentioned in the previous section, all these proposed sulfur-nitrogen reactions are hypothetical; none have been studied in the laboratory. Searches for potentially important species such as NS and HNCS at ultraviolet through microwave wavelengths (or even quoted upper limits) would greatly help in distinguishing between several possible pathways in the ammonia photochemistry. Note that these species take days or weeks to build up into potentially observable quantities. Failing such direct observational constraints, the observed lifetime of the NH_3 at the impact sites might help determine whether such proposed reactions are occurring.

One very noticeable difference between Models A & C is the apparent lifetime of NH_3 in the middle and lower stratosphere (cf. figures 6 & 7). The greater NH_3 lifetime in Model C has nothing to do with the increased abundance of oxygen compounds in that model; instead, the culprit is increased shielding by other molecules and, more importantly, by dust. Both CS_2 and H_2S absorb in the same wavelength region as NH_3. CS_2 is not abundant enough to markedly affect ammonia photolysis. If H_2S and NH_3 are

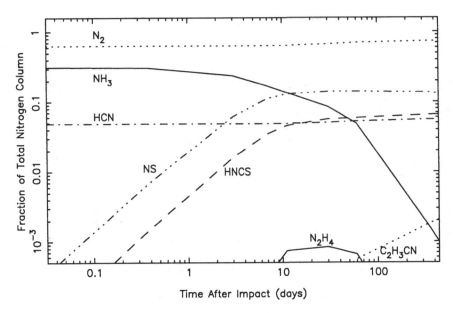

FIGURE 6. The time variation of the partitioning of nitrogen compounds in Model A (Moses *et al.* 1995b). The ordinate represents the fraction of the total nitrogen column (in terms of N atoms) that is contributed by the various molecules.

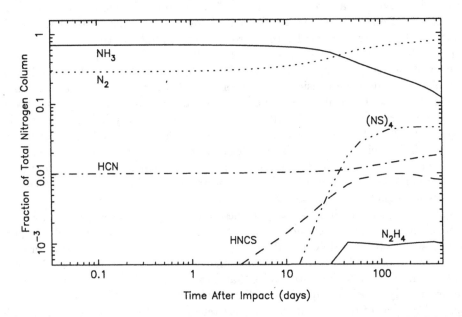

FIGURE 7. The time variation of the partitioning of nitrogen compounds in Model C (Moses *et al.* 1995c). The ordinate represents the fraction of the total nitrogen column (in terms of N atoms) that is contributed by the various molecules.

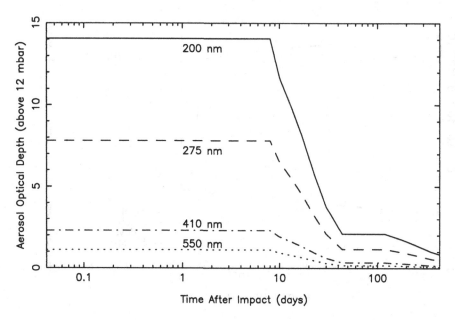

FIGURE 8. Time variation of aerosol opacity in Model C (Moses *et al.* 1995c)

co-located, as Yelle & McGrath (1996) suggest, then H_2S could be a potentially effective UV shield as long as the H_2S persists in the lower stratosphere. The dark widespread dust observed at the impact sites (*e.g.*, West *et al.* 1995) might also inhibit the penetration of ultraviolet radiation into the lower stratosphere. Several studies of the aerosol debris generated by the impacts suggest that the dust extends up to at least 0.3 mbar or perhaps even higher (*e.g.*, West *et al.* 1995, Mallama *et al.* 1995). If so, then this debris could affect ammonia photolysis.

To determine the effect of dust on the photochemistry, Moses *et al.* (1995c) add aerosol extinction to Model C and use a Mie scattering code to estimate dust opacities. The dust absorption cross sections are derived by assuming a log-normal distribution of spherical particles with a distribution width of 1.2 and a mean radius of 0.15 μm (the radius was chosen to be consistent with the dust analysis of F. Moreno *et al.* 1995 and Ortiz *et al.* 1995). The particle imaginary index of refraction was assumed to be typical of outer planetary hazes; *i.e.*, $k = \exp[-2.1 - 6.5 * \lambda(\mu m)]$ and is consistent with the derived ultraviolet imaginary refractive indices of West *et al.* (1995). The dust number densities were chosen so that the haze opacity remains approximately consistent with the West *et al.* UV-visible optical depths. The initial dust profile adopted in Model C has a constant dust density of $\sim 8 \times 10^3$ particles cm^{-3} below 10 mbar, a dust number mixing ratio of 1.8×10^{-14} between 10–10^{-2} mbar and 10^{-25} above 10^{-2} mbar. Since West *et al.* (1995) demonstrate that the dust opacity decreases with time, Moses *et al.* (1995c) allow the dust number density to be reduced over time; however, the reduction in haze abundance was instituted somewhat arbitrarily. The resulting time variation in the haze opacity is shown in figure 8. Although not shown in the figure, the ratio of the dust opacities at 275 and 893 nm derived from this haze model is not quite consistent with the West *et al.* (1995) results. In addition, the haze model of F. Moreno *et al.* (1995) and Ortiz *et al.* (1995) would not be optically thick in the ultraviolet using the optical properties extrapolated from West *et al.* (1995). Both of these results suggest that the particles are indeed larger than 0.15 μm in radius, as indicated by West *et al.*

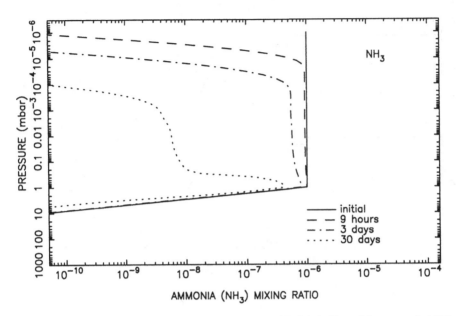

FIGURE 9. The photochemical evolution of NH_3 in Model A (from Moses *et al.* 1995a) in terms of the temporal variation of the NH_3 mixing ratio profile.

Figure 9 illustrates how the NH_3 profile varies with time in Model A. In the middle and upper stratosphere ($P < 10^{-2}$ mbar), the ammonia profile in Model C is virtually identical to that of Model A. However, absorption of solar radiation by dust keeps more NH_3 in the lower stratosphere in Model C. The haze optical depth in Model C is still ~ 1 at 1 mbar 10 days after the impacts and remains high for several weeks. Therefore, the dust provides an effective shield for the ammonia at low altitudes and prevents the NH_3 from being lost as rapidly as it normally would be. As will be discussed in Section 5, the fact that the NH_3 stays in the stratosphere in Model C much longer than is indicated by the observations suggests that the actual dust deposited by the impacts is not optically thick at millibar pressure levels in the near ultraviolet. Thus, the photochemical models can be used in conjunction with observations to place constraints on the stratospheric aerosol profile.

Hydrogen cyanide does not absorb near-ultraviolet radiation and only absorbs weakly below 190 nm. In addition, no chemical loss mechanisms are effective at removing HCN from Jupiter's stratosphere. Moses *et al.* (1995b) find that in Model A, the HCN formed during the plume splashdown is photochemically stable throughout the stratosphere, and they suggest that HCN might therefore be regarded as a good "thermometer" for the stratosphere—changes in the HCN observations might be due to temperature variations or horizontal spreading rather than to changes in total abundance. Moses *et al.* (1995c) derive a slightly different result for Model C. They see an increase in the HCN abundance over time due to the following series of reactions:

$$
\begin{array}{rcl}
NH_3 & \xrightarrow{h\nu} & NH_2 + H \\
NH_2 + S & \longrightarrow & HNS + H \\
HNS + H & \longrightarrow & H_2 + NS \\
NS & \xrightarrow{h\nu} & N + S \\
CH_3 + N & \longrightarrow & H_2CN + H \\
H_2CN + H & \longrightarrow & HCN + H_2 \\
\hline
\text{Net: } NH_3 + CH_3 & \longrightarrow & HCN + 2H_2 + H
\end{array}
$$

Because NH_3 is initially so much more abundant than HCN in Model C, and because NS photolysis has been included, the increase in HCN abundance is more noticeable in Model C than in Model A.

Molecular nitrogen was probably an important reservoir for the nitrogen after the impact and the plume re-entry shocks (Zahnle 1996, Lyons & Kansal 1996). Its presence in the post-SL9 jovian stratosphere is inferred from the fact that the O/N ratio of the observed molecules at the impact sites is greater than that of typical comets, suggesting that N_2 might make up the "missing" nitrogen component (*e.g.*, Lellouch 1996). N_2 has no efficient photochemical loss mechanisms in the jovian stratosphere. The N_2 abundance increases with time in the photochemical models due to schemes such as the following:

$$
\begin{array}{rcl}
2(NH_3 & \xrightarrow{h\nu} & NH_2 + H) \\
NH_2 + S & \longrightarrow & HNS + H \\
HNS & \xrightarrow{h\nu} & H + NS \\
NH_2 + NS & \longrightarrow & N_2 + H_2S \\
\hline
\text{Net: } 2NH_3 + S & \longrightarrow & N_2 + H_2S + 4H
\end{array}
$$

Moderate amounts of hydrazine (N_2H_4) and nitriles such as acrylonitrile (C_2H_3CN) build up at various times in the photochemical models (see figures 6 & 7). However, it's not clear that these species can build up in observable quantities.

None of the nitrogen-bearing molecules in the photochemical models are predicted to contribute much mass to the stratospheric aerosol layer. Hydrazine may be abundant enough to condense after several months, and species with uncertain vapor pressures such as $(NS)_x$ clusters or rings may also condense in the stratosphere. If ammonia is as abundant initially in the deep stratosphere as is assumed in Model C, then NH_3 is close to its saturation vapor curve in the lower stratosphere, but there are indications that the initial NH_3 abundance in Model C may be an overestimate (see Section 5).

4.3. Oxygen

Interest in the chemistry of the Earth's stratosphere has motivated many laboratory studies of oxygen photochemistry, and most of the rate constants for the oxygen reactions in the Moses *et al.* models are well known. Initial conditions for the oxygen compounds at the impact sites are less certain; CO, H_2O, and OCS are the only oxygen-bearing molecules that have been identified. Figure 10 illustrates how these and other potentially important oxygen compounds vary with time. Note that CO and H_2O are much more photochemically stable than most of the sulfur-bearing molecules or ammonia. Since Model C is the only one that contains a non-trivial initial amount of water, it will be the focus of our discussion in this section.

Water photolysis drives the oxygen photochemistry. Photolysis of H_2O occurs at short wavelengths (*e.g.*, $\lambda < 190$ nm); thus, H_2O is relatively stable in Jupiter's stratosphere. Its photolysis lifetime is almost 3 months at 10^{-3} mbar in Model C. In addition, H_2O is recycled to a large degree once it is photolyzed. The OH reacts with H_2 to reform water. However, some of the OH radicals react with S to form SO or react with SO to form

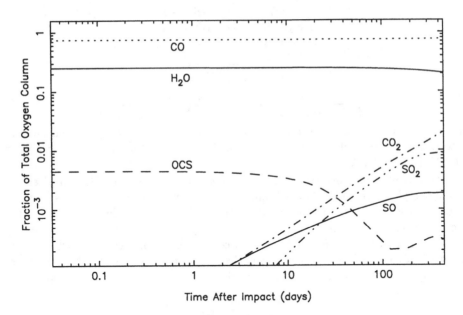

FIGURE 10. The time variation of the partitioning of oxygen compounds in Model C (Moses *et al.* 1995c). The ordinate represents the fraction of the total oxygen column (in terms of O atoms) that is contributed by the various molecules.

SO_2 so that the abundances of sulfur monoxide and sulfur dioxide increase with time. Another interesting loss process is $OH + CO \longrightarrow CO_2 + H$. This reaction is slow and barely makes a dent in the CO abundance after 1 year; however, the reaction does cause the carbon dioxide (CO_2) abundance to increase noticeably with time (see figure 10).

Theoretical models predict that large quantities of CO will form during both the initial impact explosion and the plume splashdown for virtually all possible composition, temperature, and pressure conditions assumed for the shocks (Zahnle 1996, Lyons & Kansal 1996). Carbon monoxide is photochemically stable, and Moses *et al.* (1995a,b,c) suggest that CO might be a good tracer for temperature changes or atmospheric dynamics after the impacts. In the photochemical models, CO is the ultimate repository for most of the oxygen compounds. For instance, CO is produced directly by OCS and CO_2 photolysis, and indirectly by H_2O, SO, and SO_2 photolysis. Schemes such as the following are effective at converting H_2O to CO:

$$
\begin{array}{rcl}
H_2O & \overset{h\nu}{\longrightarrow} & OH + H \\
OH + S & \longrightarrow & SO + H \\
SO & \overset{h\nu}{\longrightarrow} & S + O \\
O + CH_3 & \longrightarrow & H_2CO + H \\
H_2CO + H & \longrightarrow & HCO + H_2 \\
HCO + H & \longrightarrow & CO + H_2 \\
\hline
\text{Net: } H_2O + CH_3 & \longrightarrow & CO + 2H_2 + H
\end{array}
$$

Note that formaldehyde (H_2CO) is an intermediary in the above reaction. Moderate amounts (*e.g.*, 10^{13}–10^{15} molecules cm^{-2}) of H_2CO, methanol (CH_3OH), and nitric oxide (NO) form in the photochemical model after a week or so. Other important oxygen-bearing molecules include OCS, SO, SO_2, and $(SO)_2$. A discussion of the photochemistry of these molecules is included in Section 4.1.

None of the oxygen compounds are expected to be abundant enough to condense in the lower stratosphere. It is interesting to note, however, that the initial H_2O abundance selected for Model C puts H_2O very close to its saturation point in the lower stratosphere. No other oxygen compounds in the model are even close to saturation within a period of 15 months after the impacts.

5. Model-data comparisons

To determine if the chemical state of the atmosphere has responded as expected to the impacts, we compare the photochemical model predictions with observational data. The record of the time variation of the different observed molecules is particularly useful in determining physical and chemical conditions at the impact sites and in helping to constrain some of the uncertain initial conditions and chemical reaction schemes. Information about the temporal variation of at least 7 of the observed molecules (*e.g.*, S_2, CS_2, CS, OCS, NH_3, HCN, CO, and perhaps H_2S) has been presented in the literature. We will discuss each of these molecules and evaluate whether the photochemical models are consistent with the observations. Remember that the photochemical models are one-dimensional; because horizontal spreading is ignored, the models should always overpredict column abundances at later times.

As discussed by Noll *et al.* (1995), S_2 absorption lines were seen in the HST FOS spectra of the G impact site on July 18, 1994 ($\sim$ 3.5 hours after the G impact) and on July 21, 1994 ($\sim$ 3 days after the G impact, but only 45 minutes after the nearby S impact). The viewing geometry of the observations was not the same in the two cases, and the S_2 signature in the July 21 spectra may be due to S_2 at the G impact site, or it may be due to S_2 at the smaller, but more recent, S impact site. No S_2 absorption lines were observed on August 9, 1994 (22 days later) or at any time after July 21, 1994 (Noll *et al.* 1995, McGrath *et al.* 1995).

In the Moses *et al.* (1995a) photochemical model (Model A), the S_2 abundance drops by a factor of $\sim$ 20 in the first month after the impacts; however, most of that reduction occurs in the first day. (Note that the S_8 shown in figures 1 & 2 is directly derived from S_2). Between 3 days and 30 days, the S_2 column decreases by only a factor of $\sim$ 2. If the July 21 signature is due to the presence of S_2 at the fresh S site rather than at the 3-day-old G site, then Model A may be consistent with observations. On the other hand, if the July 21 signature is due to S_2 at the G impact site, then Model A probably overpredicts the rate of loss of S_2 at the impact sites.

A straightforward decrease in the assumed initial S_2 abundance at all altitudes in the photochemical model allows the S_2 to remain in the stratosphere for a longer period of time. However, in that situation, the S_2 probably persists too long to account for observations. For instance, in Model C, the S_2 column drops by only a factor of $\sim$3 in the first 23 days. Although no upper limits are given in Noll *et al.* (1995) or McGrath *et al.* (1995) for the August 9, 1994 data, the observations probably indicate a more dramatic decrease in the S_2 abundance than is seen in Model C.

A cautionary note needs to be inserted here with regard to the photochemical modeling of S_2. The S_2 may initially be confined to higher altitudes than is assumed in the photochemical models (see Yelle & McGrath 1995), and the photochemistry of S_2 is not well understood. More importantly, the photochemical models are diurnally averaged. S_2 and the other sulfur radicals have short photochemical lifetimes, and diurnally averaged models may not truly represent the situation at the impact sites. Further comparisons between S_2 observations and models should be delayed until more appropriate photochemical models are developed.

CS$_2$ is always seen in ultraviolet spectra of the impact sites, even 9 months after the impacts (McGrath *et al.* 1995). This result is consistent with both of the photochemical models of Moses *et al.* who find that reactions between sulfur and hydrocarbon radicals produce CS$_2$ in the mid-to-lower stratosphere several months after the impacts. Both Model A and C predict that CS$_2$ will be readily photolyzed and lost from the upper atmosphere; however, a slower rate of decline in the total CS$_2$ column is predicted in Model C relative to Model A. Model A exhibits a total drop in the CS$_2$ column of a factor of 40 in the first 30 days while Model C predicts that the CS$_2$ column would have decreased by a factor of 2.5 between the July 18 and the August 9 observations and by just a factor of $\sim$ 3 between July 18 and August 23, 1994. The actual observations seem consistent with the Model C results but inconsistent with the Model A results. Yelle & McGrath (1996) derive a decrease in the CS$_2$ column of a factor of 2–3 between July 18 and August 9, and another factor of $\sim$2 between August 9 and August 23. The persistence of CS$_2$ in the observations suggests that some shielding is going on (*e.g.*, from dust or from NH$_3$) as in Model C or that CS$_2$ is more efficiently recycled or produced than is evident in Model A.

One interesting prediction from Model C is that the CS$_2$ will end up being co-located with the stratospheric NH$_3$ weeks or months after the impacts. Yelle & McGrath (1996) suggest that the bulk of the CS$_2$ is originally located at higher altitudes than the ammonia. However, Moses *et al.* find that CS$_2$ disappears from the upper stratosphere over time but persists in a similar altitude region as the NH$_3$. It will be interesting to see whether an analysis of the FOS spectra taken in August, 1994 will support this prediction.

CS has been monitored on a regular basis at millimeter and submillimeter wavelengths from the IRAM and JCMT telescopes (Lellouch *et al.* 1995, R. Moreno *et al.* 1995, Matthews *et al.* 1995). Six to twelve months after the impacts, the CS column abundance appears to be $\sim$ 10 times less than in July 1994 (Moreno *et al.* 1995, Matthews *et al.* 1995, Lellouch 1996). If we follow the arguments of Lellouch (1996) and assume that the spots have spread by a factor of $\sim$ 80 in area in this time, then the data imply that the total CS mass may have increased by a factor of $\sim$ 8 during the 6–12 month period following the impacts.

Both Models A and C exhibit a steady increase in the CS column abundance in the weeks and months following the impacts. In Model A, a factor of 20 increase in the CS column is predicted between 1 week and 1 year after the impacts. In Model C, a factor of 80 increase is predicted between 1 week and 1 year. Both models grossly overestimate the increase of the CS column with time. Because the CS in the photochemical models is derived from S$_2$ photochemistry, the inconsistencies between models and data suggest that the models overestimate the initial S$_2$/CS ratio. In fact, Moses *et al.* (1995a,b,c) do not begin with much CS at all (see Table 1), mainly because some of the original thermochemical models (*e.g.*, Zahnle *et al.* 1995) did not favor CS production and because the initial large S$_2$ abundance in Model A allowed a substantial build-up in the CS abundance after a day or so (*i.e.*, large initial amounts of CS were not required to explain the observations). If S$_2$/CS $\lesssim$ 1 initially, then the photochemical models would not show such a dramatic increase in the CS abundance with time unless some other sulfur compound is present in much larger quantities than CS.

OCS was detected at mm wavelengths at the IRAM telescope on July 22, 1994 (Lellouch *et al.* 1995). The detection was not confirmed at the same telescope 1 month later (Lellouch 1996). Photochemical models predict that OCS is lost at a steady rate in the jovian stratosphere due to photolysis at near-ultraviolet wavelengths (with a photolysis lifetime of 24 days). Dust helps shield the OCS from photolysis in the Model C; even

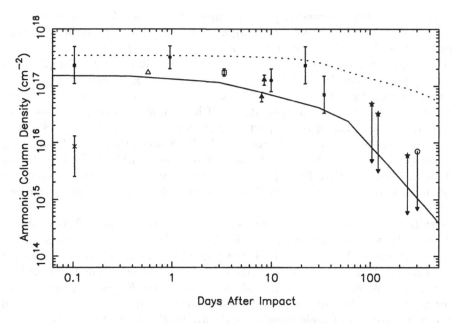

FIGURE 11. The time variation of NH₃ (after Fast *et al.* 1995). The solid line represents the photochemical model results of Model A, and the dotted line the results of Model C. Observations are indicated by individual points and associated error bars. The cross is from Atreya *et al.* (1995), the solid squares from Yelle & McGrath (1996) (where the range is calculated assuming the NH₃ is confined between 5 mbar and either 12 or 40 mbar), the open triangle from Conrath *et al.* (1995, personal communication to T. Kostiuk), the solid circles from Griffith *et al.* (1995a), the open square from Betz *et al.* (1995), the solid triangles from Kostiuk *et al.* (1995), the solid stars (upper limits) from Fast *et al.* (1995), and the open circle (upper limit) from Bézard *et al.* (1995). Note: Most of the observations are quoted for columns above 40–150 mbar; the Atreya *et al.* values are for the column above 0.3–10 mbar. Note also that observations made in the first few weeks were often centered at different impact sites.

so, the model predicts a factor of ∼ 2.6 decrease in the OCS column after 30 days. This decrease may be consistent with the observations (Lellouch 1996).

Although the H_2S detection is controversial, Yelle & McGrath (1996) suggest that H_2S is present in the HST FOS spectra 3.5 hours after the impact, but not 22 days later. This suggestion is consistent with photochemical models.

The time variation of ammonia has been monitored extensively at infrared and ultraviolet wavelengths. Figure 11 shows how the photochemical model results compare with observations. Observations taken many months after the impacts show no evidence for NH₃ in the middle and upper stratosphere (Fast *et al.* 1995, Bézard *et al.* 1995). However, NH₃ is visible in ultraviolet spectra taken 9 months after the impacts (McGrath *et al.* 1995), suggesting that NH₃ may still be enhanced in the lower stratosphere and/or upper troposphere relative to pre-impact values. This latter conclusion will not be confirmed until the data are fully analyzed. As discussed in Section 4.2, the observed time variation of NH₃ appears more consistent with Model A than Model C, suggesting that too much dust shielding has been included in the latter model. If we take the models at face value, the model-data comparisons tell us that the dust cannot be optically thick above a few mbar a week or so after the impacts. However, keep in mind that horizontal spreading has not been included in the photochemical models. It is not clear how fast the material at a few mbar to a few 10's of mbar (where most of the NH₃ is located according to the infrared observations) is spreading. If there is a factor of ~ 80 areal spreading 6–12 months after the impacts compared with the spot size during the week of the impacts, then Model C may easily be consistent with observations.

HCN, like CS, has been monitored at millimeter and submillimeter wavelengths on a regular basis (Marten *et al.* 1995, Matthews *et al.* 1995). The HCN column abundance has appeared to drop by a factor of ~ 10 in the 6–12 months following the impacts (Matthews *et al.* 1995, R. Moreno *et al.* 1995). If horizontal spreading is assumed to have caused the areal extent of the impact sites to increase by a factor of ~ 80, then the mm and sub-mm observations indicate that the HCN abundance (*e.g.*, total mass) may have increased by as much as a factor of ~ 8 over time. IRTF/IRSHELL 10-μm observations also indicate that HCN persists in the jovian stratosphere and that the total HCN mass may have increased in the 10 months following the impacts (Griffith *et al.* 1995b).

These results are consistent with the suggestion of Moses *et al.* that HCN is produced from NH_3 photolysis. The relative increase in the total HCN column depends on the initial NH_3/HCN ratio. To produce HCN, the photochemical models require a source of atomic nitrogen. In the Moses *et al.* models, N is supplied by speculative reactions involving ammonia photolysis products and sulfur, and have NS as an intermediate. Positive or negative searches for NS and other nitrogen-sulfur molecules in observational data would help constrain possible chemical pathways for the production of HCN.

The molecule with the most surprising observed temporal variation as compared with photochemical models is CO. Millimeter and submillimeter observations in the year following the impacts indicate that the impact sites have a CO column abundance that is ~ 160 times less than in July, 1994 (R. Moreno *et al.* 1995, Matthews *et al.* 1995, Lellouch 1996). When spreading is taken into account, the implied reduction in total abundance (mass) is a factor of ~ 2. Photochemical models do not reproduce a loss of this magnitude loss over the course of a year. In fact, Model C predicts that the CO abundance will increase with time due to H_2O photolysis products reacting with hydrocarbons to form CO. Lellouch (1996) discusses this problem in detail.

How could CO decrease with time while HCN and CS do not? One possible way to reconcile the problem is to assume that the bulk of the CO is located at a different altitude than that of the CS and HCN. In that case, the temperature variation and rate of horizontal spreading may be different. For instance, HCN and CS are favored by moderately shocked conditions in a "dry" plume; *i.e.*, one that has $C/O > 1$ (Zahnle 1996, Lyons & Kansal 1996). If the oxygen is derived from the comet rather than from the jovian atmosphere, then much of the CO may have come from a different region of the plume/fireball (*i.e.*, one with $C/O < 1$) and hence may have been deposited in a different altitude region during the plume splashback. Zahnle (1996) suggests that the cometary material may have been shocked to very high temperatures, and thus represents the material that has been flung the farthest and deposited the highest in the atmosphere. In addition, the molecules may be filling different areal fractions of the observed impact "scars." One can then imagine scenarios in which the CO spreads and dilutes more rapidly than the CS or HCN. One problem with such scenarios is that the altitudes derived from the millimeter observations (R. Moreno *et al.* 1995) are similar for the CO, CS, and HCN, and we know that HCN has spread over a large area (Griffith *et al.* 1995b).

Alternatively, some loss mechanism for the CO that has not been included in the Moses *et al.* photochemical models may be present in the jovian atmosphere. If the reduction in the CO abundance is due to photochemical processes, then observations should show a corresponding increase in some other oxygen compound such as CO_2, SO_2, or NO (unless the CO is converted to something that will condense in the stratosphere). No increase in any other oxygen compounds have been found, but it is not clear that anyone has been looking. Recently ($\sim$ 15 months after the impacts), IRAM observations taken under good seeing conditions showed no evidence for SO_2 in the jovian stratosphere (E. Lellouch and R. Moreno, personal communication).

6. Conclusions

The collision of comet Shoemaker-Levy 9 with Jupiter resulted in profound changes to the jovian atmosphere. To fully understand the chemical changes that occurred after the impacts, photochemical models of the post-SL9 jovian stratosphere have been developed (Moses et al. 1995a,b,c). These models are used to trace chemical changes over time and to connect observations taken days or weeks after the impacts with chemical abundances at the time of the plume splashdown.

The theoretical models indicate that the photochemistry at the impact sites is rapid and complex. The sulfur species introduced by the impacts evolve very quickly. Condensed S_8 is the main reservoir for the sulfur over time, and the aerosols should become progressively coated with sulfur compounds (e.g., S_8, H_2S_x, N–S and C–S polymers). One important prediction, that the CS abundance persists with time, seems to be supported by mm and sub-mm observations (R. Moreno et al. 1995, Matthews et al. 1995); however, the dramatic increase of the modeled CS abundance with time is not mirrored in the observations, indicating that the initial S_2/CS ratio at the impact sites was probably $\lesssim 1$. The observed abundances of CS_2, OCS, and perhaps H_2S all seem to decrease roughly as expected. Two predictions remain to be investigated: (1) unusual nitrogen-sulfur species such as NS and HNCS are predicted to become important reservoirs for both the sulfur and the nitrogen at the impact sites, and (2) SO_2 will become an important reservoir for the sulfur if H_2O was an important initial constituent at the impact sites. Positive or negative searches for these molecules in data taken months after the impacts will help distinguish between several possible chemical schemes that may have occurred at the impact sites and will help constrain the initial water abundance.

Nitrogen compounds tend to be more stable at the impact sites than sulfur compounds. NH_3 photolysis, which operates on a week time scale, drives the nitrogen photochemistry. N_2 is predicted to be the main nitrogen reservoir over time. In the photochemical models, the HCN abundance increases slightly with time due to reactions catalyzed by sulfur radicals; the HCN observations appear to be consistent with this prediction (Matthews et al. 1995, R. Moreno et al. 1995, Griffith et al. 1995b). The observed rate of decrease of NH_3 may indicate that the dust introduced by the impacts was not optically thick at a few mbar at near-ultraviolet wavelengths after the first week of the impacts. This conclusion will not be firm until we have a better handle on the amount of horizontal spreading at the impact sites.

H_2O and CO are predicted to be very stable in the jovian stratosphere. Water, which has a photolysis lifetime of months, will drive the oxygen photochemistry. If large quantities of H_2O were present throughout the impact sites, then CO_2, SO_2, SO, and NO will become important oxygen reservoirs over time. The observed decrease in the CO abundance with time (R. Moreno et al. 1995, Matthews et al. 1995, Lellouch 1996) is not consistent with photochemical models. The models may be neglecting some scheme that converts the CO to other as-yet-unobserved oxygen compounds, or the CO may have been deposited at a different location than the HCN and CS and so may have experienced a different rate of horizontal spreading.

Improved photochemical models should help resolve some of the problems with the current model-data comparisons. More realistic initial vertical profiles for the different observed species can be developed now that more sophisticated observational analyses are being published. Future photochemical models should also include parameterizations to take horizontal spreading into account and should allow for diurnal variations. More extensive thermochemical modeling should also help constrain the initial conditions required for accurate photochemical modeling. The Shoemaker-Levy 9 impacts have provided us with a unique opportunity to observe some unusual atmospheric photochemistry in action; photochemical models can be valuable tools in interpreting the chemical evolution at the impact sites.

We thank R. Yelle, K. Zahnle, E. Lellouch, and M. Allen for thorough reviews of the manuscript and many useful discussions.

REFERENCES

ATREYA, S. K., EDGINGTON, S. G., TRAFTON, L. M., CALDWELL, J. J., NOLL, K. S., & WEAVER, H. A. 1995 Abundances of ammonia and carbon disulfide in the Jovian stratosphere following the impact of comet Shoemaker-Levy 9. *Geophys. Res. Lett.* **22**, 1625–1628.

BETZ, A. L., BOREIKO, R. T., BESTER, M., DANCHI, W. C., & HALE D. D. 1995 Stratospheric ammonia in Jupiter after the impact of comet SL-9. *Bull. Amer. Astron. Soc.* **26**, 1590–1591.

BÉZARD, B., GRIFFITH, C., GREATHOUSE, T., KELLY, D., LACY, J., & ORTON, G. 1995 Jupiter ten months after the collision of comet SL9: Stratospheric temperatures and ammonia distribution. *Bull. Amer. Astron. Soc.* **27**, 1126.

BJORAKER, G. L., HERTER, T., GULL, G., STOLOVY, S. & PIRGER, B. 1995 Detection of water in the "splash" of fragments G and K of comet Shoemaker-Levy 9. In *Abstracts for IAU Colloquium 156: The Collision of Comet P/Shoemaker-Levy 9 and Jupiter*, p. 8.

BORUNOV, S., DROSSART, P., ENCRENAZ, TH., & DOROFEEVA, V. 1995 Thermochemistry in the fireball of SL9 impacts. *Bull. Amer. Astron. Soc.* **27**, 1120.

BROOKE, T. Y., ORTON, G. S., CRISP, D., FRIEDSON, A. J. & BJORAKER, G. 1995 Near-infrared spectroscopy of the Shoemaker-Levy 9 impact sites with UKIRT: CO emission from the L site. In *Abstracts for IAU Colloquium 156: The Collision of Comet P/Shoemaker-Levy 9 and Jupiter*, p. 12.

CALDWELL, J., MARTYN, M., DELRIZZO, D., ATREYA, S., EDGINGTON, S., BARNET, C., NOLL, K., WEAVER, H., TRAFTON, L., & YOST, S. 1995 Upper limits on SiO, H_2S, C_2H_2 and H_2O on Jupiter from SL9. *Bull. Amer. Astron. Soc.* **27**, 1118–1119.

CARLSON, R. W., WEISSMAN, P. R., HUI, J., SEGURA, M., SMYTHE, W. D., BAINES, K. H., JOHNSON, T. V., DROSSART, P., ENCRENAZ, T., LEADER, F. & MEHLMAN, R. 1995 Galileo infrared observations of the Shoemaker-Levy 9 G and R fireballs and splash. In *Abstracts for IAU Colloquium 156: The Collision of Comet P/Shoemaker-Levy 9 and Jupiter*, p. 15.

CONRATH, B. J., GIERASCH, P. J., HAYWARD, T., McGHEE, C. NICHOLSON, P. D. & VAN CLEVE, J. 1995 Palomar mid-infrared spectroscopic observations of comet Shoemaker-Levy 9 impact sites. In *Abstracts for IAU Colloquium 156: The Collision of Comet P/Shoemaker-Levy 9 and Jupiter*, p. 24.

COSMOVICI, C. B., MONTEBUGNOLI, S., POGREBENKO, S. & COLOM, P. 1995 Water MASER detection at 22 GHz after the SL-9/Jupiter collision. *Bull. Amer. Astron. Soc.* **27**, 1133.

FAST, K. E., LIVENGOOD, T. A., KOSTIUK, T., BUHL, D., ESPENAK, F., BJORAKER, G. L., ROMANI, P. N., JENNINGS, D. E., SADA, P., ZIPOY, D., GOLDSTEIN, J. J., & HEWEGAMA, T. 1995 NH_3 in Jupiter's stratosphere within the year following the SL9 impacts. *Bull. Amer. Astron. Soc.* **27**, 1126–1127.

GLADSTONE, G. R., ALLEN, M., & YUNG, Y. L. 1996 Hydrocarbon photochemistry in the upper atmosphere of Jupiter. *Icarus* 119, 1–52.

GRIFFITH, C. A., BÉZARD, B., KELLY, D., LACY, J., GREATHOUSE, T., & ORTON, G. 1995a Mid-IR spectroscopy and NH₃ and HCN images of K impact site. In *Abstracts for IAU Colloquium 156: The Collision of Comet P/Shoemaker-Levy 9 and Jupiter*, p. 42.

GRIFFITH, C. A., BÉZARD, B., GREATHOUSE, T., KELLY, D., LACY, J., & ORTON, G. 1995b Jupiter ten months after the collision of comet SL9: Spectral maps of HCN and NH₃. *Bull. Amer. Astron. Soc.* 27, 1126.

HEAL, H. G. 1972 The sulfur nitrides. *Adv. Inorg. Chem. Radiochem.* 15, 375–412.

KIM, S. J., RUIZ, M., RIEKE, G. H., & RIEKE, M. J. 1995 Thermal history of the R impact flare of comet Shoemaker-Levy 9. *Bull. Amer. Astron. Soc.* 27, 1120.

KNACKE, R. F., FAJARDO-ACOSTA, S. B., GEBALLE, T. R., & NOLL, K. S. 1995 Infrared spectroscopy of the R-impact of comet Shoemaker-Levy 9. *Bull. Amer. Astron. Soc.* 27, 1114.

KOSTIUK, T., BUHL, D., ESPENAK, F., ROMANI, P., BJORAKER, G., FAST, K., LIVENGOOD, T., & ZIPOY, D. 1996 Stratospheric ammonia on Jupiter after the SL9 collision. *Icarus*, submitted.

LELLOUCH, E. 1996 Chemistry induced by the impacts: Observations. *This volume.*

LELLOUCH, E., PAUBERT, G., MORENO, R., FESTOU, M. C., BÉZARD, B., BOCKELÉE-MORVAN, D., COLOM, P., CROVISIER, J., ENCRENAZ, T., GAUTIER, D., MARTEN, A., DESPOIS, D., STROBEL, D. F., & SIEVERS, A. 1995 Chemical and thermal response of Jupiter's atmosphere following the impact of comet Shoemaker-Levy 9. *Nature* 373, 592–595.

LYONS, J. R., & KANSAL, A. 1996 A chemical kinetics model for analysis of the comet Shoemaker-Levy 9 impacts with Jupiter. *Icarus*, submitted.

MAILLARD, J.-P., DROSSART, P., BÉZARD, B., DE BERGH, C., LELLOUCH, E., MARTEN, A., CALDWELL, J., HILICO, J.-C., & ATREYA, S. K. 1995 Methane and carbon monoxide infrared emissions observed at the Canada-France-Hawaii Telescope during the collision of comet SL-9 with Jupiter. *Geophys. Res. Lett.* 22, 1573–1576.

MALLAMA, A., NELSON, P., & PARK, J. 1995 Detection of very high altitude fallout from the comet Shoemaker-Levy 9 explosions in Jupiter's atmosphere. *Geophys. Res. Lett.* 100, 16,879–16,884.

MALLARD, W. G., WESTLEY, F., HERRON, J. T., HAMPSON, R. F., & FRIZZELL, D. H. 1994 NIST Chemical Kinetics Database—Version 6.0. *NIST Standard Reference Data*, Gaithersburg, MD.

MARTEN, A., GAUTIER, D., GRIFFIN, M. J., MATTHEWS, H. E., NAYLOR, D. A., DAVIS, G. R., OWEN, T., ORTON, G., BOCKELÉE-MORVAN, D., COLOM, P., CROVISIER, J., LELLOUCH, E., DE PATER, I., ATREYA, S., STROBEL, D., HAN, B., & SANDERS, D. B. 1995 The collision of comet Shoemaker-Levy 9 with Jupiter: Detection and evolution of HCN in the stratosphere of the planet. *Geophys. Res. Lett.* 22, 1589–1592.

MATTHEWS, H. E., MARTEN, A., GRIFFIN, M. J., OWEN, T., & GAUTIER, D. 1995 JCMT observations of long-lived molecules on Jupiter in the aftermath of the comet Shoemaker-Levy 9 collision. *Bull. Amer. Astron. Soc.* 27, 1121.

MCGRATH, M. A., YELLE, R. V., NOLL, K. S., WEAVER, H. A., & SMITH, T. E. 1995 Hubble Space Telescope spectroscopic observations of the Jovian atmosphere following the SL9 impacts. *Bull. Amer. Astron. Soc.* 27, 1118.

MEADOWS, V. S., & CRISP, D. 1995 Near-infrared imaging spectroscopy of the impacts of SL9 fragments C, D, G, K, N, R, V, and W with Jupiter. *Bull. Amer. Astron. Soc.* 27, 1127.

MOLTZEN, E. K., KLABUNDE, K. J., & SENNING, A. 1988 Carbon monosulfide: A review. *Chem. Rev.* 88, 391–406.

MORENO, F., MUÑOZ, O., MOLINA, A., LÓPEZ-MORENO, J. J., ORTIZ, J. L., RODRÍGUEZ, J., LÓPEZ-JIMÉNEZ, A., GIRELA, F., LARSON, S. M., & CAMPINS, H. 1995 Physical properties of the aerosol debris generated by the impact of fragment H of comet P/Shoemaker-Levy 9 on Jupiter. *Geophys. Res. Lett.* 22, 1609–1612.

MORENO, R., MARTEN, A., LELLOUCH, E., PAUBERT, G., & WILD, W. 1995 Long-term evolution of CO and CS in the Jupiter stratosphere after the comet Shoemaker-Levy 9 collision: Millimeter observations with the IRAM-30m telescope. *Bull. Amer. Astron. Soc.* **27**, 1129.

MOSES, J. I., ALLEN, M., & GLADSTONE, G. R. 1995a Post-SL9 sulfur photochemistry on Jupiter. *Geophys. Res. Lett.* **22**, 1597–1600.

MOSES, J. I., ALLEN, M., & GLADSTONE, G. R. 1995b Nitrogen and oxygen photochemistry following SL9. *Geophys. Res. Lett.* **22**, 1601–1604.

MOSES, J. I., ALLEN, M., FEGLEY, B., JR., & GLADSTONE, G. R. 1995c Photochemical evolution of the post-SL9 Jovian stratosphere. *Bull. Amer. Astron. Soc.* **27**, 1119.

NICHOLAS, J. E., AMODIO, C. A., & BAKER, M. J. 1979 Kinetics and mechanism of the decomposition of H_2S, CH_3SH and $(CH_3)_2S$ in a radio-frequency pulse discharge. *J. Chem. Soc. Faraday Trans. 1* **75**, 1868–1875.

NOLL, K. S., MCGRATH, M. A., TRAFTON, L. M., ATREYA, S. K., CALDWELL, J. J., WEAVER, H. A., YELLE, R. V., BARNET, C., & EDGINGTON, S. 1995 HST spectroscopic observations of Jupiter after the collision of Comet Shoemaker-Levy 9. *Science* **267**, 1307–1313.

ORTIZ, J. L., MUÑOZ, O., MORENO, F., MOLINA, A., HERBST, T. M., BIRKLE, K., BÖHNHARDT, & HAMILTON, D. P. 1995 Models of the SL-9 collision-generated hazes. *Geophys. Res. Lett.* **22**, 1605–1608.

ORTON, G. & 57 CO-AUTHORS. 1995 Collision of comet Shoemaker-Levy 9 with Jupiter observed by the NASA Infrared Telescope facility. *Science* **267**, 1277–1282.

SPRAGUE, A. L., BJORAKER, G. L., HUNTEN, D. M., WITTEBORN, F. C., KOZLOWSKI, R. W. H. & WOODEN, D. H. 1995. Water brought into Jupiter's atmosphere by fragments R and W of comet SL-9 *Icarus*, in press.

TURNER, B. E. 1989. Recent progress in astrochemistry. *Space Sci. Rev.* **51**, 235–337.

WEST, R. A., KARKOSCHKA, E., FRIEDSON, A. J., SEYMOUR, M., BAINES, K. H. & HAMMEL, H. B. 1995 Impact debris particles in Jupiter's stratosphere. *Science* **267**, 1296–1301.

YANG, S. C., FREEDMAN, A., KAWASAKI, M., & BERSOHN, R. 1980 Energy distribution of the fragments produced by photodissociation of CS_2 at 193 nm. *J. Chem. Phys.* **72**, 4058–4062.

YELLE, R. V., & MCGRATH, M. A. 1995 Results from HST spectroscopy of the SL9 impact sites. *Bull. Amer. Astron. Soc.* **27**, 1118.

YELLE, R. V., & MCGRATH, M. A. 1996 Ultraviolet spectroscopy of the SL9 impact sites. I. The 175–230 nm region. *Icarus* **119**, 90–111.

ZAHNLE, K. 1995 Dynamics and chemistry of SL9 plumes. *This volume.*

ZAHNLE, K., MAC LOW, M.-M., LODDERS, K., & FEGLEY, B., JR. 1995 Sulfur chemistry in the wake of comet Shoemaker-Levy 9. *Geophys. Res. Lett.* **22**, 1593–1596.

Particles in Jupiter's atmosphere from the impacts of Comet P/Shoemaker-Levy 9

By ROBERT A. WEST

Jet Propulsion Lab, California Institute of Technology, 4800 Oak Grove Drive, Pasadena, CA 91109, U.S.A.

The dark clouds that were easily seen in small telescopes after the comet impacts were caused by small particles which were deposited in Jupiter's stratosphere. Observations from the Hubble Space Telescope and from ground-based instruments at visible and infrared wavelengths indicate that the mean radius of the particles is in the range 0.1 to 0.3 μm, and the total volume of particles is approximately the same as that for a 1-km diameter sphere. In the dark core regions of freshly-formed impacts, the particles are distributed over a large vertical extent, between about 1 mb and 200 mb or deeper. The diffuse outlying haze is confined to the high-altitude end of the range. Such a distribution probably reflects different methods of emplacement of the debris as a function of distance from the impact. The color of the particles, and their volatility as required to make waves visible, suggest an organic material as the main constituent. These relatively volatile materials are thought to have condensed onto more refractory grains after the plume material cooled, some 30 minutes or more after impact. The most refractory materials expected to condense from an evolving fireball are Al_2O_3, magnesium and iron silicates, and soot, depending on the C/O ratio. A silicate spectral feature was observed, confirming that cometary material was incorporated into the grains, although silicate grains make up only 10–20% of the particle volume. After one year in Jupiter's stratosphere, the particles have spread some 20° in latitude and a significant number have sedimented into the troposphere where they are no longer visible.

1. Introduction

Impacts from fragments of comet Shoemaker-Levy 9 left visible marks on the jovian atmosphere, at wavelengths ranging from the ultraviolet to the near-infrared. The dark features (at visible wavelengths) were easily seen even in small telescopes, and were described by some observers as the most prominent markings ever seen on Jupiter. Clouds of small dust particles are largely responsible for these features.

Knowledge about the dust particles is of value for several reasons. They may give clues concerning the physics and chemistry of a high velocity impact into a deep atmosphere and subsequent fireball and plume eruption, with implications for the Earth. They constitute a considerable mass of observable material that may bear on questions surrounding the composition and chemistry of the impacting bodies and/or the chemistry of the jovian atmosphere at levels where we have no other information. Over longer time scales they serve as excellent tracers of wave motion and tracers of atmospheric mass motion in Jupiter's stratosphere where previously there were none. In this way they are analogous to sulfuric acid particles deposited in the Earth's stratosphere from volcanic eruptions, and their evolution during the first year after impact is in many ways similar to the evolution of terrestrial stratospheric particles.

This chapter will review the observations of the impact debris particles and address questions concerning their optical, physical and microphysical properties, composition, origin (whether from the comet or from Jupiter), and evolution.

2. Observations of impact debris particles

Data relevant to aerosol studies fall into four main categories. These are (1) reflectivity and emissivity at ultraviolet (UV), visible, and infrared (IR) wavelengths, (2) observations of gas composition in the impact regions, (3) the visibility of waves, and (4) measurements made as a Galilean satellite moves into or out of eclipse (Mallama *et al.* 1995). The first of these categories contributes most directly toward studies of the debris particles. West *et al.* (1995) analyzed HST data at UV, visible and in the 889-nm methane and nearby continuum filters for a variety of spots both near the time of impact, and on a global scale up to a month after the last impact. Moreno *et al.* (1995) studied ground-based CCD images between 360 and 948 nm of impact H obtained at La Palma Observatory. Ortiz *et al.* (1995) reported on observations of images and spectra in the K and H bands (1.5–2.4 μm) made at the Calar Alto Observatory. Banfield *et al.* (1995) obtained images in the spectral range 2.0–2.3 μm taken on the 5m Hale telescope at Palomar mountain. Orton *et al.* (1995a; 1995b) obtained images at several wavelengths in the same spectral region, and in the very strong 3.4-μm methane band as well from the NASA Infrared Telescope Facility. Nicholson *et al.* (1995) provided spectra of the emerging plume from the R impact in the thermal IR (8 μm to 13 μm) observed from Mt. Palomar. Rosenqvist *et al.* (1995) obtained spectra near 4 μm of the area around the W and K impact sites with the 6 m telescope of the Special Astronomical Observatory of Zelenchuk.

3. Particle optical properties, optical depth, mean radius, and total volume

At wavelengths outside the strong methane absorption bands, Particles appear as dark markings over a very broad spectral range, from the ultraviolet to 1 μm in the near-infrared. Signatures of particles are seen in Hubble Space Telescope (hereafter HST) UV images (Clarke *et al.* 1995; West *et al.* 1995) and spectra at wavelengths as short as 125 nm (Noll *et al.* 1995). Figure 1 shows the appearance of many fresh impact sites on 21 July, 1994 in the HST Wide Field/Planetary Camera 2 (hereafter WFPC2) imaged with the F255W filter which has a solar flux-weighted effective wavelength of 275 nm.

The darkest region of the planet in Fig. 1 is the cloud from the H impact near the dawn terminator. It was immediately obvious from these images that the particles reside at high altitudes in Jupiter's stratosphere, otherwise scattering from the overlying gas particles would make the limb brighter than it is. At the nadir 275-nm photons penetrate to 0.35 bar by the time scattering by H_2 and He attenuates the beam by the factor $1/e$. Near the limb where the slant path is large, the pressure level sampled by 275-nm photons is just a few mb.

The debris particles have a flat or slightly brown spectrum at visible wavelengths, out to 1 μm. Figure 2 is a half-tone version of a color composite made from HST WFPC2 images in the blue (F410M, effective wavelength 410 nm), green (F555W, 555 nm), and near-IR (F953N, 955 nm). The image was taken almost 2 hours after the impact of fragment G. The impact from fragment D is also seen in Fig. 2. Several features are to be noted. The main features are (1) dark core regions, (2) circular rings, and (3) a crescent-shaped region at larger distance from the core. A key observation, evident in Fig. 2, is that *the color of the particles in all of these regions is the same.*

The lowest reflectivities occur at the center of the impact. These core regions have a central circular shape, about 2000 km in diameter for G, with a wedge of material fanning out to the southeast.

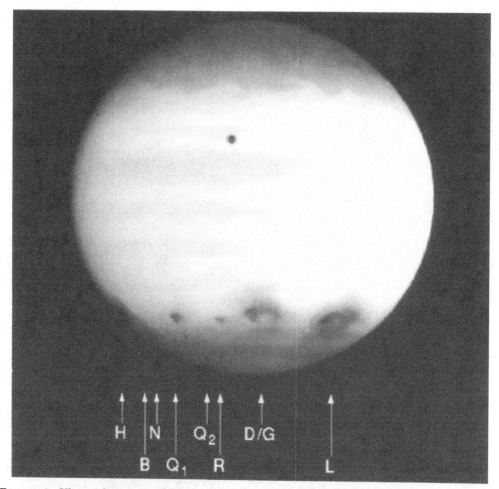

FIGURE 1. Ultraviolet image of Jupiter taken by the Wide Field Camera 2 on NASA's Hubble Space Telescope. The image shows Jupiter's atmosphere at a wavelength of 275 nm after many impacts by fragments of comet Shoemaker-Levy 9. The most recent impactor is fragment R. This photo was taken on 21 July 1994, about 2.5 hours after the R impact. A large, dark patch from the impact of fragment H is visible rising on the morning (left) side. Jupiter's moon Io is the dark spot just above the center of the planet.

A thin circular ring in Fig. 2 is centered on the core region and is observed to expand radially at a rate of about 450 km s^{-1} (Hammel *et al.* 1995; Ingersoll and Kanamori, 1995). Similar rings were seen in HST images of impacts A, E, Q1 and R. These features are also visible as bright rings in the 890-nm methane-band images, proving that they are made visible by the presence of particles in Jupiter's stratosphere. Inside of this ring is a fainter one whose expansion velocity is roughly half that of the more visible ring (Hammel *et al.* 1995, Ingersoll and Kanamori, 1995). Ingersoll and Kanamori argued that the wave speed is determined by the lapse rate in the water cloud region between 6 and 10 bars pressure, but temperature perturbations in the stratosphere are responsible for particle formation during the passage of a wave (see the chapter by Ingersoll and Kanamori in this volume). This mechanism for particle formation provides a constraint on the vapor pressure of the condensing material and hence its composition.

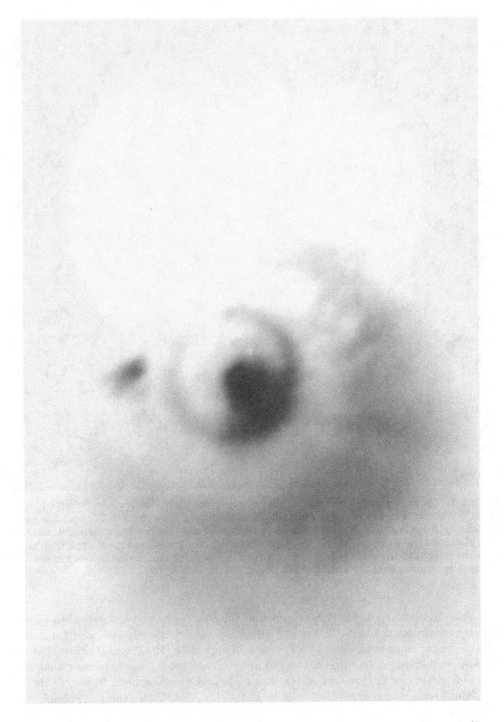

FIGURE 2. The D and G impact sites as they appeared almost 2 hours after the G impact (from West *et al.* 1995). The diameter of the wave (circular ring centered on the G impact) is about 7000 km. A color version of this figure appears in West *et al.* (1995).

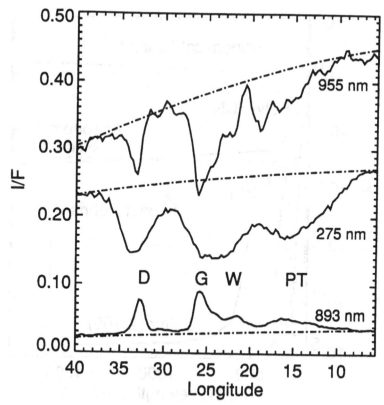

FIGURE 3. The ratio of intensity, I, observed by the WFPC2 to incident solar flux (πF) is shown for a cut across the cores of the D and G impact sites shown in Fig. 2, along a line of constant latitude at three wavelengths (from West *et al.* 1995). The dot-dash lines indicate the level of the longitudinal-average brightness of the undisturbed adjacent regions determined by fitting adjacent regions to a generalized Minnaert law: $\ln(\mu I/F) = C_0 + C_1 x + C_1 x^2 + C_1 x^3$, where $x = \ln(\mu \mu_0)$ and μ and μ_0 are the cosines of the emission angle and solar incidence angle, respectively. The C coefficients were obtained by a linear least-squares singular value decomposion algorithm. The dark cores of the D and G sites are indicated, along with the location of the wave (W) and plume terminus (PT) or optically thin crescent (from West *et al.* 1995).

The crescent of dark material lying at greater distance from the core is optically thin at visible wavelengths. Contrast from features in the troposphere can be seen through the haze. The location of the haze is consistent with the idea that particles in the outlying regions were initially ejected at large velocity from an erupting plume seen on the limb in HST and some high-resolution ground-based images (see Hammel *et al.* 1995 and the chapter by Hammel in this volume).

The magnitude of the contrast can be judged quantitatively from Fig. 3. The features are more diffuse at 275 nm compared to longer wavelength both because the optically thin regions are optically thicker at UV wavelengths, and because there is some image smear in the UV due to the long exposure time. In addition, some of the absorption in the F255W filter is due to S_2 gas (Noll *et al.* 1995). The 893 nm filter is centered on a methane band where the ambient atmosphere value for I/F is close to 0.015 at the geometry of the data plotted in Fig. 3. The much higher I/F in the cores of the spots is due to significant aerosol optical depth in the stratosphere. The particles are darker than the ambient atmosphere in the nearby 955 nm continuum band where methane

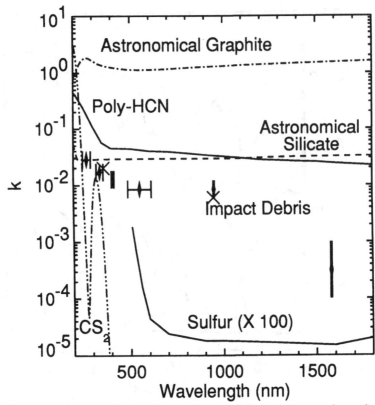

FIGURE 4. The imaginary part of the refractive index (k) derived at several wavelengths below 1 μm from HST images (West *et al.* 1995, diamonds with error bars) and ground-based images (Moreno *et al.* 1995, $\times$ symbols), and at 1.58 μm by Ortiz *et al.* 1995. The results derived by West *et al.* include bars showing the HST WFPC2 filter passbands and uncertainty in k. Also shown are values for several candidate materials (from Gustavsen, 1989; Sasson *et al.* 1985; Draine, 1985; Khare *et al.* 1994).

absorption is small. Moreno *et al.* (1995) found results from ground-based CCD images similar to those reported by West *et al.* (1995).

At longer wavelengths the contrast at continuum wavelengths is much reduced. Ortiz *et al.* (1995) reported difficulty detecting the spots in the 1.58 μm continuum band, indicating that the optical depth is smaller or the imaginary refractive index is smaller than at shorter wavelengths, or a combination of the two.

West *et al.* (1995) studied the darkest core regions of several of the fresh impact sites where the large optical depth greatly reduced contributions from the underlying atmosphere to the observed intensities. Multiple scattering models for the reflectivity and optical depth of the particles in these regions provided estimates of the imaginary part of the refractive index of the particles as well as their mean radius. Ortiz *et al.* (1995) followed a similar procedure for their near-infrared spectra of the G/D complex and the H impact site. Results are shown in Figure 4.

Mie theory for spheres was used in both studies to relate refractive index to particle optical properties. Since no observations exist which could differentiate between spherical and nonspherical particles we do not know if Mie theory is appropriate. It is quite possible that the particles are clusters of smaller monomers, as proposed for the jovian polar stratospheric aerosols (West and Smith, 1991). If that is the case the value of k derived using Mie theory, as well as the particle size, could be in error. West (1991) compared optical properties of small aggregate particles to spheres.

Even if the assumption of spherical shape is correct, the particle size distribution and the real part of the refractive index contribute to the optical properties and to the uncertainty in the derived value of k. There is not enough information to determine the particle size distribution. The first moment of the size distribution (the particle mean radius) can be estimated from the wavelength dependence of the optical depth. The most thorough studies to date are those by West *et al.* Moreno *et al.* (1995), and Ortiz *et al.* (1995), covering the spectral range from 275 nm to 1.58 μm. Rosenqvist *et al.* (1985) derived average optical depth at 3.9 μm for a 3-arcsec region centered on the K and W impact sites. They found optical depths in the range 0.07–0.15 during the first few days after the impacts.

An estimation of particle mean radius requires knowledge or an assumption of the real part of the refractive index. West *et al.* (1995) assumed the real part of the refractive index to be 1.4–1.44, consistent with many organic and some inorganic (NH_3 ice) compounds. West *et al.* were able to fit the observed I/F provided particle mean radii were in the range 0.15 to 0.3 μm, with the imaginary part of the refractive index also uncertain by a factor of 2 as shown in Fig. 4. The uncertainty in k is larger at 1.58 μm because both the optical depth and the contrast decrease at the longer wavelength. Ortiz *et al.* (1995) and Moreno *et al.* (1995) assumed the real part of the refractive index to be 1.7 and found a value of 0.15 μm for the particle radius, on the low end of the range determined by West *et al.* Mie calculations are sensitive to the product $(n-1)a/\lambda$ where n is the real part of the refractive index, a is the particle radius, and λ is the wavelength. The results for particle mean radius derived by Ortiz *et al.* and Moreno *et al.* are near the middle of the range derived by West *et al.* when scaled by the factor $(n-1)$. Although the optically thick core regions were used to determine refractive index, finite optical depth and uncertainty in the vertical location of the haze particles also contribute to uncertainty in optical depth, especially in the infrared. Additional modeling is needed to better define the values and their uncertainties.

Aerosol optical depth generated by the impacts depends strongly on location, time and wavelength. Near the optically thick core regions of a fresh impact, the contrast between the absorbing material and the ambient cloud reflectivity can change by more than a factor of 2 over a distance of 1000 km. The best ground-based images (atmospheric blur disk $\sim$ 0.4 arcsec) can resolve features as small as 1500 km. The HST Planetary Camera-2 was able to resolve (at the Nyquist frequency) features as small as 342 km. West *et al.* (1995) and Moreno *et al.* (1995) found 889-nm optical depths near 4 for the core regions of several fresh impacts. Optical depth at other wavelengths can be estimated from Fig. 5 which shows the wavelength dependence of the extinction cross section for size distributions of spheres. The ratio of optical depth between 950 nm and 275 nm changes by a factor of 3 when the particle radius changes only by a factor of 1.4 from 0.21 to 0.28 μm.

West *et al.* (1995) generated maps of the optical depth of the particles at two wavelengths (275 and 889 nm), shown in Fig. 6. The largest optical depths are greater than 12 at 275 nm one day after the last of the impacts. During the following week and month the material diffused horizontally due to the motion of vortices and the zonal wind shear. The role of small vortices is especially apparent on 23 July. The arcs of material extending to the north and south of impacts A and E near longitudes 180° and 150°, respectively appear to be following the flow pattern of nearby vortices (see the chapter by Beebe).

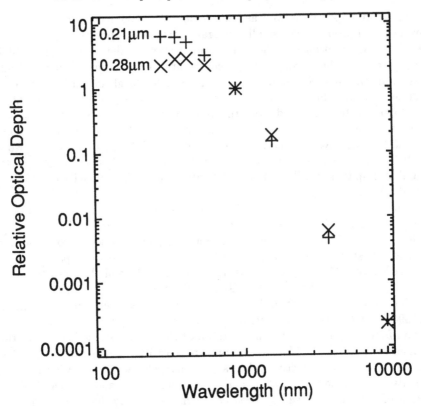

FIGURE 5. Wavelength dependence of the optical depth for particles whose mean radius is 0.21 and 0.28 μm, as indicated. Both curves are normalized to 950 nm and both were calculated with Mie theory for a real refractive index of 1.44 and for an imaginary part that varies as shown in Fig. 4 for the impact debris. The imaginary refractive index at 3.9 and 10 μm has the same value (3×10^{-4}) as that used for the point at 1.58 μm. The particle emissivity at the longer wavelengths is proportional to the imaginary part of the polarizibility of the material which is related to the imaginary refractive index. Values for the optical constants of the debris particles at wavelengths longer 1.58 μm are not known.

West *et al.* integrated the particle optical depth over 360 degrees of longitude and 17.7 degrees of latitude corresponding to the latitude bands in Fig. 6. The total area of that region is 6.48×10^9 km^2. The total integrated optical depth on 23 July, 30 July, and 24 August was 7.0, 4.6, and 3.0 (all 10^9km^2) at 275 nm. The optical depth at 893 nm on the same dates was 1.1, 1.5, and 1.5 $\times 10^9$km^2. The ratio of optical depth at the two wavelengths on the same three dates is 6.4, 3.1, and 2.0. As seen in Fig. 5, the ratio on 23 July implies the particle mean radius was 0.21 μm, while on 30 July and August it was 0.24 μm and 0.28 μm. West *et al.* argued that coagulation of particles could account for an increase in particle mean radius during the month following the impacts.

The total volume of the debris particles can be calculated from knowledge of the particle mean radius, refractive index, and optical depth. Using the values of refractive index assumed (n = 1.44) and derived (Fig. 4), West *et al.* (1995) calculated optical cross sections of particles with the mean radii mentioned above. The total number of particles is given (assuming all particles have the mean radius) by the total optical depth divided by the extinction cross section of an individual particle. The total volume is then the total number of particles times the volume of each particle. West *et al.* found total

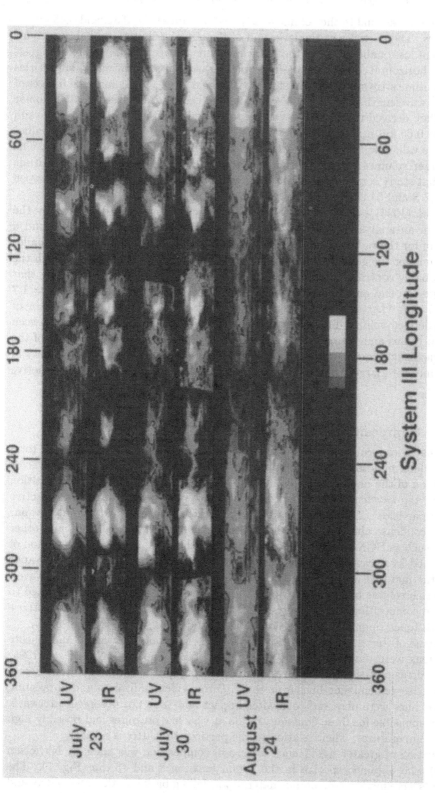

FIGURE 6. Optical depths of debris material are shown for wavelengths 275 nm and 893 nm as functions of latitude (between 36.5° S and 54.2° S) and longitude for 23 July, 30 July, and 24 August (from West et al. 1995). Each strip has latitude 36.5° S at the top and latitude 54.2° S at the bottom. The intensity steps at the bottom calibrate the colors in terms of the optical depth. The brightest intensities correspond to optical depth 12 or higher, and the lowest correspond to optical depth 0.2 (adapted from a color figure by West et al. 1995).

particle volume to be equal to the volume of a sphere of radius 0.51, 0.52, and 0.45 km on the three dates. The particles at the large end of the size distribution probably account for the apparent loss of about 30% of the particle volume between 23 July and 24 August, augmented by horizontal eddy diffusion (from vortices as discussed previously) of particles outside the latitude strip studied by West *et al.* Banfield *et al.* (1995) analyzed images of Jupiter in the wavelength range 2.0–2.35 μm taken on the 5m Hale telescope at Palomar mountain. They derived total particle volume to be equal to that of spheres with radii 0.67, 0.60 and 0.65 km in late July, mid-August, and late August of 1995. Those values are close to the values derived by West *et al.* given the uncertainties in both studies. The somewhat larger volumes derived by Banfield *et al.* may be due to the greater latitude coverage of that study compared to West *et al.* who retrieved optical depth only between latitudes 36.5° S and 54.2° S.

Moreno *et al.* (1995) estimated the volume of particulate material deposited by the H impact from an analysis similar to that done by West *et al.* (1995). They found a value of 115 m for the radius of an equal-volume sphere. That value is 80 times smaller than the volume for the globally integrated debris found by West *et al.*, even though the H impact probably contributed more than 5% of the total optical depth. Recalling that the Mie results depend on the product (n - 1)×a, and that Moreno *et al.* used n = 1.7 and a = 0.15 μm, whereas West *et al.* used n = 1.44 and a = 0.25 μm, the factor of roughly 4 in derived volume difference can be attributed to the different trades between particle mean radius and the real part of the refractive index. Since the real part of the refractive index is not known, its uncertainty leads to a factor of 4 or more uncertainty in particle volume, with larger volumes derived under assumption of smaller real refractive index.

4. Particle Composition

The solid particles formed by the impacts constitute a substantial fraction of the total debris, and a determination of their composition would be valuable, with implications for the composition of the impacting body and Jupiter's atmosphere. Clues to composition come from (1) the wavelength dependence and absolute value of the imaginary refractive index of the particles, (2) observations of gas-phase constituents such as water vapor which would condense after the initial hot phase of the fireball and fallback, and other constituents such as HCN and CS_2 which coexist with the solid, and (3) the detection of particles formed by waves which constrains the combination of mixing ratio and vapor pressure of the material in the phase transition region (the cooling phase of the wave). Chapters in this volume by Lellouch, Moses, and Ingersoll and Kanamori are devoted to the gas-phase composition and the physics of the impact-generated waves. Only those aspects which touch upon the composition of the particulates will be treated here.

At least some of the material is of cometary origin. Emission from metal and silicate atoms and ions were observed shortly after some of the impacts. Noll *et al.* (1995) reported on ultraviolet spectral features of iron, silicon, and magnesium. Roos-Serote *et al.* (1995) observed emission lines from sodium, iron, calcium, lithium and potassium. The emission lines were observed soon after impact and were not observed afterward. The atoms responsible for these emissions reside at very low pressures and possibly high (~ 1000 K) temperatures. Their relation to the particulate matter is unclear.

An observation of greater significance to aerosol composition was made by Nicholson *et al.* (1995) who obtained spectra in the region between 8 and 13 μm (Fig. 7). The measurements were made as the plume material was falling back into the atmosphere and the thermal energy generated from the process heated the gas and particles. Nicholson

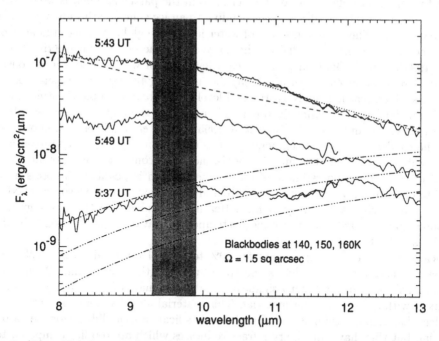

FIGURE 7. (From Nicholson *et al.* 1995) Spectra (flux) observed during the R impact on 21 July, 1995 from Palomar Observatory. The first spectrum (5:37 UT) was obtained prior to significant brightening at 10 μm. The second (5:43 UT) was obtained just prior to peak brightness, and the third (5:49 UT) during the decay phase. The dashed curve shows flux calculated for a 2000 K blackbody. The dotted curve which tracks the data at 5:43 UT is a synthetic spectrum using optical constants for "astronomical silicate" (Draine, 1985). Dot-dash curves show blackbody fluxes corresponding 140, 150, and 160 K. The stippled region between 9 and 10 μm was obscured by telluric opacity.

et al. assumed emission from methane gas could be modeled by a power-law function of wavelength. They used the data to find the best value of the exponent (-7.5), along with a temperature (370 K) and the product of the effective solid angle of the emitting dust and its optical depth at 9.7 μm. The fit to the spectrum is shown in Fig. 7.

Nicholson *et al.* derived a value of 6×10^{12} g for the total mass of silicate dust generated by the R impact. According to Banfield *et al.* (1995), the R impact contributed only 2% of the total number of particles to the globally integrated debris. Therefore the total mass of silicate material could be 3×10^{14} g, and its volume (assuming a density of 3.3 g cm^{-3}) equal to that for a sphere of radius 0.28 km, some 17% of the total particulate volume derived by West *et al.* (1995), and 10% of the total volume derived by Banfield *et al.* Such a large proportion of silicate material would be consistent with the idea that a large fraction, perhaps most, of the ejected material was of cometary origin.

Water, either from the impacting body or from Jupiter, was expected to contribute to the debris inventory, and water vapor lines were observed by ground-based observers (Meadows and Crisp, 1995) and observers on the Kuiper Airborne Observatory (Bjoraker *et al.* 1995). Water lines were observed only when the plume was visible and when temperatures were warm (1000 K or more according to Bjoraker *et al.*). Bjoraker *et al.* report masses of water for the G and K impacts to be each in the range 1.4 to 2.8×10^{12} g, corresponding to ice spheres with radii 0.07 and 0.09 km. According to Banfield *et al.* (1995) the G and K events contributed about half of the globally integrated optical depth produced by debris particles. The total water mass in the particulate debris would then be in the range 5.6 to 11.2×10^{12} g, corresponding to an ice sphere with radius in the range 0.11 to 0.14 km. The volume fraction of water ice for the globally integrated aerosol debris is then in the range 1% to 2% for the range of total aerosol volumes derived by West *et al.* (1995), and by Banfield *et al.* Such a low ratio of water to silicate is consistent with the proposal by Zahnle *et al.* (1995) the impacting bodies were water-poor comet fragments which reached the NH_4SH cloud level near 2 bars in Jupiter's atmosphere.

Spectral signatures of ammonia vapor, enhanced over the impact sites, were observed in the ultraviolet and infrared (Noll *et al.* 1995; Orton *et al.* 1995a). Ammonia would favor the condensed phase at temperatures less than about 140 K, in the pressure range between about 10 mb and 600 mb. Ammonia should be considered as a potential aerosol constituent, although no estimate of its volume fraction is possible. Aerosols near the 140 K temperature level (where P = 12 mb) and deeper may provide a source of NH_3 gas in the stratosphere lasting many months. Heating by sunlight could produce grain temperatures that are somewhat above the gas temperature at the same location, stimulating NH_3 ice sublimation.

Silicate and water ice can account for 10% to 20% of the total particle volume, but neither can account for the debris spectra at wavelengths shorter than 2 μm. Water and ammonia ice do not absorb significantly at wavelengths less than 1 μm and must be mixed with some absorbing material. This material absorbs more efficiently at short wavelengths, as indicated in Fig. 4. Graphite and silicate are candidates for the absorbing material, but they have imaginary refractive indices which are too flat compared to the derived values. Carbon disulfide was observed in UV spectra (Noll *et al.* 1995; Atreya *et al.* 1995) and appeared to be an attractive candidate for producing the particles in the waves, but it lacks significant absorption at most wavelengths. Elemental sulfur was also observed in UV spectra (Noll *et al.* 1995). Moses *et al.* (1995) and Zahnle *et al.* (1995) showed that most of the gas-phase sulfur would be converted to S_8 which forms a solid. It would therefore be expected to contribute to the aerosol mass. But it too lacks significant absorption at red wavelengths.

The interpretation of the debris spectra may not be as simple as deriving a value for the imaginary part of the refractive index as West *et al.* (1995) and others have done. If the particles are a heterogeneous mix of two or more components, the use of Mie theory for a homogeneous sphere may not be appropriate. It is likely that very small (a < 0.1 μm) graphite and silicate grains condensed early in the cooling phase of the plume and served as condensation nuclei for water ice, sulfur and organics (Field *et al.* 1995; Field, 1995; Friedson, 1995). Graphite has an absorption maximum near 250 nm. It is conceivable that a mixture of very small graphite and silicate grains with larger water ice grains could produce the observed reflectivity, although a quantitative test of that hypothesis has not been done.

The role of carbon in the impact process raises many questions. If all the carbon in the impacting body and in the jovian atmosphere which became entrained in the fireball were processed through a high temperature phase of several thousand K, and if oxygen from the comet or Jupiter were sufficiently abundant, all of the carbon would end up as CO which is stable as a gas in the atmosphere. CO was observed, in significant abundance (see the chapter by Lellouch), but observations of CS and CS_2, and HCN indicate that carbon in the form of organic material may be abundant. There is also a rationale for expecting carbon soot to be one of the refractory materials that condenses first in the cooling fireball (Friedson, 1995). The ratio of carbon to oxygen is a key parameter.

The heterogeneous nature of the complex process of fireball formation and evolution, fallback, entrainment and mixing of the comet material and jovian gas, and the chemistry of the shock and radiation environments near the eruption leave open the possibility of formation of a variety of constituents under a wide variety and rapidly evolving thermodynamic and chemical states. Field (1995) recently pointed out that refractory grains which form first as the plume cools could catalyze the formation of organics from CO reacting with H_2 through a Fisher-Tropsch process (Anders, 1971), and that water might be bound to silicates as water of crystallization as proposed by Anders for the cooling solar nebula. The relatively low temperature organic condensates and ices (water and possibly ammonia) would condense on previously formed refractory silicate and graphite nuclei and form grain mantles which contribute most to the particle volume.

Wickramasinghe and Wallis (1994) anticipated that submicron particles of organic material would be produced by the impacts. West *et al.* (1995) noted that the shape of the derived refractive index spectrum was similar to that measured for polymerized hydrogen cyanide (poly-HCN). Wilson and Sagan (1995) showed that the imaginary refractive indices of an organic residue from the Murchison carbonaceous chondrite provide an excellent match to the derived spectrum at wavelengths less than 1 μm. Noting that several organic gas-phase molecules were observed, including CS, CS_2 (Noll *et al.* 1995), and HCN (Marten *et al.* 1995), and that some organics are sufficiently volatile to exist in the gas phase as needed to produce particles in waves at temperatures near 150 K, West *et al.* proposed that an organic material rich in S and N is responsible for the color of the particles. A cornerstone of the argument is that the particles produced by the waves are observed to have the same color as the more permanent debris particles (Fig. 2), suggesting that the same material is responsible for both. The only way that could happen is to have the material which produces the color to favor the solid phase at temperatures where the long-lived particles reside (in the vicinity of 1 mb or deeper), while the particles in the waves are at a slightly higher altitude where the ambient temperature is higher and the thermodynamics favor the gas phase before and after the passage of the wave. While simple organics like CS_2 are sufficiently volatile to condense near 150 K, more complex organics are usually less volatile. Poly-HCN seems unlikely now that Ortiz *et al.* (1995) derived a low value for the imaginary refractive index at 1.58 μm (see Fig. 4), while k for poly-HCN remains high at that wavelength. A candidate or mixture with the right thermodynamic and spectral properties remains to be identified.

5. Particle Vertical Structure

The impact debris particles appear as bright features in strong methane absorption band images and as dark features in ultraviolet images (Fig. 1). Those contrasts are produced because the particles are higher in the atmosphere than the background aerosol, reflecting photons that would otherwise be absorbed in the methane bands, and absorbing photons that would otherwise be backscattered at ultraviolet wavelengths by H_2 and He molecules. The latitude of the impacts (near 43° S) was fortuitous for their visibility because Jupiter's intrinsic reflectivity in the strong methane bands is lowest at that latitude. The deeper tropospheric ammonia clouds which occupy pressure levels between about 350 and 600 mb at 43° S normally account for almost all of the reflected light at that latitude in the strong methane bands. Further to the south, beginning near 60° S, Jupiter's polar stratospheric haze darkens the planet in the UV and brightens it in the strong methane bands for the same reason the impact particles do. Except for their location closer to the equator, the impact particles are hard to distinguish from the polar haze particles, indicating similar altitudes and optical properties. West *et al.* (1992) summarized observations regarding the polar haze and argued that they extend upwards to pressure levels near 1 mb.

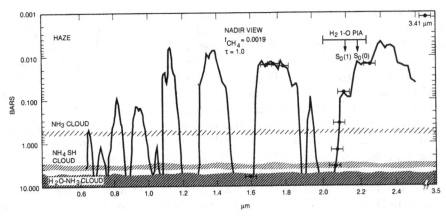

FIGURE 8. The pressure level coresponding to unit optical depth from methane and hydrogen pressure-induced absorption (PIA) are shown at visible and near-IR wavelengths, along with locations of cloud layers in Jupiter's upper troposphere and stratosphere (adapted from Baines *et al.* 1993). Points with error bars mark locations of filters used by K. H. Baines and others to observe Jupiter at the NASA Infrared Telescope Facility.

One measure of the vertical location in a plane-parallel atmosphere sampled by photons at different wavelengths is the pressure level where unit optical depth is reached for light directed toward the nadir. At that location a fraction $1/e$ of the upward-directed photons will emerge without being scattered or absorbed. When the solar incidence and viewing directions are away from the zenith and nadir, the equivalent depth of penetration is smaller by the factor $0.5(1/\mu + 1/\mu_0)$ where $\mu = \cos(\theta)$, $\mu_0 = \cos(\theta_0)$, and θ and θ_0 are the angles between the surface normal and the view and sun directions, respectively. Unit optical depth for scattering by molecular H_2 and He in HST images taken with the F255W filter with effective wavelength 275 nm (Fig. 1) occurs at 0.35 bar. At other UV wavelengths unit optical depth from gas molecules scales approximately as λ^{-4}. The wavelength dependence of methane and hydrogen gas determines the level of photon penetration in the near-infrared. Fig. 8 shows that dependence.

Methane absorption coefficients range over more than three orders of magnitude. Photons in the strong band at 3.4 μm sample the 1-mb level of the stratosphere. At that wavelength the contrast between belts and zones, and even the Great Red Spot, which is an elevated tropospheric cloud, disappears. Figures 9–11 show Jupiter at selected wavelengths which sample a range of methane absorption coefficients. The appearance of the planet after the impacts (Figs. 10 and 11) can be compared with that before the impacts (Fig. 9).

Images at 3.41 μm sample only altitudes higher than the 10-mb pressure level, and unit optical depth for vertical viewing occurs near the 1-mb level. Prior to the impacts only reflection from the high-latitude haze, and emission from the H_3^+ ion, also at polar latitudes, was seen. Shortly after the impacts reflection from haze particles near latitude 45° S is prominent (Fig. 10), proving that impact particles reside at altitudes at least as high as the 10-mb level, and perhaps much higher. Orton *et al.* (1995b) showed that only a very faint trace of particle scattering could be seen near latitude 45° S by 11 October, 1994, and none could be seen in a 3.4-μm image taken on 1 February, 1995. From these observations it appears that the e-folding time for particle sedimentation near the 1-mb level is 4–6 weeks, corresponding to the sedimentation time calculated for sphere of radius 0.07–0.1 μm and density 2 g cm^{-3} (West *et al.* 1995).

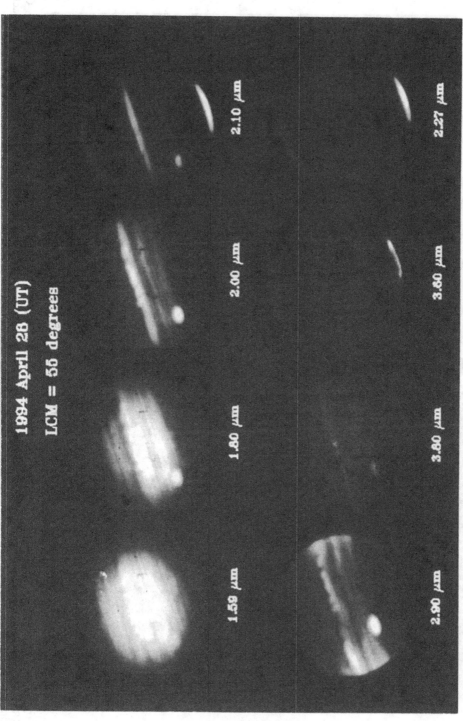

FIGURE 9. Images of Jupiter in the near-infrared methane bands observed with the NASA IRTF telescope (courtesy K. H. Baines and G. Orton. These images were taken three months prior to the comet impacts. The Great Red Spot is the bright (between wavelengths 1.8 and 2.9 μm) oval about half way between the central meridian and the morning limb.

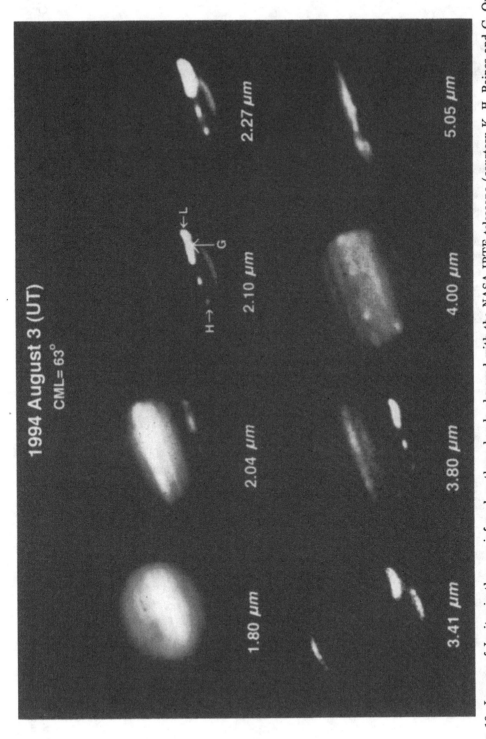

FIGURE 10. Images of Jupiter in the near-infrared methane bands observed with the NASA IRTF telescope (courtesy K. H. Baines and G. Orton, and the IRTF SL9 observing team—see Orton *et al.* 1995a). These images were taken about 10 days after the last impact. Several of the impact sites are identified on the 2.10-μm image.

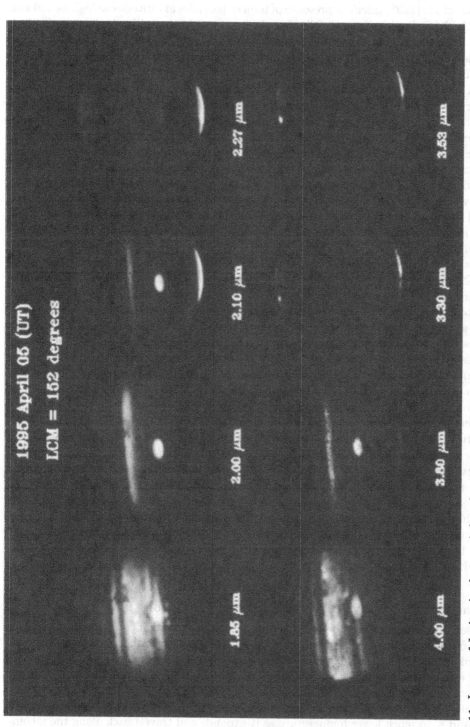

FIGURE 11. Images of Jupiter in the near-infrared methane bands observed with the NASA IRTF telescope (courtesy G. Orton and K. H. Baines). These images were taken almost 9 months after the impacts and show the remnants of the impact-generated aerosols at 2.1 and 2.27 μm.

Evidence for particles at higher altitudes comes from measurements of satellite eclipses. Mallama *et al.* (1995) infer the presence of impact particles at altitudes as high as 300 km above the 1-bar level one month following the impacts. At that altitude the pressure is a few μb. The vertical optical depth of these particles is only 0.1 at 540 nm wavelength and much less at 2–4 μm, assuming the extinction coefficient of 0.001 derived by Mallama *et al.* for the top of the haze layer applies to a 100-km thick layer. If the particles were formed during the emplacement of the fallback material they would have to be much smaller than 0.1 μm radius, since particles of radius 0.1 μm sediment out in a matter of hours at such a low pressure. Alternatively, gas-phase material at a few μb pressure may be a source of solid particles one month after the impact. Moses *et al.* (1995) and Zahnle *et al.* (1995) pointed out that S_2 gas will be converted to solid S_8 although the time scale for that process should be much shorter than one month.

Banfield *et al.* (1995) took advantage of the the broad range in depth probed by different wavelengths in the 2-μm methane spectrum to derive haze vertical structure and optical depth in late July, mid-August, and late August of 1994. Their inversion algorithm always selected the highest altitude and lowest pressure (20 mb or less) as the location of the haze.

West *et al.* (1995) examined the darkest cores of several fresh impact sites using WFPC2 images at UV (275 nm) and in the 893-nm methane band. They found that the particles need to extend up to pressures of a few mb or less to produce the low reflected UV intensity at the limb. But material confined to pressures less than about 100 mb produced a center-to-limb behavior in the methane band that was not consistent with observations. The observations required that a significant optical depth of particles extending to pressure levels as high as 200 mb. Baines *et al.* (1995) and Orton *et al.* (1995a) found that such a model was consistent with reflectivity in the near-infrared methane bands. Baines *et al.* derived particle column number density of $3 \times 10^8 \mathrm{cm}^{-2}$ for core of the G site on 3-August-1994. Models with aerosols extending down to 400 mb constructed by both West *et al.* and Moreno *et al.* (1995) were able to fit HST and ground-based data.

Vertical structure models point to two types or regions formed from the impacts. In dense cores the particles extend over many scale heights, from 200 mb pressure or deeper to a few mb pressure or less. In the diffuse outlying regions the particles cover a much larger areal extent but their optical depth is smaller than in the cores and they are confined to altitudes higher than the 20-mb level. This picture is consistent with the idea that particles in the outlying regions were emplaced from above by fallback from the plume, but that particles in the core region were emplaced both from above and below by plume fallback and by turbulent entrainment and buoyancy of the hot rising fireball.

6. Particle Microphysical Processes and Evolution

Observations of plumes shortly after impact, and observations of ultraviolet, visible, and infrared emissions by hot plume material (see the chapters in this volume by Hammel, Chapman, and Nicholson) have confirmed and enlarged upon many aspects of models of the events surrounding the impacts (see chapters by Crawford, MacLow and Zahnle). The main features in common to all the models of the largest impacts include an incoming bolide which creates a shock wave and an eruptive plume which cools from several thousand K to much lower temperature as it expands and travels back along the incoming path to altitudes near 3000 km, followed by fallback into the upper atmosphere and heating upon re-entry. The process is heterogeneous, with temperature, pressure and composition all functions of location and time. Models differ in terms of the depth of penetration attained, and the fractions of the impacting body and Jupiter gas which was ejected in the plume, and to what distances from the impact site various parcels came to rest. How did grains form during this process, and what may have been their size distribution and composition as functions of time?

Field *et al.* (1995) and West *et al.* (1995) were the first to examine the possible compositions and microphysical processes which shaped the grain properties during the early stages of the formation of the haze clouds. Both studies pointed out that silicates are expected to form the early condensate if cometary material was present, and observations by Nicholson *et al.* (1995) confirmed that premise.

The most recent and thorough work to date was done by Friedson (1995) who looked at two possible compositions for the hot initial fireball and studied the formation of Al_2O_3, silicate, and soot grains. Friedson's work builds on previous models of grain formation in stellar atmospheres. The following remarks are based on his work.

The ratio of carbon to oxygen determines whether Al_2O_3 or soot will form the refractory grains which condense first from the hot vapor. If Al_2O_3 grains appear they will serve as condensation nuclei for less refractory materials like metallic iron, and magnesium and iron silicates. Friedson argued that the environment within each fireball may have been sufficiently heterogeneous that both types of grains formed in different locations. SiC or SiS may also have been abundant if the ratios of carbon or sulfur to silicon atoms were favorable.

The most refractory grains should form by homogeneous nucleation from the vapor phase. The process is complex because the free energy of the small atomic clusters are thought to dominate the nucleation process, and the formalism surrounding homogeneous nucleation is based on a free energy which is not easily calculated for such small particles. Formation of soot is further complicated by the empirical finding from flame studies that small soot particles tend to agglomerate to larger sizes rather than by direct growth of particles from the vapor. The chemistry of soot formation is complex, beginning with C_2H_2 reacting with aromatic hydrocarbons in such a way that the C/H ratio is enhanced as the molecular weight increases. Many organic molecules, including polycyclic aromatic hydrocarbons, and even fullerenes may be important in grain formation.

The main events governing particle microphysics as described by Friedson are shown in Figure 12. The process of mantling of Al_2O_3 by magnesium and iron silicates was omitted from the figure because it occurs very shortly after formation of the Al_2O_3 grains. The pressure/temperature model used in the calculation was taken from Zahnle *et al.* (1995). The temperature at 20 s after impact is 3400 K. Grain formation occurs when T $\sim$ 2000 K. Mantling by more volatile species (organics and ice) occurs at much lower temperature. Volatiles which may condense during the initial cooling phase of the plume are expected to vaporize upon re-entry into the atmosphere, and then re-condense after the parcels cool to near-ambient temperature.

During the hours to months following the impacts the particles may undergo additional growth by coagulation or from gas to particle conversion, and their numbers in the stratosphere will decrease by sedimentation to the troposphere where rapid vertical mixing will remove them. Time scales for sedimentation and coagulation as a function of pressure are shown in Fig. 13. West *et al.* (1995) noted that the increase in particle mean radius during the month following the impacts (from 0.21 to 0.28 μm) was consistent with the idea that particle coagulation was taking place in the dense core regions. West *et al.* and Banfield *et al.* (1995) reported a decrease in total particle volume after one month, indicating sedimentation is also taking place.

The disappearance of particles from the few-mb level or higher as seen in the 3.4-μm methane band images two months after impact (Orton *et al.* 1995b) is consistent with

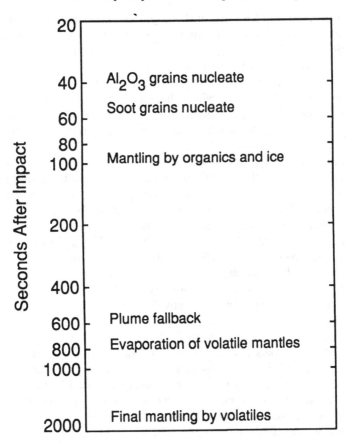

FIGURE 12. Particle microphysical processes during the first 2000 s in an evolving fireball/plume as discussed by Friedson (1995).

particle radius greater than about 0.1 μm. Few particles with radii near 0.3 μm or larger should reside at pressure levels less than 100 mb after one year. Deeper in the atmosphere vertical mixing from the troposphere increases rapidly, removing particles. Haze particles can be seen in images taken about a year after the impacts (see Fig. 11), although their number density is clearly much lower than it was a month after impact. Part of the decrease in surface density, perhaps a factor of 2–3 is due to horizontal spreading, but a significant part is due to sedimentation. The haze is most visible in the images at 2.1 and 2.27 μm which sample pressures in the range 10 to 100 mb. The most recent images show particles reaching latitudes near 20° S. It is difficult to tell how far toward the south pole the particles may have reached because of the strong gradient in Jupiter's polar stratospheric haze. As the particles continue to disperse they will tell us much about meridional transport in the jovian stratosphere.

7. Summary

After initial analysis of some of the voluminous data obtained on particles created by the SL9 impacts, a consensus view is emerging regarding the broad questions of particle origin, composition, and evolution. Cometary material is responsible for a significant portion of the particle total volume, although silicate material accounts for only 10–20%

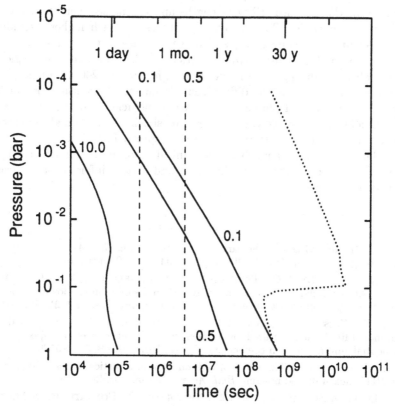

FIGURE 13. Time scales for sedimentation and coagulation of impact particles are shown as a function of pressure and particle mean radius (assuming spherical shape) from West *et al.* (1995). Solid curves indicate sedimentation times for particles of the indicated radius. Dashed curves indicate coagulation times for an initial population of 0.1 and 0.5 μm particles. The initial particle number density for the two cases was taken to be 570 and 23 cm^{-3}, respectively, consistent with particle optical depths determined for dense core regions by West *et al.* The dotted curve indicates the eddy diffusive time scale.

of the particle volume. The plumes on the limb were made visible by refractory grains which include silicates but may also contain aluminum oxide and soot. Onto these grains condensed more volatile material, including a small amount for water ice and probably organics, with ammonia possibly contributing as well. The color of the particles and their ability to form during the passage of a stratospheric wave suggest some kind of volatile organic material, but a specific candidate with both the spectral and thermodynamic properties has not been identified.

The total volume of particles (equal to that of a sphere with diameter approximately 1 km) is consistent with expectations of many of the hypervelocity impact models. The particles behave optically and dynamically like spheres with radii in the range 0.1 to 0.3 μm, although they may be aggregates of smaller spheres. Their color is homogeneous on a global scale, a puzzling fact since many heterogeneous processes were operative over a wide range of distance and time scales during the impact, rebound, and fallback phases of the events. This heterogeneity may be reflected in the internal heterogeneity of particle composition, but we have no observations which bear on that question.

Most of the particles at locations distant from the dark core regions were emplaced near the 1 mb pressure level, with an uncertainty of a factor of about 3, constrained by reflectivity in UV and near-IR wavelengths and by the sedimentation time scale for a 0.25-μm radius particle. The report of particles at much higher altitudes implies a source (gas-to-particle conversion) and/or the particles at very high altitudes must have radii much smaller than 0.1 μm. Particles within the dark core regions extend over a large altitude and pressure range, from a few mb to 200 mb or deeper. Such a distribution probably reflects the different emplacement mechanisms. Particles distant from the cores were part of plume fallback, while particles near the center could come from entrainment of upward flowing gas near the shock tube, thermal upwelling, as well as fallback. As the debris pattern continues to spread in latitude over the course of many months, it is behaving much like volcanically-produced sulfuric acid particles in the earth's stratosphere and is providing new and unique information on transport in the jovian stratosphere.

REFERENCES

ANDERS, E. 1971 Meteorites and the early solar system. *Ann. Rev. Astron. Astrophys.* **9**, (eds. L. Goldberg, D. Layzer, & J. Phillips), pp. 1–34. Ann. Reviews, Inc.

ATREYA S. K., EDGINGTON S G., TRAFTON L. M., CALDWELL J. J., NOLL K. S., & WEAVER, H. A. 1995 Abundances of ammonia and carbon-disulfide in the jovian stratosphere following the impact of comet Shoemaker Levy 9. *Geophys. Res. Lett.* **22**, 1625–1628.

BAINES, K. H., WEST, R. A., GIVER, L. P., & MORENO, F. 1993 Quasi-random narrowband model fits to near-infrared low-temperature laboratory methane spectra and derived exponential-sum absorption coefficients. *J. Geophys. Res.* **98**, 5517–5529.

BAINES, K. H., *et al.* 1994 The effect of SL9 on Jupiter's vertical aerosol structure: Results from IRTF near-infrared imaging *Bull. Amer. Astron. Soc.* **26**, 1591.

BANFIELD, D., GIERASCH, P., SQUYRES, S., NICHOLSON, P., CONRATH, B., & MATTHEWS, K. 1995 2μm spectrophotometry of jovian stratospheric aerosols—scattering opacities, vertical distributions and wind speeds. submitted for publication (*Icarus*).

BJORAKER, D., STOLOVY, S. R., HERTER, T. L., GULL, G. E., & PIRGER, B. E. 1995 Detection of water after the collision of fragments G and K of comet Shoemaker-Levy 9 with Jupiter. submitted for publication (*Icarus*).

CLARKE, J., *et al.* 1995 HST far-ultraviolet imaging of Jupiter during the impacts of comet Shoemaker-Levy 9. *Science* **267**, 1302–1307.

CONRATH, B. J., GIERASCH, P. J., & LEROY, S. S. 1990 Temperature and circulation in the stratosphere of the outer planets. *Icarus* **83**, 255–281.

TABULATED OPTICAL PROPERTIES OF GRAPHITE AND SILICATE GRAINS 1985 *Astrophys. J. Supp. Ser.* **57**, 587–594.

FIELD, G. B. 1995 Dust at the SL9 impact sites. In *Abstracts for IAU Colloquium 156: The collision of Comet P/Shoemaker-Levy 9 and Jupiter.*

FIELD, G. B., TOZZI, G. P., & STANGA, R. M. 1995 Dust as the cause of spots on Jupiter. *Astron. and Astropys. Lett.* **294**, L53–L55.

FRIEDSON, A. J. 1995 Refractory grain formation in Shoemaker-Levy 9 fireballs. submitted for publication (*Icarus*).

GIERASCH, P. J., B. J., & MAGALHÃES, J. A. 1986 Zonal mean properties of Jupiter's upper troposphere from Voyager infrared observations. *Icarus* **67**, 456–483.

GUSTAVSEN, R. L. 1986 Ph.D. Thesis, Washington State University

HAMMEL, H. B., *et al.* 1995 HST Imaging of atmospheric phenomena created by the impact of comet Shoemaker-Levy 9. *Science* **267**, 1288–1296.

INGERSOLL, A. P. & KANAMORI, H. 1995 Waves from the collisions of comet Shoemaker-Levy 9 with Jupiter. *Nature* **374**, 706–708.

KHARE, G. N., SAGAN, C., REID, THOMPSON, W. R., ARAKAWA, E. T., MEISSE, C. & TU-MINELLO, P. S. 1994 Optical properties of poly-HCN and their astronomical applications. *Canadian J. Chem.* **72**, 678–694.

MALLAMA, A., NELSON, P., & PARK, J. 1995 Detection of very high altitude fall-out from the comet Shoemaker-Levy 9 explosions in Jupiter's atmosphere. *J. Geophys. Res.* **100**, 16,879–16,884.

MARTEN A., et al. 1995 The collision of comet Shoemaker-Levy 9 with Jupiter: Detection and evolution of HCN in the stratosphere of the planet. *Geophys. Res. Lett.* **22**, 1589–1592.

MEADOWS, V. & CRISP, D. 1995 Impact plume composition from near-infrared spectroscopy. In *Proceedings of the European SL-9/Jupiter Workshop* (eds. R. West and H. Böhnhardt), pp. 239–244. European Southern Observatory.

MORENO, F., MUÑOZ, O., MOLINA, A., LÓPEZ-MORENO, J. J., ORTIZ J. L., RODRÍGUEZ, J., LÓPEZ-JIMÉNEZ, A., GIRELA, F., LARSON, S. M., & CAMPINS, H. 1995 Physical properties of the aerosol debris generated by the impact of fragment-H of comet-P Shoemaker-Levy 9 on Jupiter. *Geophys. Res. Lett.* **22**, 1609–1612.

MOSES, J. L., ALLEN, M., & GLADSTONE, G. R. 1995 Post-SL9 sulfur photochemistry on Jupiter. *Geophys. Res. Lett.* **22**, 1597–1600.

NICHOLSON, P. D., GIERASCH, P. J., HAYWARD, T. L., MCGHEE, C. A., MOERSCH, J. E., SQUYRES, S. W., VAN CLEVE, J., MATTHEWS, K., NEUGEBAUER, G., SHUPE, D., WEINBERGER, A., MILES, J. W., & CONRATH, B. J. 1995 Palomar observations of the R impact of comet Shoemaker Levy 9. 2. Spectra. *Geophys. Res. Lett.* **22**, 1617–1620.

NOLL, K. S., MCGRATH, M. A., TRAFTON, L. M., ATREYA, S. K., CALDWELL, J. J., WEAVER, H. A., YELLE, R. V., BARNET, C., & EDGINGTON, S. 1995 HST spectroscopic observations of Jupiter after the collision of comet Shoemaker-Levy 9. *Science* **267**, 1307–1313.

ORTIZ, J. L., MUÑOZ, O., MORENO, F., MOLINA, A., HERBEST, T. M., BIRKLE, K., BÖHNHARDT, H., & HAMILTON, D. P. 1995 Models of the SL-9 collision-generated hazes. *Geophys. Res. Lett.* **22**, 1605–1608.

ORTON, G. et al. 1995a Collision of comet Shoemaker Levy 9 with Jupiter observed by the NASA Infrared Telescope Facility. *Science* **267**, 1277–1282.

ORTON, G. et al. 1995b Some Results from the NASA Infrared Telescope Facility Shoemaker-Levy 9 observing campaign. In *Proceedings of the European SL-9/Jupiter Workshop* (eds. R. West and H. Böhnhardt), pp. 123–128. European Southern Observatory.

ROOSSEROTE, M., BARUCCI, A., CROVISIER, J., DROSSART, P., FULCHIGNONI, M., LECACHEUX, J., & ROQUES, F. 1995 Metallic emission lines during the impacts L and Q_1 of comet P Shoemaker Levy 9 in Jupiter. *Geophys. Res. Lett.* **22**, 1621–1624.

ROSENQVIST, J, BIRAUD, Y. G., CUISENIER, M., MARTEN, A., HIDAYAT, T., CHOUNTONOV, G., MOREAU, D., MULLER, C., MASLOV, I., ACKERMAN, M., BALEGA, Y., & KORABLEV, O. 1995 Four micron infrared observations of the comet Shoemaker-Levy 9 collision with Jupiter at the Zelenchuk Observatory: Spectral evidence for a stratospheric haze and determination of its physical properties. *Geophys. Res. Lett.* **22**, 1585–1588.

SASSON, R., WRIGHT, R., ARAKAWA, E. T., KHARE, B. N. & SAGAN, C. 1985 Optical properties of solid and liquid sulfur at visible and infrared wavelengths. *Icarus* **64**, 368–374.

WEST, R. A. 1991 Optical properties of aggregate particles whose outer diameter is comparable to the wavelength. *Appl. Optics* **30**, 5316–5324.

WEST, R. A. & SMITH, P. H. 1991 Evidence for aggregate particles in the atmospheres of Titan and Jupiter. *Icarus* **90**, 330–333.

WEST, R. A., FRIEDSON, A. J., & APPLEBY, J. F. 1992 Jovian large-scale stratospheric circulation. *Icarus* **100**, 245–259.

WEST, R. A., KARKOSCHKA, E., FRIEDSON, A. J., SEYMOUR, M., BAINES, K. H., & HAMMEL, H. B. 1995 Impact debris particles in Jupiter stratosphere. *Science* **267**, 1296–1301.

WICKRAMASINGHE, N. C. & WALLIS, M. K. 1994 Submicron dust and the collision of comet SL-9 with Jupiter. *Astrophys. and Space Sci.* **219**, 295–301.

WILSON, P. D., & SAGAN, C. 1995 Chemistry of the Shoemaker-Levy 9 jovian impact blemishes: Indigenous cometary vs. shock-synthesized organic matter. In *The collision of comet P/Shoemaker-Levy 9 and Jupiter* (eds. K. Noll *et al.*) Cambridge.

ZAHNLE, K., MACLOW, M. M., LODDERS, K., & FEGLEY, B. 1995 Sulfur chemistry in the wake of comet Shoemaker Levy 9. *Geophys. Res. Lett.* **22**, 1593–1596.

Jupiter's post-impact atmospheric thermal response

By BARNEY J. CONRATH

Laboratory for Extraterrestrial Physics, NASA Goddard Space Flight Center, Greenbelt, MD 20771, USA

Measurements of thermal emission in spectral regions, ranging from the near-infrared to mm wavelengths provide information on the atmospheric thermal structure over impact sites from μbar levels in the upper stratosphere down to the upper troposphere. Systematic time series of observations relevant to this entire height range over individual spots do not exist. However, by piecing together information at different times from various spots, it is possible to obtain a provisional, semi-quantitative picture of the behavior of the thermal structure over a typical impact site. Immediately after fall-back of the ejecta plume, the upper stratosphere is heated to $\sim$ 600–1300 K above ambient temperature. The amplitude of the temperature perturbation diminishes with increasing depth in the atmosphere, but even in the upper troposphere a temperature increase of a few kelvins is observed. Initially, the upper stratosphere cools very rapidly with time scales of tens of minutes, presumably the result of strong radiative cooling associated with the high temperatures. After the initial cooling, all levels continue to cool at slower rates with time scales of a few days; however, this is still very rapid compared to radiative cooling of the ambient atmosphere. Enhancements in infrared opacity necessary to produce the cooling radiatively do not appear to be viable, suggesting that dynamical effects may play a dominant role. Possible mechanisms include horizontal mixing with the ambient atmosphere and adiabatic cooling produced by upward motion associated with an anticyclonic vortex. Many questions remain concerning the thermal structure above the impact sites; these are being addressed through ongoing data analysis and modeling efforts.

1. Introduction

The impacting fragments of comet P/Shoemaker-Levy 9 produced significant local perturbations to Jupiter's atmospheric thermal structure. Detailed study of the behavior of the temperature structure of the impact sites can provide information on the radiative and dynamical properties of these sites and may ultimately prove to be diagnostic of the ambient state of Jupiter's upper troposphere and stratosphere.

In this review, we attempt to synthesize a picture of the spatial and temporal behavior of a typical impact site by examining selected observations relevant to the temperature structure. Ideally, one would like to have a time series of observations of a particular site covering a broad height range. Unfortunately, such a systematic set of measurements does not exist because of many practical considerations including observational constraints and the fact that as the sequence of impacts progressed, many sites overlapped and merged with other sites. Therefore, it is necessary to piece together data from various spots at various times, obtained with a variety of observational techniques.

In the following Section, we first review selected observations relevant to the thermal structure of the various impact sites and their ambient surroundings. In Section 3, we attempt to combine information based on these observations to reconstruct the behavior of thermal structure associated with a "generic spot" as a function of atmospheric pressure level and time after impact. The possible atmospheric radiative and dynamic effects implied by this behavior are considered in Section 4. Finally, the present status of our understanding and some remaining questions are summarized in Section 5.

2. Observations

In this section, we summarize selected observations that are relevant to achieving an understanding of the behavior of the atmospheric thermal structure from the time of fall-back of the ejecta plume until several weeks following impact. This discussion is necessarily based primarily on preliminary reports by various groups of observers. Undoubtedly, an improved understanding of the thermal behavior will emerge in the future as detailed intercomparisons of the data sets are made, and more refined modeling is carried out. We will first examine observations that provide information on atmospheric temperature at and immediately following fall-back. Then we will discuss observations taken during the hours and days following the impacts that are relevant to the thermal behavior over a range of atmospheric levels from the upper stratosphere to the troposphere.

2.1. *Temperatures immediately after fall-back*

Observations in the near-infrared of several sites show evidence of major increases in stratospheric temperatures. Shortly after the impact of fragment H, Encrenaz *et al.* (1995a; 1995b), using the IRSPEC imaging spectrometer on the New Technology Telescope of the European Southern Observatory in Chile, observed strong brightening in the 3.5 μm spectral region. This is a region of strong methane absorption, and, consequently, is usually quite dark under normal Jovian conditions. The solar flux is absorbed with little backscattered radiation present, and at nominal stratospheric temperatures, thermal emission is very small. On this basis, the observed post-impact brightening has been interpreted as thermal emission from methane resulting from an enhanced upper stratospheric temperature. Examples of two of the spectra, the first taken at maximum brightening 14 minutes after impact and the second approximately 6 minutes later, are shown in Fig. 1. By fitting parameterized models to the measurements, Encrenaz *et al.* infer a temperature of 750 ± 100 K near the 0.01 mbar level at the time of maximum emission. By the time of the spectrum shown in Fig. 1b, the temperature had decreased to about 640 K, consistent with a very rapid initial cooling.

Maillard *et al.* (1995) obtained measurements in the 1.6 to 4.7 μm spectral region using a Fourier Transform Spectrometer at the Canada-France-Hawaii Telescope. Approximately 10 minutes after the impact of the C fragment, they observed strong methane emission consistent with temperatures between 750 and 1500 K at levels between 0.1 and 0.01 mbar. They also observed CO emission from the L site at 4.7 μm 4.5 hours after the impact. Model calculations suggest a temperature of 274 ± 10 K. This is found to correspond to approximately the 2 μbar level if the CO mixing ratio profile derived by Lellouch *et al.* (1995) from millimeter observations of the G site is assumed. CO emission from the L site at 4.7 μm was also observed by Brooke *et al.* (1995) about 4 hours after impact using the CGS4 spectrometer on the United Kingdom Infrared Telescope (UKIRT). Knacke *et al.* (1995), also using the UKIRT, infer initial temperatures near 2000 K from 2.2– 2.4 μm spectroscopy of the R event.

Crisp & Meadows (1995) obtained measurements with the Infrared Imaging Spectrometer on the Anglo-Australian Telescope that indicate strong near-infrared CO line emission from several sites, apparently associated with fall-back events. Rotational temperatures in excess of 2000 K were inferred; however, the emission decayed rapidly, and could no longer be detected by about 30 minutes after impact.

Emission in the 3.4 μm region has been directly observed by NIMS on Galileo during the G and R events (Carlson *et al.* 1995). Under the assumption that the emission originates from a continuum of particulates, Carlson *et al.* infer a temperature immediately after fall-back of the ejecta of 1000 K or less.

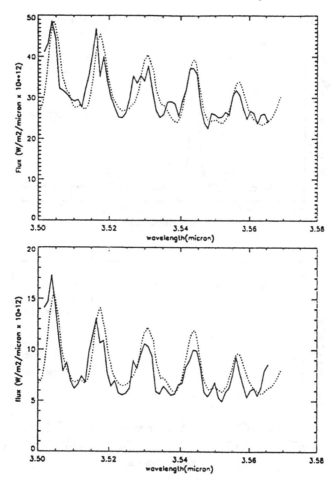

FIGURE 1. Spectra obtained with the IRSPEC imaging spectrometer at the European Southern Observatory. Measurements are shown as solid curves and model fits as broken lines. The spectrum in the upper panel was acquired 14 minutes after impact H, and the model fit was obtained assuming a temperature of 750 K near 0.01 mbar. The lower panel shows a spectrum taken 6 minutes later, along with a model spectrum calculated assuming a temperature of 640 K near 0.01 mbar. Note the change in flux scale between the upper and lower panels. (From Encranez *et al.* 1995.)

Observing with the MIRAC2 mid-infrared array camera on the IRTF, Friedson *et al.* (1995) found enhanced emission at 7.85, 10.3, and 12.2 μm in the time period immediately following the impact of fragment R. From these data, it is not possible to uniquely infer both the angular extent of the source and the temperature. However, for a source 1900 km in diameter, Freidson *et al.* estimate a temperature of at least 1350 K, while if the source were a factor of two larger, the inferred temperature would be 800 K. In the nominal Jovian atmosphere, these measurements would be indicative of temperatures in approximately the 1–10 mbar region. The presence of possible additional gaseous and particulate opacity associated with the impact would tend to move the region of sensitivity toward lower pressures.

The measurements summarized above pertain to several different spots, are in some cases from different spectral regions, and were interpreted under various modeling as-

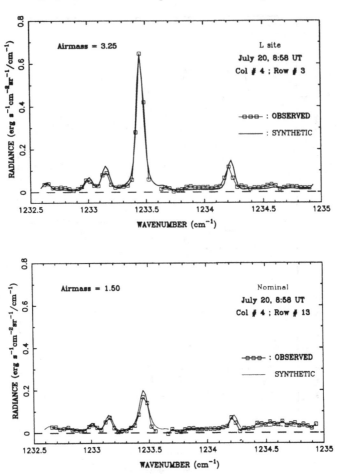

FIGURE 2. Irshell spectra obtained from the L impact site (upper panel) and an adjacent area (lower panel). Both observed and synthetic spectra are shown as indicated. A strong methane absorption lines is seen at 1233.455 cm^{-1} with weak lines at 1233.147 and 1234.226 cm^{-1}. The measurements were acquired approximately 11 hours after impact. (From Bézard *et al.* 1995.)

sumptions so it is difficult to synthesize a completely coherent picture at this time. However, it does appear that the upper stratospheric temperatures were strongly perturbed by the fall-back or "splash", perhaps ranging from about 700 to 2000 K in the region between approximately 1 mbar and a few μbars. The heating was followed by a very rapid cooling with temperatures probably dropping over 100 K in a matter of minutes.

2.2. *Irshell observations*

Next, measurements relevant to the behavior of temperatures following the initial cooling phase will be considered. We shall begin in the upper stratosphere and work our way downward.

Measurements with the Irshell, a mid-infrared high spectral resolution imaging spectrometer, have been used by Bézard *et al.* (1995) to infer information on the atmospheric temperatures above about the 0.1 mbar level. Spectra obtained from the L site and an adjacent region approximately 11 hours after impact are shown in Fig. 2. Within the narrow spectral region covered there is both a strong and a weak methane line. The strong

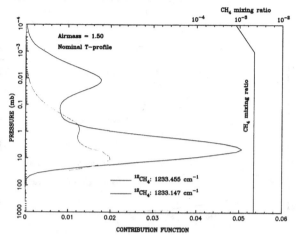

FIGURE 3. Contribution functions for a strong and a weak methane absorption line. The relative contributions of each atmospheric level to the radiance at the top of the atmosphere are shown. The radiance from the strong line at 1233.455 cm^{-1} contains contributions from both the upper and lower stratosphere, while that from the weak line at 1233.147 cm^{-1} includes contributions primarily from the lower stratosphere. The methane profile used in the models is shown in the right-hand side of the figure. (From Bézard *et al.* 1995.)

line shows a much more intense emission on the site than off, while the enhancement of emission from the weak line on the site is significantly less. By combining measurements from these two lines, Bézard *et al.* were able to infer information on the behavior of the temperature perturbation as a function of pressure level. The contribution functions for these two lines are shown in Fig. 3.

The strong line emission includes contributions from both the lower and upper stratosphere, while for the weaker line, the contributions are mostly from the lower stratosphere. This permits a crude separation of information on the temperature perturbation of the upper stratosphere from that of the lower stratosphere. Model temperature structures that provide good fits to the spectra are shown in Fig. 4. Because of the limited information, the details of the inferred structure are highly non-unique and two families of solutions are shown. However, the results clearly constrain the primary temperature perturbation to lie mainly above the 1-mbar level. The maximum temperatures permitted by the weak line observations are also shown. Results were also obtained by Bézard *et al.* for the K site 23 hours after impact showing a weaker perturbation. The measurements appear to be consistent with a continued cooling of the sites after the fall-back, but at a much lower rate than that initially observed immediately after the splash.

2.3. *IRAM Observations*

Strong emission by CO in the millimeter wavelength region was observed with the IRAM telescope by Lellouch *et al.* (1995). The spots were not spatially resolved because of the relatively large beam size of the telescope, and most of the observations pertain to the G-Q-R-S complex. The emission was observed to decay rapidly over several days time. In the interpretation of these data, it is not possible to uniquely determine both the CO abundance and atmospheric temperature. Based on considerations of the CO chemistry, it was assumed that the CO abundance remained constant over the period of the observations. Under this assumption, it was found that the temporal evolution of thermal structure could be modeled as shown in Fig. 5. The temperature profile labeled

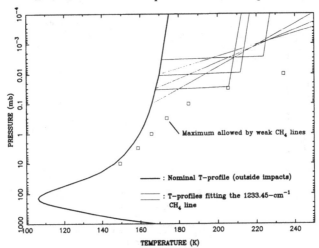

FIGURE 4. Model temperature profiles for the L impact site and adjacent region approximately 11 hours after impact. The heavy solid curve is the nominal profile that fits the spectrum from the adjacent region. The thin curves represent families of profiles that all fit the spectrum from the impact site equally well. The square symbols represent the highest temperatures permitted by the weak methane lines. (From Bézard *et al.* 1995.)

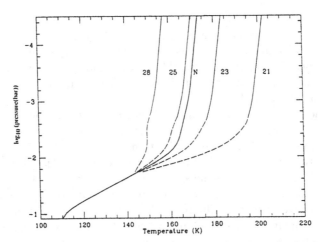

FIGURE 5. Model temperature profiles used to fit measurements obtained with the IRAM telescope. The curves are labeled with the July 1994 date on which they were taken, and the curve labeled N is taken to be the nominal profile. The measurements are sensitive to levels only above about the 1 mbar level. (From Lellouch *et al.* 1995.)

N was taken as nominal. The measurements are sensitive primarily to the atmospheric region above approximately 1 mbar, and little detailed vertical structure information is available. The numbered profiles correspond to the dates in July 1994 on which the observations were made. A very rapid cooling is noted with a characteristic time scale of a few days. Remarkably, the upper stratosphere appears to cool significantly below the assumed nominal temperature.

2.4. *SpectroCam-10 observations*

Observations of impact sites were made in the middle infrared by Nicholson *et al.* (1995a; 1995b) using SpectroCam-10, an imaging array spectrometer on the 5-meter Hale tele-

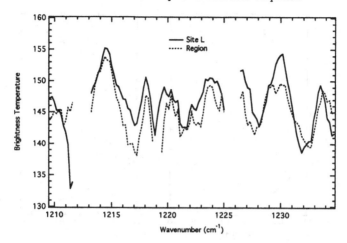

FIGURE 6. Measurements obtained with SpectroCam-10 on the 5-meter Hale telescope at Palomar. The solid curve is a spectrum taken on the L site approximately 4 hours after impact, and the broken curve is a spectrum from an adjacent region. The primary absorber in this region is methane. (From Conrath *et al.* 1995.)

scope at Palomar. Both low resolution (10 cm^{-1}) spectra in the 8–12 μm region and high resolution (0.5 cm^{-1}) spectra in selected narrow intervals were acquired. An example of high resolution spectra between 1210 and 1240 cm^{-1} of the L site and an adjacent region is shown in Fig. 6. This spectral region is dominated by methane emission from the stratosphere and is sensitive to temperature in a broad layer between approximately 1 and 10 mbar. Comparison of the two spectra show enhanced emission over the L site compared to the adjacent region. These measurements were combined with low resolution spectra to retrieve estimates of temperature structure in the 1–10 mbar region and near the tropopause and upper troposphere (Conrath *et al.* 1995). The results are shown in Fig. 7. Although extremely crude in terms of vertical structure information, the results indicate a temperature perturbation above the 10 mbar level of about 5 K. There may be a slight warming in the upper troposphere, but the apparent warming at deeper levels is an artifact of the retrieval since the measurements do not contain information on these levels. The results were obtained about 4 hours after the impact.

2.5. *MIRAC2 observations*

Data obtained with MIRAC2, an imaging middle infrared array camera at the IRTF, have been used to infer information on the thermal structure of several impact sites (Orton *et al.* 1995a). Measurements in a spectral interval centered at 7.85 μm are sensitive to a relatively thick layer centered near the 10 mbar level. From these images, temperature perturbations over the site relative to the surrounding regions were inferred. Site E on July 18 was found to have a perturbation of about 1.5 K which decreased to less than 0.5 K in about 3 days. A thermal perturbation of 2 K was observed over site L on July 20, but had decayed to about 1 K 19 hours later, and was no longer detectable when the site was observed again on July 28. Observations of the Q1+R site 10 hours after the R impact indicated an temperature enhancement relative to the surrounding region of 3–4 K.

Orton *et al.* (1995b) have used images at 13.0, 17.8, 20.2, and 20.8 μm to retrieve information on thermal perturbations in the upper troposphere. Temperature retrievals from these data are shown in Fig. 8 for site K on July 28 and sites G and L on July 21.

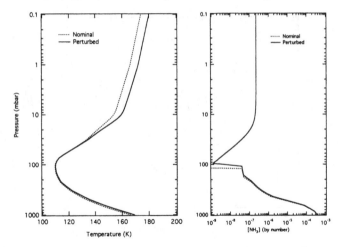

FIGURE 7. Profiles of temperature (left panel) and ammonia (right panel) retrieved from SpectroCam-10 measurements. The broken curves labeled nominal are for the region adjacent to the impact site and the solid curves are from measurements on impact site L. The retrievals are meaningful only in the 1–10 mbar region and in the upper troposphere. (From Conrath *et al.* 1995.)

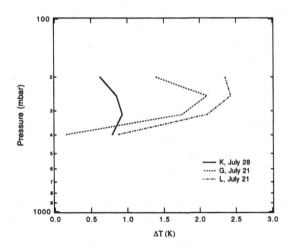

FIGURE 8. Tropospheric temperature retrievals obtained from MIRAC2 measurements. The retrievals are expressed in terms of a temperature perturbation over a site relative to an adjacent region. Results for the the K, G, and L sites are shown. The measurements used for the K site were taken July 28 and those for the G and L sites were obtained on July 21. (From Orton *et al.* 1995.)

Temperature perturbations are shown relative to surrounding regions. They appear to be confined to the atmospheric levels above approximately 400 mbar.

3. Synthesis of results

In an attempt to organize the thermal structure data, we schematically show the various results discussed above in Fig. 9. The temperature perturbations are shown as temperature differences ΔT relative to the nominal jovian atmosphere. In most cases the nominal thermal structure is taken to be that in unperturbed regions near the impact

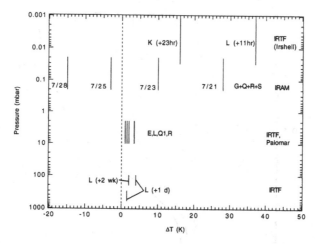

FIGURE 9. Schematic summary of temperature inferences. The approximate vertical range of sensitivity of each determination is shown as a vertical line. The results are expressed as temperature perturbations ΔT over the sites relative to the ambient atmosphere. The site to which each result pertains is indicated along with either the date of the observation or time after impact. The facility used is indicated on the right of the figure.

sites. The approximate pressure ranges to which the results apply are shown. Temporal variations are indicated where available. The temperature perturbations of 600 to 1000 K observed immediately after fall-back of the impact-generated plumes are not included in this figure.

Unfortunately, a time-history of the thermal perturbation associated with any one particular spot at all relevant atmospheric levels does not exist. In the absence of such direct information, we have attempted to combine the available measurements from several spots shown in Fig. 9, and postulate the thermal behavior of a "generic" spot as shown in Fig. 10. However, the over-all picture that seems to emerge is one of very strong heating of the upper stratosphere by the impacting ejecta plume as it falls back into the atmosphere. This is followed by a very rapid drop in temperature with a characteristic time scale of the order of minutes. Presumably this is due to strong radiative cooling associated with the very high initial temperatures. With decreasing temperatures the cooling rate decreases; several hours after impact, the characteristic cooling time in the upper stratosphere is of the order of 2–3 days. It is this phase of the temporal evolution that is depicted schematically in Fig. 10. While the detailed vertical structure of the temperature perturbation has not been well established, it does decrease significantly with increasing pressure in the lower stratosphere. The indicated "overshoot" in cooling with the temperature near 0.1 mbar dropping below the nominal value is based on the IRAM observations and their model-dependent interpretation.

The behavior of the temperature structure between 10 and 100 mbar, as indicated in the figure, is largely an interpolation between higher levels and the troposphere since none of the available measurements is very sensitive to this portion of the atmosphere. The upper troposphere appears to cool somewhat more slowly than the stratosphere. The cause of the tropospheric temperature perturbation is not well understood at the present time. Possibilities include local heating at the time of impact, downward penetration of heating following ejecta plume fall-back, or a buoyant plume rising adiabatically as the result of impact perturbations at deeper levels. Further investigations of the horizontal

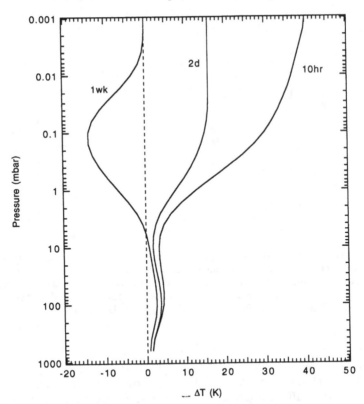

FIGURE 10. Thermal behavior of a "generic" spot. This cartoon of the vertical structure and temporal behavior of the thermal perturbation over a spot relative to the ambient atmosphere is based on the results summarized in Fig. 9. The curve labels indicate time after impact.

spatial extent of the tropospheric perturbations relative to those in the stratosphere may lead to a better understanding of this phenomenon.

4. Radiative and dynamic effects

The behavior of the thermal structure over the impact sites raises questions concerning radiative and dynamic processes that may be associated with the sites and their temporal evolution. The very rapid initial cooling rates in the upper stratosphere can be understood, at least qualitatively, in terms of the initial very high temperatures. However, several hours after impact when the upper stratospheric temperatures have returned to within a few tens of K of equilibrium, apparent cooling times of a few days are observed, which are long compared to those of the initial cooling, but are still short compared to those associated with the ambient jovian atmosphere. This appears to be true at all levels, including the more slowly cooling upper tropsophere.

An example of a radiative time constant calculated for the ambient jovian atmosphere is shown as a function of pressure level in Fig. 11 (Flasar, 1989). Although the time constant is not shown for pressures less than 10 mbar, it will continue to increase with decreasing pressure. From this we see that the ambient radiative time constant ranges from about 600 days near the tropopause to over 1000 days in the upper stratosphere. If the rapid cooling is radiative, then an increase in infrared opacity above ambient is

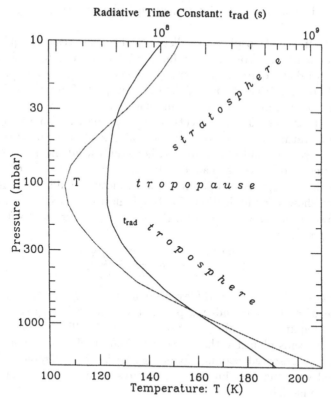

FIGURE 11. Radiative time constant or radiative relaxation time (t_{rad}) as a function of atmospheric pressure level for the nominal jovian atmosphere. The temperature profile used to calculate the time constant is labeled T. The time constant is expressed in seconds. (From Flasar, 1989.)

required. This could result from either an increased gas opacity or increased particulate opacity.

We can make a rough estimate of the increase in infrared opacity needed to yield the necessary radiative time constant by considering a gray atmospheric emitting layer at temperature T_0 with a pressure difference across the layer Δp and an optical thickness $\Delta \tau$. The radiative time constant is then given approximately by

$$t_{rad}^{-1} \sim \frac{4g\sigma T_0^3}{c_p} \frac{\Delta \tau}{\Delta p} \qquad (4.1)$$

where g is gravitational acceleration, σ is the Stefan-Boltzmann constant and c_p is the specific heat of the atmosphere (see for example Andrews *et al.* 1987). For $p = 1$ mbar and $T_0 = 160$ K, a radiative time constant $t_{rad} \sim 1$ day requires a gray optical thickness $\Delta \tau > 1$. Such a large additional optical depth from either particulates or enhanced gaseous absorption does not seem to be consistent with mid-infrared spectral measurements (Conrath *et al.* 1995). If the additional opacity were due to particulates associated with the impact sites, it is likely that they would also produce heating due to absorption of solar energy, and this would partially offset any enhanced infrared cooling, further compounding the problem. Thus, it appears unlikely that radiative cooling alone can account for the observed cooling rates near 10 mbars and deeper. This conclusion is supported by detailed, non-gray radiative transfer calculations carried out by Wang *et al.* (1995).

At higher levels in the atmosphere, smaller optical thicknesses are required to produce the necessary cooling rates. For example, for $p = 0.1$ mbar and $T_0 = 180$ K, a radiative relaxation time of 1 day requires a gray optical thickness of only ~ 0.05. This is probably near the upper limit of particulate optical depth in this part of the atmosphere that would be permitted by the mid-infrared spectral measurements.

It appears likely that dynamical effects may be primarily responsible for the rapid decay of the temperature perturbations. Horizontal mixing of the perturbed region with the ambient atmosphere could play a role. A horizontal wind shear of the order several tens of meters per second over a scale of 10^4 km might be capable of producing such mixing. Adiabatic cooling associated with upward motion could also contribute. Some spots apparently show some indication of anticyclonic vorticity (Beebe 1995). If such is the case, an upward motion of the central region would be expected with an associated adiabatic cooling given by

$$\left(\frac{\partial T}{\partial t} \right)_{ad} \sim -w \left(\frac{\partial T}{\partial z} + \frac{g}{c_p} \right). \tag{4.2}$$

For a cooling rate $(\partial T/\partial t)_{ad} \sim 1$ K/day, we find a vertical velocity $w \sim 0.5$ cm s^{-1}. For a spot size of $\sim 10^4$ km and a depth of the order of one pressure scale height, mass continuity requires a radial outflow of ~ 2.5 m s^{-1} (250 km day^{-1}). From these considerations, it seems plausible that some combination of dynamic effects is probably responsible for the observed cooling, following the brief very rapid radiative cooling phase associated with the high initial temperatures. This can be confirmed only with more quantitative modeling.

5. Summary

Several sets of thermal emission data, acquired during the fragment impacts and in the days immediately following the impacts, have been used to extract information on the perturbed atmospheric temperature structure over the sites. The measurements range from the near-infrared to millimeter wavelengths. Collectively, these measurements yield some information on the atmosphere between a few μbars and about 500 mbar. No uniform sets of measurements exist for an individual spot that can provide a complete time history of the thermal behavior at all atmospheric levels. We can only attempt to combine information from various sites at various times to obtain a semi-quantitative picture of the behavior of a "typical" spot.

In the time period immediately following the fall-back of the ejecta-plume, or the so-called splash, temperatures in the upper stratosphere were ~ 600–1300 K above ambient. The amplitude of the temperature perturbations decreased with increasing depth in the atmosphere as might be expected if the atmospheric heating is primarily due to the fall-back of material. However, temperatures a few Kelvins higher than normal were observed all of the way down into the upper troposphere to about the 500 mbar level. The thermal perturbations in the upper stratosphere initially decayed very rapidly with characteristic time scales of tens of minutes. This rapid change presumably resulted from very strong radiative cooling associated with the high initial temperatures. As the temperatures dropped, the cooling rate also decreased. Several hours after impact, temperatures in the 0.001 to 0.1 mbar region were a few tens of K above ambient, while near 10 mbar, perturbations of only 3–4 K were observed. Temperatures then continued to decrease in these regions with time scales of the order of days. In the upper troposphere, somewhat longer cooling times were noted. However, in all cases the cooling times appear to have been at least two orders of magnitude shorter than estimates of radiative cooling times for the ambient jovian atmosphere.

It is difficult to account for cooling times of the order of days entirely in terms of enhanced radiative cooling resulting from increased infrared opacities. Estimates of the required equivalent gray opacities suggest values so large that they may be inconsistent with middle infrared spectral observations. More detailed radiative transfer calculations support the conclusion that radiative cooling cannot be the primary factor, at least in the lower statosphere and upper troposphere; dynamic effects must therefore be significant. Candidates include horizontal mixing with the surrounding environment and adiabatic cooling, possibly associated with the formation of an anticyclonic vortex. A more detailed examination of the dynamics associated with the sites is needed.

Observations of the thermal structure over the impact sites have raised many questions that remain to be answered. What processes control the vertical structure above a spot? Is the structure diagnostic of spot dynamics? What are the detailed mechanisms that produce rapid stratospheric cooling? What role do particulates play in the structure and evolution of a site? What is responsible for the thermal perturbations in the upper troposphere? On-going data analysis and modeling should provide some answers to these and related questions in the future.

REFERENCES

ANDREWS, D. G., HOLTON, J. R., & LEOVY, C. B. 1987 *Middle Atmosphere Dynamics*. Academic Press.

BEEBE, R. 1995 Growth and disperison of the Shoemaker-Levy 9 impact features from HST imaging. This volume.

BÉZARD, B., GRIFFITH, C. A., KELLY, D., LACY, J., GREATHOUSE, T., & ORTON, G. 1995 Mid-IR high-resolution spectoscopy of the SL9 impact sites: temperature and HCN retrievals. Paper presented at IAU Colloquium 156: The Collision of Comet P/Shoemaker-Levy 9 and Jupiter, Baltimore, Maryland, 9–12 May.

BROOKE, T. Y., ORTON, G. S., CRISP, D., FRIEDSON, A. J., & BJORAKER, G. 1995 Near-infrared spectroscopy of the Shoemaker-Levy 9 impact sites with UKIRT: CO emission from the L site. Paper presented at IAU Colloquium 156: The Collision of Comet P/Shoemaker-Levy 9 and Jupiter, Baltimore, Maryland, 9–12 May.

CARLSON, R. W., WEISSMAN, P. R., SEGURA, M., HUI, J., SMYTHE, W. D., JOHNSON, T., BAINES, K. H., DROSSART, P., ENCRENAZ, TH., LEADER, F. 'E., & THE NIMS SCIENCE TEAM 1995 Galileo infrared observations of the Shoemaker-levy 9 G impact fireball: a preliminary report. *Geophysical Research Letters* **22**, 1557–1560.

CONRATH, B. J., GIERASCH, P. J., HAYWARD, T., MCGHEE, C., NICHOLSON, P. D., & VAN CLEVE, J. 1995 Palomar mid-infrared spectroscopic observations of comet Shoemaker-Levy 9 impact sites. Paper presented at IAU Colloquium 156: The Collision of Comet P/Shoemaker-Levy 9 and Jupiter, Baltimore, Maryland, 9–12 May.

CRISP, D. & MEADOWS, V. 1995 Near-infrared imaging spectroscopy of the impacts of SL9 fragments C, D, G, K, N, R, V, and W with Jupiter. Paper presented at IAU Colloquium 156: The Collision of Comet P/Shoemaker-Levy 9 and Jupiter, Baltimore, Maryland, 9–12 May.

ENCRENAZ, TH., SCHULZ, R., STUWE, J. A., WIEDEMANN, G., DROSSART, P. & CROVISIER, J. 1995 Near-ir spectroscopy of Jupiter at the time of comet Shoemaker-Levy 9 impacts: emissions of CH_4, H_3^+, and H_2. *Geophysical Research Letters* **22**, 1577–1580.

FLASAR, F.M. 1986 In *Time Variable Phenomena in the Jovian System* (eds. M. Belton, G. Hunt, & R. West) Spec. Publ. 494, National Aeronautics and Space Administration, Washington, DC, p. 324.

FRIEDSON, A. J., HOFFMANN, W. F., GOGUEN, J. D., DEUTSCH, L. K., ORTON, G. S., HORA, J. L., DAYAL, A., SPITALE, J. N., WELLS, W. K., & FAZIO, G. G. 1995 Thermal infrared lightcurves of the impact of comet Shoemaker-Levy 9 fragment R. *Geophysical Research Letters* **22** 1569–1572.

KNACKE, R. F., FAJARDO-ACOSTA, GEBALLE, T. R., & NOLL, K. S. 1995 Near-infrared spectroscopy of the R impact site of comet Shoemaker-Levy 9. Paper presented at IAU Colloquium 156: The Collision of Comet P/Shoemaker-Levy 9 and Jupiter, Baltimore, Maryland, 9–12 May.

LELLOUCH, E., PAUBERT, G., MORENO, R., FESTOU, M. 'C., MARTEN, A., DESPOIS, D., STROBEL, D. F., & SIEVERS, A. 1995 Chemical and thermal effects in Jupiter after comet impacts from millimetre observations. *Nature* **373**, 592.

MAILLARD, J.-P., DROSSARD, P., BÉZARD, B., DE BERGH, D., LELLOUCH, E., MARTEN, A., CALDWELL, J., HILICO, J.-C., & ATREYA, S. K. 1995 Methane and carbon monoxide infrared emissions observed at the Canada-France-Hawaii Telescope during the collision of comet SL-9 with Jupiter. *Geophysical Research Letters* **22**, 1573–1576.

NICHOLSON, P. D., GIERASCH, P. J., HAYWARD, T. L., McGHEE, C. A., MOERSCH, J. E., SQUYRES, S. W., VAN CLEVE, J., MATTHEWS, K., NEUGEBAUER, G., SHUPE, D., WEINBERGER, A., MILES, J. W., & CONRATH, B. J. 1995a Palomar observations of the R impact of comet Shoemaker-Levy 9: I. Light curves. *Geophysical Research Letters* **22**, 1613–1616.

NICHOLSON, P. D., GIERASCH, P. J., HAYWARD, T. L., McGHEE, C. A., MOERSCH, J. E., SQUYRES, S. W., VAN CLEVE, J., MATTHEWS, K., NEUGEBAUER, G., SHUPE, D., WEINBERGER, A., MILES, J. W., & CONRATH, B. J. 1995b Palomar observations of the R impact of comet Shoemaker-Levy 9: II. Spectra *Geophysical Research Letters* **22**, 1617–1620.

ORTON, G. AND 57 OTHERS 1995a Collision of comet Shoemaker-Levy 9 with Jupiter observed by the NASA Infrared Telescope Facility. *Science* **267**, 1277–1282.

ORTON, G., SPITALE, J., FRIEDSON, J., YANAMANDRA-FISHER, P., BAINES, K., HOFFMANN, W., DAYAL, A., DEUTSCH, L., & HORA, J. 1995b Spatial variation and time dependence of the temperature structure of impact sites. Paper presented at IAU Colloquium 156: The Collision of Comet P/Shoemaker-Levy 9 and Jupiter, Baltimore, Maryland, 9–12 May.

WANG, Y., NOLL, K. S., & YELLE, R. 1995 Thermal effects from minor constituents in Jupiter's upper atmosphere after SL9. Paper presented at IAU Colloquium 156: The Collision of Comet P/Shoemaker-Levy 9 and Jupiter, Baltimore, Maryland, 9–12 May.

Growth and dispersion of the Shoemaker-Levy 9 impact features from HST imaging

By RETA F. BEEBE

Department of Astronomy, New Mexico State University, P.O. Box 30001/Dept. 4500,
Las Cruces, NM 88003, USA

The Hubble Space Telescope Wide Field Planetary Camera 2 imaging data provide the highest spatial resolution of individual Shoemaker-Levy 9 impact sites. Analysis of images obtained with the F410M filter yielded horizontal translation rates of tropospheric cloud structures and the east-west components have been interpreted as zonal winds which vary with latitude. When the tropospheric zonal winds between $-60°$ and $-30°$, which were derived from the SL9 images, are compared with Voyager data there are no discernible changes in the magnitude or latitudinal positions of wind minima and maxima. This result provides additional evidence of the long-term stability of the zonal winds. Changes in individual sites during a two week period in July 1994 have been mapped. Their evolution is consistent with zonal winds decreasing with height and it provides evidence that local circulation associated with isolated weather systems perturbs the lower stratosphere.

---

1. Introduction

On July 16, 1994 at $21^h 30\text{--}51^m$ the first multicolor images revealed the site of the A fragment impact of Comet P/Shoemaker-Levy 9 (SL9) as it rotated into view about 1.5 hours after it formed. The lack of color dependence and the resulting orientation and morphology of the ejecta blanket had not been anticipated. The blowout region was located more to the east than expected and dark rings and crescent-shaped structures centered on the impact site were observed, but the most obvious aspect of site A was the dark core (see the chapter by Hammel). Due to the fact that the dark debris were in the stratosphere, the extent and contrast of the extended ejecta were strongly enhanced near the limb and terminator. This aspect of the SL9 sites is unique in the history of visual observations of Jupiter (Reese, 1994).

This study utilizes HST/WFPC2 SL9 Campaign imaging data to derive tropospheric winds from $-60°$ to $-30°$ latitude and compares them to Voyager results obtained from data with the same spatial resolution (unless otherwise specified all latitudes are planetographic and all longitudes are in System III—see section 2 for definitions). The temporal changes of the impact sites are interpreted relative to this underlying tropospheric wind field. The effects of local cyclonic and anticyclonic storm systems are considered in an effort to determine the extent to which the tropospheric wind field and the vertical propagation of organized storm systems influence stratospheric circulation.

Temporal variability of the reflectivity and longitudinal drift rates of large, long-lived cloud systems had been well documented long before the SL9 fragments impacted Jupiter (Peek, 1958; Smith & Hunt, 1976; Beebe & Youngblood, 1979; Smith, et al. 1979a & 1979b; Beebe, Orton & West, 1989). Quantification of variations of the magnitude of the zonal winds and latitudinal positions of the jets (maxima and minima of the east-west component of the tropospheric winds) was hampered by the fact that coherent cloud systems, which can be resolved in groundbased observations, tend to interact with the local wind field, rotate about their centers and translate eastward or westward at a rate equivalent to the zonal wind at the central latitude. Consequently, historical zonal wind profiles (east-west wind component versus latitude) (Peek, 1958; Chapman, 1969; Smith & Hunt, 1976; Beebe & Youngblood, 1979) had lower wind speeds associated with the zonal wind maxima and minima than those obtained from analysis of the more highly resolved Voyager imaging data where cloud markers that drifted with the winds could be resolved (Ingersoll, *et al.* 1981; Limaye, *et al.* 1982; Limaye, 1986; Beebe, Orton & West, 1989).

Because there is no terrestrial equivalent to the strong westerly (west-to-east) winds that dominate cloud structure at low Jovian latitudes, considerable effort has been expended to understand global circulation (Ingersoll & Cuzzi, 1969; Williams & Wilson, 1988; Dowling & Ingersoll, 1989; Stone, 1976; Dowling, 1993; Marcus, 1993). Much of the temporal information concerning long term stability of winds has been derived from images spanning the visible range (0.4 to 0.7 micron) of the spectrum. Broadband transmission filters, which sound down to the ammonia cloud deck (0.5 to 0.7 bars) have generally been used to record temporal changes in the cloud deck. Because these filters do not vertically sample the atmosphere, our knowledge of variation of winds with altitude within Jupiter's atmosphere is limited. The rate of decrease of zonal winds with altitude above the visible cloud deck has been derived from interpretation of the Voyager infrared spectroscopic data (Gierasch, Conrath & Magalhaes, 1986). In addition, stratospheric and upper tropospheric circulation has been modeled by West, Friedson & Appleby (1992). Although it is understood that the magnitude of zonal winds must go to zero relative to the rate of rotation of the bulk of the planet, there are few observational constraints which define the depth to which zonal winds penetrate. Ingersoll, Beebe, Conrath & Hunt (1984) have argued in favor of deep penetration of the winds in order to explain Saturnian winds, which, at the level of the visible cloud deck, blow from west to east over most of the planet.

Throughout the Voyager era, detectors capable of high spatial resolution were limited to visible wavelengths. A review by West, Strobel & Tomasko (1986) summarizes the extent to which these data constrained models of the upper atmosphere. Development of infrared sensitive detectors and narrow-band interference filters has allowed the strong frequency dependence of the methane gas to be utilized to selectively sound the atmosphere above the ammonia cloud deck (Orton, *et al.* 1994). Low IR emissivity tends to correlate with low reflectivity of clouds at visible wavelengths; however, infrared monitoring reveals nonseasonal variability and a lack of correlation between IR emission and low cloud reflectivity within cyclonic regions, such as the South Equatorial Belt and North Temperate Belt, which display aperiodic convective disturbances.

Gierasch & Conrath (1993) point out that, until probes have measured conditions below the ammonia cloud deck at a sufficient number of latitudes, the atmospheric energy balance will not be well understood. Although Mac Low and Zahnle (see their chapters) claim that comet fragments may have penetrated no further than 200 millibars, hopefully the December 7, 1995 entry of the Galileo probe is the first of a series of future probes to sound down through the water cloud at various latitudes. In the meantime, the Hubble Space Telescope (HST) Wide Field Planetary Camera 2 (WFPC2), Faint Object Camera, Faint Object Spectrograph and Goddard High Resolution Spectrograph provide ultraviolet capabilities and spatial resolution more than ten times better than can be obtained from Earth. These instruments allow sensing of the upper atmosphere over a wavelength range from 0.12 to 1.0 microns. Unlike the Voyager high resolution data, which spanned a few months, HST (with high resolution cameras and spectroscopic instruments) has a nominal expectancy of about 15 years, more than a Jovian year.

The WFPC2 allows two modes of operation; planetary camera mode (PC) samples at 0.0455 arcsec/pixel and wide field mode (WF), using detector 3, samples at 0.0995

FIGURE 1. Latitude-Longitude map of impact sites. This map extends from $-65°$ to $5°$ in latitude and (left to right) from $205°$ through $0°$ to $-35°(325°)$ W longitude. These data were obtained on July 23, 1994. The A, E, H, Q1, R, G-S and L sites are visible from left to right in the projection.

arcsec/pixel (Burrows, *et al.* 1994; Holtzman, *et al.* 1995). For the SL9 data, this leads to a pixelation scale at the sub-spacecraft point on Jupiter of 170 and 375 km, respectively (although the resolution of the PC mode of WFPC2 is somewhat poorer than the pixelation scale, Wiener deconvolution of F410M images with a theoretical point-spread function increased the contrast but neither positively enhanced the resolution nor negatively introduced systematic structure in the data). Figure 1 illustrates the effective resolution of these two modes of operation. This latitude-longitude map is composed of WF images (left) and PC (right) images projected to optimize the PC resolution. The images were enhanced by removing large scale limb-darkening with a Minnaert filter (Minnaert, 1941) where

$$B(\mu,\mu_0) = B_0(\mu,\mu_0)^k/\mu, \qquad 0.0 < k < 1.0 \qquad (1.1)$$

is the estimated surface brightness expressed as a function of the cosine of the local emergent, μ, and incident, μ_0, angles and B_0 is the value when μ and μ_0 equal 0. The value of k can be determined by a least squares fit of the log of equation 1.1 and although this function has no physical significance, it is an effective scene flattener. The brighter the limb of the planet, the smaller the k value will be; for example, $k = 0.5, 0.6$ and 1.0 are effective for the F336W, F410M and 890-nm methane filters, respectively. After the large scale limb darkening was removed, the contrast of the remaining high frequency component of the images was enhanced and maps like the one shown in Figure 1 were constructed. In regions where images obtained in consecutive HST orbits overlap, selection of the brightness value from the image with maximum value of $(\mu \times \mu_0)$ yields the most consistent longitudinal contrast. In the case of a remote planet like Jupiter, the incident sunlight and emergent viewing angle are similar, therefore this criterion selects regions closest to the central meridian or samples the underlying clouds through minimum path lengths of the overlying hazes and Rayleigh scatterers.

The following analysis and discussion is based on imaging data, obtained by the HST/SL9 Campaign. Two major constraints limit interpretation of the evolution of individual impact sites. Because there is considerable vignetting in the methane filter, the WF mode was frequently used to obtain photometrically calibrated global coverage. In addition, the telescope was shared with other non-SL9 programs during the week of impacts, resulting in less than optimal (10-hr. intervals) sampling. Observing constraints were such that there was 38 to 42 min. of observing time during each 96 min. orbit. WFPC2 is an all purpose camera, optimized for imaging faint objects, therefore it does not have a fast readout option. Nominal setup and readout time is about 3 minutes per

Image ID	Time d/hr/min	Mode	CM(sys III)[†]	Elapsed Time (hr)	Elapsed Rotations
u2fi0h05	17/17/09	WF	120.33	0.00	0.00
u2fi0w05	19/09/20	PC	137.67	40.18	4.05
u2fi1c05	21/09/35	PC	87.54	88.43	8.91
u2fi1m05	22/06/34	WF	128.54	109.41	11.02

[†] System III latitude is defined as $L(t) = 217.956 + 870.536$ deg/day$(t-t_0)$, where $t_0 = 0$ hr UT Jan 1, 1965 in Riddle, A. C. & Warwick, J. W. 1976 Redefinition of System III Longitude, *Icarus* **27**, 457–459.

TABLE 1. July 1994 images used to measure the zonal winds

exposure; and, depending on the filters and whether new guide stars had to be acquired, 9–13 images were obtained within a given orbit. Although images with filters F953N, FQCH4N15 (889 nm), F673N, F555W (PC), F547M (WF), F410M, F336W, F255W, F218W are available (where W, M and N indicate wide, medium and narrow bandpasses and the three digits the effective wavelength of the filter in nm), in this analysis the 410, 889, 336, and 255-nm images have been utilized to define the extent of the impact sites and the 410-nm images to measure horizontal wind fields.

2. Zonal winds

A set of images obtained with the F410M filter was selected to measure the zonal (east-west) tropospheric winds. This filter is similar to the Voyager 2 violet filter, used by Ingersoll, *et al.* (1981) to establish the average zonal wind as a function of latitude. Table 1 lists the characteristics of the images. Although a sequence of 4 images separated in time by equal increments of 9.925 hrs. (the rotation period of Jupiter) would be ideal, the following table, containing frames selected to span a selected range of longitudes, illustrates the degree of compromise that was required to utilize allocated observing time to attain the goals of all campaign participants while providing continuing observing time for the rest of the astronomical community.

Images in Table 1 were cleaned to remove cosmic ray impacts and a least squares fit of the limb was used to determine the center of the planet. At the time of observations, the range to the planet was 764,450,000 km, subspacecraft latitude was $-3.00°$, subspacecraft longitude is tabulated as the CM (central meridian) in Table 1, subsolar latitude is $-3.12°$ and subsolar longitude is 10–11 degrees less than the CM (east). Polar (R_p) and equatorial radii (R_e) were assumed to be 66,854 and 71,492 km, respectively. Large scale limb darkening was removed, using the Minnaert function with $k = 0.6$. Maps centered on the same central latitude and longitude were constructed from all four images. In an effort to select cloud features that are true markers of the local winds, software which allowed sequential display of the mapped time sequence was utilized and cloud features were selected based on their evolution in the time sequence. Because desirable wind tracers in rapidly moving cloud systems evolved rapidly, their horizontal displacement was determined in the last two frames (separated by 21 hours) and the resulting displacement vector reduced to zonal and meridional components.

Measured east-west rates of translation of selected clouds are plotted in Figure 2 as a function of planetographic latitude. Planetographic latitude is defined as the angle of intersection of a local normal with the equatorial plane of the planet. On an elliptical

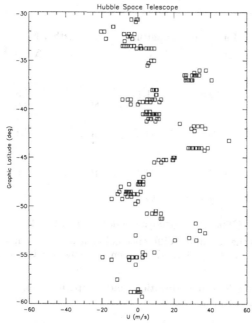

FIGURE 2. Zonal wind as a function of planetographic latitude.

planet such as Jupiter, the normal to a local equipressure surface (corresponding roughly to the cloud deck) passes through the center of the planet only at 0° and 90°. Planetocentric latitude is defined as the angle subtended between the equatorial plane and a vector passing through the planet center and a local point on the planet. Although the magnitudes of planetographic (θ_g) and planetocentric (θ_c) latitude converge at 0° and 90°, planetographic latitude is larger at all other latitudes and the relation between the two is:

$$tan(\theta_g) = (R_e/R_p)^2 tan(\theta_c) \qquad (2.2)$$

and $(R_e/R_p)^2 = 1.14356$ for Jupiter.

Using a least squares fitting routine, the planet center can be located to within 0.02 degrees in longitude and latitude if sufficient limb is present in the images. Pixelation of images in PC mode is such that a displacement of 1 pixel/20 hrs corresponds to 3–4 m s^{-1}, thus the measuring error associated with individual points in Figure 2 is on the order of 8 m s^{-1}. Dispersions greater than this at a given latitude are due to the local turbulent flow.

When the resulting wind profile is compared to Voyager results (Ingersoll, *et al.* 1981), the magnitude of the winds and latitudes of minima and maxima are the same to within the measuring error for latitudes ranging from −60° to −30°. Although the White Ovals (FA, DE and BC—left to right in Figure 1) have drifted closer together than they have ever been since their formation in 1938 and the number of small anticyclonic systems at −40° (north of the impact sites in Fig. 1) is now 6 compared to 12 during the Voyager era, the zonal wind component as a function of latitude is unaltered. This implies that the zonal wind profile is not readily altered by typical tropospheric changes and that Jovian meteorology differs greatly from that on Earth. Unlike terrestrial storm systems, Jovian meteorology is dominated by the strong east-west flow. Tropospheric cloud systems which are generated in response to convective disturbances cannot migrate north or south but are trapped and driven by the strong, nearly constant, east-west wind field.

Site	Latitude centric	Latitude graphic	Longitude[†] Sys. III °	Other Nearby Sites	Δ long[‡] °	Δ t hrs.
B	−42.8	−46.6	71.1	Q1	−4.8	89.37
				N	2.0	79.57
D	−43.3	−47.1	33.5	G	−6.1	19.71
				S	0.5	99.45
N	−43.4	−47.3	73.1	B	−2.0	−79.57
Q2	−44.7	−48.5	47.5	R	−3.9	9.80

[†] Hammel, *et al.* (1995). Longitudes are determined to ±2°

[‡] Δ longitude is the estimated difference in longitude from impact center to impact center and Δ t is the time elapsed between impacts of sites within 10 degrees of specified site.

TABLE 2. Impact locations and competing impact sites

3. Temporal behavior of SL9 impact sites

SL9 impact sites can be divided into classes based on the size of the site. The sizes vary from sites that are barely detectable in the PC frames (subtending 170 km/pixel at the subspacecraft point) to site L, with visible ejecta extending from −65° to −30° latitude when the site is viewed near the terminator (sunset line) 4 days after impact (see 3.3.2). Because evolution of a site will depend on conditions of the initial impact as well as interaction with the local wind field after the ejecta enter the lower stratosphere and troposphere, several examples from the small, intermediate and large sites (Hammel, *et al.* 1995) are presented to illustrate the degree of variability in the evolution of impact sites.

3.1. *Small sites*

Impact sites associated with faint fragments (Weaver, *et al.* 1995) that can be characterized with the WFPC2 data are B, D, Q2 and N. Although these features are difficult to resolve in the WF mode, their color dependence in images obtained with filters that sound down to the ammonia cloud deck is similar to that of larger sites, and, like larger sites, they tend to be bright in the 889-nm images. Whether the observed structures of these small sites have formed from ejected material or residue from the bolide that was generated when the fragment entered the atmosphere is not clear (see chapter by Hammel). The morphology and time dependence of these features are reviewed in this section.

3.1.1. *Position of impacts relative to the wind field*

All four fragments entered the atmosphere in a region of cyclonic wind shear. Table 2 summarizes the impact locations. Relative to the System III rotation rate (870.536° day^{-1}), the zonal winds (see Fig. 2) have latitudinal shears of 6.67 m s^{-1} deg^{-1} from −49.0° to −47.5°, 3.2 m s^{-1} deg^{-1} from −47.5° to −45.5°, and more than 10 m s^{-1} deg^{-1} from −45.5° to −43.0° latitude. Thus B, D and N entered the atmosphere in a region with little latitudinal shear in the zonal winds, while Q2 impacted in a region where the shear is a factor of 2 larger.

λ_{eff} nm	Frame ID	Time dd hh mm	Lat	Long	μ_0	μ	Lat	Long	μ_0	μ
				B Site				*D Site*		
225	u2fi0g07	17 15 39	−46.8	72.5	0.72	0.70	−47.6	33.5	0.61	0.67
336	u2fi0g06	17 15 33	−46.8	72.5	0.71	0.68	−47.5	34.3	0.63	0.68
410	u2fi0g05	17 15 30	−47.0	72.5	0.70	0.67	−47.4	34.1	0.64	0.69
889	u2fi0g01	17 15 19					−47.3	33.9	0.68	0.71
410	u2fi0v01	19 07 46					−47.5	34.4	0.50	0.59
				N Site				*Q Site*		
410	u2fi0c05	21 09 35	−48.1	71.5	0.68	0.70	−48.1	46.8	0.54	0.62
889	u2fi0c01	21 09 24	−47.9	71.6	0.70	0.71	−48.3	47.0	0.59	0.65

TABLE 3. July 1994 Observational parameters of visible centers small sites

3.1.2. *The reflectivity and dispersal of small sites*

From Table 2 it is apparent that site B is the only small site that is located in an undisturbed region for more than 20 hours. However, the B fragment generated the smallest impact site, which is not bright in the 889-nm filter. Figure 3 is a map of the B and D sites constructed from (top to bottom) F255W, F336W, F410M and 889-nm methane images. Although site B shows no brightening at 889 nm, a comparison of site B with surrounding dark tropospheric features in 255, 336 and 410-nm images is consistent with dark material located higher in the atmosphere than the other dark features, where it is not as affected by overlying aerosol and Rayleigh scattering. Figure 4 contains 410-nm images where the top image is the third image in Figure 3. The second image is a 410-nm image 42.3 hours later and the bottom two are 410 and 889-nm images obtained about 90 hours after those in Figure 3, when the N and Q2 fragments are present. Here both sites are visible as faint bright spots in the 889-nm image and comparison of the B and N sites indicates that B was no longer visible when N entered. (Note: color dependence implies the dark particles were at high altitudes where zonal wind velocities should be small. Evaporation or growth of particles may affect the optical depth of the debris.) Table 3 lists image ID, time of observation, latitude and longitude of the center of the observed sites, and cosine of incident and emergent angles of site centers (the optical path is inversely proportional to the cosine of the angles for the optically thin case).

The G impact occurred on July 18 at 7:35 UT, contaminating the small D site in the 2 lower frames in Figure 4. Comparison of the top 3 frames in Figure 4 indicates that site D dispersed westward, which is consistent with the sense of the underlying tropospheric winds (see Fig. 2) and indicates that some of the dark material was deposited or had migrated down to the lower stratosphere. In the two bottom frames in Figure 4 site Q2 is east and site N is west of the intermediate Q1 site. Site R is east of Q2. The N and Q2 fragments impacted on July 20 and during the following three day period of the HST/SL9 intensive observing program they changed very little. Thus, these limited observations indicate that only D, the largest of the four sites, perturbed by the nearby impact of G, has dark material extending low enough into the stratosphere to interact with the tropospheric winds during this period.

3.2. *Intermediate Sites*

Table 4 contains impact parameters for the individual intermediate sites. The fact that R and S arrived later than G and impacted less than 17 degrees west of this large site

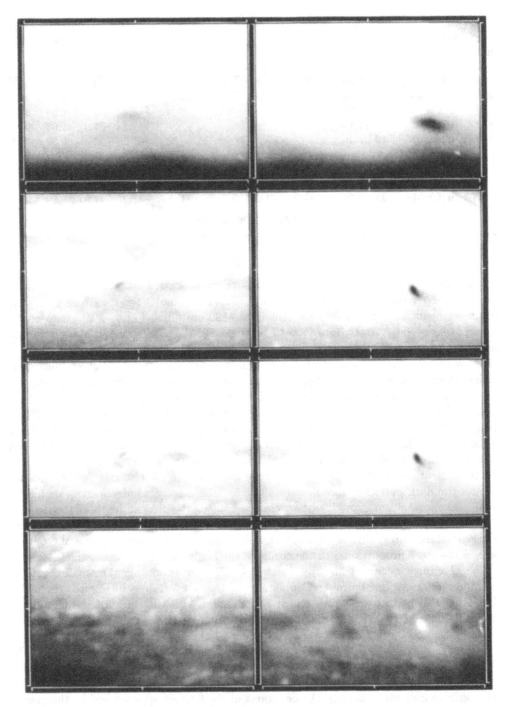

FIGURE 3. Comparison of the B site in filters with effective wavelengths of (top to bottom) 255, 336, 410, and 889 nm.

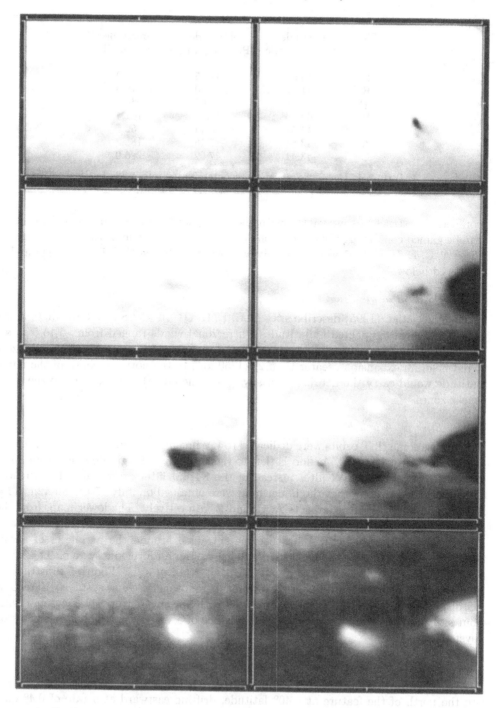

FIGURE 4. Comparison of sites B, D, N, and Q2 in filters with effective wavelengths of 410 (top three) and 889 nm (bottom).

Site	Latitude Planetocentric	Latitude Planetographic	Longitude System III
S	−43.91	−47.75	34.0
R	−44.17	−48.00	46.8
Q1	−43.41	−47.02	66.3
H	−43.66	−47.27	101.4
E	−44.54	−48.14	152.5
A	−43.54	−47.15	186.3
C	−43.41	−47.02	225.0

TABLE 4. Impact location of uncontaminated intermediate sites

confuses the structures associated with these fragments to the point that little insight can be gained concerning the interaction of these sites with the undisturbed environment. Sites A, C, E, H, and Q1 were relatively well observed during the SL9 campaign and are reviewed below.

3.2.1. *Position in the zonal winds*

Hammel, *et al.* (1995) describe sites A, C, E, H, Q1, R, and S as intermediate in size. These fragments impacted at latitudes where zonal winds ranged from −5 to 5 m s^{-1}, however, if the ejecta from the sites were at a level where it was sensitive to the zonal wind field, an expansion greater than 2° (1750 km) to the north or south of the impact latitude would carry the debris into eastward (to the north) or westward (to the south) shear zones.

3.2.2. *Sites C, A and E*

Because some HST orbits were dedicated to UV imaging, in order to obtain the earliest combined view of sites C, A, and E (left to right), latitude-longitude maps from images obtained with the F336W filter instead of the F410M filter are presented in Figure 5. The data used to construct the upper map were obtained on July 17 at 18h 42m, 11.47, 22.48, and 3.50 hrs. after impacts A, C, and E, respectively. The lower map, from data obtained on July 23 at 14h 42m UT, illustrates the evolution of the features over a five day period. Note that site A has become entrained around the perimeter of a white cloud at −51° latitude that has the morphology of an anticyclonic system. This system is moving eastward at a rate of about 1.6° day^{-1} (16 m s^{-1}). Figure 5 illustrates the sensitivity of evolution of impact sites to local tropospheric cloud systems. Here C and E dispersed in a manner that is consistent with the mean tropospheric zonal winds, but site A is drastically modified by the presence of a well-developed anticyclonic storm system.

3.2.3. *Site H*

Evolution of the H site has been modified by the presence of an anticyclonic system to the north of the feature at −40° latitude, drifting eastward at a rate of 0.4° day^{-1} (4 m s^{-1}) relative to System III. Figure 6 contains a sequence of map projections that reveals the development of the site during its first 5 days. Corresponding information on times of observation is tabulated in Table 5.

Maps in Figure 6 are centered on −48° latitude and 99.5° longitude and extend ±15° in both latitude and longitude. The average zonal wind field has been utilized to compute the actual longitudinal translation of the cloud deck that would occur between the times

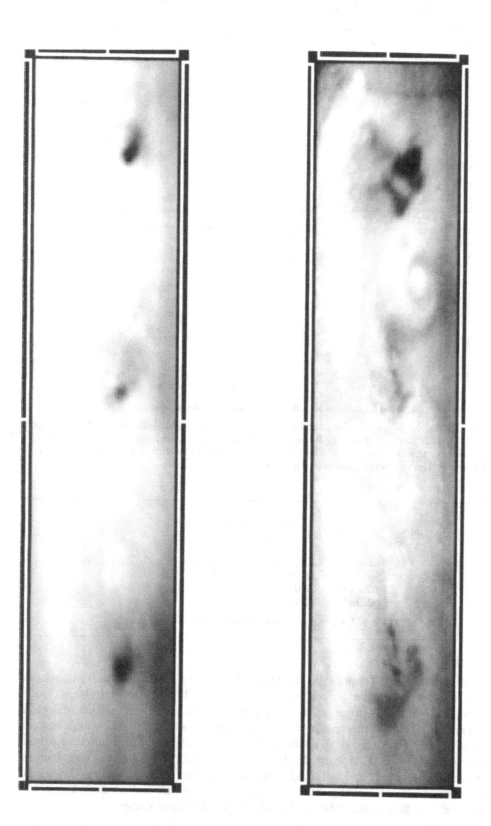

FIGURE 5. Evolution of intermediate sites C, A, and E. The two maps were generated from images obtained on July 17, when all sites were less than 24 hrs. old, and from a second image 5 days later.

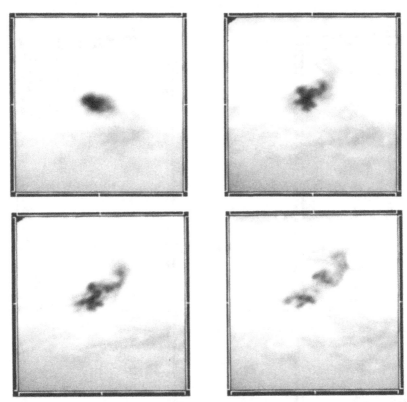

FIGURE 6. Evolution of site H. The maps at upper left, upper right, lower left and lower right are 12.2, 62.0, 81.5, and 111.9 hours after impact of the fragment, respectively. The dotted line superimposed on the lower right map represents tropospheric cloud displacement over 52 hrs., the difference in time between the upper right and lower right figures.

ID	Time dd hh mm	Δ t since impact hrs
u2fi0v05	19 07 46	12.2
u2fi1c05	21 09 35	62.0
u2fi1l05	22 04 58	81.5
u2fi1u05	23 11 26	111.9

TABLE 5. Observational parameters of Site H images

of acquisition of u2fi1c05 (upper right) and u2fi1u05 (lower right). The winds range from -10 to 15 m s^{-1} in this region and during the intervening 50 hours the underlying clouds translated as much as 10° in longitude. In general, average meridional motion of Jovian clouds is not detectable at this spatial resolution unless a closed turbulent cyclonic system is present. In this case, though, a northward velocity of 25 m s^{-1} is required to attain the observed expansion. This figure illustrates that the dark ejecta become entrained in lower stratospheric winds and, although some entrainment occurred around the bright anticyclonic feature, some of the dark material continued northward where it was swept into the westward flow around the northward edge of the storm center.

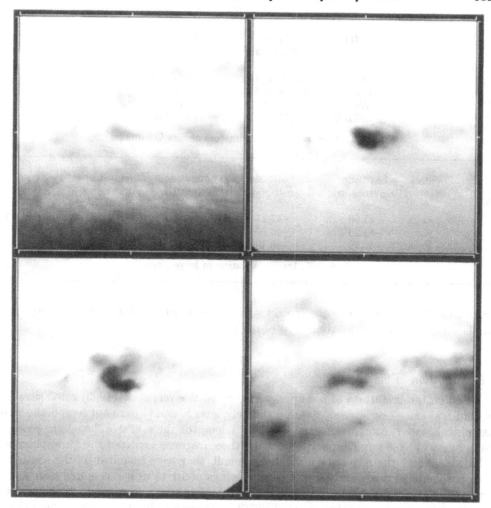

FIGURE 7. Development of the Q1 site over more than a month.

3.2.4. *Site Q1*

Figure 7 shows the longterm development of site Q1, an intermediate with no nearby anticyclonic systems at the time of impact. These maps are centered on −47.5° latitude and 64° longitude. They extend ±15° in both latitude and longitude. During the pre-impact phase two dark spots at the latitude of the impact extended from 57.1° to 51.4° (eastern spot) and 67.3° to 63.25° (western spot). During the week that the SL9 impacts occurred, these features changed very little, indicating minimum turbulence associated with typical tropospheric low albedo regions at this latitude. The time sequence (upper left, upper right, and lower left) reveals a small white oval surrounded by a dark collar which was translating eastward at a rate of 3.5° day^{-1} (34 m s^{-1}). Passage of this feature perturbed the structure of the Q1 ejecta. On August 23 a white oval was located north-west of site Q1 (lower right). This anticyclonic structure was centered at −40.5° latitude and 72.3° longitude. It was translating eastward at a rate of 0.61° day^{-1} (6.8 m s^{-1}). This was the same white oval that was interacting with site H on July 23. During the first week site Q1 was not greatly perturbed by local storm systems. By August 23, it

ID	Time m dd hh mm	Δ t since impact hrs
u2fi0v05	07 19 07 46	−60.6
u2fi1c05	07 21 09 35	13.4
u2fi1t05	02 23 09 49	61.5
u2fi2n05	08 23 02 54	798.8 (33.3 days)

TABLE 6. Observational parameters of Site Q1 images

Site	Latitude Planetocentric	Latitude Planetographic	Longitude System III	Time of Impact dd hh mm
G	−43.66	−47.27	26.8	18 07 35
K	−43.29	−46.90	282.6	19 10 31
L	−42.79	−46.40	351.6	19 22 21

TABLE 7. Impact location of large sites

was extending to the west of the impact sight in agreement with reduced stratospheric winds that reflect the mean zonal flow of the troposphere.

3.3. *Large sites*

3.3.1. *Position in the zonal winds*

Based on the brightness of pre-impact fragments, Weaver, *et al.* (1995) rated sites G, K, and L as class 1 or large sites. Unfortunately, sites K and G were not temporally well sampled; however, comparisons can be made among the three. Table 7 contains parameters related to the large impacts. Although these fragments entered the atmosphere at a latitude where the mean zonal winds are small, they were bounded on the south by a tropospheric jet with maximum westward winds of 10–15 m s^{-1} centered near 48.5° latitude. Because the ejecta have been thrown high into the atmosphere above aerosol and Rayleigh scatterers (Munoz, *et al.* 1996), obscuration due to comet ejecta will be particularly sensitive to effective path length through the absorbers and most visible near the planet limb (note: this is markedly different from Jovian tropospheric features, which are dark at wavelengths between 0.5 and 1 micron. They brighten near the limb or terminator where high altitude scattering greatly reduces contrast with surrounding clouds). The top right and left images in Figure 8 show site L rotating onto the western limb of the planet and approaching the eastern terminator, respectively. The lower images are of site K (left) crossing the limb and site G (right) approaching the limb. It is apparent that material ejected from site L spans latitudes from −65° to −30°. Material that is ejected further south than −50° or that extending northward would be entrained in an eastward flow on return to the lower stratosphere or troposphere. Comparison of sites K and L indicates that K ejecta were not as wide spread as those from the site L and, although the R and S impacts have entered near site G, the total latitudinal dispersion is not greater than that from site L.

3.3.2. *Site L*

Figure 9 shows three stages of development of site L. The site was observed 22.6 (left), 32.1 (center), and 81.9 (right) hours after impact. The dotted profile on the right map shows translational motion of tropospheric features drifting with the mean zonal

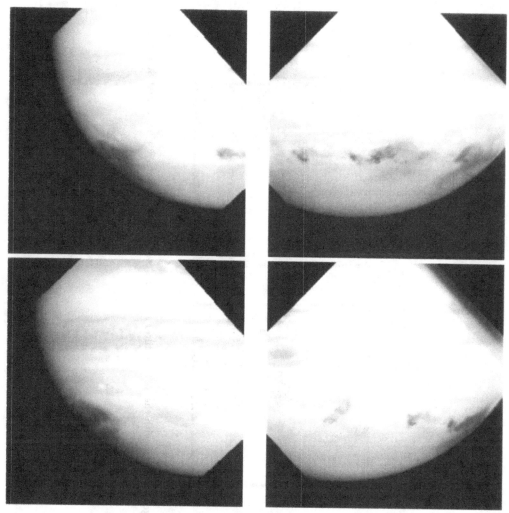

FIGURE 8. Terminator and limb views of sites G, L and K.

winds that would occur during 49.8 hours that have elapsed between the center and right images. Comparison of the three images suggests rotation about the impact center. The measured translation of dark structures around the westernmost perimeter of the darkest region implies an anticyclonic rotation with perimeter velocities on the order of 45 m s^{-1}. At the same time, a dark region to the southeast of the impact center expands eastward 1.2° of longitude over this 82-hour interval.

Table 8 contains the observational parameters for site L. Comparison of these parameters with those in Table 5 reveals that both sites H and L were observed about 81 hours after impact. Figure 10 is a comparison of site L (upper) and site H (lower). The left images show the impact sites before arrival of the L and H fragments, the central (410-nm) and right (889-nm) images compare the sites 81.5 hours after impact. Note that in the case of site L, the dark streak was not disturbed by impact L. There is a well-formed cyclonic storm center to the north of the site and the dark material does not encroach into it. Simon & Beebe (1996) measured winds of about 45 m s^{-1} around the northern perimeter of this system. The return westward flow along its southern edge was less

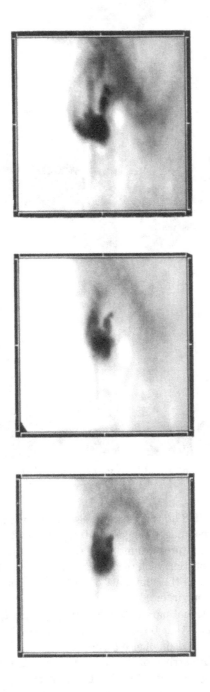

FIGURE 9. Evolution of site L. These 3 images, obtained with the 410-nm filter show the evolution of the site during its first 82 hours.

ID	Time	Δ t since impact
	dd hh mm	hrs
u2fi6803	20 20 56	22.6
u2fi1905	21 06 28	32.1
u2fils05	23 08 12	81.9

TABLE 8. Observational parameters of Site L images

organized but of similar magnitude. Such a system should have a central subsidence, and conservation of mass flow would require upwelling around the perimeter. Observed morphology of site L suggests that this local upwelling constrains northward expansion of the ejecta. Because the local flow is westward, this local storm cell could contribute to the counterclockwise rotation of more southerly dark material.

A bright oval centered at −51° latitude drifts eastward at a rate of about 1.6° day^{-1} (16 m s^{-1}). During the time since impact it has shifted eastward 5.5° relative to the impact site. The central upwelling associated with this anticyclonic system would be associated with subsidence around the perimeter of the cloud system. This would encourage concentration of dark particles along the southern perimeter of the impact site near −49° latitude.

The fact that the underlying dark tropospheric streak (Figure 10—upper left) appears undisturbed limits the extent to which the L impact has disturbed the local troposphere and is consistent with either shallow penetration of the impactor or a clean entry and exit of a more deeply penetrating fragment. Failure of the dark material to disperse according to the predicted pattern inferred by the mean tropospheric zonal flow, along with the constrained nature of the core of the impact site and apparent rotational speed, are consistent with a rising central region. However, the fact that the expanding stratospheric material encounters a westward flow to the north and a well-formed anticyclonic system is passing to the south can impose the observed structure on the expanding ejecta. Comparison of the 410 and 889-nm images of site H in Figure 10 (lower central and right images, respectively) with those of site L (upper center and right images) when both sites were about 82 hours old does not infer stronger upwelling at site L. In addition, visibility of anticyclonic storm centers in the 889-nm images (upper and lower right) in Figure 10 demonstrates that these systems are elevated and their associated circulation could influence dispersion of the dark ejecta.

The manner in which intermediate sites A and H responded to nearby anticyclonic systems is consistent with this interpretation of the evolution of site L. Inspection of the H impact site in Figure 10 (lower left) reveals that the fragment entered a local turbulent cyclonic system. The flocculent structure of site H in Figure 6 is consistent with local entrainment into the vertical motions within the system.

3.3.3. *Sites G and K*

The well-known "Bull's eye" was imaged on July 18, 1.5 hr. after impact (Ingersoll & Kanamori, 1996). Due to the fact that fragments R and S entered the atmosphere near site G, evolution of the original site is confusing but comparison with the ejecta blanket from sites L and K in Figure 8 indicates that ejecta from the G complex were more extensive than that from K.

From the time of impact of the K fragment until July 23, when a set of images containing sites L, K and G was obtained, no clear view of site K was acquired with the WFPC2.

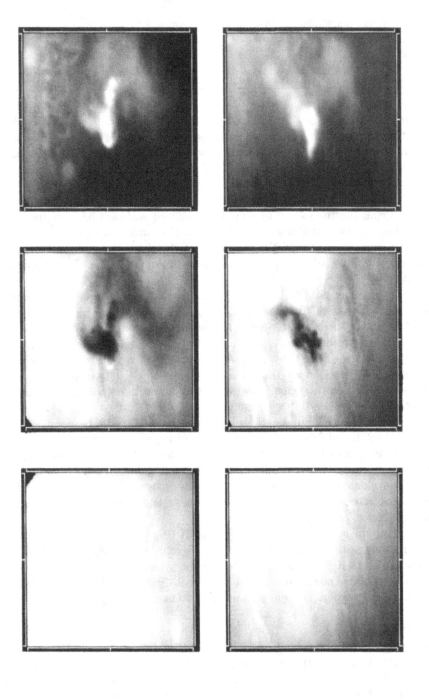

FIGURE 10. Comparison of local cloud structure and vertical development of sites L and H. The left maps show the sites before impact and the center and right maps are from 410-nm and 889-nm images about 82 hours after impact.

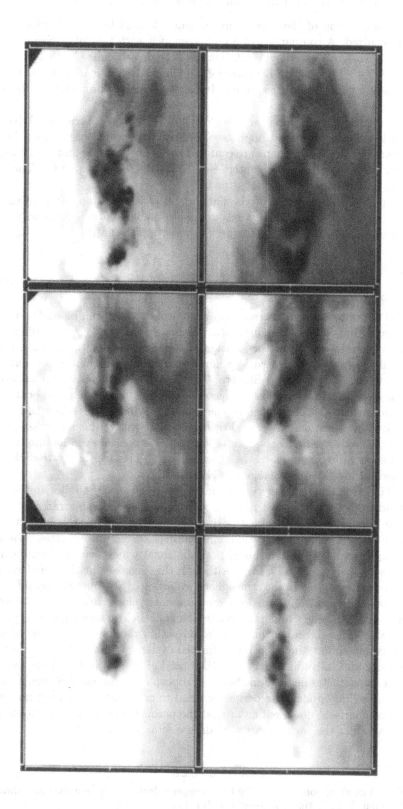

FIGURE 11. Dispersion of sites K, L and G. K is on the left, L is at center and G is on the right.

By the time of acquisition of the July 23 image more than 92 hours had elapsed since the K impact occurred. A comparison of a limb view of K with that of L (See Figure 8) indicates that K had a less extended ejecta blanket. Simon & Beebe (1996) have shown that, during the first 92 hours, site K evolved in a manner that indicated that it was not strongly influenced by local weather systems. It was dispersing as predicted by the mean zonal flow.

Figure 11 compares the evolution of sites K (left) and G (right) sites with site L (center) during the week following the final impacts. The top maps were obtained on July 23 and the lower on July 30. These maps are centered at $-47.5°$ latitude and at the impact longitudes, 283° (left—site K) 352° (center—site L) and 27° (right—site G). The maps extend $±15°$ in latitude and $±20°$ in longitude.

During this period the dispersal of the dark material is consistent with subsidence into the region where the mean zonal winds dominate the circulation. Similarity of the dispersal of passive site K and site L is further evidence that the anticyclonic morphology of site L was not internally driven.

4. Conclusions

SL9 WFPC2 data provide a high resolution wind profile, which can be compared with that derived from the Voyager imaging data. Although the wind profile indicates no significant differences in magnitude or location of the wind maxima and wind minima, the longitudinal spacing associated with cyclonic-anticyclonic weather systems at $-41°$ latitude is a factor of 2 larger for the SL9 era. This result has significant implications. It supports the argument that the winds are deep-seated, driven by an internal heat source and that the meteorology is constrained by the prevailing winds. Morphology differences may be generated by variations in vertical transport rates within this stable zonal flow.

Morphology of small sites indicates that dark material is present even for site B, which is not detectable as a bright patch in the 889-nm image. This can imply that the dark material formed during the bolide stage as well as later.

Evolution of individual impact sites is complex and varies from site to site. Initial ejecta blankets extend from $-65°$ to $-30°$ latitude and contain considerable radial structure relative to the impact site. Although the sites are modified by nearby impacts, they are much more influenced by local cyclonic and anticyclonic storm systems. Gierasch, Conrath & Magalhaes (1986) have shown that the zonal winds decrease with altitude within 2 to 3 scale heights (50–75 km). Although this decrease in the zonal winds is expected due to frictional drag in the atmosphere, evolution of SL9 impact sites infers that circulation associated with well-formed cyclonic and anticyclonic storm centers does not decrease as rapidly with height and has considerable influence on developing morphology of the impact site. This indicates that circulation within the lower stratosphere would deviate from the mean flow above both well-formed cyclonic and anticyclonic storms.

The complexity and latitudinal extent of the initial ejecta blanket and the extent of dispersal of the impact sites depend on the local tropospheric weather systems. Cyclonic systems, with upwelling around the perimeter, tend to constrain and inhibit northward expansion. In comparison, downwelling around the perimeter of anticyclonic systems to the north and south of the sites entrains the dark ejecta.

REFERENCES

BEEBE, R. F. & YOUNGBLOOD, L. A. 1979 Pre-Voyager velocities, accelerations and shrinkage rates of Jovian cloud features. *Nature* **280**, 771–772.

BEEBE, R. F., ORTON G. S. & WEST R. A. 1989 Time-variable nature of the Jovian cloud properties and thermal structure: an observational perspective. In *Time-Variable Phenomena in the Jovian System* (ed. M. J. S. Belton, R. A. West and J. Rahe). NASA SP-494, 245–288.

BURROWS, C. J., CLAMPIN, M., GRIFFITHS, R. E., KRIST, J., & MACKENTY, J. W. 1994 WFPC2 instrument handbook, version 2.0. Space Telescope Science Institute.

CHAPMAN, C. R. 1969 Jupiter's zonal winds: variation with latitude. *J. Atmos. Sci.* **26**, 986–990.

DOWLING, T. E. & INGERSOLL, A. P. 1989 Jupiter's Great Red Spot as a shallow water system. *J. Atmos. Sci.* **46**, 3256–78.

DOWLING, T. E. 1993 A relationship between potential vorticity and zonal wind on Jupiter. *Science* **50**, 14–22.

GIERASCH, P. J., CONRATH, B. J., & MAGALHAES, J. A. 1986 Zonal mean properties of Jupiter's upper troposphere from Voyager infrared observations. *Icarus* **67**, 456–483.

GIERASCH, P. J. & CONRATH, B. J. 1993 Dynamics of the atmosphere of the outer planets: post-Voyager measurement objectives. *J. Geophys. Res.* **98**, 5459–5469.

HAMMEL, H. B., BEEBE, R. F., INGERSOLL, A. P., ORTON. G. S., MILLS, J. R., SIMON, A. A., CHODAS, P., CLARKE, J. T., DE JONG, E., DOWLING, T. E., HARRINGTON, J., HUBER, L. E., KARKOSCHKA, E., SANTORI, C. M., TOIGO, A., YEOMANS, D. & WEST, R. A. 1995 Hubble Space Telescope imaging of Jupiter: atmospheric phenomenon created by the impact of comet Shoemaker-Levy 9. *Science* **267**, 1288–1296.

HOLTZMAN, J., HESTER, J. J., CASERTANO, S., TRAUGER, J. T., WATSON, A. M., BALLESTER, G. E., BURROWS, C. J., CLARKE, J. T., CRISP, D., EVANS, R. W., GALLAGHER III, J. S., GRIFFITHS, R. E., HOSSEL, J. G., MATTHEWS, L. D., MOULD, R. J., SCOWEN, P. A., STAPELFELDT, K. R., & WESTPHAL, J. A. 1995 The performance and calibration of WFPC2 on the Hubble Space Telescope. *PASP* **107**, 156–178.

INGERSOLL, A. P. & CUZZI, J. N. 1969 Dynamics of Jupiter's cloud bands. *J. Atmos. Sci.* **26**, 981–985.

INGERSOLL, A. P. BEEBE, R. F., MITCHELL, J. L., GARNEAU, G. W., YAGI, G. M., & MULLER, J. P. 1981 Interactions of eddies and mean zonal flow on Jupiter as inferred from Voyager 1 and 2 images. *J. Geophys. Res.* **86**, 8733–8743.

INGERSOLL, A. P., BEEBE, R. F., CONRATH, B. J. & HUNT, G. E. 1984 Structure and dynamics of Saturn's atmosphere. In *Saturn* (ed. T. Gehrels & M. S. Matthews), pp. 195–238. Arizona Press.

INGERSOLL, A. P. AND KANAMORI, H. 1996 Two waves from SL9 to probe the Jovian water cloud. *Icarus* Special SL-9 Issue (in press).

LIMAYE, S. S., REVERCOMB, H. E., SROMOVSKY, L. A., KRAUSS, R. J., SANTEK, D. A., SUOMI, V. E., COLLINS, S. A., & AVIS, C. C. 1982 Jovian winds from Voyager 2, part I: zonal mean circulation. *J. Atmos. Sci.* **39**, 1413–1432.

LIMAYE, S. S. 1986 Jupiter: new estimates of the mean zonal flow at the cloud level. *Icarus* **65**, 335–352.

MARCUS, P. S. 1993 Jupiter's Great Red Spot and other vortices. *Ann. Rev. Astron. Astrophys.* **31**, 523–73.

MINNAERT, M. 1941 The reciprocity principle in lunar photometry. *Astrophys. J.* **93**, 403–410.

MUNOZ, O., MORENO, F. AND MOLINA, A. 1996 Aerosol properties of debris from fragments E/F of comet Shoemaker-Levy 9 *Icarus* Special SL9 Issue (in press).

ORTON, G. S., FRIEDSON, A. J., YANAMANDRA-FISHER, P. A., CALDWELL, J., HAMMEL, H., BAINES, K. H., BERGSTRALH, J. T., MARTIN, T. Z., WEST, R. A., VEEDER JR., G. J., LYNCH, D. K., RUSSELL, R., MALCOM, M. E., GOLISCH, W. F., GRIEP, D. M., KAMINSKI, C. D., TOKUNAGA, A. T., BARON, R., HERBST, T. & SHURE, M. 1994 Thermal maps of Jupiter: Spatial organization and time dependence of tropospheric temperatures, 1980–1993. *Science* **265**, 625–631.

PEEK, B. M. 1958 *The Planet Jupiter.* Faber and Faber.

REESE, E. R. (Private communication with longterm BAA & ALPO observer)

SIMON, A. A. AND BEEBE, R. F. 1996 Jovian tropospheric features—wind field, morphology and motion of long-lived systems. *Icarus* Special SL-9 Issue (in press).

SMITH, B. A. & HUNT, G. E. 1976 Motions and morphology of clouds in the atmosphere of Jupiter. In *Jupiter, The Giant Planet* (ed. T. Gehrels) pp. 564–585. Arizona Press.

SMITH, B. A., SODERBLOM, L. A., JOHNSON, T. V., INGERSOLL, A. P., COLLINS, S. A., SHOEMAKER, E. M., HUNT, G. E., MASURSKY, H., CARR, M., DAVIES, M. E., COOK, A. F., BOYCE, J. M., DANIELSON, G. E., OWEN, T., SAGAN, C., BEEBE, R. F., VEVERKA, J., STROM, R. G., MCCAULEY, J. F., MORRISON, D., BRIGGS, G. A. & SUOMI, V.E. 1979a The Jupiter system through the eyes of Voyager 1. *Science* **204**, 951–972.

SMITH, B. A., SODERBLOM, L. A., BEEBE, R. F., BOYCE, J. M., BRIGGS, G. A., CARR, M., COLLINS, S. A., COOK, A. F., DANIELSON, G. E., DAVIES, M. E., HUNT, G. E., INGERSOLL, A. P, JOHNSON, T. V., MASURSKY, H., MCCAULEY, J. F., MORRISON, D., OWEN, T., SAGAN, C., SHOEMAKER, E. M., STROM, R., SUOMI, V. E. & VEVERKA, J. 1979b The Galilean satellites and Jupiter: Voyager 2 imaging science results. *Science* **206**, 927–950.

WEAVER, H. A., A'HEARN, M. F., ARPIGNY, C., BOICE, D. C., FELDMAN, P. D., LARSON, S. M., LAMY, P., LEVY, D. H., MARSDEN, B. G., MEECH, K. J., NOLL, K. S., SCOTTI, J. V., SEKANINA, Z., SHOEMAKER, C. S., SHOEMAKER, E. M., SMITH, T. E., STERN, S. A., STORRS, A. D., TRAUGER, J. T., YEOMANS, D. K., & ZELLNER, B. 1995 The Hubble Space Telescope (HST) observing campaign on Comet Shoemaker-Levy 9. *Science* **267**, 1282–1287.

WEST, R. A., FRIEDSON, A. J. & APPLEBY, J. F. 1992 Jovian large-scale stratospheric circulation. *Icarus* **100**, 245–259.

WEST, R. A., STROBEL, D. F. & TOMASKO, M. G. 1986 Clouds, aerosols and photochemistry in the Jovian atmosphere. *Icarus* **65**, 161–217.

WILLIAMS, G. P. & WILSON, R. J. 1988 The stability and genesis of Rossby vortices. *J. Atmos. Sci.* **45**, 207–241.

Waves from the Shoemaker-Levy 9 impacts

By ANDREW P. INGERSOLL
AND HIROO KANAMORI

Division of Geological and Planetary Sciences, California Institute of Technology, Pasadena,
CA 91125, USA

Images of Jupiter taken by the Hubble Space Telescope (HST) reveal two concentric circular rings surrounding five of the impact sites from comet Shoemaker-Levy 9 (SL9). The rings are visible 1.0 to 2.5 hours after the impacts. The outer ring expands at a constant rate of 450 m s^{-1}. The inner ring expands at about half that speed. The rings appear to be waves. Other features (diffuse rings and crescent) further out appear to be debris thrown out by the impact. Sound waves (p-modes), internal gravity waves (g-modes), surface gravity waves (f-modes), and rotational waves (r-modes) all are excited by the impacts. Most of these waves do not match the slow speed, relatively large amplitude, and narrow width of the observed rings. Ingersoll and Kanamori have argued that internal gravity waves trapped in a stable layer within the putative water cloud are the only waves that can match the observations. If they are correct, and if moist convection in the water cloud is producing the stable layer, then the O/H ratio on Jupiter is roughly ten times that on the Sun.

1. Introduction

Much of what we know about the interior of the Earth has come from the study of seismic waves—a branch of seismology. Recently, much has been learned about the interior of the Sun from helioseismology. Now, the SL9 impacts give us an opportunity to do jovian seismology. The waves probe Jupiter's atmosphere to depths that cannot be reached by remote-sensing instruments. Their speed of propagation reveals properties of the medium through which they are passing. The speed is a definite, measured quantity—perhaps the most well-determined property of the atmosphere to emerge from the SL9 impacts. The interpretation, in terms of composition and structure of Jupiter's atmosphere, depends on knowing what kind of waves were seen and what levels they are sounding.

Waves may be classified by the different restoring forces that cause them to propagate. Sound waves have compressibility as the restoring force—the increase in pressure that accompanies an increase in density. Sound waves are also known as acoustic waves, and when trapped inside a star they are known as p-modes. Surface waves have gravity as the restoring force. In astronomy they are known as f-modes. The dispersion relation—the relation between frequency and wavelength—is the same for the surface of a star as for the surface of a deep ocean. The density structure of the ambient medium does not enter. Internal gravity waves have buoyancy as the restoring force—the difference between a parcel's own weight and that of the ambient fluid it displaces. Inertia-gravity waves are simply gravity waves modified by rotation, in which case Coriolis forces provide an additional restoring force. In astronomy, these are the g- and r-modes. Rossby waves are a special set of r-modes that have vortex tube stretching as the primary restoring force: Parcels displaced across topographic contours or across latitude circles develop anomalous vorticity and are sent spinning back. Steady geostrophic flow is a limiting case of Rossby waves. The flow does not cross topographic contours or latitude circles, and the frequency is zero.

Further classification is possible depending on whether one is using ray theory or normal modes to describe the waves. Seismologists have a scheme based on ray theory that keeps track of the reflections and transmissions at internal interfaces (*e.g.,* the core-mantle boundary) and internal reflections off the surface. The direct ray is a special case because it propagates without reflection, usually from one point on the surface to another point on the surface. Normal mode theory provides an alternate classification. In general, each ray can be viewed as a superposition of normal modes, and vice versa. But special rays like the direct ray, and low-order modes—those with a small number of nodes on the surface and interior of the sphere, generally have very different properties.

Certain modes are trapped within horizontal layers. For example, the sound channel in the ocean is an acoustic waveguide—a layer where the speed of sound is a minimum with respect to depth. Near-horizontal rays are refracted back toward this layer by the gradients in the speed of sound. Normal modes centered on this layer have exponential tails in the high-velocity layers above and below. Jupiter's atmosphere contains a similar waveguide, centered at the tropopause where the temperature and hence the speed of sound is a minimum. The atmosphere may also contain a waveguide for internal gravity waves, which can be trapped within a stably stratified layer (one where entropy per unit mass increases with altitude) if the layers above and below are neutral (entropy per unit mass independent of altitude).

A wave is a propagating disturbance. The speed of propagation is larger than the speed of the fluid. An atmospheric wave becomes visible if a vapor condenses to form a cloud which then evaporates as the wave moves past. Advection, on the other hand, is moving fluid. The advancing front becomes visible if it carries particles with it. Both waves and advection from the SL9 impacts might manifest themselves as expanding circular rings surrounding the impact sites.

Circular wave-like features were seen in Hubble Space Telescope (HST) images (Hammel *et al.* 1995) and in Earth-based images at 3.42 μm in the strong ν_3 band of methane (McGregor *et al.* 1995). During the first few hours, the rings spread out from the impact sites like ripples in a pond. The observations are described in §2. We also discuss the possibility that the expansion of the rings is due to outward motion (advection) rather than wave propagation. The *p*-modes, *f*-modes, and *r*-modes are discussed in §3. These wave types do not fit the observations, because they have either the wrong speed of propagation, the wrong waveform, or the wrong amplitude. The only waves that fit are *g*-modes of a special type—those trapped within a stable layer at the level of the jovian water cloud. These waves are discussed in §4. Fitting to the observations has important implications about the abundance of water on Jupiter. The implications and a comparison of results from the Galileo probe are described in §5.

2. Observations

Hammel *et al.* (1995) analyzed the atmospheric waves seen in the HST images. Figure 1 is an orthonormal projection of an image taken in a filter at 555 nm of the G impact site. A colored version of the same figure appeared in West *et al.* (1995) and as the cover photo accompanying Ingersoll & Kanamori (1995). Hammel *et al.* (1995) show single-filter images of rings around other impact sites. In Fig. 1, the time interval between the impact and the exposure is 109 min. North is at the top. The radius of the prominent dark ring is 3700 km. A smaller dark ring, of radius 1750 km, is faintly visible. The two rings appear to be concentric, although the smaller ring is partially obscured by the prominent dark spot to the southeast of center. The dark streak outside and to the west of the ring is the site of the D impact, which occurred 20 hours earlier.

Figure 2, which is re-drawn from Hammel *et al.* (1995), shows measurements of the radii of the circular rings that were seen spreading outward from the impact sites. The

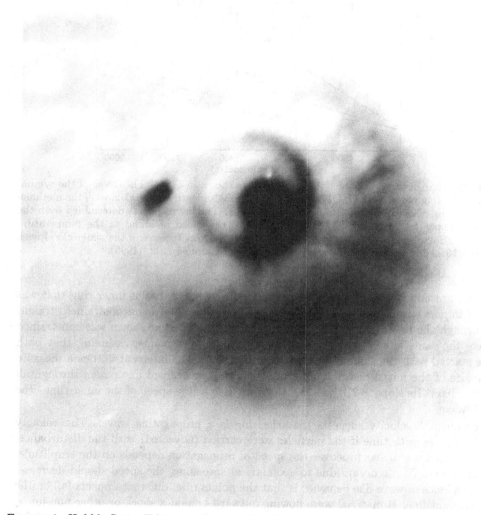

FIGURE 1. Hubble Space Telescope image of the G impact site. The image was taken through the 555 nm filter 109 minutes after the impact, and has been map-projected to show the view from a point directly over the site. The radius of the prominent dark ring is 3700 km.

measurements cover the period from 1 hour to 2.5 hours after the impacts. Five impact sites—A, E, G, Q1, and R—were observed during this time, and the data from the different sites are superposed. Other impact sites were observed one or more jovian rotations later (ten hours or more after the impacts), and no rings were visible.

There are two sets of points in Fig. 2. The upper points lie on a well-defined straight line with slope 450 m s^{-1} and intercept 500 km (radius $r = 500$ km at time $t = 0$). This is the "fast" wave, whose speed is 450 m s^{-1}. The other set of points are clustered near

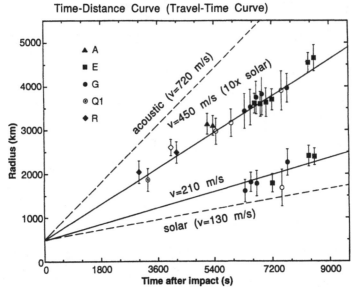

FIGURE 2. Time-distance curve for the circular rings shown in Fig. 1. The shape of the symbol identifies the fragments A, E, G, Q1, and R. Open symbols identify data taken with the methane (889 nm) filter. Closed symbols are for all other filters. The intercept was determined from the fit to the upper set of points. "Acoustic" refers to the speed of sound at the temperature minimum. "Solar" and "10×solar" refer to the speed of a gravity wave in the water cloud with solar and 10×solar abundance of water. Redrawn from Hammel *et al.* (1995).

$r = 2000$ km and $t = 7000$ sec. These are measurements of a faint inner ring that was seen for impact sites E and G. The speed is hard to measure. The unconstrained straight line through the lower points has a negative intercept. The line shown was constrained to have the same intercept as the line through the upper points, assuming that both waves started with the same radius, 500 km, at $t = 0$. The intercept is thus a measure of the size of the initial disturbance. Note that the constrained line passes through all the error bars. Its slope is 210 m s^{-1}, which we take as the speed of the inner ring—the "slow" wave.

The constant velocity suggests that the ring is a propagating wave. The velocity would decrease with time if the particles were carried (advected) with the disturbance. Advection is a non-linear process—the speed of propagation depends on the amplitude. And as the amplitude decays due to geometrical spreading, the speed should decrease. Other evidence of wave-like behavior is that the points from different impacts fall on the same straight line. If material were flowing outward behind a shock or other non-linear disturbance the speed would depend on the energy of impact, and the points would fall on different lines. For a linear wave the speed depends only on properties of the medium.

Zahnle (1995) discusses advection as a possible source of the ring. In an axisymmetric numerical model of the impact, particles that were near the axis before the impact spread out into a flat disk just above the tropopause. The average speed is close to 450 m s^{-1} for the first 1000 seconds, but then the spreading stops. The final radius of the disk is 500 km, which is only 12% of the maximum radius that was observed (Fig. 2).

Zhang & Ingersoll (1995) use a similarity solution (Zel'dovich & Raizer 1967) to reproduce Zahnle's results. The parcel of heated air—the mushroom cloud—rises under its own buoyancy to where its mean entropy matches that of the environment. There, just above the tropopause, the air spreads horizontally, forming a flat disk—a concave lens-shaped structure symmetric about a vertical axis. The outer radius r_0 varies as t^n and the speed varies as t^{n-1}, where n is a constant to be determined. To conserve volume, the thickness h varies as $t^{-2n}f(r/r_0(t))$, where f is a function to be determined. From dimensional analysis or from the Bernoulli and hydrostatic equations, the speed is proportional to Nh_0, where N is the Brunt-Väisälä frequency of the ambient atmosphere and h_0 is the thickness at the outer edge. Equating powers of t in the two expressions for the speed yields $n = 1/3$, so that $r_0 \propto t^{1/3}$. Such a power law does not fit the data in Fig. 2, so Zhang & Ingersoll (1995) conclude that the rings cannot be advection of particles associated with the spreading lens. Advection might occur during the first 1000 seconds when the dark material spreads to a radius of 500 km, but subsequent spreading at constant speed must be due to propagation.

Other features of the G impact site, including the large diffuse crescent outside the ring, are discussed by Hammel *et al.* (1995), McGregor *et al.* (1995), Pankine & Ingersoll (1995), and others. The outer edge of the crescent is about 13,000 km from the impact site, which is four times the maximum height of the plumes. Hammel *et al.* (1995) argue that the crescent is impact debris from the plumes and that the outer edge corresponds to plume material that was launched upward at a 45° angle with respect to the vertical. One hour after the impacts the crescent is still spreading, but at a slower rate than that implied by the mean radius divided by the time. Zahnle (1995) argues that this spreading is due to sliding of the debris along the top of the atmosphere.

Earth-based imaging in the strong methane band at 3.42 μm reveals a large ring outside the crescent (McGregor *et al.* 1995). The material must be high in the atmosphere—higher than the material observed in the HST images. The inner and outer radii of the ring are 15,000 and 18,000 km, respectively. Like the crescent, the ring continues to expand for 2 hours after the impacts, but at a slower rate than that suggested by the mean radius divided by the time. McGregor *et al.* (1995) discuss two possibilities—that the ring is sliding debris and that it is a propagating wave. The measured expansion velocity modestly exceeds the sound speed in the undisturbed stratosphere, suggesting the possibility of a weak shock wave. Numerical simulations, reviewed by Zahnle (1995), show that sliding does occur. McGregor *et al.* (1995) regard the large ring as a challenge to our understanding. We shall focus our discussion on the smaller, slower rings that appear in the HST images.

The brown ring material is similar to the impact debris. HST images in the 889 nm band suggest that it is located in the stratosphere somewhere between the 1 mb and 300 mb levels (Hammel *et al.* 1995; West *et al.* 1995). West *et al.* (1995) argue that the particles are created by condensation and destroyed by evaporation as the wave moves past. This requires that the particles be volatile. West *et al.* (1995) cite polymerized HCN as an interesting candidate for the composition of the brown material. Zahnle (1995) regards S_8 as the most likely candidate, but he is skeptical of all condensates. Finding a volatile material that condenses at just the right temperature and pressure is difficult. From a chemical point of view, it is easier to postulate that the brown material is refractory—either carbon dust or silicate dust from the comet that is advected with the fluid. But then one has to explain why the material forms a narrow ring that spreads at constant speed. We shall proceed under the assumption that the rings represent a propagating wave, even though the composition of the ring material is uncertain.

3. Surface waves, Acoustic waves, and Rossby waves

3.1. *f-modes*

The f-modes are surface gravity waves. As with all waves considered here the fluid motion is adiabatic, so the entropy of each fluid parcel is constant. For f-modes, the density, pressure, temperature, and all other thermodynamic variables are constant as well (Gill 1982; Goldreich & Kumar 1990). Therefore the f-modes can only be detected by measuring displacement, velocity, or acceleration. This implies that the rings seen in the HST images are not f-modes, since HST is sensitive mainly to temperature. That is, if the particles are a condensate they are probably responding mostly to the Lagrangian temperature perturbation (that experienced by a parcel) as the wave moves past. For f-modes this perturbation is zero, so the wave should be invisible in HST images.

For wavelengths that are short compared to the radius of Jupiter, the speed of propagation (phase speed) of f-modes is $\sqrt{g/k}$, where k is the horizontal wavenumber. This dispersion relation is the same as that for waves on the surface of a deep ocean. The type of fluid and its thermodynamic state do not enter, because the perturbations to the state variables are zero. The waves are highly dispersive. Those whose group velocity matches the 450 m s^{-1} speed observed in HST images have a wavelength of order 200 km. The speed is greater for longer wavelengths. Kanamori (1993) computed surface waves as a superposition of f-modes excited by momentum transfer from the comet impact.

3.2. *p-modes*

Several authors (Deming 1994; Gough 1994; Lognonné *et al.* 1994) have calculated the shape of the pulse propagating outward from a hypothetical impact site. Figure 3, reprinted from Lognonné *et al.* (1994), shows the particle displacement over a period of several days at two locations, 90° and 180° away from the impact point. From top to bottom, the first and third curves are for f-modes only. The other two curves include both f-modes and p-modes—those waves that propagate into the interior of the planet. The low-frequency p-modes are trapped in the interior by the cold surface layers; the high-frequency p-modes leak energy into the stratosphere and are dissipated.

For a spherical non-rotating planet, all the waves converge on the antipode (the point 180° from the impact site), though not at the same time. Lognonné *et al.* (1994) assume an impact energy of 5×10^{27} erg, and obtain displacements of order 100 m at the antipode and 4 m at the 90° point. The multiple packets in Fig. 3 represent waves that have gone multiple times around the planet. Scattering and absorption by jovian winds and turbulence are not included in the calculation. Clearly, the packets shown do not match the waves seen in the HST images. They propagate too fast and they are spread out too far. Except near the antipodes, their amplitudes are likely to be smaller than the normal jovian weather noise.

3.3. *Normal mode oscilllations*

Waves that go several times around the planet can be regarded as superpositions of global normal modes (Lee *et al.* 1994; Mosser *et al.* 1995). Each p-mode has two angular order numbers, which give the number of zero-crossings in latitude and longitude, respectively, and one radial order number. The f-modes have only the two angular order numbers. The frequency of the mode is a unique function of these numbers. In principle, the modes with low angular order numbers should be visible from Earth, since Earth-based telescopes can resolve features that are less than 1/50 the diameter of Jupiter. In practice, the amplitude must be large enough to stand out above the fluctuations associated with jovian weather. Such large amplitudes are unlikely, although several searches were made.

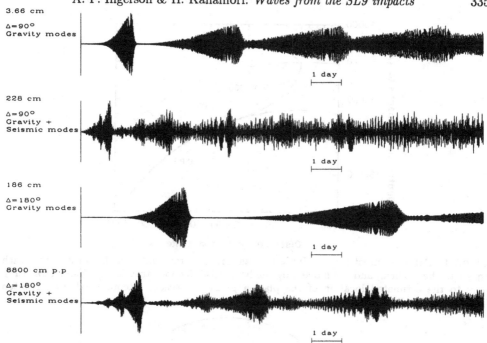

FIGURE 3. Vertical displacement as a function of time at locations 90° and 180° from the impact site. Displacement amplitude is given at left. Time increases to the right. The vertical line marks the time of impact. The first and third plots show surface gravity waves. The second and fourth plots show a superposition of surface gravity waves and acoustic waves. Reproduced from Lognonné *et al.* (1994).

According to Mosser *et al.* (1995), "Eight months after the impact of comet SL9 on Jupiter, it is still impossible to say whether the attempt to observe seismic waves excited by the impacts has been successfully conclusive."

3.4. Direct ray

Much of the interest in p-modes stems from a desire to learn about Jupiter's interior. The waves made up by superposition of high-frequency p-modes can be interpreted as seismic rays. Figure 4, reprinted from Hunten *et al.* (1994), shows the paths taken by rays launched at various angles with respect to the vertical. Only those launched nearly straight down can reach the plasma phase transition (PPT), the level below which hydrogen is a metal. Rays that are launched at other angles are refracted upward to the surface. The figure shows only the direct rays, which in seismological terms correspond to p-waves. The curves are labeled by the distance in km at which the ray re-surfaces.

Since all the rays start out at the same time but reach the surface at different times, the disturbance will appear to spread outward from the impact site. The position of the disturbance at various times after the impact is shown in Fig. 5, which is taken from Marley (1994). After 75 minutes, the disturbance radius is comparable to the radius of Jupiter—about 71,400 km. The apparent speed is much faster than the speed measured by HST (Fig. 2), suggesting that the observed waves are not related to the direct p-wave. This conclusion is consistent with estimates by Marley (1994) that the wave amplitudes are greater than 0.1 K only for impacts larger than about 10^{28} ergs. This energy is larger than most estimates, so the direct p-wave is unlikely to be seen against the jovian weather background.

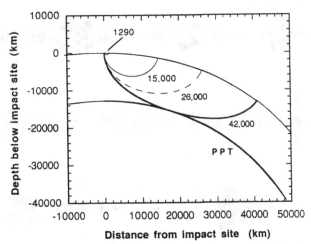

FIGURE 4. Paths of direct rays in Jupiter's interior. The rays differ in the initial angle with respect to the vertical, and are labeled by the horizontal distance traveled, in kilometers. PPT refers to the estimated location of the plasma phase transition. Reproduced from Hunten *et al.* (1994).

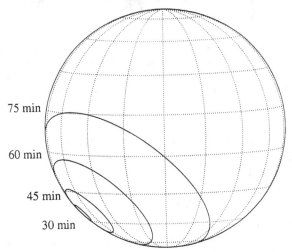

FIGURE 5. Intersection of the acoustic wave pulse (Fig. 4) with the surface, as a function of time from impact. The figure shows the appearance of the planet at various times if the amplitude were large. Reproduced from Marley (1994).

3.5. *Acoustic waveguide*

The speed of sound is proportional to $T^{1/2}$, where T is temperature. The minimum temperature is found at the tropopause, where the pressure P is 100 mb and $T = 110$ K. Figure 6 shows three profiles measured by the *Voyager* radio occultation experiment (Lindal *et al.* 1981). For Jupiter's hydrogen-helium atmosphere, the minimum speed of sound is 770 $\mathrm{m\,s^{-1}}$, which is too fast to match the observed speed (Fig. 2).

It is nevertheless useful to ask why sound waves were not observed. The tropopause acts as an acoustic waveguide: Sound waves can propagate horizontally over large distances because they do not lose energy through vertical propagation. Figure 7, reprinted from Collins *et al.* (1995), shows the normal modes of the acoustic waveguide. The dashed curve is the speed of sound, which varies with depth. The solid curves are the normal

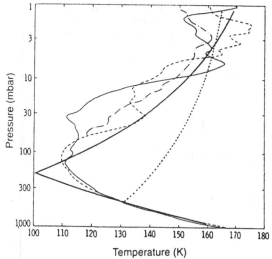

FIGURE 6. Temperature profiles in Jupiter's atmosphere. The light solid curve, the long dash curve, and the dot-dash curve are measured by the *Voyager* radio occultation experiment (Lindal *et al.* 1981). The heavy solid curve and the short dash curve are those used by Ingersoll & Kanamori (1995). Reproduced from Ingersoll & Kanamori (1995).

modes. The lowest mode (on the left) is a single peak centered on the temperature minimum; its amplitude falls off exponentially above and below this level. Both the phase and group speeds of this mode are near 770 $m s^{-1}$. Higher modes have higher speeds. For a source near the tropopause most of the energy goes into the lowest mode. For a source at deeper levels the energy is distributed among all the modes, and the amplitude far from the source region is small.

Figure 8, reprinted from Ingersoll & Kanamori (1995), shows the amplitude of the acoustic wave 1.5 hours after the impact for point sources at various pressure levels. The amplitude in this case is the Lagrangian temperature perturbation—the temperature change experienced by a parcel at the 126 mb level. The source is a 10^{27} erg pulse of heat at $r = 0$, $t = 0$. This is the kinetic energy of a solid sphere of ice of diameter 0.5 km, travelling at the impact speed of the comet. This energy is consistent with current estimates for the larger fragments. As shown in Fig. 8, the response drops as the depth increases. For a source at 0.05 bar the amplitude of the response is ±35 K. For a source at 5 bar the amplitude is ±0.65 K, and for a source at 10 bar the amplitude is ±0.20 K. From the fact that sound waves were not seen, Ingersoll & Kanamori (1995) conclude that the energy release was deep.

Another possibility is that the level of observation is high. A condensate is required to make the waves visible. Perhaps this condensate forms only above the 1 mb level— 100 km above the tropopause. There the amplitude of the lowest mode is small (Fig. 7), so the wave would be invisible to HST. Some combination of a source at low altitude and a condensate at high altitude is perhaps the most likely explanation for the absence of sound waves.

3.6. *r-modes and steady geostrophic motion*

Effects of rotation become important for low frequencies and long times. Lee & Van Horn (1994) estimate the amplitude of inertial oscillations (*r*-modes) in the interior of the planet, where buoyancy effects are negligible, and conclude that the oscillations would be unobservable unless the impact energy were greater than 10^{30} ergs. Ingersoll *et al.* (1994)

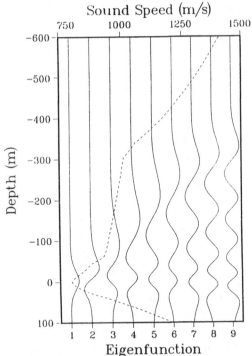

FIGURE 7. Sound speed (dashed curve) and eigenfunctions (solid curves) vs depth for acoustic waves trapped in the sound channel at the tropopause temperature minimum. Reproduced from Collins *et al.* (1995).

show that inertial oscillations become the largest mode in the atmosphere one day or more after the impact. They argue that the rings are inertia-gravity waves trapped in the water cloud, where buoyancy effects are important. As discussed in § 7.3 of Gill (1982), the waves are dispersive. The short waves propagate away and leave the long waves behind. The latter are the inertial oscillations with frequency $f = 2\Omega\sin\phi$ where ϕ is latitude and Ω is the rotation rate of the planet. During the first 2.5 hours the effects of rotation are negligible, and the waves are indistinguishable from pure gravity waves unmodified by rotation.

Steady geostrophic flow is the limiting case of a low-frequency inertial mode (Greenspan 1968). On a rotating planet, such a flow can be excited by an initial transient. Gill (1982) calls this "The Rossby Adjustment Problem," and develops the theory for a single horizontal layer of fluid. Ingersoll *et al.* (1994) extend the theory to three dimensions, giving expressions for the steady circular vortex that remains after the inertia-gravity waves have propagated away. Harrington *et al.* (1994) obtained a steady vortex in their numerical experiments. The ultimate fate of the vortex depends on the ambient zonal flow. The HST observations do not distinguish between impact debris and impact-produced vortices. Hammel *et al.* (1995) do not report evidence of vortex-type motion at the impact sites, but another look at the images is warranted.

4. Gravity Waves in the Water Cloud

The key parameter for internal gravity waves (*g*-modes) is N, the Brunt-Väisälä frequency of the atmosphere. This is the maximum frequency of a displaced parcel oscillating under its own buoyancy. N^2 is proportional to the difference between the lapse rate of

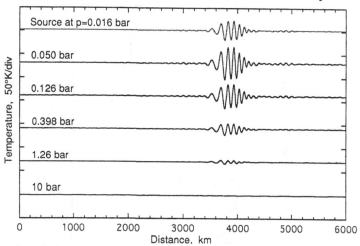

FIGURE 8. Profiles of the acoustic wave amplitude as a function of distance, for sources at different altitudes. The time is 1.5 hours after the impact of a 10^{27} erg body. "Amplitude" refers to the temperature change experienced by a parcel at the 126 mb level. The amplitude scale is 50 K per division, as shown at left. Reproduced from Ingersoll & Kanamori (1995).

temperature and the adiabatic lapse rate. When temperature falls off more slowly than the adiabatic lapse rate, N^2 is positive. Then the atmosphere is said to be stable—the parcel executes sinusoidal oscillations about its equilibrium altitude.

Figure 9, reprinted from Ingersoll *et al.* (1994), shows the vertical profile $N(P)$ from 1 mb to 10 bar. The thin wiggly line is computed from an average of the three *Voyager* radio occultation curves shown in Fig. 6. The data end at pressures slightly less than 1 bar, and the curve has been extended downward along a moist adiabat. The basic assumption is that moist convection makes the atmosphere stable, and that N^2 is proportional to the difference between a virtual moist adiabat and a dry adiabat. Observations in the Earth's tropics (Emanuel 1994, pp. 472–491) show that the mean density structure is close to that of a virtual moist adiabat, which is the density of a rising parcel undergoing latent heat release and change of molecular weight as the condensate falls out. This mean density structure is maintained by convection in isolated "cores" of cumulonimbus clouds that occupy only 0.1% of the area (Riehl & Malkus 1958; Palmén & Newton 1969). Outside the cores the atmosphere is unsaturated—the relative humidity is less than one. So a wave that does not cause the atmosphere to saturate will propagate dry adiabatically. Thus the atmosphere is neutral with respect to moist convection but is stable with respect to large-scale waves.

In Fig. 9 the water cloud base is at 5 bar. This is the saturation point in a "solar composition" atmosphere—one where the O/H ratio of Jupiter's interior is the same as that on the Sun (Gautier & Owen 1989). Since water is the main oxygen-bearing molecule, the O/H ratio sets the water abundance. N^2 is large at cloud base and falls abruptly to zero below it. Ingersoll & Kanamori (1995) use simple analytic functions fitted to the *Voyager* profile in the stratosphere and the moist adiabat in the water cloud, for O/H ratios up to 10 times solar. The fitted function for solar O/H is shown in Fig. 9. If we define e_{H_2O} as the O/H abundance on Jupiter divided by the O/H abundance on the Sun, then "solar" composition corresponds to $e_{H_2O} = 1$.

Energy from the impacts tends to excite two kinds of internal gravity waves, corresponding to the two regions with large N^2 in Fig. 9. The amplitude of the stratospheric gravity wave (SGW) compared to that of the tropospheric gravity wave (TGW) depends

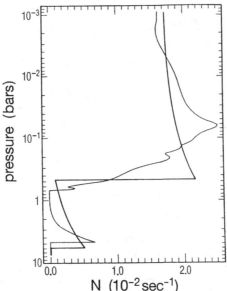

FIGURE 9. Profiles of the Brunt-Väisälä frequency in Jupiter's atmosphere. The light wiggly curve is derived from the *Voyager* radio occultation data (Fig. 6), with a moist adiabatic extrapolation below the 1 bar level. The heavy smooth curve is derived from an analytic fit to the temperature data. Reproduced from Ingersoll *et al.* (1994).

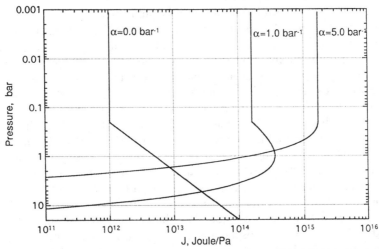

FIGURE 10. Vertical profiles of heating used by Ingersoll & Kanamori (1995) in their calculation of inertia-gravity waves from the SL9 impacts. The ordinate is the energy density in J Pa^{-1}, and the integral of each curve with respect to pressure is 10^{27} erg. Reproduced from Ingersoll & Kanamori (1995).

on the vertical distribution of heating during the impact. Figure 10 shows three heating profiles used by Ingersoll & Kanamori (1995). The parameter α has units of inverse pressure. Large α corresponds to heating at small pressure, and vice versa. Heating is zero below cloud base, which in this case is at 20 bar, corresponding to 10 times the solar abundance of water. The units of heating are J Pa^{-1}, and the integral of each curve with respect to pressure is 10^{27} erg. This is the total impact energy for this simulation.

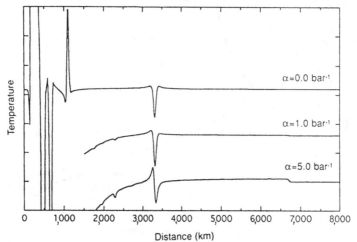

FIGURE 11. Profiles of gravity wave amplitudes as a function of distance, for the three different heating profiles of Fig. 10. "Amplitude" is the temperature change experienced by a parcel at the 45 mb level. The time is 2 hours after the impact of a 10^{27} erg body. Units are 1 K per division on the vertical axis. The pulse at 3300 km is the fundamental gravity wave mode in the water cloud. The pulse at 1100 km is the next higher mode. The bump at 6700 km in the $\alpha = 5$ bar^{-1} curve (heating at high altitude) is the gravity wave in the stratosphere. Reproduced from Ingersoll & Kanamori (1995).

Figure 11, which is also reprinted from Ingersoll & Kanamori (1995), shows three snapshots of internal gravity waves corresponding to the three heating profiles of Fig. 10. The ordinate is the Lagrangian temperature perturbation at the 45 mb level. The abscissa is radial distance from the impact point, and the time is two hours (7200 s) after the impacts. The model uses the analytic temperature profile represented by the heavy solid line in Fig. 6. The extension to the base of the water cloud at $P = 20$ bars is shown in Fig. 12. This model, with $e_{H_2O} = 10$, is the one that matches the 450 m s^{-1} speed of the rings observed in the HST images. As shown in Ingersoll *et al.* (1994), the speed varies as the square root of the water abundance and little else. For $e_{H_2O} = 1$ the speed is 130 m s^{-1}.

In Fig. 11, the pulse at $r = 3300$ km is the TGW travelling at 450 m s^{-1}. Although the wave has most of its energy in the water cloud, it leaks energy into the stratosphere where we can observe it with HST. As in Fig. 2 of Ingersoll *et al.* (1994), the disturbance amplitude is considerably larger at 45 mb than it is at 1 bar or at the NH$_3$ cloud tops near 500 mb. As pointed out by Achterberg & Ingersoll (1989), the disturbance grows in amplitude and oscillates with height in the stratosphere. Allison (1990) showed that the vertical wavelength roughly matches that observed in the radio occultation profiles (Fig. 6), provided the water abundance is 2–3 times solar.

The response function also increases with height in the stratosphere: For a given input energy the amplitude of the TGW is greater when the source is at high altitude than when the source is at low altitude. The same is true of atmospheric tides on Earth, which are internal gravity waves forced by solar heating: Absorption of small amounts of sunlight in the mesosphere by water vapor and ozone makes a large contribution to the tidal pressure oscillation at the ground (Chapman & Lindzen 1970).

In Fig. 11 the disturbance at $r = 6700$ km is the leading edge of the SGW. Its speed is 930 m s^{-1} and is very hard to change. We have run cases with e_{H_2O} ranging from 0.5 to 10, and we have tried putting all the heat at one level, varying the pressure of the level

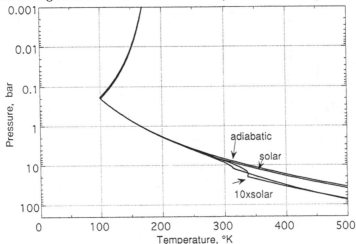

FIGURE 12. Profiles of virtual temperature Tm_{dry}/m for the no-water (adiabatic), solar, 5×solar, and 10×solar cases used by Ingersoll and Kanamori (1995) in their gravity wave calculations, where m is the molecular mass of the mixture and m_{dry} is the molecular mass of the atmosphere without condensate.

from 10 mb to 10 bar. We have increased the pressure at the tropopause by a factor of three (Fig. 6, dashed line), and we have varied T_∞—the temperature at the top of the stratosphere. The speed of the SGW depends only on T_∞, varying as $T_\infty^{1/2}$. It does not match the 450 m s^{-1} observed speed for any reasonable profile. And in all cases the amplitude of the SGW is less than that of the TGW. We conclude that the observed waves are TGWs, not SGWs.

This conclusion is supported by the model described in Harrington *et al.* (1994). Running their model with 40 vertical layers gives results almost identical to those in Fig. 11. With 40 vertical layers the numerical model successfully mimics the dissipation of wave energy propagating up from the troposphere (Lindzen *et al.* 1968). With just 5 vertical layers the model gives spurious refections from the upper layer, which reduces the speed of the SGW.

The disturbance at $r = 1100$ km in Fig. 11 is the second mode of the tropospheric waveguide. For this temperature profile (Figs. 6 and 12), the mode travels at 150 m s^{-1}, one-third the speed of the first mode. As described in Ingersoll *et al.* (1994), the first mode spans one quarter-cycle and the second mode spans three quarter-cycles from cloud base to tropopause. The two modes might be related to the outer and inner rings, although the ratio of the two speeds (3 to 1 in the model) does not exactly match the observed ratio (about 2.15 to 1).

5. Implications and conclusions

Other inferences about jovian water tend to support $e_{H_2O} \geq 1$, but there are major uncertainties. Bjoraker *et al.* (1986) looked at both the high-resolution spectra taken from the Kuiper Airborne Observatory and the lower-resolution spectra taken by the *Voyager* IRIS. They concluded that $e_{H_2O} \sim 1/50$. Analysing just the *Voyager* data, Carlson *et al.* (1992) conclude $e_{H_2O} \geq 1$. They argue that $e_{H_2O} = 10$ fits the data equally well (see their Fig. 9). The high-resolution data are best for isolating individual spectral lines, particularly the weak ones. With *Voyager* IRIS one is limited to strong lines only. However the main difference between the two models concerns assumptions made about scattering in the clouds.

Voyager and Earth-based observations (Gautier & Owen 1989) give $e_{CH_4} \sim e_{NH_3} \sim 2$ and $e_{He} \sim 0.7$. Analysis of the gravitational harmonics of Jupiter (Hubbard 1989) suggests at least a 5-fold enrichment of heavy elements like O, C, and N. This enrichment drops by a factor of two if Jupiter has a subadiabatic (radiative) zone in the 1–40 kbar pressure range, as suggested by Guillot *et al.* (1994).

The Galileo spacecraft encountered Jupiter on December 7, 1995. The probe sent back useful data from above cloud tops down to the 20 bar level. Preliminary results suggest that water is present at less than solar abundance. The mass spectrometer found less water than the solar value, and the temperature profile measured by the atmospheric structure instrument was closer to a dry adiabat than to a moist adiabat. Also, the nephelometer found almost no cloud particles at the level of the purported water cloud, and the net flux radiometer found low opacity in the 5-micron band where water vapor is a dominant absorber.

Thus the preliminary Galileo results do not support the inferences made from studying the SL9 waves. There are a number of possibilities. First, the water vapor abundance may vary from place to place (the SL9 fragments entered the atmosphere at a latitude of $-44°$, and the Galileo probe entered at 6.5°). Second, the identification of the ring as a wave in the jovian water cloud may be flawed. And third, the Galileo results may change after further calibration of the instruments.

If there is a deep stable layer extending down below 10 bars over most of the planet, then our best guess is that this layer is maintained by moist convection, with $e_{H_2O} \sim 10$. Such a result has important implications regarding the origin of Jupiter, its internal history, and the chemistry and meteorology of its current atmosphere.

Ingersoll acknowledges support from NASA's Planetary Atmospheres Program, grant NAGW-1956, and from NASA's Space Telescope Science Institute; Kanamori acknowledges support from NSF grant EAR-92 18809.

REFERENCES

ACHTERBERG, R. K. & INGERSOLL, A. P. 1989 A normal-mode approach to jovian atmospheric dynamics. *J. Atmos. Sci.* **46**, 2448–2462.

ALLISON, M. 1990 Planetary waves in Jupiter's equatorial atmosphere. *Icarus* **83**, 282–307.

BJORAKER, G. L., LARSON, H. P. & KUNDE, V. G. 1986 The gas composition of Jupiter derived from 5-μm airborne spectroscopic observations. *Icarus* **66**, 579–609.

CARLSON, B. E., LACIS, A. A. & ROSSOW, W. B. 1992 The abundance and distribution of water vapor in the jovian troposphere as inferred from *Voyager* IRIS observations. *Astrophys. J.* **388**, 648–668.

CHAPMAN, S. & LINDZEN, R. S. 1970 *Atmospheric Tides*. Gordon and Breach Science Publishers.

COLLINS, M., MCDONALD, B. E., KUPPERMAN, W. A. & SIEGMANN, W. L. 1995 Jovian acoustics and comet Shoemaker-Levy 9. *J. Acoust. Soc. Amer.* **97**, 2147–2158.

DEMING, D. 1994 Prospects for jovian seismological observations following the impact of comet Shoemaker-Levy 9. *Geophys. Res. Lett.* **21**, 1095–1098.

EMANUEL, K. A. 1994 *Atmospheric Convection*, Oxford University Press.

GAUTIER, D. & OWEN, T. 1989 The composition of outer planet atmospheres. In *Origin and Evolution of Planetary and Satellite Atmospheres* (ed. S. K. Atreya, J. B. Pollack & M. S. Matthews), pp. 487–512. University of Arizona Press.

GILL, A. E. 1982 *Atmosphere-Ocean Dynamics*, Academic Press.

GOLDREICH, P. & KUMAR, P. 1990 Wave generation by turbulent convection. *Astrophys. J.* **363**, 694–704.

GOUGH, D. O. 1994 Seismic consequences of the Shoemaker-Levy impact. *Mon. Not. R. Astron. Soc.* **269**, L17–L20.

GREENSPAN, H. P. 1968 *The Theory of Rotating Fluids*, Cambridge University Press.

GUILLOT, T., CHABRIER, G., MOREL, P. & GAUTIER, D. 1994 Nonadiabatic models of Jupiter and Saturn. *Icarus* **112**, 354–357.

HAMMEL, H. B., BEEBE, R. F., INGERSOLL, A. P., ORTON, G. S., MILLS, J. R., SIMON, A. A., CHODAS, P., CLARKE, J. T., DE JONG, E., DOWLING, T. E., HARRINGTON, J., HUBER, L. F., KARKOSCHKA, E., SANTORI, C. M., TOIGO, A., YEOMANS, D. & WEST, R. A. 1995 HST imaging of atmospheric phenomena created by the impact of comet Shoemaker-Levy 9. *Science* **267**, 1288–1296.

HARRINGTON, J., LEBEAU, R. P., BACKES, K. A. & DOWLING, T. E. 1994 Dynamic response of Jupiter's atmosphere to the impact of comet Shoemaker-Levy 9. *Nature* **368**, 525–527.

HUBBARD, W. B. 1989 Structure and composition of giant planet interiors. In *Origin and Evolution of Planetary and Satellite Atmospheres* (ed. S. K. Atreya, J. B. Pollack & M. S. Matthews), pp. 539–563. University of Arizona Press.

HUNTEN, D. M., HOFFMANN, W. F. & SPRAGUE, A. L. 1994 Jovian seismic waves and their detection. *Geophys. Res. Lett.* **21**, 1091–1094.

INGERSOLL, A. P., KANAMORI, H. & DOWLING, T. E. 1994 Atmospheric gravity waves from the impact of comet Shoemaker-Levy 9 with Jupiter. *Geophys. Res. Lett.* **21**, 1083–1086.

INGERSOLL, A. P. & KANAMORI, H. 1995 Waves from the collisions of comet Shoemaker-Levy 9 with Jupiter. *Nature* **374**, 706–708.

KANAMORI, H. 1993 Excitation of jovian normal modes by an impact source. *Geophys. Res. Lett.* **20**, 2921–2924.

LEE, U. & VAN HORN, H. M. 1994 Global oscillation amplitudes excited by the Jupiter-comet collision. *Astrophys. J.* **428**, L41–L44.

LINDAL, G. F., WOOD, G. E., LEVY, G. S., ANDERSON, J. D., SWEETNAM, D. N., HOTZ, H. B., BUCKLES, B. J., HOLMES, D. P., DOMS, P. E., ESHLEMAN, V. R., TYLER, G. L. & CROFT, T. A. 1981 The atmosphere of Jupiter: An analysis of the *Voyager* radio occultation measurements. *J. Geophys. Res.* **86**, 8721–8727.

LINDZEN, R. S., BATTEN, E. S. & KIM, J. W. 1968 Oscillations in atmospheres with tops. *Mon. Wea. Rev.* **96**, 133–140.

LOGNONNÉ, P., MOSSER, B. & DAHLEN, F. A. 1994 Excitation of jovian seismic waves by the Shoemaker-Levy 9 cometary impact. *Icarus* **110**, 180–195.

MARLEY, M. S. 1994 Seismological consequences of the collision of comet Shoemaker-Levy/9 with Jupiter. *Astrophys. J.* **427**, L63–L66.

McGREGOR, P. J., NICHOLSON, P. D. & ALLEN, M. G. 1995 CASPIR observations of the collision of comet Shoemaker-Levy 9 with Jupiter. *Icarus*, in press.

MOSSER, B., GALDEMARD, P., JOUAN, R., LAGAGE, P., MASSE, P., PANTIN, E., SAUVAGE, M., LOGNONNÉ, P., GAUTIER, D., DROSSART, P., MERLIN, P., SIBILLE, F., VAUGLIN, I., BILLEBAUD, F., LIVENGOOD, T., KÄUFI, H. U., MARLEY, M., HULTGREN, M., NORDH, L., OLOFSSON, G., ULLA, A., BELMONTE, J. A., REGULO, C., ROCA-CORTEZ, C., SELBY, M., RODRIGUEZ-ESPINOSA, J. M. & VIDAL, I. 1995 Seismic studies of Jupiter at the time of SL-9 impacts. In *European SL-9/Jupiter Workshop, Garching, 13–15 February 1995* (eds. R. M. West & H. Boehnhardt), pp. 397–402. European Southern Observatory Conference and Workshop Proceedings.

PALMÉN, E. & NEWTON, C. W. 1969 *Atmospheric Circulation Systems*, Academic Press.

PANKINE, A. A. & INGERSOLL, A. P. 1995 Ejecta patterns of the impacts of comet Shoemaker-Levy 9. *Bull. Amer. Astron. Soc.* **27**, 76.

RIEHL, H. & MALKUS, J. S. 1958 On the heat balance in the equatorial trough zone. *Geophysica (Helsinki)* **6**, 503–537.

WEST, R. A., KARKOSCHKA, E., FRIEDSON, A. J., SEYMOUR, M., BAINES, K. H. & HAMMEL, H B. 1995 Impact debris particles in Jupiter's stratosphere. *Science* **267**, 1296–1301.

ZAHNLE, K. 1995 Dynamics and chemistry of SL9 plumes. This volume.

ZEL'DOVICH, YA. B. & RAIZER, YU. P. 1967 *Physics of Shock Waves and High-Temperature Hydrodynamic Phenomena, Volume II*, Academic Press.

ZHANG, H. & INGERSOLL, A. P. 1995 Rings in HST images of SL9 collision with Jupiter: Waves or advection? *Bull. Amer. Astron. Soc.* **27**, 76.

Jovian magnetospheric and auroral effects of the SL9 impacts

By WING-HUEN IP

Max-Planck-Institut für Aeronomie, D-37191 Katlenburg-Lindau, Germany

The collisions of comet Shoemaker-Levy 9 with Jupiter have produced many surprising auroral and magnetospheric phenomena. The energy released during the passage of the cometary dust comas through the jovian magnetosphere and at atmospheric explosion could lead to impulsive particle acceleration, enhanced radial diffusive transport, and the establishment of field-aligned current systems connecting the comet impact sites to their respective magnetic conjugate points. Some of the observed effects such as the abrupt increase of decimetric radio emission, the excitation of infrared H_3^+ emissions and mid-latitude auroral emission in the ultraviolet, could be interpreted within the framework of these mechanisms. Several auroral features like the X-ray outbursts and short-term variations in the UV emissions are more puzzling and require further observation of jovian auroral dynamics in these wavelength ranges in coordination with the Galileo mission.

The important thing is to be there when the picture is painted.

—John Minton

1. Introduction

The collisions of Comet Shoemaker-Levy 9 with Jupiter in the third week of July 1994 have opened a new chapter in cometary physics and the study of Jupiter. The hyper-velocity impacts of these comet nuclei at Jupiter caused many spectacular effects in the upper atmosphere and ionosphere. Very dynamic phenomena were also observed in the Jovian auroral zones and magnetosphere during the impact week. In a certain sense, the study of the magnetospheric and auroral effects caused by the atmospheric impacts of Comet Shoemaker-Levy 9 is complicated. Not only were the energetic phenomena observed closely coupled to the energy release processes in the jovian upper atmosphere and ionosphere, they could also be affected by the plasma interaction of the dust coma as well as the global dynamics of the jovian magnetosphere. Both of these topics are new in the sense that we have only an incomplete understanding of the physical effects involved. Furthermore, unlike the spectacular fireballs created by atmospheric explosions, it is difficult in some cases to identify the observed auroral features, without any ambiguity, as being the signatures associated with comet impacts. With this said, a comprehensive review of the major results will be attempted by taking a three-step approach. In the first part, we shall briefly describe the general aspects of the Jovian magnetosphere and the cometary dust coma which are important in data interpretation. This is followed by a summary of observations relevant to auroral and magnetospheric effects made at different wavelengths (*e.g.*, radio, infrared, UV and X-rays). Theoretical ideas and models proposed to explain the many exciting and challenging observations described in preliminary reports are then described.

1.1. *The jovian magnetosphere*

In the case of the jovian magnetosphere, perhaps the most important things to know are that Jupiter has a rather fast rotation period (P = 9 hr 55 min 29.7 s) and that its

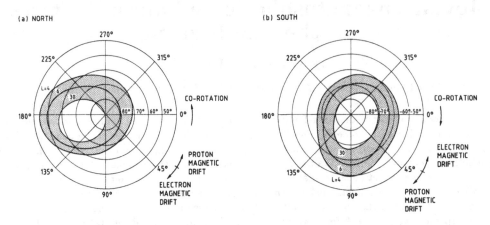

FIGURE 1. L = constant drift shell contours in the northern hemisphere (a) and southern hemisphere (b) of Jupiter. Adapted from Acuna *et al.* (1983).

magnetic field at the planetary surface is the strongest among all planetary objects. A combination of the fast rotation and a large intrinsic magnetic field results in a basic difference between the jovian magnetosphere and Earth's magnetosphere. That is, while the solar wind interaction plays a dominant role in determining the overall plasma flow pattern in the terrestrial case, the jovian magnetosphere is mainly controlled by the corotational electric field so that the plasma flow velocity in a large part of the magnetosphere is in the azimuthal direction (Brice & Ioannidis 1970; Belcher 1983).

Another important point is that the chemical composition of the jovian magnetosphere is a mixture of the solar wind plasma and the oxygen and sulfur ions with their origin traced back to the volcanic gas of Io. A thermal plasma disc of O^+, O^{2+}, S^+, S^{2+} and S^{3+} ions forms in the vicinity of Io. This plasma structure, called Io plasma torus (IPT), has it peak density of 3000 electrons cm^{-3} at $L \approx 5.6$ (L is the equatorial radial distance in units of the jovian radius, R_J) and is one of the most prominent features in the jovian system in optical (Na D lines and SII at 6716 Å and 6731 Å) and UV radiation (Broadfoot *et al.* 1979; Brown *et al.* 1983). The radiative energy of the UV emissions from the jovian aurora and the IPT is derived from the mass loading effect of the newly ionized ions (Broadfoot *et al.* 1979; Dessler 1980) and probably also from other sources (Shemansky 1988).

X-ray emissions from the polar regions were monitored by the Einstein and ROSAT space observatories long before the SL9 impacts (Metzger *et al.* 1983; Waite *et al.* 1994). The X-ray radiation shows a significant level of north-south asymmetry with strong enhancement near the north pole. Furthermore, such radiation, which is believed to be generated by the K-shell emission from the precipitating oxygen and sulfur ions, has a significant rotational dependence with its brightness peaking near the System III Longitude $(\lambda_{III})180° - 200°$ (Waite *et al.* 1994, 1995). To a certain extent, this has to do with the fact that Jupiter's magnetic dipole moment is inclined to the planetary rotation axis as well as offset from the planet center (Acuna *et al.* 1983). As a result, the northern aurora exposes itself most to Earth viewing when the central meridian longitude is at $\lambda_{III} \sim 180°$; a similar effect occurs for the southern aurora when $\lambda_{III} \approx 90°$ (see Fig. 1).

A similar property of longitudinal variation is shared by the UV and infrared emissions from the jovian aurora (Herbert *et al.* 1987; Livengood *et al.* 1990; Miller *et al.* 1995). The imaging observations of H_3^+ auroral emission in the infrared (Connerney *et al.* 1993)

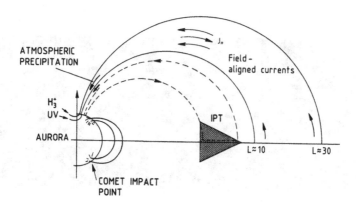

FIGURE 2. A sketch of the magnetic field line connection from different parts of the jovian auroral zone to the magnetosphere.

and of H and H_2 emissions at far ultraviolet wavelengths by HST (Gerard *et al.* 1994a) have also shown that both of the discrete H_3^+ and UV aurora zones could be mapped to a magnetospheric region at $L > 30$ (or even until reaching the magnetopause) while the more diffuse UV emissions detected by IUE and Voyager is mapped to the outer IPT region—where pitch angle scattering of the energetic ions could have the effect of producing the auroral emissions equatorward of the H_3^+ and the discrete UV aurora (Livengood 1990, Gerard 1994a). During its high latitude crossing of the jovian system, the Ulysses spacecraft observed a system of field-aligned currents connecting the polar auroral zones to the outer magnetosphere at $L \approx 20$–30 with a total current of 10–90×10^6 A (Dougherty *et al.* 1993). The measurements of field-aligned beams of energetic electrons and ions by the particle instruments on Ulysses (Lanzerotti *et al.* 1993) are also consistent with the scenario that the jovian auroras are excited by atmospheric precipitation of the current-carrying particles. Prangé *et al.* (1995a) reviewed the collisional excitation effects of the H_2 Werner bands in the FUV and concluded that the energy of the primary population of precipitating particles is likely to be $\approx 40 - 150$ keV for electrons and 2–10 MeV/nucleon for ions. Figure 2 illustrates the present understanding of the regions where auroral emissions have their sources.

Also of special interest to the present discussion is that long term monitoring of the jovian UV auroral activity by IUE has established that the average auroral brightness profile in general does not vary by more than a factor of 2-3. Because of the lack of spatial resolution, the IUE observations could not provide information on whether any of the brightness variations are associated with morphological changes in localized areas. A recent coordinated observation program of simultaneous measurements using IUE and the Faint Object Camera on HST recorded a sudden brightness change in the northern aurora by a factor > 10 over a time period of < 20 hours (Gerard *et al.* 1994b). The maximum value of the total H_2 emission (≈ 6 MR) and the short time scale suggested that disruption of the global current system of the jovian magnetosphere because of an interplanetary disturbance might have been the underlying cause of this event. According to Gerard *et al.* (1994b), an interesting consequence of the large energy deposition rate (≈ 1 W/m^2) from particle precipitation is that strong thermospheric winds could be generated because of the intense atmospheric heating. This effect will be a familar theme in our discussions of the SL9 atmospheric impacts.

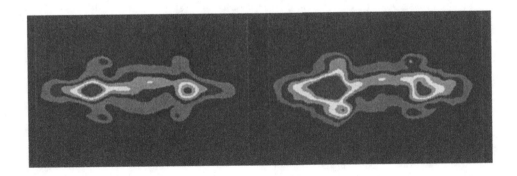

FIGURE 3. VLA 20 cm radio images of Jupiter at $\lambda_{III} \approx 118°$ at two 1994 UT dates:
(a) June 24:03 h UT (before the comet impacts), and (b) July 20:03 UT (during the impact
week). The two images are on the same intensity scale. The peak value in image (a) is ≈ 1700 K
and ≈ 2500 K in image (b). The full beam width at half power is $\approx 0.3R_J$. From de Pater
et al. (1995).

Jupiter has long been known to be a strong source of decimetric and decametric radio
emissions (see Goldstein and Goertz 1983; de Pater and Klein 1989). As a result, the
information concerning their short-term and long-term variations is also most complete
and telling as far as the responses to the SL9 impacts are concerned. The decimetric
emissions from the synchrotron radiation of relativistic electrons in the inner trapped
radiation belt are particularly well documented. As depicted in Fig. 3, the 2D intensity
map shows a maximum at $L \approx 1.5$ with the brightness concentrated near the equatorial
region. This pancake-like structure originates from the flattening of the pitch angle distribution of the MeV electrons as a consequence of their inward diffusion while conserving
the first adiabatic invariant. The displacement and tilting of the planetary magnetic
moment leads to a rocking of the synchrotron radiation structure as Jupiter rotates. Detailed comparisons of such pre-impact data and measurements obtained during and after
the impact week will prove to be very valuable in analyses of magnetospheric effects.

1.2. *The dust coma*

The best source of information on the properties of the dust coma of SL9 have come
from the HST Wide Field Planetary Camera observations by Weaver *et al.* (1995). In
the inner coma the brightness profiles display a $\rho^{-1.16}$ dependence out to 1″ (where ρ
is the radial distance from the comet nucleus) which means that there might have been
continuous dust emission from the nucleus surface. These observers estimated that the
dust production rate should be on the order of 5 kg s^{-1} for some of the brighter nuclei.
Only upper limits of the corresponding gas production rates were given. For example,
according to H. Rickman (priv. comm., 1995) the production rate of CO—which was
possibly the dominant sublimating gas—was found to be below that of P/Schwassmann-
Wachmann 1 at 5.8 AU the value of which is $Q(CO) \sim 5 \times 10^{28}$ molecules s^{-1} (Senay and
Jewitt 1994; Crovisier *et al.* 1995), and the upper limit on the H_2O production rate was
found to be $\sim 10^{27}$ molecules s^{-1} by Weaver *et al.* (1995). Thus, it is by no means certain
that the dust coma must be accompanied by any significant level of gas sublimation. As
an alternative, an electrostatic blowoff mechanism could play a role in ejecting small dust
particles from the bare nucleus (Mendis *et al.* 1981; Ip 1984).

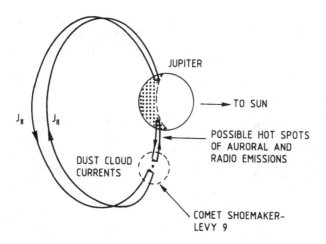

FIGURE 4. A global structure of the field-aligned current system that might have been established between the coma (or the atmospheric impact site) of comet SL9 and the jovian ionosphere. From Ip and Prangé (1994).

Because the cometary dust has a very fragile structure, it is possible that electrostatic force on charged grains could lead to fragmentation when subject to the surface charging effect of energetic electrons (Hill & Mendis 1980). It was on this basis that Dessler & Hill (1994) suggested that as the SL9 comet comas entered the jovian magnetosphere, the electrostatic fragmentation process could make the dust coma visible because of the enhanced cross sectional area of the dust fragments. However, no significant changes to the coma structures were observed except for a short burst of Mg^+ emission followed by a three-fold increase in continuum emission when the G fragment was 54 R_J away from Jupiter (Weaver *et al.* 1995). The brightness increase in continuum emission lasted about 8 minutes and then returned to the quiescent level 20 minutes after the outburst. This time variation might in part be due to the dispersal of the cloud of fragmented dust of submicron size by the Lorentz force. For particles of micron size in the extended coma the Lorentz force would still be effective in causing some deviations of their orbital motion from that of the comet nucleus as it moves toward Jupiter (Horanyi 1994). A careful study of such a dynamical effect is still to be carried out.

The dust coma could interact with the jovian magnetosphere in several ways depending on the effective conductivity of the charged dust cloud and other factors. If the dust coma is partially conducting because of the development of an ionosphere, the magnetospheric interaction will be cometary-like and a current system will be generated between the comet and the jovian ionosphere possibly triggering decametric emissions (Kellogg 1994; Ip and Prangé 1994). On the other hand, if the relative motion between the charged dust particles and the magnetospheric plasma is unimpeded inside the dust coma, the corresponding differential motion between the negatively charged dust and the corotating ions would also lead to a field-aligned current system as shown in Fig. 4. It is in this manner that magnetospheric disturbances in the southern hemisphere associated with a comet impact could be transmitted to the magnetic conjugate point in the northern hemisphere along the magnetic field line.

To summarize, some of the auroral and magnetospheric effects are expected to be arranged in terms of field-aligned current systems connecting the impact site to its magnetic

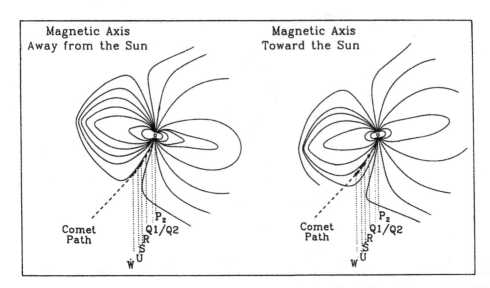

FIGURE 5. A sketch of the orbital spiral traced by comet SL9 in the rotating jovian magnetosphere. Before its final atmospheric impact, the comet and its dust coma could sweep across a
wide range of L shells covering regions with closed or open magnetic field lines. From Prangé
et al. (1995b).

conjugate point. However, such magnetic connection is complicated by the inclination
and rotation of the jovian magnetosphere. For example, the position of the comet fragments could alternate between the region of closed magnetic field lines and the region with
the planetary magnetic field lines opened to the interplanetary magnetic lines (Prangé
et al. 1995a). These two basic configurations are depicted in Fig. 5. As we shall see
some observed features could indeed be described by the simple picture outlined here.
However, there are many unexpected surprises—as is typical of magnetospheric plasmas.

2. Observations

The world-wide coverage of the magnetospheric effects of the SL9 impacts has been
quite phenomenal and extremely fruitful. These comprehensive data sets dealing with a
variety of temporal variations will be very useful in the diagnosis of SL9-related effects
and the physics of the inner radiation belt of Jupiter. But it is really the fortuitous combination of the coordinated microwave measurements by several major radio telescopes,
observations of the H_3^+ emission at 3.5 μm by infrared telescopes (*i.e.*, UKIRT at Mauna
Kea and the 3.5-m NTT at La Silla), ultraviolet observations using the International Ultraviolet Explorer (IUE) satellite, the Extreme Ultraviolet Explorer (EUVE), the Hubble
Space Telescope, and finally the X-ray observations by ROSAT which provides much of
the synergism. If the SL9 impacts were to have taken place five years earlier, our information on the magnetospheric and auroral effects would have been limited to microwave
and IUE observations and perhaps nothing else. With this in mind, we will summarize
the highlights of these very unique results.

2.1. *Microwave observations*

As discussed in the Introduction, the synchrotron radiation of relativistic electrons in the energy range of 10–20 MeV peaks at about 1.5 R_J; this means the narrow jovian ring located between 1.72 and 1.81 R_J would cause absorption collision and energy degradation of these relativistic electrons as they diffuse inward from larger radial distance (de Pater and Goertz 1990). Besides the particulate matter in the jovian ring, magnetospheric electrons could be depleted by dust particles injected from the comet comas into the jovian magnetosphere. Thus if the collisional interaction of magnetospheric electrons with the comet dust is significant, a reduction of microwave radiation would probably be observable (de Pater 1994; Ip 1994). Even though closer scrutiny of the mass budget showed that the increase in the dust population, if any, would be quite small and little decrease in the synchrotron radiation would be expected (Dessler and Hill 1994), the general expectation before the comet impacts was still that some reduction in the microwave emissions might occur. The fact that sharp increases in the brightness fluxes were seen at different wavelengths throughout the impact week therefore came as a major surprise (Dulk *et al.* 1995; Leblanc and Dulk 1995; de Pater *et al.* 1995; Klein *et al.* 1995). According to Klein *et al.* (1995) such changes were unprecedented in the 23-year history of the NASA-JPL Jupiter patrol. There is, therefore, little doubt that they are impact-related. A summary of the temporal variations of the brightness fluxes between 6-cm and 90-cm wavelengths is given in Fig. 6. The radio emissions all tend to reach peak values at the end of the impact week. The increases were wavelength dependent, with the amplitude varying from about 10% at 70–90 cm to about 45% at 6 cm.

It is important to note that the decay time scales (t_d) for the brightness flux increases were also found to be different for different wavelengths. For the first few months, the 6-cm component had $t_d \approx 250$ days, the 13-cm had $t_d \approx 125$ days and $t_d \approx 76$ days for the 21-cm component (Klein *et al.* 1995). As a result, most of the microwave emissions have not returned to pre-impact values several months later except for the 70–90 cm component, which appeared to drop below its preimpact value forty days after the impact week; curiously enough, there also appeared to be a drop in the brightness flux immediately after the impact week as observed by the Westerbork Synthesis Radio Telescope (WSRT).

Leblanc & Dulk (1995) reported that there existed a significant longitudinal asymmetry in the brightness increase at least until July 28 with most of the enhancement confined to one hemisphere with $\lambda \approx 100°–240°$, *e.g.*, the active sector of the jovian magnetosphere (Vasyliunas and Dessler 1981). On the other hand, de Pater *et al.* (1995) concluded that the impact-induced asymmetry lasted no more than four days on the basis of the VLA observations. The dispersion time scale of 10 MeV electrons as a result of azimuthal drift is on the order of two days. A confinement of the brightness enhancement to the magnetic active sector for a time interval much longer than this drift period would require pitch angle scattering of the trapped electrons and/or a continuous injection of new electrons into this region (I. de Pater, priv. comm. 1995). A detailed analysis of the time variations of the longitudinal asymmetry of the microwave emission enhancement will be essential to the understanding of the source region and storage of the relativistic electrons generated by the cometary impacts.

A comparison of the brightness fluxes at different times shows that the spectral shape of the electron energy distribution might have been hardened or alternatively, the pitch-angle distribution of the MeV electrons was isotropized following the comet impacts (Bolton and Thorne 1995; de Pater *et al.* 1995). Before the impacts, the radio spectrum follows a simple power law of $S \sim \nu^{-\alpha}$, with $\alpha \approx 0.10$ at $\nu > 20$ cm and $\alpha \approx 0.4$ at

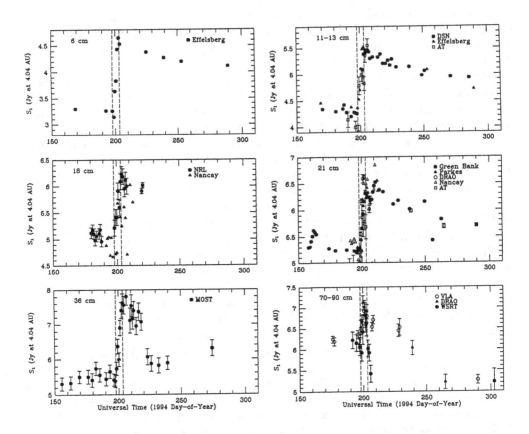

FIGURE 6. Jupiter's nonthermal flux densities between 6 cm and 70–90 cm. From de Pater
et al. (1995).

$\lambda < 20$ cm. After the impact week, there was a conspicuous maximum at 36 cm and
$\lambda \approx 0.30$ for $\lambda > 20$ cm.

The observed effect of radio spectrum hardening could be caused by several mechanisms
including (a) impact-driven radial diffusion of the radiation belt electrons initially located
at larger radial distance, (b) pitch-angle scattering of the pre-existing electron population
to larger pitch angles at which mirroring points the corresponding synchrotron radiation
would be intensified, and (c) *in-situ* acceleration of electrons to energies > 10 MeV
(see Fig. 7). In the first case, electron energization because of conservation of the first
adiabatic invariant (*i.e.*, $E^2/B = \text{const.}$) during the impulsive inward diffusion could
be a very effective mechanism in producing an increase in the synchrotron radiation (Ip,
1995a). Because the radiation intensity $S \approx E^2 B^2$ and hence $S \approx 1/L^9$ in a dipole field
geometry; an inward shift of the relativistic electrons by just $\Delta L \approx 0.05$ at $L \approx 1.5$ would
lead to an increase of the brightness flux by as much as 30%. This effect is consistent with
the decrease of the radial separation between the intensity peaks from 2.92 R_J to 2.82 R_J
observed by Leblanc and Dulk (1995) at 13 cm. The 20 cm VLA images also showed
an inward movement of the intensity peak near $\lambda_{III} \sim 60°$–$110°$ by 0.2 R_J (de Pater
et al. 1995).

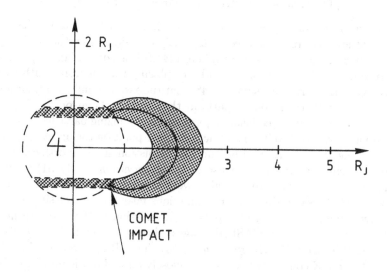

FIGURE 7. A schematic view of the magnetic flux tube with enhanced neutral wind dynamo effect and possibly pitch angle scattering process caused by the comet impact-driven plasma wave turbulence.

Conservation of the first adiabatic invariant would require that the pitch angle distribution become more and more flattened as the radiation belt particles diffuse inward to regions of stronger magnetic field. Thus, we would expect a higher concentration of the synchrotron radiation near the magnetic equator if the impact-driven radial diffusion mechanism is the cause of the increase in the microwave radiations. However, Klein *et al.* (1995) reported that the observed change in the magnetic latitude beaming curves during the impact week suggests an increase in the emission at higher magnetic latitudes. For this to happen, significant level of pitch angle scattering of the locally trapped electrons [*e.g.*, mechanism (b)] would have to take place (Bolton and Thorne 1995). Because the determination of beaming curves could be affected by the longitudinal asymmetry of the microwave brightness distribution detected by de Pater *et al.* (1995) and Leblanc & Dulk (1995), a detailed examination of the 2D images of the radio maps would be required to separate these two effects.

Finally, *in-situ* electron acceleration in the magnetic flux tubes connected to the locations of atmospheric explosions in Jupiter remain a possibility (Brecht *et al.* 1995). This particular mechanism would probably instigate a sequence of step function-like increases in the microwave emissions following individual comet impacts. Examination of high time resolution data of radio measurements would be very useful in resolving this particular issue (see Section 3.3).

2.2. *Infrared observations*

Several programs of infrared observations during the comet impact week were planned. Of major importance in the study of jovian auroral activities are (a) the imaging and spectroscopic measurements at 3.53 μm of the H_3^+ emission by the University College London group (Miller *et al.* 1995) using the CGS4 on the United Kingdom Infrared Telescope and the NSFcam imager on NASA's Infrared Telescope Facility at Mauna Kea, (b) the near-IR spectrometric measurements using the 3.5-m ESO New Technology Telescope at La Silla (Schulz *et al.* 1995), and (c) the CASPIR on the ANU 2.3 m telescope at the Siding Spring Observatory, Australia (McGregor *et al.* 1995).

The ESO measurements in the 3.5 μm region (3.501 μm–3.566 μm, $\lambda/\Delta\lambda \sim 1700$) between July 22 and 31, 1994, showed intense H_3^+ emissions over the positions of the comet impact sites in the southern hemisphere (44° S latitude). Mapping of the jovian disk was performed by Schulz *et al.* (1995) by displacing the slit from south to north in several steps. Anomalous H_3^+ emissions at 44° N latitude were found with their longitudes in good correlation with their counterparts in the southern hemisphere (Fig. 8). This effect showed that after a comet collision, a population of ionizing particles was injected at the impact site and travelled to its magnetic conjugate point on the opposite hemisphere. What is interesting is that the mid-latitude H_3^+ emission in the northern aurora was considerably intensified at the later phase of the observations (July 31) until the flux values of H_3^+ in the northern auroral region were around ten times higher than in the southern aurora; at the same time, the H_3^+ emissions near the impact sites had faded to a relatively low value. Discrete bright spots of IR emission in the northern hemisphere were also observed in the 3–4 μm CASPIR images taken at the Siding Spring Observatory after the K and G impacts (McGregor *et al.* 1995). The IR spot observed in the image made 45 min after the K collision appears to be closely related to the intense UV emission arcs detected by HST at nearly the same time interval (Clarke *et al.* 1995); and both emissions triggered by the comet impact displayed rather short time scale ($\approx$ one hour) in the brightness decrease. The transient nature of the mid-latitude auroral activity detected by CASPIR is therefore not consistent with the long-term effect reported by Schulz *et al.* (1995). This discrepancy remains to be resolved by detailed analysis of the impact-related IR obesrvations and continuous monitoring of the H_3^+ emission in the jovian disk.

Before comet impacts, the north/south (N/S) ratio of the H_3^+ auroral emissions ratio was $\approx$ 1–3, which is within the norm of the average auroral variabilities. Sudden unusual variations of the auroral emission were reported by all three groups. For example, the UKIRT and IRTF observations by Miller *et al.* (1995) showed that between 25 and 27 July the N/S ratio was seen to increase to $\approx$ 6–10 before returning to the average value of about 3. This change was caused in part by the enhancement of the northern emission and in part by a suppression of the southern one. Another important result from Miller *et al.* (1995) concerns the fact that very strong longitudinal variations were also found. The northern emission reached the highest intensity at λ_{III} between 90° and 270° while the southern emission generally leveled off to values below the pre-impact intensities. To explain this remarkable north-south asymmetry, Miller *et al.* (1995) suggested that it could be due to the reduction of the ionospheric Pedersen conductivity in the southern auroral zone because of the injection of a large amount of H_2O and CO molecules and dust of cometary origin (Cravens 1994). This would effectively short-circuit the field-aligned current flow between the equatorial plasma sheet and the southern hemisphere but would double the electric current going into the northern part and hence the temporal variation of the N/S ratio of the H_3^+ auroral emissions. This scenario can be further tested by searching for the putative H_3O^+ ions (at 2.9 μm) in the near-IR spectra and detailed aeronomical model calculations taking into consideration the vertical distribution of dust and H_2O and CO molecules in the jovian upper atmosphere. It is however important to note that, as will be discussed later, energetic events characterized by ultraviolet and X-ray emissions associated with the P2 and Q1/Q2 impacts have been detected in the northern auroral zone. Such effect could potentially play an important role in the generation of the H_3^+ emission in the northern aurora.

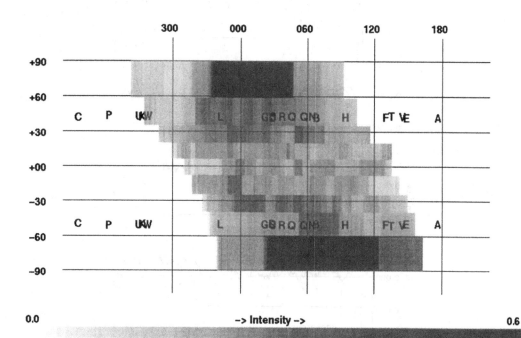

FIGURE 8. A map showing the spatial distribution of the H_3^+ line (3.528–3.536 μm) on the jovian disk on July 24, 1994. The map was constructed by shifting the slit in 9 steps from the south pole to the north pole. The System III longitudes of the impact sites are specified in the figure. This observation was made using the IRSPEC spectrometer at the 3.5-m New Technology Telescope at ESO, La Silla. From Schulz *et al.* (1995).

2.3. *Ultraviolet observations*

The IUE was employed to monitor the auroral activities of Jupiter during the impact week (Ballester *et al.* 1995). Three observational sessions were scheduled for the K, S and P2 fragments each with an integration time $\sim$ 20 minutes centered around the impact time. (Strong emissions were detected only for the K and S impacts). The position of the aperture was centered at the impact sites in the southern hemisphere. The exact dimension of the source region of H Lyman alpha and H_2 emission features is limited by the full-width-half-maximum (FWHM $\approx$ 5″) of the SWP camera. For the K impact, which has the brightest UV emissions, the H_2 feature at 1610 Å appeared to be point-like along dispersion. A diameter of 2″ (7330 km) for the emission area and an effective exposure time of 10 minutes were used to estimate the lower limit of the surface brightnesses. The derived value of a few tens of kR for the hydrogen emissions is in fact as large as the normal jovian auroral brightnesses (Livengood *et al.* 1990). The lower limit of the radiative energy output was estimated to be about 10^{21} ergs. Because of the long integration time ($\approx$ 10 minutes) these UV emissions which must be of a transient nature might have a much greater brightness reaching MR levels at the initial phase of the mid-latitude auroral activities.

From fits to observed spectra (see Figure 9) with electron excitation models, it was tentatively concluded that the H_2 emissions might have been caused by collisional excitation by electrons with energies $\gtrsim$ 20 eV (Ballester *et al.* 1995). [It is however probable

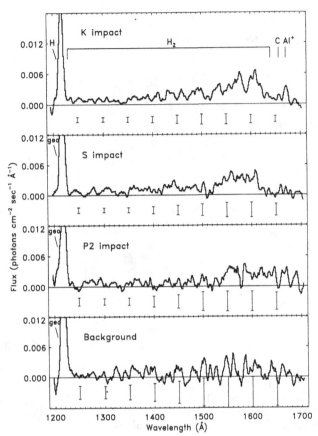

FIGURE 9. Representative IUE spectra for the K, S and P2 impacts and background observations, respectively from top to bottom. The absolute spectral fluxes are lower limits evaluated with the full exposure times. From Ballester *et al.* (1995).

that the primary precipitating particles could have much higher energies.] In addition to the H Lyman alpha and the H_2 Lyman- and Werner-band emissions, spectral features of C 1657 A and possible Al^+ 1671 A could be tentatively identified. Both C and Al atoms could have originated from the cometary material deposited in the jovian atmosphere.

It was unfortunate that the HST observation of the K impact was not scheduled at the same time as the IUE measurements. Otherwise, the immediate response of the jovian atmosphere at the impact site as well as at the magnetic conjugate point in the northern hemisphere would have been registered. Two HST observations were made 47 and 57 minutes after the K impact (Clarke *et al.* 1995). Figure 10 shows one of the FUV images of discrete mid-latitude auroral emissions of about 280 kR near the approaching limb. The two northern emission arcs covering (a) 51° N (latitude), 257°–277° (longitude) and (b) 56° N, 238°–258° are the most conspicuous features. Some faint emission could also be identified just south of the K impact site with the latitude 54° S, 275° and 52° S, 280°. The north and south UV emissions appeared to be connected by magnetic field geometry. However, it is important to note that no emission was detected at the northern conjugate point (at 38° N latitude and 269° longitude) of the K impact site (at 43.8° S latitude and 279° longitude).

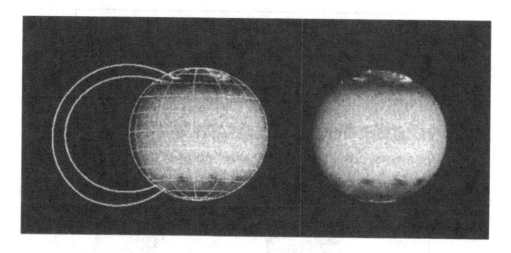

FIGURE 10. A WFPC2 FUV images from HST taken 47 and 57 min after the K impact (assumed at 10:24 UT), showing auroral emissions at lower latitudes than normally observed and apparently associated with the K impact event. Magnetic field lines from the O_6 model are overplotted to show the connection of the northern emission centers to near the K impact region in the southern hemisphere. The integration time of the image is 400 sec. From Clarke *et al.* (1995).

The cause of the observed spatial displacement is uncertain. There are three possibilities. First, the neutral wind disturbances which could be instrumental in driving a field-aligned current (FAC) system could propagate towards the poles bringing with them the FAC footprints (Hill and Dessler 1995). Second, if a population of energetic ions are accelerated in the magnetic flux tube connected to the K impact site, their subsequent interaction with the jovian magnetosphere could lead to a radial transport and diffusion of the particles. Finally, the particle acceleration effect need not be confined to the magnetic flux tube connected to the K impact site; that is, plasma disturbances could be generated along the path of the comet fragment which swept through a range of magnetic latitudes just before atmospheric collision. In this way, excitation of UV emissions at somewhat higher latitudes would be expected. There are indeed some indications of the latter process. According to Clarke *et al.* (1995) and Prangé *et al.* (1995b), two spots of significant UV emission (brightnesses $\approx 200 - 240$ kR) were detected near the southern polar cap at the time of the P2 impact. These UV spots were detected twice at 20 minutes interval (hence the term "blinking aurora") at 20.590 (*i.e.*, July 20. 14:10) and 20.603 before the actual P impact at 20.615 (Figure 11). At the times of the blinking auroral emissions, the P fragment was still about one jovian radius away from the planetary surface. By using a set of jovian magnetic field models, Prangé *et al.* (1995b) showed that these UV spots could possibly be magnetically connected to the Q1 and Q2 fragments which were about 7 R_J away from Jupiter. Consequently, these authors suggested that interaction of the dust coma of the Q1/Q2 fragment with the jovian magnetosphere could be responsible for the excitation of these UV emissions.

Even though the HST observations of the blinking aurora could possibly be interpreted in terms of dust coma interaction, caution must be excercised. This is because we are

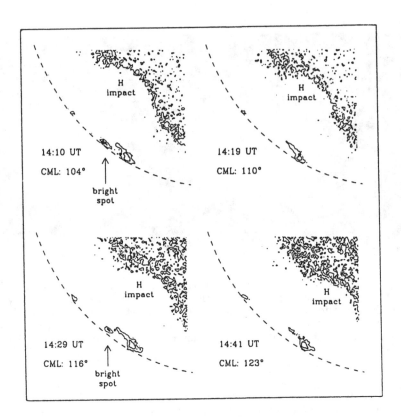

FIGURE 11. Four consecutive views of Jupiter's south pole from the FUV images taken with the HST WFPC2 on July 20, 1994. Isocontours are about 100 and 170 kR. The dashed line has been adjusted on the auroral oval limb gradient and is not the planet disk limb. Note the unusual bright spot on images (a) at 14:10 and (c) at 14:29 amid the emission features associated with the permanent auroral oval. From Prangé *et al.* (1995b).

still at a very early stage in the study of the morphology of the jovian aurora. In the few observational opportunities by HST, rather dynamic changes of the auroral features were found with the sporadic formation of auroral arcs and patchy structures in the polar cap region presumably in response to the solar wind condition (Gerard *et al.* 1994b; Clarke *et al.* 1995; J. E. P. Connerney, priv. comm., 1995). There thus exists the possibility that the presence of the "blinking aurora" might just be part of the normal variation of the jovian auroral activity and not necessary related to the comet impacts. The discovery of these interesting time-variable features during the SL9 impacts hence underlines the need for more HST observations so that the statistical behaviour of the jovian polar aurora can be better understood.

2.4. *X-ray observations*

The maximum surface brightness of the UV emissions following the K impact was 250 kR in the first HST observation at 47 min and 180 kR at 57 min after the atmospheric collision (Clarke *et al.* 1995). This implies a time scale of about 10 min for an exponential decay of the UV emission. The initial UV emission immediately after the comet im-

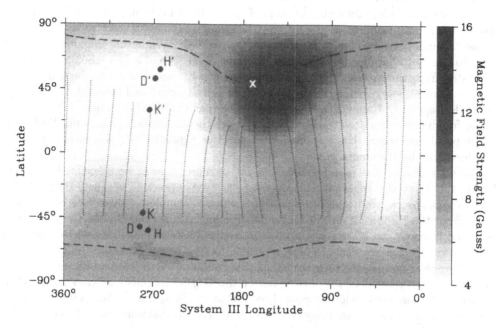

FIGURE 12. Locations of the K impact and its conjugate point (K′) in the northern hemisphere, the low-latitude UV emissions detected by HST (D/H and D′/H′), and the X-ray outburst (X) in the O4 magnetic model. The dotted curves (.....) represent magnetic field lines connecting from the comet impact points at latitude = 45° S and different System III longitudes to the corresponding conjugate points. The footprints of the Io flux tube are indicated by the dashed curves (– – –).

pact could therefore be as high as 1200 kR, a value compatible with the lower limit determined by the IUE observations of Ballester *et al.* (1995). This also means that the total radiative energy output could be on the order of 10^{22} ergs and the corresponding mechanical energy input ten times larger. The peak power is therefore on the order of 10^{12}–10^{13} W. However, an even more energetic event was observed by ROSAT. Using the high-resolution imager (HRI) of ROSAT, Waite *et al.* (1995) monitored the X-ray emission of Jupiter during the impact week. X-ray outbursts were detected in the northern hemisphere in time intervals associated with the K and P2 impacts (Figure 11). These events will be discussed in more detail in the following.

As described earlier, previous observations by the Einstein Observatory and ROSAT favoured the statistical argument that the normal X-ray emissions from the jovian aurora could be the result of K shell emission from precipitating energetic ($\geq$ 400 keV/amu) sulfur and oxygen ions (Metzger *et al.* 1983; Waite *et al.* 1994). From a comparison of the BATSE measurements on the Compton Gamma Ray Observatory with the ROSAT results in the same time frame, Waite *et al.* (1995) reached the tentative conclusion that the heavy ions (instead of the 1–10 MeV electrons) were also responsible for the excitation of the X-ray emission at the K impact. In these circumstances, an input power of 2×10^{13} W and 4×10^{22} ergs of particle energy would be required to produce the K emission which lasted about 30 sec. [A similar energy budget holds for the P2 emission.] Where did these energetic ions come from? How could they be generated in such a short time and then disappear just as quickly? To address these issues we must first identify the locations of these X-ray emissions.

The nominal pointing accuracy of HRI is 5–6 arcseconds. However, the identification and location of visible stars within the field of view of ROSAT during the SL9 observations permitted an improvement of the pointing accuracy to $3''$. This led to the positioning of the K impact burst to 50° N latitude and $\lambda_{III} = 170°$. While this emission peak was close to the footpoint of the Io flux tube (latitude = 50° N and $\lambda_{III} = 186°$), it was well separated from the magnetic conjugate point of the K impact site. As for timing, the impact K event was registered three minutes before the predicted atmospheric collision of the K fragment. However, the crossing of the magnetic field lines near Io's orbit by the K fragment took place at about 14 min before the atmospheric explosion. Such lack of temporal and spatial correlations has no obvious explanation if this X-ray emission was produced by the comet impact.

A similar difficulty is found with the X-ray outburst said to be associated with the P2 impact. First, the brightness enhancement detected between 15:07–15:40 UT (July 20) was seen at $\lambda_{III} = 180°$ and 70° N latitude which is completely disconnected magnetically from the P2 impact site in the southern hemisphere. Second, the X-ray emission occurred between the P2 and the Q1/Q2 impacts. There is therefore uncertainty as to which fragment (in any) should be responsible for the X-ray emission. As a matter of fact, Prangé et al. (1995b) suggested that this energetic event was caused by the magnetospheric interaction of the dust coma of the Q1/Q2 fragment as they approached Jupiter. The magnetic field mapping showed that the footprint of the Q1/Q2 fragment should be located in the proximity of $\lambda_{III} \sim 150°$–160° at this time interval, the putative interconnection with the x-ray burst an intriguing possibility.

The fact that both the K and P2-related X-ray bursts appeared at $\lambda_{III} \sim 180°$ is noteworthy. Since the ROSAT X-ray data obtained in previous years indicate that the X-ray aurora displays a pattern of longitudinal asymmetry similar to that of the UV and IR aurorae with a peak brightness near $\lambda_{III} = 180°$–200° (Waite et al. 1994), the locations of the two X-ray events suggest that they might not be directly related to the SL9 impacts, but rather only reflecting the "regular" pattern of the X-ray auroral emission. The seeming confusion in the timing and location of the outbursts has reinforced the idea that the ordinary X-ray aurora might appear in short-lived bursts with the occurrence frequency peaking near $\lambda_{III} \sim 180°$–200°. A definite answer would require long-term monitoring of the jovian X-ray aurora with high time and spatial resolutions and good pointing.

2.5. *The Io plasma torus*

Before the comet impacts several predictions were made that the jovian ionosphere would be significantly modified so that the corotational motion of the Io plasma torus (IPT) would display certain anomalous features such as a slowing down of the plasma flow (Cravens 1994). Furthermore, the possible injection of cometary dust grains into the jovian magnetosphere could lead to the addition of metallic ions to the IPT as well as absorption of the magnetospheric charged particles (Horanyi 1994; de Pater 1994; Herbert 1994). Several observational programs in the optical and UV wavelength ranges were thus planned to detect these predicted effects. In essence, no exceptional changes which could be related to the comet impacts were found. For example, the EUVE observations of the IPT luminosity reported by McGrath et al. (1995) have been shown to be consistent with no temporal changes during the impact week (Hall et al. 1995). High spectral-resolution ground-based observations by Brown et al. (1995) in a three-week period covering the SL9 impacts have shown that the day-to-day variations in the ion densities, ion temperature, and plasma rotation velocities were not atypical of the time changes in the data obtained by identical method and instrument in previous years.

We are therefore left with the impression that the IPT was not disturbed by the comet impacts to any significant level. Of course, the question remains as to why the energetic events leading to the X-ray bursts in the magnetic flux tube threading through the IPT and/or the enhanced radial diffusion did not leave any observable signatures.

3. Theories

Most of the magnetospheric and auroral effects produced by the SL9 impacts were not foreseen. This should not come as a surprise because of our lack of precise knowledge of the jovian magnetosphere and plasma-dust interaction which might have played an important role. In this regard, the observations of the comet impact effects have in fact provided us with a unique opportunity to test some of the basic theories which could be applied to the impact effects and vice versa. With this in mind, we shall consider several related theoretical topics, namely, the radial diffusion, current systems and particle acceleration mechanisms in their simplified forms.

3.1. *Radial diffusion*

The interplanetary electric field plays a dominant role in defining the convection pattern of the terrestrial magnetosphere. As a result, the radial diffusion of charged particles in the trapped radiation belt is caused by the fluctuations of the electrostatic field which, in turn, is controlled by the solar wind interaction (Schulz and Lanzerotti 1974). In the case of Jupiter, because of its large intrinsic magnetic field and rapid rotation the magnetospheric plasma flow pattern is largely determined by the corotational electric field. For this reason, the electrostatic field variations driven by the neutral wind turbulence in the upper atmosphere become important in defining the random walk motion of the magnetospheric particles. For relativistic electrons with slow gradient drift, the diffusion motion is related to the electrostatic potential fluctuations $\Delta\phi$ with time scale comparable to the orbital period of the energetic electrons in the dipole field by the following expression (Brice and McDonough 1973):

$$D_{LL} = 6.7 \times 10^{-8} L^3 \left(\frac{\Delta\phi_o}{m_e c^2} \right)^2 R_J^2 / \text{hr} \tag{3.1}$$

where $\Delta\phi = \Delta\phi_o \cos\theta$ and θ is the latitude, $m_e c^2$ is the electron rest mass. It is interesting to note that data analyses of the Pioneer 10/11 and Voyager 1/2 measurements of the energetic charged particles suggested an empirical expression of $D_{LL} = D_o L^3$ with $D_o = 4 \times 10^{-9} R_J^2$ s^{-1} (Thomsen 1979; Hood 1993; de Pater and Goertz 1994). Such an experimental value of D_o is compatible with $\Delta\phi/m_e c^2 \approx 16$ in the above equation.

For an electron to drift from dawn to dusk through a longitudinal angle of 180° in a drift shell of radial distance L, the effective electric field would be $< E_n > \approx \Delta\phi_o/\pi L R_J$. Because the impact points of comet Shoemaker-Levy 9 correspond approximately to $L \approx 2$ in a centered dipole field (see later discussion), the required change in the neutral wind speed $V_n (= E_n/B)$ is on the order of 110 m s^{-1} per 5 hr. This means that if a transient neutral component could be generated as a result of atmospheric explosions following the comet impacts and that the wind magnitude at the magnetic flux tube of $L \approx 2$ could reach a value of about 100–200 m s^{-1}, the radial diffusion coefficient in the vicinity of the L shell encompassing the impact sites could accordingly be increased by a factor of two (see Fig. 7). By assuming a step-function-like change in the value of D_o:

$$D_o = 9.3 \times 10^{-9} \quad \text{for} \quad 1.5 < L < 2.5$$
$$4.6 \times 10^{-9} \quad \text{elsewhere} \tag{3.2}$$

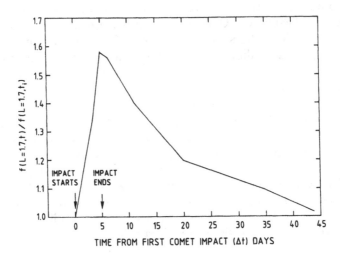

FIGURE 13. The time history of the normalized phase space density of energetic electrons (and the corresponding flux-density of synchrotron radiation) at $L \sim 1.7$ according to the impact-driven radial diffusion model.

and the further assumption that immediately after the last comet impact on July 22, the radial diffusion coefficient would return to the normal value given by $D_o = 4.6 \times 10^{-9}$, it can be shown that the phase space density of the MeV electrons at $L \approx 1.7$ would increase by 60% (Ip 1995a). An example of the time history of the flux-density of the synchrotron radiation is illustrated in Fig. 13. The time scale for the decay of the flux density enhancement could be estimated to be about 40 days. In this simple model treating the one-dimensional diffusion problem, the betatron acceleration effect as a result of the conservation of the first adiabatic invariant was not taken into account. In a more realistic approach, both radial diffusion and the energy variations should be considered simultaneously (see de Pater and Goertz 1994). The energy-dependence and hence frequency-dependence of the time variations of the microwave emissions could then be compared with the observations described in Sec. 2.

Another important fact, of course, is that the displacement of the dipole moment from the planetary center could have caused additional temporal and longitudinal variations. As depicted in Fig. 14, magnetic field lines connected to comet impact sites with different λ_{III} values could map to very different radial distances at the magnetic equator. For example, the magnetic flux tube of the A impact reaches a maximum distance of 1.5 R_J while that of the Q impact has a value of about 2 R_J. This means relativistic electrons injected inward after the A impact would probably cause immediate enhancement of the synchrotron radiation near the brightness center of the microwave emission. On the other hand, there would be a diffusion time scale of about 10 days for the electrons injected at $L \approx 2$ to reach $L = 1.5$ (by which time the azimuthal drift effect would have dispersed the electrons longitudinally around Jupiter). This might be one possible explanation why a strong concentration of microwave emissions was observed during the impact week (Dulk *et al.* 1995; de Pater *et al.* 1995).

3.2. *Current systems*

The momentum and energy in a magnetospheric system can be transferred from one region to another via electric current flows. If a current system could be established

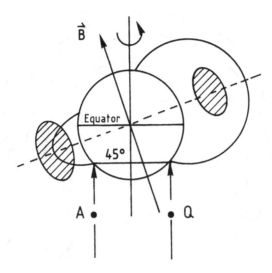

FIGURE 14. A schematic diagram showing the synchrotron radiation belts and the L-shells in the meridian plane of the magnetic field. The latitude line of 45° S is at the footprint of $L \approx 1.5$ at $\lambda_{III} \sim 200°$ (for impact A), and at the $L \approx 2.5$ and $\lambda_{III} \sim 20°$ (for impact Q). From Dulk and Leblanc (1994).

between the comet coma and the jovian ionosphere, it is conceivable that the kinetic energy from the interaction of the comet with the magnetosphere could be dissipated in the topside ionosphere in the form of auroral or radio wave emission. Following this line of argument, Farrell *et al.* (1994) and Ip and Prangé (1994) have independently suggested that the relative motion between the charged dust particles in the cometary coma and the corotating plasma could generate a field-aligned current system mapping from the comet to the two footpoints of the magnetic flux tube (see Fig. 4). Prangé *et al.* (1995b,c) have further suggested that such a process could be a source mechanism for the occurrence of the "blinking aurora" in the northern hemisphere observed by HST. It should be noted that, in addition to the possibility of field-aligned currents, MHD waves and plasma waves might be generated by the passage of the charged dust cloud. This aspect of dust-plasma interaction could indeed be of potential importance.

At atmospheric explosion, a substantial amount of the impact energy of the nucleus fragment was released in the form of shock wave and large-scale wind motion. The frictional interaction between the neutral wind and the ionized gas in the vicinity of the impact site (which might also be instrumental in causing an enhanced level of radial diffusion) could then act as an ionospheric dynamo. Following this line of thinking, Hill & Dessler (1995) have suggested that the atmospheric expansion associated with the impact plume of the K fragment could maintain a field-aligned current system connecting the dynamo region to the conjugate foot point of the corresponding magnetic flux tube. Kinetic energy released in the comet impact could therefore be transferred to the opposite hemisphere which acted as a load in the circuit. If the parallel current density $(j_\parallel)$ exceeds a certain value, voltage drops or electrostatic double layers would form along the magnetic field lines. Electrons and ions could thus be accelerated to high energies causing ultraviolet and X-ray emissions (if the voltage drop is large enough) at atmospheric precipitation—just as in the case of the terrestrial aurora. In this way, the particle

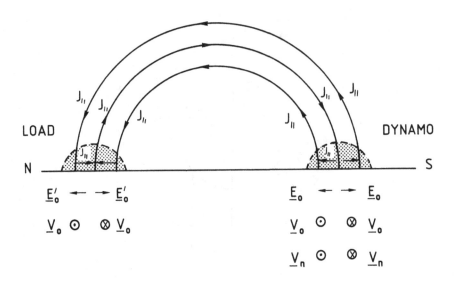

FIGURE 15. An idealized model of the magnetic field coupling between the dynamo region to the energy dissipation region in the opposite hemisphere. The ionospheric dynamo mechanism is driven by a neutral wind shear of velocity (V_n). The maximum energy transfer is reached when the plasma derift velocity (V_o) is half the value of V_n.

acceleration effect could play an important role in the dissipation of the energy stored in the current system. The IUE observed intense UV emissions at the K and S impacts with radiative energies in the H_2 Lyman- and Werner-band and Lyman-alpha exceeding 10^{21} ergs (Ballester *et al.* 1995). Also the mid-latitude UV emissions observed between 47 and 57 min after the K impact by HST (Clarke *et al.* 1995; Prangé *et al.* 1995a) have a total radiative power output on the order of 3×10^{11} W (J. T. Clarke, priv. comm., 1995). The questions are then, what might be the required magnitude of the current flow and whether this current could maintain a voltage drop large enough to excite the observed UV emissions.

The actual situation is rather complicated because of the three dimensional geometry and time evolution of the neutral wind system (see Hill and Dessler 1995). To make some order of magnitude estimates, we shall consider the highly simplified case of a steady state atmospheric dynamo mechanism in which shear flows with wind speeds (V_n) of 100–200 m s^{-1} move across the footpoint of a dipole field (Fig. 15). Following the analytical treatment of Kan *et al.* (1983) the field-aligned current density can be expressed as

$$j_\parallel \approx 4 \times 10^{-4} \left(\frac{\Sigma_p}{1 \ mho} \right) \cdot \left(\frac{V_n}{100 \ m \ s^{-1}} \right) \cdot \left(\frac{B_o}{8 \ G} \right) \cdot \left(\frac{200 \ km}{W} \right) A \ m^{-2} \qquad (3.3)$$

where Σ_p is the height-integrated Pedersen conductivity, B_o the surface magnetic field at the impact point and W is the typical width of the magnetic flux tube. Thus, the maximum value of the total parallel current would be as much as 1.6×10^{10} Å if the arc length is $l \approx 2 \times 10^7$ m.

In a dipole magnetic field geometry, the upward (from equator to pole) FAC ($j_\parallel$) is limited by the loss cone effect. That is, charged particles with pitch angles larger than the loss cone will be mirrored back to the equatorial region before reaching the dynamo or the load region near the planetary surface. To overcome such loss cone limitation, a voltage drop (Φ) along the magnetic field line must be developed to accelerate electrons so that their pitch angles could be reduced. From a consideration of the current flow of a thermal electron population with Maxwellian velocity distribution, a $j_\parallel - \Phi$ relation has been derived by Knight (1973) to be

$$j_\parallel(o) = \left(\frac{B_o}{B_m} \right) \frac{e n_e V_{th}}{\sqrt{2\pi}} \left[1 - \frac{\exp(-\delta\Phi/KT_e)}{1+\delta} \right] \qquad (3.4)$$

where B_m is the magnetic field at the equator, $j_\parallel(0)$ is the parallel current density in the dynamo region, n_e and T_e are the number density and temperature of the ionospheric electrons in the magnetic flux tube, and the electron thermal velocity is $V_{th} = (KT_e/m_e)^{\frac{1}{2}}$. Finally, $\delta = B_m/(B_o - B_m)$ is the numerical factor to account for the convergence of the magnetic field lines.

The combination of Eqs. (3.3) and (3.4) therefore provides a method to estimate the Φ value to be maintained by the impact-driven dynamo mechanism. As an example, for $B_o = 8G$, $B_m = 0.5G$, $n_e = 10^9$ cm^{-3}, $T_e = 20$ eV, we find that a voltage drop $\Phi \approx 70$ V would be established if $j_\parallel^{(0)} \approx 0.4$ mA/m^2. Since the energy dissipation rate due to particle acceleration is on the order of $P = j_\parallel \Phi A_\parallel$ where $A_\parallel$ is the effective area of the FAC we have $A_\parallel \approx 9 \times 10^{12}$ m^2 from the HST observations and therefore $P \approx 3 \times 10^{12}$ W for the K impact (assuming an energy conversion factor of 10). As a result, our simplified consideration shows that the atmospheric dynamo effect is, in principle, a viable mechanism for the generation of the mid-altitude auroral emissions associated with the K impact.

To explain the morphology of the two bright UV arcs in the northern hemisphere and the two other arcs in the southern hemisphere after the K impact, Hill & Dessler (1995) suggested that these are the result of the "snowplow" effect of the impact plumes as they re-entered atmosphere south of the impact site. Because the re-entry speed was supersonic, a bow shock will form just ahead of the plume material which was separated from the ambient jovian atmosphere by a contact surface. In the reference frame of the impact plume, the jovian atmosphere will appear as a subsonic flow, which was diverted in the lateral direction after passing through the bow shock [Fig. 16(a)]. The shear velocity field in the lateral flow around the impact plume would then set up a pair of FAC of opposite directions [see Fig. 16(b)].

3.3. Particle acceleration

The ionospheric dynamo effect and the associated FAC systems are only one of the potential excitation mechanisms of the FUV mid-latitude auroral emissions and the impact-related infrared H$_3^+$ emissions. Other processes such as pitch angle scattering of the pre-existing charged particle population and/or in situ acceleration could also play a role. This is particularly true in the case of magnetospheric phenomena involving MeV electrons (*i.e.*, the synchrotron radiations in decametric wavelengths) and the X-ray bursts associated with the K and P2 impacts.

The time scale for strong pitch angle scattering of energetic ions is given as (Gehrels and Stone 1983)

$$\tau_s = \frac{\tau_b/2}{\alpha_o^2} \approx 5\mu^{-\frac{1}{2}} L^{\frac{11}{2}} [4 - 3/L]^{\frac{1}{2}} \quad \text{sec} \qquad (3.5)$$

where τ_b is the particle bounce period, α_o is the equatorial loss cone of the non-relativistic particles with magnetic moment (μ) in MeV/amu. If oxygen and sulfur ions with energies > 400 keV/amu were responsible for the X-ray emissions of the K impact (Metzger

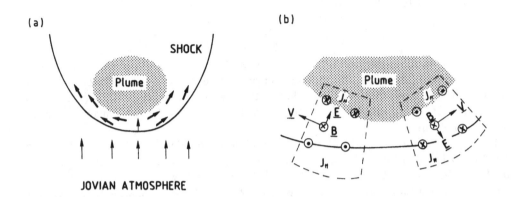

FIGURE 16. The Hill-Dessler model of the field-aligned current system driven by the atmospheric reentry of the "K" impact plume (a) and the symmetric pair of dynamo regions (b). From Hill and Dessler (1995).

et al. 1983; Waite *et al.* 1994; 1995), the corresponding value of T_s at $L \approx 6$ would be ~ 28 hours which is far too long in comparison with a time duration of just 30 seconds for the K event. Therefore, pitch angle scattering of energetic ions would not explain the X-ray outbursts even if the energy budget could be met by the trapped ion population. The pitch angle scattering time scale for MeV electrons is much shorter. But several major obstacles can be immediately identified. First, for electrons to supply the radiative power of 10^{10} W the corresponding mechanical power would have to be a factor of 10^6 larger. This amounts to a total energy of 10^{26} ergs which is a significant fraction ($\sim 10\%$) of the total energy in the jovian magnetosphere according to the Dessler-Parker relation.

In their preliminary report, Waite *et al.* (1995) pointed out that the location of the X-ray emission associated with the K impact was rather close to the Io flux tube which is known to be a source of decametric radio emissions (Goldstein and Goertz 1983) and infrared emission of the H_3^+ ions (Connerney *et al.* 1993). In addition, an intense flux of electrons with energies ~ 0.5 MeV was detected near the L shell of Io by the Trapped Radiation Detector on Pioneer 11 (Fillius 1976). However, it is not clear why the K impact at $\lambda_{III} \sim 270°$ should have caused an energetic event at $\lambda_{III} \sim 180°$, not to mention their separation in latitudinal angle.

As mentioned earlier, the X-ray emission presumably associated with the P2 impact provides yet another major puzzle. If this emission was situated in the polar cap region with magnetic field lines interconnected with the interplanetary magnetic field, it would be difficult to relate it to the P2 impact in the closed field region in the southern hemisphere. A careful magnetic field mapping might show that the X-ray hot spot could in fact be mapped back to a conjugate point in the southern hemisphere. The large angular separation between this X-ray hotspot and the conjugate footprint of the P2 or the Q1/Q2 fragment nevertheless remains problematic. In order to clarify this, it is essential to make detailed surveys of the jovian X-ray emissions using ROSAT and future spaceborne X-ray telescopes to document the spatial and temporal correlations with Io and the System Longitude λ_{III}. Probably after we have learned more about the statistical behaviour of the jovian X-ray aurora (do the emissions appear in localized regions and in bursts of short-duration?) we will be in a better position to pinpoint the origin of the impact-related events.

The interaction of the dust coma or the atmospheric impact of a comet fragment could cause impulsive disturbances to the magnetic flux tubes so that MHD waves could be excited to force random scattering of the charged particles. In this way, stochastic acceleration could be responsible for producing a population of energetic ions in the disturbed magnetic flux tubes. It is, therefore, possible that the UV emissions associated with the K impact could be related to this effect.

Brecht *et al.* (1995) have investigated electron acceleration by a collisionless shock wave generated by the comet impact in the jovian atmosphere. From the application of a theoretical model estimating the time required to accelerate electrons in the terrestrial inner Van Allen belt by the shock wave generated in the Starfish high altitude nuclear test, it was found that the mean time required to accelerate jovian magnetospheric electrons in the synchrotron radiation region from several keV to tens of MeV is < 1 second. However, it is not certain whether the plasma environment surrounding the atmospheric fireball and subsequently the impact plume would permit the formation of the MHD shock as required in this model. Furthermore, de Pater *et al.* (1995) noted that there was no significant correlation between the longitudinal distribution of the synchrotron radiation enhancement and the impact sites.

4. Future prospects

Observations of the many facets of auroral and magnetospheric effects generated by the comet impacts have focused our attention on many hidden issues and provided us with a unique opportunity to examine the problems in a new light. For example, we have invoked radial diffusion, current systems, and stochastic acceleration to explain the excitation of anomalous UV and X-ray emissions observed. Many of these processes in the jovian magnetosphere are not well understood. The jovian system is known to be able to accelerate electrons and ions to high energies (MeV) very efficiently. But there is no definite idea of the basic mechanism(s). We know also very little about the electric current systems connecting the jovian ionosphere to the corotating magnetosphere on a global scale. As for radial diffusion, there exists no comprehensive theory describing the transport and energization of relativistic electrons in the inner magnetosphere which is characterized by strong longitudinal asymmetry. The important measurements of the impact-related microwave emissions described in Sec. 2.1 have demonstrated its need. We are, therefore, still at the very beginning of the understanding of the working of the jovian magnetosphere. The new measurements to be carried out by the Galileo mission in 1996 and 1997 will be most timely and useful in this respect.

At the same time, the continuing efforts in data analysis and interpretations of the observations of SL9-related effects will certainly bring more insights into the ionospheric and magnetospheric responses. Many interesting topics such as dust ablation in the high-altitude atmosphere (Moses 1992) and impact-induced aeronomical processes as discussed in Cravens (1994) will be investigated further. Detailed study will also be given to the excitation mechanisms of the mid-latitude UV aurora and the infrared H_3^+ emissions by theoretical modelling. As for the X-ray aurora, it is hoped that beyond the ROSAT era the XMM mission of ESA and other Earth-orbiting X-ray telescopes could be utilized to observe the corresponding spectral features and spatial and temporal variations. The

ROSAT detection of the X-ray outbursts associated with the K and P2 impacts are most tantalizing.

Finally, the HST observations by Gerard *et al.* (1994a,b) and Clarke *et al.* (1995) have shown what a wealth of information is contained in the UV auroral emissions. Only by engaging in a systematic survey of the very dynamical behaviour of the jovian aurora, which is mapped to a wide range of plasma regions, will we be able to formulate a more complete picture of the jovian magnetosphere and hence its responses (*i.e.*, the blinking aurora) to comet impacts or external disturbances of solar wind origin. Therefore, the saga of SL9 will continue.

It has been a great thrill to be able to participate in the magnetospheric study of the SL9 impact event at first hand. I must therefore thank Renée Prangé for drawing my attention, in the summer of 1993, to the pending comet collisions when I was still in Taiwan and for involving me in her HST observation project. I am also grateful to many colleagues who have provided me with useful information, stimulating discussions, and valuable advice in the preparation of this work. They are Nick Achilleos, Gilda Ballester, Scott Bolton, John Clarke, Imke de Pater, Alex Dessler, George Dulk, Sam Gulkis, Doyle Hall, Yolande Leblanc, Rita Schulz, Joachim Stüwe, Nick Thomas, and Hunter Waite. The assistance and efforts by Inge Gehne, Karin Peschke and Kim Schumann in the preparation of this manuscript are very much appreciated. I thank them all. The present work is supported in part by the Galileo Project of DARA. The quotation of John Minton is from the book "What Mad Pursuit" by Francis Crick.

REFERENCES

ACUNA, M. H., BEHANNON, K. W. & CONNERNEY, J. E. P. 1983 Jupiter's magnetic field and magnetosphere. In *Physics of the Jovian Magnetosphere*, (ed. A. J. Dessler), pp. 1–50, Cambridge University Press.

BALLESTER, G. E., HARRIS, W. M., GLADSTONE, G. R. et al. 1995 Far-ultraviolet emissions from the impact sites of comet P/Shoemaker-Levy 9 with Jupiter. *Geophys. Res. Lett.* **22**, 2425.

BELCHER, J. W. 1983 The low-energy plasma in the Jovian magnetosphere, in *Physics of the Jovian Magnetosphere*, (ed. A. J. Dessler) pp. 68–105, Cambridge University Press.

BOLTON, S. J. AND THORNE, R. M. 1995 Assessment of mechanisms for Jovian synchrotron variability associated with comet SL-9. *Geophys. Res. Lett.* **22**, 1813.

BRECHT, S. H., PESSES, M., LYON, J. G., GLADD, N. T. AND McDONALD, S. W. 1995 An explanation of synchrotron radiation enhancement following the impact of Shoemaker-Levy 9 with Jupiter. *Geophys. Res. Lett.* **22**, 1805.

BRICE, N. AND IOANNIDIS, G. A. 1970 The magnetospheres of Jupiter and earth. *Icarus* **13**, 173.

BRICE, N. AND McDONOUGH, T. R. 1973 Jupiter's radiation belts. *Icarus* **18**, 206.

BROADFOOT, A. L., BELTON, M. J. S., TAKACS, P. Z., et al. 1979 Extreme ultarviolet observations from Voyager 1 encounter with Jupiter. *Science* **204**, 979.

BROWN, M. E., MOYER, E. J., BOUCHEZ, A. H., AND SPINRAD, H. 1995 Comet Shoemaker-Levy 9: No effect on the Io plasma torus, *Geophys. Res. Lett.* **22**, 1833.

BROWN, R. A., PILCHER, C. B. AND STROBEL, D. F. 1983 Spectrophotometric studies of the Io torus. In *Physics of the Jovian Magnetosphere*, (ed. A. J. Dessler, pp. 197–225. Cambridge University Press.

CLARKE, J. T., PRANGÉ, R., BALLESTER, G. E., et al. 1995 Hubble Space Telescope far-ultraviolet imaging of Jupiter during the impacts of comet Shoemaker-Levy 9. *Science* **267**, 1302.

CONNERNEY, J. E. P., BARON, R., SATOH, T. AND OWEN, T. 1993 Images of excited H_3^+ at the foot of the Io flux tube in Jupiter's atmosphere. *Science* **262**, 1035.

CRAVENS, T. E. 1994 Comet Shoemaker-Levy 9 impact with Jupiter; Aeronomical predictions. *Geophys. Res. Lett.* **21**, 1075.

CROVISIER, J., BIVER, N., BOCKELÉE-MORVAN, D., COLOM, P., JORDA, L. AND LELLOUCH, E. 1995: Carbon monoxide outgassing from comet P/Schwassmann-Wachmann 1. *Icarus* **115**, 213.

DE PATER, I. 1994 The effect of comet Shoemaker-Levy 9 on Jupiter synchrotron radiation. *Geophys. Res. Lett.* **21**, 1071.

DE PATER, I. AND GOERTZ, C. K. 1990 Radial diffusion of energetic electrons and Jupiter's synchrotron radiation. I. Steady state solution. *J. Geophys. Res.* **95**, 39.

DE PATER, I. AND GOERTZ, C. K. 1994 Radial diffusion models of energetic electrons and Jupiter's synchrotron radiation. II. Time variability. *J. Geophys. Res.* **99**, 2271.

DE PATER, I. AND KLEIN, M. J. 1989 Time variability in Jupiter's synchrotron radiation. In Proc. Conference on *Time Variable Phenomena in the Jovian System*, pp. 139–150.

DE PATER, I., HEILES, C., WONG, M., *et al.* 1995 Outburst of Jupiter's synchrotron radiation following the impact of comet P/Shoemaker-Levy 9. *Science* **268**, 1879.

DESSLER, A. J. 1980 Mass-injection rate from Io into the Io plasma torus. *Icarus* **44**, 291.

DESSLER, A. J. AND HILL, T. W. 1994 Some interactions between dust from comet Shoemaker-Levy 9 and Jupiter. *Geophys. Res. Lett.* **21**, 1043.

DOUGHERTY, M. K., SOUTHWOOD, D. J., BALOGH, A. AND SMITH, E. J. 1993 Field-aligned currents in the jovian magnetosphere during the Ulysses flyby. *Planet. Space Sci.* **41**, 291.

DULK, G. A. AND LEBLANC, Y. 1994 Changes in jupiter's synchrotron radiation belts during and after SL-9 impacts. In Proc. European SL-9/Jupiter Workshop (eds. R. West and H. Böhnhardt), p. 381. ESO.

DULK, G. A., LEBLANC, Y., AND HUNSTEAD, R. W. 1995 Flux and images of Jupiter at 13, 22 and 36 cm before, during and after SL-9 impacts. *Geophys. Res. Lett.* **22**, 1789.

FARRELL, W. M., KAISER, M. L., DESCH, M. D. AND MACDOWALL, R. J. 1994 Possible radio wave precursors associated with the comet Shoemaker-Levy 9/Jupiter impacts. *Geophys. Res. Lett.* **21**, 1067.

FILLIUS, R. 1976 The trapped radiation belts of Jupiter. In Jupiter (ed. T. Gehrels), pp. 896–927. University of Arizona Press.

GEHRELS, N. AND STONE, E C. 1983 Energetic oxygen and sulfur ions in the Jovian magnetosphere and their contribution to the auroral excitation. *J. Geophys. Res.* **88**, 5537.

GERARD, J. C., DOLS, V., PRANGÉ, R., AND PARESCE, F. 1994a Morphology and time variation of the Jovian far uv. *Planet. Space Sci.* **42**, 905.

GERARD, J. C., GRODENT, D., PRANGÉ, R., *et al.* 1994b A remarkable auroral event on Jupiter observed in the ultraviolet with the Hubble Space Telescope aurora: Hubble Space Telescope observations. *Science* **266**, 1675.

GOLDSTEIN, M. L. AND GOERTZ, C. K. 1983 Theories of radio emissions and plasma waves. In Physics of the Jovian Magnetosphere pp. 317–352. Cambridge University Press.

HALL, D. T., GLADSTONE, G. R., HERBERT, F., LIEU, R., AND THOMAS, N. 1995 Io torus EUV emissions during the comet Shoemaker-Levy 9 impacts. *Geophys. Res. Lett.* **22**, 3441.

HERBERT, F. 1994 The impact of comet Shoemaker-Lev 9 on the Jovian magnetosphere. *Geophys. Res. Lett.* **21**, 1047.

HERBERT, F., SANDEL, B. R. AND BROADFOOT, A. L. 1987 Observations of the jovian UV aurora by Voyager. *J. Geophys. Res.* **92**, 3141.

HILL, J. R. AND MENDIS, D. A. 1980 On the origin of striae in cometary dust tails. *Astrophys. J.* **242**, 395.

HILL, T. W. AND DESSLER, A. J. 1995 Midlatitude Jovian aurora produced by the impact of comet Shoemaker-Levy 9. *Geophys. Res. Lett.* **22**, 1817.

HOOD, L. L. 1993 Long-term changes in jovian synchrotron radio emission: intrinsic variations or effects of viewing geometry. *J. Geophys. Res.* **98**, 5769.

HORANYI, M. H. 1994 New Jovian rings. *Geophys. Res. Lett.* **21**, 1039.

IP, W.-H. 1994 Time variations of the Jovian synchrotron radiation following the collisional impacts of comet Shoemaker-Levy 9. *Planet. Space Sci.* **42**, 527.

IP, W.-H. 1995a Time variations of the Jovian synchrotron radiation following the collisional impacts of comet Shoemaker-Levy 9: 2. flux enhancement induced by neutral atmospheric turbulence. *Planet. Space Sci.* **43**, 221.

IP, W.-H. 1995b On particle acceleration by the impact-driven field-aligned current systems. *Icarus*, submitted.

IP, W.-H. AND PRANGÉ, R. 1994 On possible magnetospheric dust interactions of comet Shoemaker-Levy 9 at Jupiter. *Geophys. Res. Lett.* **21**, 1051.

KAN, J. R., AKASOFU, S.-I. AND LEE, L. C. 1983 A dynamo theory of solar flares. *Solar Phys.* **84**, 153.

KELLOGG, P. J. 1994 Plasma effects on the interaciton of a comet with Jupiter. *Geophys. Res. Lett.* **21**, 1055.

KLEIN, M. J., GULKIS, S., BOLTON, S. J. 1995 Changes in Jupiter's 13-cm synchrotron radio emission following the impact of comet Shoemaker-Levy 9. *Geophys. Res. Lett.* **22**, 1797.

KNIGHT, S. 1973 Parallel electric field. *Planet. Space Sci.* **21**, 741.

LANZEROTTI, L. J., ARMSTRONG, T. P., MACLENNAN, C. G., *et al.* 1993 Measurements of hot plasmas in the magnetosphere of Jupiter. *Planet. Space Sci.* **41**, 893.

LEBLANC, Y. AND DULK, G. A. 1995 Changes in brightness of Jupiter's radiation belt at 13 and 22 cm during and after impacts of comet SL-9. *Geophys. Res. Lett.* **22**, 1793.

LIVENGOOD, T. A., STROBEL, D. F. AND MOOS, H. W. 1990 Jupiter's north and south polar aurorae with IUE data. *J. Geophys. Res.* **95**, 10375.

McGRATH, M. A., HALL, D. T., MATHESON, P. L., *et al.* 1995. Response of the Io plasma torus to comet Shoemaker-Levy 9. *Science* **267**, 1313.

McGREGOR, P. J., NICHOLSON, P. D. AND ALLEN, M. G. 1995 CASPIR observations of the collision of comet Shoemaker-Levy 9. *Icarus* submitted.

MENDIS, D. A., HILL, J. R., HOUPIS, H. F., AND WHIPPLE, E. C. 1981 On the electrostatic charging of the cometary nucleus. *Astrophys. J.* **249**, 787.

METZGER, A. E., GILMAN, D. A., LUTHEY, J. L. *et al.* 1983 The detection of x-rays from Jupiter. *J. Geophys. Res.* **88**, 7731.

MILLER, S., ACHILLEOS, N., DINELLI, B. M., *et al.* 1995 The effect of the impact of comet Shoemaker-Levy 9 on Jupiter's aurora. *Geophys. Res. Lett.* **22**, 1629.

MOSES, J. 1992 Meteoroid ablation in Neptune's atmosphere. *Icarus* **99**, 368.

PRANGÉ, R., REGO, D. AND GERARD, J. C. 1995a Auroral Lyman alpha and H2 bands from the giant planets 2. Effect of the anisotropy of the precipitating particles on the interpretation of the "color ratio." *J. Geophys. Res.* **100**, E4, 7513.

PRANGÉ, R., ENGLE, I., CLARKE, J. T., *et al.* 1995b Auroral signature of comet SL9 in the Jovian magnetosphere. *Science* **267**, 1317.

PRANGÉ, R., ENGLE, I., DUNLOP, M., MAURICE, S., AND REGO, D. 1995c Magnetic mapping of auroral signatures of comet SL9 in the Jovian magnetosphere. *Geophys. Res. Lett.*, submitted.

SENAY, M. AND JEWITT, D. 1994 Coma formation driven by carbon monoxide release from Comet Schwassmann-Wachmann, I., *Nature* **371**, 229.

SHEMANSKY, D. E. 1988 Energy branching in the Io plasma torus: The failure of neutral cloud theory. *J. Geophys. Res.* **93**, 1773.

SCHULZ, M. AND LANZEROTTI, L. J. 1974 Particle Diffusion in the Radiation Belt. Springer-Verlag.

SCHULZ, R., ENCRENAZ, TH., STÜWE, J. A., AND WIEDEMANN, G. 1995 Near-IR emissions in the upper Jovian atmosphere after SL9 impact: Indications of possible northern counterpart. *Geophys. Res. Lett.*, **22**, 2421.

THOMSEN, M. F. 1979 Jovian magnetosphere-satellite interactions: aspects of energetic charged particle loss. *Rev. Geophys. Space Phys.* **17**, 369.

VASYLIUNAS, V. M. AND DESSLER, A. J. 1981 The magnetic anomaly model of the Jovian magnetosphere: A post-Voyager assessment. *J. Geophys. Res.* **86**, 8435.

WAITE, J. H. JR., BAGENAL, F., SEWARD, C. *et al.* 1994 ROSAT observations of the Jupiter aurora. *J. Geophys. Res.* **99**, 14799.

WAITE, J. H. JR., GLADSTONE, G. R., FRANKE, K., *et al.* 1995 ROSAT observations of X-ray emissions from Jupiter during the impact of comet Shoemaker-Levy 9. *Science*, **268**, 1598.

WEAVER, H. A., A'HEARN, M. F., ARPIGNY, C. *et al.* 1995 The Hubble Space Telescope oberving campaign on comet P/Shoemaker-Levy 9. *Science* **267**, 1282.